# Nihilism Inc.

## Environmental Destruction and the Metaphysics of Sustainability

# *Nihilism Inc.*

## *Environmental Destruction and the Metaphysics of Sustainability*

**Arran Gare**

**ECO-LOGICAL PRESS**

**1996**

NIHILISM INC. Environmental Destruction and the Metaphysics of Sustainability



First published in 1996 by Eco-Logical Press.

Editorial Address:
Eco-Logical Press
99 Woronora Crescent
Como, NSW, Australia 2226



Gare, Arran
NIHILISM INC. Environmental Destruction and the Metaphysics of Sustainability.
Includes Bibliography and index, 438 p.
1996.
ISBN 1 876236 00 0

Printed in Australia by Panther Publishing & Printing

The author gratefully acknowledges the encouragement and support of Richard Sylvan whose tragic death in June, 1996 has robbed philosophy of one of its great thinkers.

# CONTENTS

# PREFACE

In this work I have attempted to explain the failure of humanity to effectively confront the global environmental crisis, and thereby to reveal what is required to overcome it. It is argued that the destruction of the environment on a global scale is the final outcome of the expansion and domination of the world by European civilization, a civilization which is inherently destructive and implicitly nihilistic. The failure to respond to a crisis threatening all humanity and most other forms of life is symptomatic of this nihilism. Environmentalists have failed because they have not fully appreciated the nature of this civilization, an appreciation which requires an understanding of the history and dynamics of European culture and its offshoots from Ancient Greece to the present. The first part of the book analyses the origins and dynamics of Western civilization to reveal the orgins of nihilism and to show how in the modern world nihilism has come to be embodied by institutions and individuals, while the second part is devoted to analysing Marxism, Russian culture and the Soviet Union.

The focus on culture does not exclude concern with the economic determinants of environmentally destructive behaviour. The work is designed to facilitate a deeper understanding of these determinants. But it is also designed to show why orthodox Marxist analyses of these determinants is not enough and why existing forms of Marxism cannot provide an effective response to, or alternative to, capitalism. Only by focussing on culture is it possible to reveal the possibilities open to people to create a social order free of the destructive imperatives of modern civilization. The final part of the work offers the philosophical foundations necessary for an alternative, environmentally sustainable civilization.

Research on Russian culture and Soviet Marxism has enabled me to situate the central ideas of this book within a tradition originating in the work of Aleksandr Bogdanov. Bogdanov not only conceived humans as part of and within nature and recognized the environmental limits to economic activity, but also argued for the central role of culture in the dynamics of history and in the creation of a new social order. The *Proletkul't* movement inspired and led by him after the Bolshevik revolution represented and still symbolizes the alternative to both capitalism and the centralized system of State control forged by Lenin and Stalin. The tradition of thought originating in this movement encompasses the pioneering efforts of Joseph Needham to combine Marxism and process philosophy and to transcend the limitations of European culture in his monumental study of science and civilization in China, and links the achievements of more recent philosophers, scientists and environmentalists inspired by process philosophy such as Milic Capek, Ivor Leclerc, Ilya Prigogine, C.H. Waddington, Brian Goodwin, David Bohm, John Cobb Jr and Herman Daly.

This work was conceived in courses of lectures on environmental philosophy given at the University of Queensland and the University of Western Australia, and while assisting Professor Douglas Weiner teach a course on conservation in USA and USSR at Harvard University. An earlier version has been published previously in two separate books: *Nihilism Incorporated: European Civilization and Environmental Destruction* and *Beyond European Civilization: Marxism, Process Philosophy and the Environment*. Writing and revising this work has taken place over an extended period, and research has brought me into contact with specialists in a wide range of disciplines. This forced me to question anew the nature of inquiry, of questioning the past and constructing historical narratives, and of understanding and describing other cultures and societies. This in turn has created another problem - to what audience is this work addressed? Initially I struggled for a compromise between readability for a general audience and the standards of intellectual rigour set by the diverse and to some extent competing disciplines whose domains I had invaded. I came to realize that it is the acceptance of and accommodation to existing audience differentiations which fractures efforts to confront the problems of society, that it is essential to work towards the creation of an audience which destroys such differentiation if a movement able to solve these problems is to be created. This book is meant as an attempt to undermine institutionalized audience differentiation.

For their encouragement and support I am grateful to my colleagues in Perth and Melbourne, and beyond these cities, to Charles Birch, Robert Cohen, Val Plumwood, Valeria Russo, Richard Sylvan and Douglas Weiner. I am particularly thankful to Douglas Weiner for sharing his vast knowledge of Russian environmentalism and for his guidance in my research on Russian and Soviet culture. I am also indebted to a number of institutions. These include the Western Australian Society for the History and Philosophy of Science for a research fellowship to Curtin University, the Australian-American Educational Foundation for a Fulbright Post-doctoral Fellowship, the Center for the Philosophy and History of Science, Boston University for hosting my stay in USA., and the Department of Philosophy and Cultural Inquiry at Swinburne University for their support for this project. Finally I am grateful to Richard and Louise Sylvan for their help in publishing this book. I dedicate the book to my father Frank Gare and to my wife, Jennifer, and to the memory of my mother, Nene Gare.

Arran Gare
14 July 1996.

# INTRODUCTION

Mexico City was described in a lead article in *Time* magazine. The article began:

> When the ragged and exhausted Spanish conquistadors first beheld the lake-encircled capital of the Aztecs one November morning in 1519, they were stunned by its grandeur. A shining metropolis of some 300,000 people, far larger than any city in Europe, Tenochtitlan displayed immense stone temples to the gods of rain and war and an even more immense royal palace, where Aztec nobles stood guard in jaguar-head helmets and brightly feathered robes. In the nearby marketplace, vendors offered an abundance of jungle fruits and rare herbs and skilfully wrought creations of silver and gold. 'The magnificence, the strange and marvellous things of this great city are so remarkable as not to be believed,' Hernando Cortés wrote back to the imperial court of Charles V. 'We were seeing things,' Bernal Díaz del Castillo recalled in his memoir of the Spanish invasion, 'that had never been heard of or seen before, nor even dreamed about.'
>
> A newcomer today is more apt to arrive by air, and before he even glimpses the dried-up bed of Lake Texacoco, now edged with miles of slum hovels, the first thing he sees is the almost perpetual blanket of smog that shrouds the entire city. It is an ugly grayish brown. There is something strangely sinister about it - a cloud of poison. The pilot orders the seat belts tightened and announces an imminent descent into the murk and filth.[1]

The lakes and flower gardens of Tenochtitlan have disappeared. In the last quarter-century Mexico City has lost 75% of its woodland, and 14 million saplings planted between 1976 and 1982 withered and turned yellow within a few years. Drawing water from the subsoil has caused parts of the city to sink, sometimes up to 30 feet. Caged birds placed in the middle of the city die within two hours and birds, their lungs laden with lead and cadmium, have begun to drop from the sky. 70% of newborn babies now have high lead content in their blood. The level of ozone tripled between 1986 and 1988. Pollution is killing 30,000 children each year through respiratory and gastro-intestinal diseases, and may be responsible for the deaths of 100,000 people a year. The vast majority of Mexico City's population of more than 20 million, the descendants of the Aztecs, are living in the appalling poverty of the slums, more than 3 million without sewage facilities.

Mexico City symbolizes the success with which Western European civilization has conquered and subjugated almost every other civilization and culture over the past five hundred years, the impoverishment of the survivors of this conquest, and the future in store for the world if this civilization continues to progress on its present path. It symbolizes the strengths and weaknesses of this civilization: the prodigious developments in technology, and the systematic blindness to, and incapacity to deal with, the destructive side-effects of these developments. This failure is most fully manifest in the destruction of the natural environment and the oppressive environments people are creating for themselves.

Environmental degradation is the most important complex of problems ever confronted by humanity. Humans are interfering with the world's ecosystems so severely that they are beginning to undermine the conditions for their own continued existence. They are polluting the air, the oceans and the land. They are rapidly exhausting the reserves of minerals and destroying the

---

1. Otto Friedrich, reported by Ricardo Chavira and David DeVoss, 'A Proud Capital's Distress: Overcrowded, polluted, corrupted, Mexico City offers the world a grim lesson', *Time*, August 6, 1984, pp.14-21.

resources of the world on which civilization depends, while destroying other life forms on a massive scale. At the same time humans are increasingly enclosing themselves in built environments which isolate them and fragment their lives, destroy their health and reduce them to either the dehumanized instruments of military-industrial complexes, or to abject poverty. The problem of the environment is also the problem of over-population, the disproportionate consumption of resources by Western nations and the starvation of those in the Third World who lose out in the struggle for the remainder. If present trends continue the total destruction of civilization is probable within a few hundred years - and the extinction of the human species is a real possibility.

This situation also presents the greatest intellectual challenge of the era. It undermines the traditional idea of economic progress - the ultimate evaluative concept and the virtual *telos* of European civilization. It brings into question the economic, legal, political and ethical institutions of modern societies and the modes of thought on which they are based, including the natural and social sciences and the institutions supporting them. In doing so, it opens up the most fundamental questions about human existence: the nature of knowledge and value, of meaning and rationality, and of the significance of life itself. Confronting the environmental crisis requires a complete review of the way we think of ourselves and our place in the world.

But in presenting these social and intellectual challenges to humanity, environmental problems reveal not only the total inadequacy of prevailing modes of thought, but also the nihilistic attitudes which dominate the modern world. They reveal the imperviousness of governments to the obvious irrationalities of the present economic order and their unwillingness to do anything to seriously tackle the world's long-term problems. They reveal a political world dominated by politicians, corporation chiefs, bureaucrats, media, military and intelligence moguls who are devoid of ideals and principles and for whom the struggle for power has become mere sport. The recent patina of concern for the environment shown by opportunistic politicians can hardly be taken seriously. They remain committed to economic growth, deregulation of the market and consumerist values which together make increasing environmental destruction inevitable. The attitude of corporation chiefs to environmental problems is illustrated by the announcement of British Petroleum that to refurbish its image as an environmentally sensitive corporation it will paint its service stations a more conspicuous green. The environmental crisis also reveals widespread indifference by most people in the affluent countries to the damage they are doing to nature, to the poor of the world and to future generations. It makes apparent their fascination with military power and their readiness to support oppression to force others to bear the costs of their own life-styles. And it reveals the poverty of the academic world, a world in which education is being reduced to vocational training, knowledge to a commodity and all critical discourse is being eliminated. Most academics do not even question the fragmentation of inquiry and the noise explosion engendered by the publish or perish syndrome, though this is now burying knowledge rather than advancing it. And what may be worse, where fundamental intellectual and social problems are excluded from consideration in mainstream academic life because they cannot be encompassed within established disciplinary boundaries.[2] Such pusillanimity reflects the tacit acceptance of the prevailing world-orientation according to which the world itself is devoid of meaning, life is just a struggle for survival and for power in which the destruction of the weak is inevitable, knowledge is simply a means to control the world, and the only real values in life are survival, pleasurable stimuli and entertaining distractions. The belief that there is something more noble to life has lost its foundations. As Nietzsche wrote of the modern predicament: 'the highest values devaluate themselves. The aim is lacking; "why" finds no answer.'[3] This nihilism must be confronted if humanity is to meet the challenge of the environmental crisis.

---

2. Increasingly, natural scientists are addressing the issues, but social scientists, economists and philosophers who have taken up environmental issues have been marginalized by their academic colleagues.

3. Friedrich Nietzsche, *The Will to Power*, tr. Walter Kaufmann, N.Y.: Vintage, I, 2, p.9.

In this work it will be argued that environmental problems and the nihilism underlying the failure to confront them are manifestations of basic deficiencies in the world-orientation which dominates throughout the world. The roots of these deficiencies will be shown to lie in metaphysical notions that originated in Ancient Greece, were developed in medieval Europe, incorporated into mechanistic materialist science, assumed by economic theory and institutionalized in capitalist society. With the development of capitalism and the elaboration of mechanistic materialism into evolutionary theory, Social Darwinism and information theory, these notions have come to inform almost all the practices of those people who now dominate the world. They underlie the concepts in terms of which people define themselves, their relationships to each other, to society and to nature. They provide the basis on which people make their decisions about how to live and what to do. In this way they largely have come to constitute the existing social order so that people are enmeshed in a framework of defective concepts which defines their reality and limits their comprehension: they have great difficulty in perceiving or thinking about anything not intelligible in terms of these concepts. It is not only that these concepts have blinded people to the intrinsic value and fragility of their world, though this is important. By disorienting them and frustrating their potentialities, they have also engendered aggression, nihilistic violence and destructive social dynamics which exceeds the comprehension of most people. Environmental problems reveal the deep-rooted nature of these deficient metaphysical notions.

The main opposition to Western culture is still Marxism (although Islamic fundamentalism is becoming increasingly important). Marxists, it will be argued, have identified the main causes of global environmental destruction in 'commodity fetishism', the autonomous dynamics of the market and the domination of people politically and ideologically to maintain and extend these dynamics. It is these which have produced a global system of environmental exploitation. Environmental degradation is the second and ultimate contradiction of capitalism, impelling people in their struggle for livelihoods to participate in the destruction of the conditions not only of capitalism, but of humanity itself, together with most other forms of life. Marxists are correct to argue that if the world is to have any future, the market must be replaced, or at least subordinated, to some organization or complex of organizations which evaluate life in terms transcending exchange value. However the former Soviet Union and other communist countries also generated massive environmental problems of their own, and societies so oppressive and corrupt that most people now believe that there is no alternative to capitalism. So while Marxism explains the dynamics of environmental destruction, in its present form it neither overcomes the nihilistic destructiveness of Western culture nor provides a foundation for a less environmentally destructive socio-economic formation.

Revealing the defective assumptions underlying the environmental crisis both in the West and in the former Soviet Union will reveal the need for a more radical revolution than envisaged by orthodox Marxists. What is required is a 'metaphysical revolution'. However before the fundamental reorientation required to overcome the environmental crisis can be seriously entertained or even properly understood, it is necessary to comprehend the metaphysical assumptions which dominate the existing social order and which dominated the former Soviet Union. People take such assumptions so much for granted that, until these are exposed, it is impossible to comprehend that the world could be understood differently. Questioning these assumptions presents difficulties unlike those confronting any other intellectual pursuit. Societies have developed in such a way that the forms of thought which have been embodied in all their major social institutions presuppose their validity. To challenge the prevailing metaphysical assumptions and to attempt to develop an alternative metaphysics is to set out on tasks which have been more than simply censured. This enterprise is now barely acknowledged to have any meaning at all, and the forms of reasoning associated with it have almost no acknowledged status.

This presents the problem of where to begin. The approach adopted has been to assume the validity of the new metaphysics as a starting point and to analyse the ideas and modes of thought dominating the modern world and the former Soviet Union from this perspective. The prevailing metaphysical assumptions and their deficiencies are exposed by using the new metaphysics as a framework for analysing environmental problems, the deficiencies of prevailing ethical and political thinking (manifest in the intellectual responses to these problems), and the relationship between modes of thought, social dynamics and environmental destruction in the evolution of Western European civilization. It is also argued from this perspective that Marx's ideas cannot be dissociated from metaphysics, that they are confused and inconsistent because Marx drew on ideas from different metaphysical traditions without clarifying his own metaphysical commitments. While in the early years of the Russian revolution some Marxists accentuating radical aspects of Marx's thought promoted a spectacularly successful environmentalist movement, Russians were predisposed to adopt other aspects of Marx's work. Marxism became the vehicle through which Russians assimilated the Western orientation towards technological domination of the world to their culture. As Soviet Marxism crystallized in the 1930's, and those Marxists who had promoted the more radical version of Marxism were suppressed, the early successes of environmentalists were obliterated. However the radical Marxists provided a vision of what could be achieved if Marx's ideas were consistently rethought on the basis of new metaphysical foundations.

The new metaphysics is expounded, elaborated and defended in the final part of this work. A dialectical theory of knowledge, in which the goal of disciplined inquiry is taken to be understanding, is defended, showing the indissociable relationship between science and metaphysics. A categorial scheme is outlined and it is shown how this generates an alternative grand research programme for the sciences. In terms of this scheme, the world is understood as a process of creative becoming continually generating emergent processes. Humanity itself is then represented and explained as a complex of emergent processes, thereby resolving the most important problems in the philosophy of mind and philosophical anthropology: the relationship between mind and body, consciousness and the world, thought and action, freedom and determination, and the individual and society. This conception of humanity is used to formulate a new ethical and political philosophy and the foundations for a reflexive, critical science of humanity - designed to reveal to people what they are and what role they can take in the creation of the future. It is shown how this will enable people to transcend the prevailing nihilism, to effectively confront the environmentally destructive tendencies of society and to create new forms of relationships between people and between humans and the rest of nature. In this way this metaphysics is offered as the foundation for an alternative culture to oppose and to replace the culture which underlies the existing economic, social and political world order; the foundation for a new, environmentally sustainable civilization.

# 1

# ECOCIDE AS APPLIED NIHILISM

This chapter will explore the dimensions of the contemporary environmental crisis. This is a more difficult task than it appears. The view of environmentalists was summed up by Thomas Sancton when he argued in *Time* magazine:

> Let there be no illusions. Taking effective action to halt massive injury to the earth's environment will require a mobilization of political will, international cooperation and sacrifice unknown except in wartime. Yet humanity is in a war right now, and it is not too draconian to call it a war for survival. It is a war in which all nations must be allies.[1]

But this is not the view of the business leaders, politicians, bureaucrats and technocrats who dominate the world. Such people are more inclined to believe, as Herman Kahn and his colleagues argued in opposition to environmentalists, that: '200 years ago almost everywhere human beings were comparatively few, poor and at the mercy of the forces of nature, and 200 years from now, we expect, almost everywhere they will be numerous, rich and in control of the forces of nature.'[2] Establishing that there is an environmental crisis is not simply a matter of pointing to the facts to refute this view. The way the situation is interpreted is largely an expression of people's basic assumptions about the nature of the world and their place within it.

According to the assumptions of most people in positions of power, there cannot be an environmental crisis. There can only be more or less efficient control of nature, and separate, isolated environmental problems which can be treated independently of each other. Since life is assumed to be essentially a struggle against others for survival, problems are taken to be of significance only when someone's own interests are affected. The destruction of other businesses, other people and other species is just part of life, part of economic, political and evolutionary progress in which the weak and inefficient are being weeded out by the strong and efficient. Those who see a global environmental crisis on the other hand, do so because they are tacitly rejecting such assumptions. Environmentalists tend to see ecosystems not just as groups of individual organisms but as fragile communities. They tend to see a global politico-economic order threatening the stability of the world ecosystem as a whole. More significantly, despite the prevailing culture, many environmentalists see intrinsic value in the world: in non-human life, in the lives of the impoverished in the peripheries of the world economy, in human potentialities beyond the capacity to survive and consume, and in the future of humanity and of life on earth. They refuse to accept the nihilistic implications of the prevailing culture. As such they are seen as immature, wooly-minded idealists unable to face reality.

Between these poles lie a multiplicity of perspectives on the environment associated with different standpoints, different social and cultural contexts and different life experiences. Environmental problems are defined differently through different social systems and discursive formations - scientific, political, economic, educational and legal, and through different media, quite apart from how they are defined through different cultures and from the standpoints of different regions, different nations and different social classes. Such a multiplicity of perspectives highlights the problematic status not only of any claim that there is a global environmental crisis,

---

1. Thomas A. Sancton, 'Planet of the Year', *Time*, Jan.2, 1989, p.14.

2. Herman Kahn, William Brown and Leon Martel *The Next 200 Years: A Scenario for America and the World*, London: Associated Business Programs, 1977, p.1.

but also the 'God's eye' perspective assumed by any effort to characterize the global situation.[3] It has to be recognized that the very idea of a global environmental crisis is a social construct produced by a certain class of people from particular social, cultural and institutional standpoints.

That the global environmental crisis is a construct from particular standpoints will be accepted here, not in order to subvert its claim to truth, but to show how it is possible to achieve a global perspective from particular standpoints, and to defend such a construct as essential to humanity's continued self-creation as a viable process within nature. To do this, and to reveal the nature and extent of environmental problems, an orientation to the world in terms of which the aspiration to achieving a global perspective and the concerns and values of the environmentalists can be justified will be presupposed, while the situated nature of all such aspirations, concerns and values will be acknowledged. The environmental situation will be described from this orientation, bringing together the variety of issues which have aroused the concerns of people from all walks of life. The picture which will be conveyed is one in which, as Richard Barnet put it: 'There is a misfit between politics and the natural order which neither economists nor scientists nor corporate executives nor government bureaucrats quite understand.'[4] It is a world in which people throughout the world are living and acting rationally in terms of prevailing assumptions; but in doing so are producing effects far beyond their intentions. They remain blind to these effects because they continue to interpret the world in terms of these assumptions.

## *The Degradation of Non-Human Life*

Such effects are clearly evident in the degradation of non-human life. Nearly forty per cent of the earth's land-based photosynthetic activity is now devoted to human needs or has been destroyed by human activity.[5] Wilderness areas consisting of unique species and ecosystems are being destroyed to make way for domesticated forms of life, while domesticated forms of life are being denatured through breeding in order to eliminate features not useful to humans. Agriculture throughout the world is being reorganized into large scale, highly mechanized agribusinesses which subordinate everything to the goal of maximising profits. 'Economic progress' is rapidly leading to the state of the world predicted by J.S. Mill:

> ...with nothing left to the spontaneous activity of nature; with every rood of land brought into cultivation, which is capable of growing food for human beings; every flowery waste or natural pasture ploughed up, all quadrupeds or birds which are not domesticated for man's use exterminated as his rivals for food, every hedgerow or superfluous tree rooted out, and scarcely a place left where a wild shrub or flower could grow without being eradicated as a weed in the name of improved agriculture.[6]

Farms are becoming factories in which '[a]nimals are treated like machines that convert low-priced fodder into high-priced flesh, and any innovation that results in a cheaper "conversion-ratio" is liable to be adopted.'[7] Exemplifying this, the three billion chickens killed each year in USA have been bred for easy packaging and are grown factory fashion in filthy, highly overcrowded conditions. The control of life to maximise profits is accelerating with the development of genetic engineering and the construction of 'trans-genic' animals: animals in which genetic material from

---

3. On this see Niklas Luhmann in *Ecological Communication*, tr. John Bednarz, Cambridge: Polity Press, 1989.

4. Richard J. Barnet, *The Lean Years*, London: Abacus, 1981, p.16.

5. Peter M. Vitousek et. al., 'Human Appropriation of the Products of Photosynthesis', *Bioscience*, June, 1986.

6. J.S. Mill, *Principles of Political Economy* [1848 and later] *Collected Works*, Vol. III, Toronto: University of Toronto Press, 1965, p.756.

7. Peter Singer *Animal Liberation: A New Ethics for Our Treatment of Animals* N.Y.: Avon Books, 1975, p.94.

one breed of animal is spliced onto the genetic material of others. And it is not only farm animals that are caught up in the industrial system. In the United States alone up to 200 million animals, including 250,000 monkeys and apes, are killed annually in experiments testing the toxicity of new chemicals.[8] As a consequence, chimpanzees are now threatened with extinction.

All forms of life, human and non-human, which do not serve the immediate purposes of agribusiness or which compete with it are being displaced or destroyed. There has always been extinction of species. However no species in the past has had as big an impact on other species as humans. Between 1600 and 1900 increased human activity led to the known extinction of roughly one species every four years.[9] Since then the rate of extinction has accelerated alarmingly. As Paul and Anne Ehrlich argued, 'only in the last half century has it become clear that humanity has been forcing species and populations to extinction at a rate greatly exceeding that of natural attrition and far beyond the rate at which natural processes can replace them.'[10] It has been estimated that by the year 2000 up to 20% of all the species which existed in 1975 will have been irretrievably lost.[11] If present trends continue the result will be a biological catastrophe greater than all the mass extinctions of the geological past, including that which led to the extinction of the dinosaurs.

## *The Limits of Resources*

While efforts to dominate nature increase, people are increasingly being denied the minimum requirements for life. More than a billion people suffer from serious hunger or malnutrition for at least part of the year, with infants, growing children and pregnant mothers being the most affected.[12] Although it is very difficult to estimate precisely, somewhere around 50,000 people now die each day as a result of malnutrition. Those who survive malnutrition are frequently permanently affected by it. The poor of the world are also facing a 'firewood crisis,' and firewood now consumes a third of their incomes.[13] This situation is worsening.

Deprivation is a consequence of the general destruction of the world's resources: arable land, vegetation, water, fish, minerals and energy. The effect of farming has been to reduce the fertility of the land and destroy vegetation. It has been estimated that, against the 1,500 million hectares of land currently in use for crop production, nearly 2 billion hectares have been lost in historical times.[14] Much of this has been due to soil erosion. K.W. Butzer has estimated that in 150 years the agricultural soil resources of USA have been cut by half, and in some areas such as Oklahoma, a single generation sufficed to destroy almost 30% of the soil matter.[15] However this is insignificant by comparison with the rate of soil destruction in the Third World, and, world-wide, net annual soil loss is now some 26 billion tons.[16] Along with this, 80% of the dry rangelands, 60% of the rain fed croplands, and a third of all irrigated lands on earth are affected by the march of the

---

8. Richard Ryder, 'Experiments on Animals' in *Animal Rights and Human Obligations* Tom Regan and Peter Singer eds, New Jersey: Prentice-Hall, 1976, pp.33-47, p.34.

9. Norman Myers, *The Sinking Ark: A New Look at the Problem of Disappearing Species*, Oxford, N.Y.: Pergamon Press, 1979, p.4f.

10. Paul and Anne Ehrlich, *Extinction*, London: Victor Gollancz, 1982, p.xiii.

11. *The Global 2000 Report to the President*, Gerald O. Barney, Study Director, Harmondsworth: Penguin, 1982, p.3.

12. Lester R. Brown and Erik P. Eckholm 'Man, Food, and Environment' in William W. Murdoch, ed., *Environment: Resources, Pollution and Society*, Sunderland Mass.: Sinauer, 2nd ed. 1975, pp.67-94, p.70.

13. On the firewood crisis, see Bina Agarwal, *Cold Hearths and Barren Slopes*, London: Zed Books, 1986.

14. See Essam El-Hinnawi and Manzur-Ul-Haque Hashmi, *Global Environmental Issues*, for the U.N. Environment Programme, Dublin: Dublin: Tycooly, 1982, p.194f.

15. K.W. Butzer, 'Accelerated Soil Erosion: A Problem of Man-Land Relationships' in *Perspectives on the Environment*, Ian R. Manners and Marvin W. Mikesell eds, Washington: Assoc. of American Geographers, 1974, pp.57-78, p.71.

16. Lester Brown et. al. eds, *State of the World: 1988*, N.Y.: Norton & Co., p.6.

deserts. 45 million square kilometres are in immediate danger of turning into desert, putting the lives of 850 million people at risk.[17] And about half the world's irrigated land has been damaged to some degree by salinity, alkalinity and waterlogging.[18] 27 million hectares of agricultural land were lost to production in 1980 and this will increase to 100 million hectares per year by 2000.[19] As a consequence of such degradation, grain production has peaked. While between 1950 and 1984 grain production increased by 3% a year, between 1984 and 1989 it increased by only 1 percent, despite billions of dollars invested in agriculture and a 14% increase in the use of fertilizers. At the same time the accelerating destruction of forests is threatening the livelihood of the 140 million people now living in and around closed forests.[20]

Pesticides, fertilizers and modern breeding practices which have been hailed as the saviours of humanity in the struggle for food are now beginning to threaten its supply. In recent years there has been some relief from famine with the cultivation of hybrid strains of rice which produce yields on average three times greater than the old strains. This is the so-called Green Revolution. But this rice produces grain which is deficient in protein, it requires the continual application of fertilizer, it is exceptionally vulnerable to pests and requires enormous quantities of pesticides, and since the same rice strain is now cultivated throughout entire countries there are likely to be years in which crops sustain catastrophic pest damage. Fertilizers are now acidifying the soil and crop yields have started to decline as a consequence. If the same methods of agriculture continue to be used, the soil will be so damaged that there will be lower yields than with the original crops. By then, however, the original strains will no longer be cultivatable, even if they are preserved: old strains are dying out, eroding the genetic potential of crops to adapt to new situations which might arise in the future. Farmers who formerly raised fish in their rice paddies are now deprived of this protein source by the high levels of pesticide in the water. Pesticides are also less effective and increasingly destructive in their side-effects, not only in rice paddies, but in agriculture throughout the world. They are disrupting ecological systems, destroying natural predators and weakening the natural defences of crops, while the pests themselves are becoming immune to the pesticides. Thus in the United States, despite a ten fold increase in the use of pesticides, there was a two fold increase in crop losses due to insects between the 1940's and the 1970's.[21] And monocultural cropping, together with the reduction of genetic diversity through controlled breeding, has increased the likelihood of devastating crop disease.

Despite such problems people still hold out hope for some technological fix, most recently from biotechnology. However genetic engineering has proved more difficult to apply to agriculture than its champions expected, and there is little reasons to believe that expectations will ever be fulfilled. As Lester Brown and John Young have noted:

> As recently as 1984, one writer predicted that "in 5 to 10 years, Saudi Arabia may look like the wheat fields of Kansas." The unfortunate reality in 1989 - when Kansas lost over a third of its wheat crop to drought - was that the fields of Kansas came to resemble the still fallow Saudi Arabian desert.[22]

---

17. From the first report of the Drylands Project, United Nations Environmental Programme, 1987; reported by Julian Cribb, 'Drylands dilemma is a heritage of disaster', *The Australian*, Wednesday, July 22, 1987, p.16.

18. *The Global 2000 Report to the President*, p.33.

19. El-Hinnawi and Hashmi, *Global Environmental Issues*, p.194.

20. See *The Vanishing Forests: A Report for the Independent Commission on International Humanitarian Issues*: London, Zed Books, 1986, p.16.

21. See J.C. Harsanyi, 'Rule Utilitarianism and Decision Theory', *Erkenntnis*, Vol.II, 1977, pp.25-54.

22. Lester R. Brown and John E. Young, 'Feeding the World in the Nineties' in *State of the World 1990*, Lester R. Brown et.al. eds, Sydney: Allen & Unwin, 1990, p.68.

Further disruption of agriculture will occur as the effects of greenhouse gases come to be felt. Increasing temperatures will also raise the sea level from somewhere between half a metre and two metres over the next century. This will affect 5 million square kilometres, an area encompassing one third of the world's cropland and home to a billion people.[23] A three metre rise would be enough to flood almost all of Bangladesh, (which in 1988 suffered devastating floods due to overclearing of trees in the catchment areas of the Ganges and Brahmaputra Rivers).

Development of cities, the growth of population and new forms of agriculture and industry are producing a critical shortage of water in many parts of the world, including some areas within the affluent nations. While water resources are being developed in many places the benefits from this will be short-lived and the effects frequently deleterious in the long run. A considerable amount of the water now being exploited throughout the world derives from underground aquifers which are being used up faster than they are being replenished.[24] Dams frequently silt up, while irrigation tends to salinate the soil. The damming of the Nile illustrates this. Unlike the irrigation from the Tigris and Euphrates Rivers in ancient Babylonia which led to salination of the soil and the consequent destruction of this civilization, irrigation from the Nile has been conducted for thousands of years without deleterious effects because flood waters have been allowed to run off irrigated land. The construction of the Aswan Dam has changed this and irrigation is now leaving deposits of salt. This is exacerbated by the residues of the expensive fertilizers which are now required because silt is deposited in the dam rather than on the land. Furthermore less water is now available for irrigation because a third of the water in the dam evaporates, the reduction of fresh water entering the Mediterranean Sea has resulted in the destruction of the Egyptian sardine fishery, and irrigation canals have brought about a proliferation of the snails which carry schistosomiasis. The quality of much available water in the world is decreasing because of pollution and salination, and the major problem remains of getting water to where it is needed. The prospects of overcoming this water shortage through desalination and purification of local water or by towing icebergs from the South Pole are limited because of the costs in work, energy and materials required for this purpose.

The oceans which in the past have been regarded as a vast under-utilized resource are now beginning to reveal their limits. Despite developments in technology and greater efforts to catch fish, catches have been falling since 1970.[25] The improved technology is also driving some important species of fish, such as the bluefin tuna, to extinction. The Peruvian anchovetta, which once provided about 10 million tonnes of fish a year, has yielded almost nothing since a final, disastrously large harvest in 1970-71. The vast schools of herring have all but disappeared from northern European waters, and the breeding grounds of fish such as estuaries and coastal marshlands are being destroyed all around the world. Pollution is also affecting catches. There has been so much pollution in the seas off the coast of USA that fish catches have dropped by a third, and those caught are often riddled with toxic chemicals and suffering from finrot, a new disease where the fins of contaminated fish erode and their internal organs disintegrate. On the other side of the Atlantic poison algae, fed by chemical fertilizers washed into the sea by northern European rivers, killed millions of fish as it drifted along the Danish, Swedish and Norwegian coasts in 1988. Another by-product of pollution has been a continuing fall off in the level of oxygen in the Baltic Sea. If such effects spread, and they are likely to with such developments as off-shore oil wells, the effects on ocean life could be disastrous.

---

23. Jodi l. Jacobson, 'Holding Back the Sea', in *State of the World 1990*, Lester R. Brown et. al. eds, Sydney: Allen & Unwin, pp.79-97.

24. *The Global 2000 Report to the President*, Ch.9

25. *The Global 2000 Report to the President*, p.105; John Gulland, 'The Harvest of the Sea' in Murdoch *Environment*, 2nd ed. pp.167-189; and 'How to Fish' in *The Economist*, 10-16 December, 1988, pp.95-98.

## *The Threat to the World Ecosystem*

Humanity's transformations of nature are now beginning to interfere with the complex self-stabilizing processes of the world ecosystem through which optimum conditions for life, particularly human life, are maintained, and on which its stability and resilience depends. It has been convincingly argued by James Lovelock in his book *Gaia* that the whole of life on earth functions as an ecosystem which maintains the conditions for its continued existence:

> ...the entire range of living matter on Earth, from whales to viruses, from oaks to algae, could be regarded as constituting a single living entity, capable of manipulating the Earth's atmosphere to suit its overall needs and endowed with faculties and powers far beyond those of its constituent parts.[26]

The original reasons for proposing this hypothesis were the evidence that over 3,500 million years the earth's climate has changed little despite changing output from the sun, changing surface properties of the earth and changing composition of the atmosphere. Also the chemical composition of the atmosphere bears no relation to the expectations of steady-state equilibrium but, as with the climate, is maintained at an optimum value for life. For instance the surface temperature of the earth has been maintained at a constant temperature from between 15°C and 30°C (or between 288°A and 303°A) despite increases in the sun's radiation over the last 4 billion years of somewhere between 1.3 and 3.3 times.[27] Lovelock pointed out that relatively small departures (in absolute terms) from these optimum levels would have had disastrous consequences for life.

If the world can be considered to be one large ecosystem, then it can be expected that it will have many of the characteristics of particular ecosystems. Studies of these have shown them to be self-stabilizing with varying degrees of resilience.[28] Interference with ecosystems might appear to have no appreciable effects while undermining this resilience. Only when there is a sudden collapse of the ecosystem is this weakness revealed. This was dramatically illustrated in the case of Lake Erie which was transformed from an ecosystem comprising a large and varied population of fish to a simple ecosystem consisting of a far smaller variety of fish in a very short time period.[29] As C.S. Holling wrote of this:

> Whatever the specific causes, it is clear that the precondition for collapse was set by the harvesting of the fish, even though during a long period there were no obvious signs of problems. The fishing activity, however, progressively reduced the resilience of the system, so that when the inevitable unexpected event occurred, the populations collapsed.[30]

This same fate is now befalling the world-ecosystem: its resilience is being reduced, paving the way for collapse, or at least, radical modification.

The stability of the world ecosystem, as with most ecosystems, depends on its diversity.[31] Yet the domestication of nature, mainly by agriculture and forestry, is destroying complex ecosystems and replacing them with simple ones. These are frequently unstable and give way to deserts. In *The*

---

26. J.E. Lovelock *Gaia: A New Look at Life on Earth* Oxford, O.U.P. 1979, p.9.

27. James Lovelock and Sidney Epton, 'The Quest for Gaia' in *New Scientist*, 6th February, 1975.

28. C.S. Holling 'Resilience and Stability of Ecosystems' in *Evolution and Consciousness*, Erich Jantsch and Conrad H. Waddington eds, Reading, Mass.: Addison-Wesley, 1976, pp.73-92.

29. This collapse is described by Barry Commoner in *The Closing Circle: Confronting the Environmental Crisis*, London: Jonathon Cape, 1971, Ch.5.

30. Holling, p.79.

31. James Lovelock has shown why diversity results in stability in *The Ages of Gaia*, Oxford: O.U.P., 1988, Ch.3, 'Exploring Daisyworld'.

*Global 2000 Report to the President* it was estimated that each year 8 million hectares of cropland and grassland are being reduced to barren wasteland, while forests are now disappearing at a rate of 18-20 million hectares a year, an area about half the size of California, out of the 2,600 million hectares of closed forest remaining.[32] Satellite photography now shows that the situation is much worse: 30 million hectares of forest are being destroyed annually. About two-thirds of this is rain-forest which contain 40% of the world's species. Furthermore, despite some efforts to reverse the trend, the rate of forest destruction is increasing.

Humans are also beginning to affect the world ecosystem by increasing the temperature of the atmosphere.[33] The single most important cause of this is the increasing level of carbon dioxide in the atmosphere. Fossil fuel combustion and destruction of forests acting as carbon "sinks" are the most important causes of this. It is now believed that a doubling of the concentration from the background 270 ppm will increase the average temperature of the earth by 2.0°C. Prior to 1974 the $CO_2$ concentration was increasing at a rate of 4% a year. This was reduced to 1% a year by huge increases in the price of oil, the spread of nuclear power stations and a global recession, but by 1988 this had again risen to 3.7%. Other greenhouse gasses - methane, chlorofluorocarbons and nitrous oxide, are between them responsible for as much heat retention as $CO_2$. Methane levels are increasing at a rate of 1.1% a year, and its concentration has increased from 0.7 to 1.68 parts per million over the past three hundred years.[34] The heating process associated with the greenhouse effect was hidden in the Northern Hemisphere between 1940 and 1970 by the cooling effect of volcanic activity. However since 1970 temperatures have continued the upward trend which began in the 1880's when the levels of carbon dioxide first begun to increase significantly. The 1980's was the warmest decade on record, and the US Environmental Protection Agency has warned that the greenhouse effect will be felt even more acutely in the 1990's.[35] The greenhouse effect is expected by most scientists to increase temperatures throughout the world by between 2.5°C and 5.5°C over the next hundred years - a change greater than the change between the last ice age and the present. However the change could be greater than this, as the heating of the atmosphere will destroy large tracts of forest and affect humus in the soil, which could release more $CO_2$ into the atmosphere than will be released by burning fossil fuels. At the same time the amount of $CO_2$ absorbed by the oceans will decrease from 40% to 25%.[36] It is important to note that this change in temperature will be unequally distributed. Only slight changes will occur at the equator, while towards the poles the temperatures will increase by three times the average. This will radically change weather patterns. The 1988 drought in USA, which can be directly attributed to the greenhouse effect, foreshadows permanent reductions in rainfall, hotter summers and colder winters in the temperate regions of the world, including both USA and Europe.[37] And the greenhouse effect is only part of the problem. Humans are increasing their generation of heat at an exponential rate. While this is of no great significance at present, it has been estimated by Robert Ayers and Allen Kneese that if energy emission continued to increases at the present rate, in 250 years it would be

---

32. *The Global 2000 Report to the President*, p.2 and 118.

33. On this see John Griffin, *Future Weather: Carbon Dioxide, Climate and the Greenhouse Effect*, [1982], Harmondsworth: Penguin, 1983; and *Hole in the Sky: Man's Threat to the Ozone Layer*, N.Y.: Bantam, 1988, Ch.5.

34. *New Scientist*, 30th June, 1988, p.45.

35. If $CO_2$ were ever to reach 1% of the atmosphere the increase in temperature would no longer be linear, and the temperature on earth would rapidly increase to boiling point. See Lovelock, *Gaia*, p.45.

36. See Fred Pearce, 'Forgotten feedback disrupts the greenhouse', *New Scientist*, 10 Dec. 1988, p.24, reporting on the work of Gundolf Kohlmaier.

37. Ian Anderson, 'Greenhouse warming grips American corn belt', *New Scientist*, 30th June, 1988, p.35, and Gribbin, *The Hole in the Sky*, p.98ff.

equivalent to 100% of the absorbed solar flux. This alone would increase the earth's temperature by 50°C.[38]

At the same time the amount of ozone in the upper atmosphere, which at present is shielding the earth's surface from ultra-violet radiation, has been reduced by 2 per cent, and a large hole has appeared over the South Pole and a smaller hole over the North Pole. This is almost certainly due to the increasing levels of chlorofluorocarbons, methyl chloroform, carbon tetrachloride and nitrous oxide in the atmosphere.[39] Some idea of the effect this will have can be gained from the way wild animals, domestic animals and humans in Tierre del Fuego, the southernmost part of South American, have been blinded. But apart from damaging humans and other forms of life, the destruction of ozone is likely to completely destabilize the world's climate. It is hard to estimate the extent of the threat posed by this development. John Gribbin who had published a book in 1982 arguing that there are far more important things to worry about than the greenhouse effect,[40] published another in 1988 on the effect of ozone depletion in which he completely reversed his position. In this he pointed out that the depletion of ozone together with the greenhouse effect will result in a drop in temperature of the upper stratosphere by 30°C. The effects of this when combined with heating in the lower atmosphere are likely to be dramatic. As Gribbin wrote:

> Changes in the circulation of the atmosphere over the Southern Hemisphere are now clearly implicated in establishing conditions that allow chlorine compounds to produce a dramatic decrease in ozone concentrations over Antarctica each spring. The 'worst case' scenario that might result from this would be if the depletion of ozone itself caused a strengthening of the atmospheric conditions that set up the chemical containment vessel. Such a positive feedback could change the climate of at least the Southern Hemisphere and perhaps the whole globe, into a state that has never been experienced by human beings.[41]

Since the records of past climates frozen in the Greenland ice sheet reveal that the last the steady weather of the last 10,000 years, the period of the emergence and development of civilizations, has been highly abnormal, even freakish, it is unlikely future climates will be favourable to humanity.[42]

All this is associated with an increasingly rapid destruction of non-human forms of life. This destruction of vegetation parallels and interacts with the levels of pollution in the atmosphere. With less vegetation there is a reduced rate of absorption of pollutants. This is particularly true of carbon dioxide since rainforests have always played a major role in removing carbon from the air and burying it. The destruction of the Amazon forest alone is estimated to be responsible for 10% of $CO_2$ emitted into the atmosphere each year. Pollution in turn interferes with the growth of vegetation. The most significant manifestation of atmospheric pollution is the acid rain of Northern Europe and North America.[43] The U.S. Forest Service has found that the annual growth of yellow pines declined by 30 to 50% between 1955 and 1985, and this seems to be a typical effect of increased pollution on temperate forests.[44] In Europe, where pollution is greater, the forests are dying. The destruction of rainforests is reducing the cloud cover, reducing the amount of heat

---

38. *Economic and Ecological Effects of a Steady State*, Resources for the Future Reprint No. 99, December 1972, p.16, cited by Robert Heilbroner in *An Enquiry Into The Human Prospect*, N.Y.: Norton & Co., 1975, p.51.

39. See Richard S. Stolarski, 'The Antarctic Ozone Hole', *Scientific American*, Jan. 1988, Vol.258. No.1, pp.20-27, and Gribbin, *The Hole in the Sky*.

40. In *Future Weather* Gribbin concluded that: 'The puzzle of future weather is fascinating, but it is nothing compared with the real problems facing the world today, and it would be a mistake to divert too much attention to the possibilities of either a greenhouse Earth or a Little Ice Age.'(p.253).

41. Gribbin, *The Hole in the Sky*, p.152f.

42. Research on this was reported in *The Age* (Melbourne), Mon.19, Feb., 1996.

43. See *The State of the Environment*, Paris: OECD Report, and Fred Pearce, *Acid Rain*, Harmondsworth: Penguin, 1987.

44. Brown et. al. *State of the World 1990*, 1990, p.6.

reflected from the earth. Consequently this destruction of life is even more significant than it appears to be, and could prevent an adequate response on the part of the world ecosystem to the rising temperatures.

These are only the most dramatic examples of humanity's destruction. It is unknown what long term effects the build up in nature of chemically stable pesticides will have.[45] One effect has been the 'feminization' of males of all species, which now have lower sperm counts and smaller penises.[46] It has been estimated that 25 per cent of all the DDT produced is now in the ocean,[47] and it has been shown that synthetic chemicals in widespread use affect the species composition of plankton. Plankton, the foundation of the ocean ecosystem, is also under threat from the depletion of ozone, as it is destroyed by ultra-violet light. This will affect the oceanic food chain and further reduce the absorption of $CO_2$; and it could also affect the role of plankton in the ocean's governance of the cycles that rule the biosphere.[48] And as James Lovelock has noted, it is unobserved areas which are likely to be most important to the future of life on earth:

> The really critical areas which need careful watching are more likely to be in the tropics and the seas close to the continental shores. It is these regions, where few do watch, that harmful practices may be pursued to the point of no-return before their dangers are recognized; and so it is from these regions that unpleasant surprises are most likely to emerge. Here man may sap the vitality of Gaia by reducing productivity and by depleting key species in her life-support system; and he may then exacerbate the situation by releasing into the air or the sea abnormal quantities of compounds that are potentially dangerous on a global scale.[49]

The final state produced by this destabilization is likely to be very unfavourable to most existing life forms - including humans. In 1988 Lovelock remarked to John Gribbin: 'People sometimes have the attitude that "Gaia will look after us." But that's wrong. If the concept means anything at all, Gaia will look after *herself*. And the best way for her to do that might well be to get rid of us.'[50]

## *The Exhaustion of Reserves*

The extraordinarily rapid increase in consumption of non-renewable mineral reserves has been well documented.[51] Between 1954 and 1980 Americans consumed more minerals than the rest of mankind in all previous history.[52] Each person in the US now requires 40,000 lbs of minerals a year, Europeans and Japanese are beginning to consume reserves at a similar rate, and there are a number of newly industrializing nations making their own demands on the world's reserves. At this rate it seems inevitable that problems will arise.

However there is considerable dispute about the significance of this. Preston Cloud estimated in 1973 that by the year 2042, current rates of consumption would have exhausted half the presently

---

45. Rachel Carson's book *Silent Spring*, N.Y.: Fawcett Crest, 1962 remains a good account of the danger of pesticides.

46. On this, see Gail vines, 'Some of our Sperm are Missing', *New Scientist*, No. 1992, 26 August, 1995, pp. 22-25.

47. Barry Commoner, *The Closing Circle: Confronting the Environmental Crisis*, London: Jonathon Cape, 1972, p.227

48. William Ophuls, *Ecology and the Politics of Scarcity*, San Francisco: Freeman, 1977, p.24

49. Lovelock, *Gaia*, p.121.

50. Gribbin, *The Hole in the Sky*, p.161.

51. See Earl Cook, 'Limits to Exploitation of Nonrenewable Resources' in Herman E. Daly, ed., *Economics, Ecology, Ethics: Essays Toward a Steady-State Economy*, San Francisco: Freeman, 1980, pp.82-99.

52. Richard Barnet, 'The Minerals Plunder' in *The National Times* July 6 to 12, 1980, p.43f. reprinted from the *New Yorker*. See also Richard Barnet, *The Lean Years*, London, Abacus, 1980, esp. Ch.5.

known recoverable reserves of half the world's now useful metals.[53] Discovery of further reserves have shown this date to be mistaken, and it is being argued by a number of people that the earth is so resourceful that there will never be a minerals shortage. The most notable proponents of this view have been Herman Kahn and his colleagues.[54] They have pointed out that new reserves have been discovered far beyond what had been anticipated in the early 1970's.[55] However increased reserves alone are of little significance if there is an exponential rate of increase in their consumption. Their basic argument is that there cannot be shortage because minerals are so abundant. It is simply a matter of developing the technology to exploit the diverse forms, extracting minerals from lower concentrations and digging deeper into the earth's crust; and in those rare cases where minerals are approaching exhaustion, of working out how to substitute different minerals. And, they argue that ultimately ore could be extracted even from sea water and high grade rocks. In relation to energy it is admitted that there is an imminent shortage of petroleum, but it is argued that substitutes for this will be found. When all the possible sources of energy are considered: other fossil fuels, geothermal, solar, nuclear fission and fusion, it is clear that there can be no greater problem than the development of the technology to exploit them.

This interpretation of the situation ignores the increased work and energy required to extract minerals, the lower efficiency of substitutes, and the effect of increased demand for work, energy and substitutes occurring simultaneously. Only six minerals are abundant in the earth's crust, and the concentrating processes which have produced easily exploitable ore bodies only occur near the surface.[56] Once these are used up, there is an enormous leap downwards in the grade of ore, and to obtain mercury, tin, zinc and other scarce metals from rocks as suggested by Kahn would require a staggering amount of effort, energy and environmental destruction. To obtain 400 tons of zinc (US annual demand is 1,300,000 tons) would require the processing with perfect efficiency of 5 million tons of rock.[57] The absurdity of anticipating being able to substitute for scarce metals is evident from the table of possible substitutions listed by Kahn. He proposes zinc as a substitute for lead, though zinc is scarcer than lead; and at the same time, tungsten as a substitute for molybdenum and molybdenum as a substitute for tungsten.[58]

The question of energy is more complex, since it is usable energy that matters, and diverse uses require diverse forms. Thus while as much energy arrives in the form of sunlight every two weeks as is contained in the entire reserves of fossil fuels, it is difficult to use this.[59] The reserves of the most useful form of energy, easily transportable liquid, are running out. If petroleum continues to be consumed at the present rate reserves will be used up in 110 years if all hypothetical and speculated reserves prove to exist.[60] Optimism about energy supplies are generally a manifestation of the failure to take into account the complete context of the problem.[61]

One proposal for obtaining liquid fuel is through the liquefaction of coal, but quite apart from the increased pollution this would produce it would also require the use of astronomic amounts of

---

53. Preston Cloud, 'Mineral Resources in Fact and Fancy', in Herman E. Daly, ed., *Toward a Steady-State Economy*, San Francisco: Freeman, 1973, pp.50-75, p.74.

54. See for instance Julian Simon and Herman Kahn *The Resourceful Earth*, Oxford: Blackwell, 1984.

55. Kahn et.al. *The Next Two Hundred Years*, Ch.4.

56. Ophuls, *Ecology and the Politics of Scarcity*, p.67.

57. Ibid. p.69.

58. Kahn et. al. *The Next 200 Years*, p.87.

59. Nicholas Georgescu-Roegen 'Selections from "Energy and Economic Myths"' in Daly *Economics, Ecology, Ethics*, pp.61-81, p.70.

60. William Ophuls, *Ecology and the Politics of Scarcity*, p.88. *The Global 2000 Report* gives a lower figure (see p.190), while a higher figure is given by P.R. Odell and K. E. Rosing in *The Future of Oil*, 2nd ed., London: Kogan Page, 1983.

61. Certainly a more optimistic stance is justified than prevailed in the early 1970's. See Peter R. Odell, 'Draining the World of Energy' in R.J. Johnston and P.J. Taylor, *A World in Crisis? Geographical Perspectives*, Oxford, Basil Blackwell, 1986 for a more recent account of the issue.

water. Another suggestion is that crops be developed for the production of ethanol, but studies of this have shown that it requires as much energy to produce energy in this way as is yielded.[62] Furthermore, enabling the rich to continue driving their cars by this method involves converting land from food production to fuel production, which inevitably results in more starvation. This has occurred in Brazil where the production of alcohol from sugar cane has also led to the pollution of rivers, and peasants forced off their lands are destroying the Amazon forests in their struggle for survival.

Where energy in general is concerned it is still thought by many that nuclear fuel is the answer. The fire at Chernobyl nuclear power plant in April, 1986 should dispel any illusions about the safety of this form of energy. This led to the release of 50 million curies of radioactive iodine and 6 million curies of radioactive caesium in the first week after the disaster, while another 6 million curies of radioactive strontium were released within the borders of the Soviet Union.[63] It is now thought that up to 10,000 people have died as a result of exposure to such pollution. Despite raising the 'permissible' annual dosage of radiation to 20 times the top international level of 0.5 rems, 20,000 sq km of agricultural land north and east of Chernobyl have had to be abandoned for normal use. This will not be a short term measure, as radioactive strontium which binds strongly to the soil has a half-life of 30 years and will contaminate the environment for centuries. Outside the former Soviet Union record levels of radioactive caesium have been found in wild animals, lake fish, wild mushrooms and forest berries. Apart from such dangers, without breeder reactors there will be a serious shortage of usable uranium within a few decades, and if breeder reactors are used there will be all the problems associated with the production of plutonium which can be used in nuclear bombs, which is highly toxic and which has a half life of 25,000 years. There are also shortages of helium and water for cooling. Helium is in short supply, and projections into the future have suggested that in the US, if the nuclear path had been adopted, by the year 2000 water equal to between one third and a half of all the fresh-water runoff in USA would have been required for cooling.[64] If nuclear fusion is ever effectively mastered, this will produce even more dangerous radioactive wastes than nuclear fission reactors. Filling the energy needs of Britain would produce 100,000 tons of unsafe waste each year.[65]

Kahn and his colleagues have posed the problem of mineral depletion and energy shortages in a way that skirts around the real issues. These have been stated succinctly by Earl Cook:

> To society ... the profit from mining (including oil and gas extraction) can be defined either as an energy surplus, as from the exploitation of fossil and nuclear fuel deposits, or as a work saving, as in the lessened expenditure of human energy and time when steel is used in place of wood in tools and structures.... The ultimate limit to exploitation of earth resources then is the limit of net energy profit (or work savings).[66]

Energy and work costs for the recovery of minerals have begun to increase with great rapidity as decreasing grades of ore are having to be used.[67] This increase has been disguised until recently because of the decreasing costs of energy used. Now that the limits of exploitation of energy itself has been reached, and now that the costs of recovery for some minerals are moving up the steeper parts of their exponential curves, the limits to exploitation of reserves of minerals are beginning to

---

62. See Juan Martinez-Alier, *Ecological Economics*, Oxford: Basil Blackwell, 1987, pp.21-26.

63. See Zhores Medvedev, *Gorbachev*, Oxford: Blackwell, 1987, pp.259-267 for an account of this.

64. Ophuls, *The Politics of Scarcity*, p.91.

65. As reported by Britain's National Radiological Protection Board (NRPB). Reported in *The Economist*, Vol.305, No.7530-31, 26 Dec. 1987 - 8 Jan. 1988, p.106.

66. Earl Cook, 'Limits to Exploitation of Nonrenewable Resources', *Science*, Vol.191, 20th February, 1976, pp.677-82, p.677.

67. Ibid. p.97

be recognized. Underlying this problem is the inevitability of failure in attempting to solve resource shortages through improved technology alone. According to the second law of thermodynamics, all creation of useful order destroys at least an equal amount of such order. Improved access to solar or nuclear energy will not solve the problem because such energy cannot offset the dispersion of minerals and the pollution associated with refining ores and using machinery. And as Georgescu-Roegen pointed out, 'since no practical procedure is available at human scale for transforming energy into matter ... accessible material low entropy is by far the most critical element from the bioeconomic viewpoint.'[68]

## *Population Growth and Inequitable Resource Use*

Both a cause and effect of environmental problems, the human population of the world is growing inexorably. Throughout human history the rate of population growth has been accelerating. This is clearly evident from the rapid reduction in doubling times for the world population throughout history:[69]

Doubling Times For Human Population Growth

| Date | Est. World Population | Years required to double |
|---|---|---|
| 8000 B.C. | 5 million | |
| 1650 B.C. | 500 million | 1,500 |
| 1850 A.D. | 1,000 million | 200 |
| 1930 A.D. | 2,000 million | 80 |
| 1975 A.D. | 4,000 million | 35 |

Globally, there has been scarcely any reduction in these trends, as is evident from figures on world population growth by decade between 1950 and 1990:

| Year | Population (billion) | Total increase (million) | Annual increase (million) |
|---|---|---|---|
| 1950 | 2.515 | | |
| 1960 | 3.019 | 504 | 50 |
| 1970 | 3.698 | 679 | 68 |
| 1980 | 4.450 | 752 | 75 |
| 1990 | 5.292 | 842 | 84 |

While the extent of global problems can be judged by considering aggregate figures, this obscures the real nature of the problem. Population growth is not uniform throughout the world. Affluent countries are often characterized by negative population growth, while it is in the poor peripheries of the world that population is growing most rapidly. These different rates of population growth are not unconnected. The global economy functions as a system consisting of economic core zones competing with each other and increasing their flow-throughs of energy and materials with low population growth, and peripheral regions of the world economy with high rates

68. Nicholas Georgescu-Roegen, 'Selections from "Energy and Economic Myths"', in Herman E. Daly ed. *Economics, Ecology, Ethics*, p.69f.

69. Paul Ehrlich, Anne Ehrlich, Robert Holdren, *Human Ecology: Problems and Solutions*, San Francisco: Freeman, 1973.

of population growth yielding resources to core zones.[70] In this competition the core zones are increasing their power to dominate and exploit the peripheries, and inevitably compete more intensely for control over their resources and reserves, while the peripheries themselves are progressively impoverished as their resources are destroyed and reserves exhausted. The disruption, impoverishment and consequent insecurity of people in the peripheries impels them to have as many children as the can.[71]

The system as a whole engenders more intense exploitation of the environment. The search for new reserves and resources has driven European imperialism since the fifteenth century. It led to the European drive for total world domination in the nineteenth century and has culminated in the global power conflicts of the twentieth century. It is impossible to account for the First and Second World Wars and the subsequent Cold Wars except in terms of the expanding economies of the core zones of the world economy requiring more resources and reserves than available within their borders. In the first half of the twentieth century, Germany, Japan and Italy, as latecomers to industrial capitalism, were struggling to gain secure access to external resources and reserves. The recognition of the imminent shortage of these by the powerful nations of the world since the Second World War has been responsible for the increasing levels of military expenditure and growing oppression throughout the world. Today, however, the struggle for control of resources in the peripheries of the world economy has replaced old fashioned imperialism centred on the control of territory. This neo-imperialism has frequently resulted in the destabilisation of reformist governments and the imposition of corrupt military or semi-military dictatorships in Third World countries.

This is what has been referred to in the United States as defence of the 'Free World'. But 'free' here refers to the freedom of enterprise, of American based transnational businesses to extract raw materials or produce commodities for the world market; freedom to exploit the available human and natural resources and to destroy the environment.[72] It has nothing to do with democratic governments and has included some of the world's most barbarously oppressive regimes. Kennedy's ambassador to Brazil and later Assistant Secretary of State for Inter-American Affairs, Lincoln Gordon, described the overthrow of the democratic government of Brazil in 1964 and its replacement by a military dictatorship as 'the single most decisive victory for freedom in the mid-twentieth century' and 'one of the critical points of inflection in mid-twentieth century world history.'[73] That this regime subsequently murdered its effective critics and initiated the onslaught on the forests of the Amazon basin which is today the single most important threat to the world ecosystem meant nothing to the defenders of 'freedom'.

Eighty five percent of materials consumed in the world economy are now used by less than 20% of the world's population. USA imports more than 95% of its mica, strontium, cobalt, manganese and titanium and over 80% of its aluminium, asbestos, platinum and tin. It imports more than half of 23 of 38 basic minerals. The EEC countries and Japan import even higher percentages of their minerals, mostly from the Third World. The developed countries import more than twice the value of food from the underdeveloped countries as they export, even without taking into account the fish taken by the wealthy nations from the waters surrounding Third World countries. Generally the food exported by the poor countries is of a higher nutritional value than that imported. Three-

---

70. On this see Stephen Bunker, *Underdeveloping the Amazon*, [1985], Chicago: The University of Chicago Press, 1988, esp. Ch.1.

71. That this is the situation has been argued by Richard Newbold Adams, *The Eighth Day*, Austin: University of Texas Press, 1988, p.133ff. and p.242.

72. Noam Chomsky designated this the Fifth freedom: 'the freedom to rob and exploit ... the one that really counts.' *Turning the Tide: US Intervention in Central America and the Struggle for Peace*, Boston: South End Press, 1985, p.47. See also Noam Chomsky, *The Culture of Terrorism*, London: Pluto Press, 1988.

73. Cited by Noam Chomsky in *Towards a New Cold War*, N.Y.: Pantheon, c.1983, p.24.

quarters of the fruit and vegetables consumed in the US come from the Third World.[74] And the First World is importing huge amounts of timber and forest products, destroying Third World forests at ever increasing rates.

The USA is not the only country responsible for exploiting Third World countries. Western European nations, and France in particular, are actively involved in supporting oppressive governments in Africa to maintain control over their investments. France maintains a standing army for just this purpose, and in 1977 and 1978 provided logistic support to Belgian and Moroccan troops who had been sent to prop up the corrupt government of Zaïre, the source of many of France's minerals. However, since the Second World War it has been the United States which has been most responsible for oppressing people in the Third World. Its actions directed towards undermining democratic governments have not been *ad hoc* responses to particular situations but have been part of a long term strategy worked out during and after the end of the Second World War. Realizing that they would be the dominant power, US policy planners worked out what areas of the world would need to be controlled in order to ensure their supply of resources. These included most of the world. US planners also realized that they would be in competition with indigenous populations for these resources. One of the principle architects of this was George Kennan, the Director of the Policy Planning Staff of the Department of State. His ideas and basic commitments can now be more clearly understood in the light of recently declassified planning documents. Kennan, the inspiration behind the policy of 'containment' of communism, based his policy recommendations on the belief that '[i]ndustrial capacity, together with access to raw materials necessary to sustain it, was the key to power in the world'.[75] Breaking with the tradition based on universalist principles, that is, the concern with international justice which had been characteristic of the administration of Franklin Roosevelt, he argued in a top secret document PPS 23 in 1948:

> [W]e would be better off to dispense now with a number of the concepts which have underlined our thinking ... We should dispense with the aspiration to "be liked" or to be regarded as a repository of a high-minded international altruism. We should stop putting ourselves in the position of being our brothers' keeper ... We should cease to talk in vague and - for the Far East - unreal objectives such as human rights, the raising of the living standards, and democratization. The day is not far off when we are going to have to deal in straight power concepts. The less we are then hampered by idealistic slogans, the better... We should make a careful study to see what parts of the Pacific and Far Eastern world are absolutely vital to our security, and we should concentrate our policy on seeing to it that those areas remain in hands which we can control or rely on.[76]

In carrying out this policy the US government engineered or supported the destruction of relatively democratic governments in Iran, Guatemala, Ecuador, the Congo, Brazil, Dominica, Peru, Indonesia, Cambodia, Ghana, Chile, Greece and Timor in their efforts to keep these nations

---

74. These figures are given by F.E. Trainer, *Abandon Affluence*, London: Zed Books, 1985, pp.130f and 142ff.

75. John Lewis Gaddis, 'The Strategy of Containment', in Thomas Etzold and John Lewis Gaddis, eds. *Containment: Documents on American Policy and Strategy, 1945-1950*, N.Y.: Columbia University Press, 1978, pp.25-37, p.27.

76. From an excerpt in Etzold and Gaddis, *Containment*, p. 227. The full document can be found in *Foreign Relations of the United States*, (FRUS) 1948, I (part 2). It is discussed along with other documents of a similar ilk by Noam Chomsky in *Turning the Tide*, pp. 47ff. Kennan was eased out of his position in 1953 because of his lack of toughness towards the Communists and replaced by Paul Nitze, and recently, expressed concern about the possibility of a nuclear war. For a full study of Kennan and his policies, showing Kennan's concern with controlling resources, see Barton Gellman, *Contending with Kennan: Towards a Philosophy of American Power*, N.Y.: Praeger Press, 1984. It is noteworthy that this, like other studies of Kennan and Kennan's own memoirs do not refer to PPS 23.

in the Free World.[77] Since the early 1970's a major concern of the United States government has been that the sources of raw materials within the dollar zone are rapidly being exhausted so that the US has had to rely on countries which are considered unstable.[78] So while the most well known instances of US intervention have been into Central and South America and South East Asia, USA is becoming increasingly involved in Africa. As Fred Halliday noted in 1982.

> Western and Japanese dependence on raw material imports from the Third World is, moreover, by no means confined to petroleum. The strategic minerals essential to military production are also mainly extracted from underdeveloped areas - in this case, above all, Southern Africa. The tougher attitude of the Reagan Administration towards the advance of national liberation forces in this region finds part of its explanation in Washington's resolve to keep a firm protective grip on these minerals, whether cobalt (Zaïre), chromium (Zimbabwe), uranium (Namibia), or manganese (Gabon, South Africa): not to speak of the gold mines of the Rand itself.[79]

It is the tensions generated between and within nations by this struggle for increasingly scarce resources which underlies the growth in military spending throughout the world.[80] In 1990 some $US700 billion dollars was spent on the military, about 7% of the world's G.N.P. (in the run-up to the first two world wars it was never more than 3%). The military employed 60 million people and the research efforts of half the world's scientists. In absolute terms, the US spends more than any other country on the military, and in 1984 64% of its scientific research and development expenditure was on armaments. The ratio of military expenditure to fixed capital investment in the United States was then 46 to 100.[81] Scarce reserves are increasingly being used up in the struggle for their control.

However the First World does not dominate the Third World by sheer force. They could not succeed in their exploitation of the Third World without support from within. The success with which the US governments have been able to prosecute this policy, to undermine democratic governments and install repressive regimes, has been due to the increasing numbers of people in the Third World willing to act on their behalf in opposition to their own people and in return for a share of the spoils. Such support has been forthcoming as the resource crisis has made it increasingly obvious that it is impossible for all countries to enjoy the affluence of First World nations. As Dudley Seers noted in his analysis of the relationship between the present economic crisis, the crisis of resources, and the political situation within the Third World:

> Whereas dictatorships were rare in the mid-1960's, they are now very common. By 1980 there were over fifty governments in the world dominated by the military, of which the great majority were described as 'repressive'... The explanation seems to be, in brief, that the bureaucrats, traders, and white-collar (as well as blue-collar) employees in the modern sector, public and private, have become increasingly determined that they and their children shall continue to enjoy the modern lifestyle, largely imported, whatever the brutality and whatever the inflows of aid and private capital needed to ensure this.[82]

---

77. For the global role of the C.I.A. see William Blum, *The CIA: A Forgotten History*, London: Zed Books, 1986.

78. See Yann Fitt, Alexandre Faire and Jean-Pierre Vigier, *The World Economic Crisis: U.S. Imperialism at Bay*, tr. Michael Pallis, London: Zed Press, 1980.

79. Fred Halliday, 'The Sources of the New Cold War', in New Left Review ed. *Exterminism and Cold War*, Verso, 1982, pp.289-328, p.316.

80. This is argued by Noam Chomsky, 'Strategic Arms the Cold War and the Third World' and Fred Halliday in 'The Sources of the New Cold War', in *Exterminism and Cold War*, pp.223-236 and 289-238.

81. Chomsky, *Towards a New Cold War*, p.32, citing Seymour Melman.

82. Dudley Seers, *The Political Economy of Nationalism*, Oxford: Oxford University Press, 1983, p.6f.

Since this was written, a facade of democracy has been resurrected in many Third World nations, particularly in South America. However this merely disguises the extent to which real policy choices have been denied to the populations of these countries.[83]

Political oppression does not end with the subjugation of people in the Third World. A new dimension to the struggle for resources is emerging with what amounts to a deliberate exclusion by affluent nations of large proportions of their own populations from participation in their economies. This is evident in the consistent, if unsteady growth in unemployment in the OECD nations since the early 1970's (from 5 million in 1967 to 32 million in 1982) associated with the promotion of precisely those economic theories and strategies which led to the Great Depression of the 1930's. As Dudley Seers again points out:

> No evidence whatever is put forward for the assumption that Northern governments wish to reduce unemployment, even in their own countries. The governments concerned have justified monetarist policies as being designed to 'fight inflation' and set the stage for economic growth. But this does not quite ring true. Would any sane government with that ultimate objective forgo for years much investment in their industries, on which eventually growth depends, just to reduce by a few points the annual rise in the consumer price index - to rates which they may not be able to maintain anyway? An alternative hypothesis would be that to run the neo-colonial system is very difficult now, without heavy unemployment. It would not be necessarily ignorant or short-sighted for governments in the North to calculate (though it might be unwise to state publicly) that a rise of even 5% in the national products of OECD countries would make OPEC and its member governments far too powerful, and also lead to sharply higher prices for metals (as well as agricultural products).[84]

Economic policies which have made life insecure for the majority of young people in advanced Western nations have contributed to negative population growth.

## *The Urbanization of Humanity*

Corresponding to the degradation, destruction and exhaustion of the natural environment, people are enclosing themselves in built-up environments which are destroying their health, their human potential, and in particular, their capacity to respond to the world's environmental problems. One of the major recent changes in the world has been the growing proportion of people who live in cities, and the increasing size of these cities. This is not simply a function of population growth but is also a consequence of the destruction of rural communities as land is taken over by large scale agribusinesses. Between 1800 and 1950 the world's population increased by a factor of 2.6 while the number of people living in settlements over 20,000 increased by a factor of 23.[85] In 1850 there were only four cities in the world with more than a million people. In 1950 there were about a hundred such cities, and by 2000 there will be over 1000 cities of this magnitude.[86] In 1950, 29% of the world's population lived in urban settlements. This had increased to 39% in 1975 and it is projected that this will approach 50% by 2000. The cities in less developed countries in

---

83. For an analysis of the present state of the world, see Noam Chomsky, *World Orders, Old and New*, London: Pluto Press, 1994.

84. Dudley Seers, *The Political Economy of Nationalism*, Oxford: Oxford University Press, 1983, p.158. See also G.F. Ray, 'Europe's Farewell to Full Employment?' in Daniel Yergin and Martin Hillenbrand, *Global Insecurity*, Penguin: Harmondsworth, 1983.

85. *The Global 2000 Report to the President* p.241ff.

86. Gwen Bell and Jacqueline Tyrwhitt 'Introduction' to *Human Identity in the Urban Environment*, Gwen Bell and Jacqueline Tyrwhitt, eds, Harmondsworth: Penguin, 1972, p.15.

particular are expected to grow rapidly between now and 2000. Then, Mexico City will house 32 million persons and Sao Paulo more than 26 million. The growth of cities is putting extreme pressure on sanitation, water supplies, health care, food, shelter and jobs. Some idea of life in these cities can be gained by considering one example: Calcutta. Here 70% of families live in one-room houses, half the houses have no indoor toilets, there is only one water faucet for every 25 slum dwellings and 600,000 people are without housing altogether and live and die on the streets.[87] The condition of life in such cities is likely to get worse in the future. Most population growth will occur in the slums and shanty-towns.

Such environments are associated with increasingly high levels of pollution.[88] While there is some overlap, the problems of pollution can be divided between those associated with the general environment and those associated with places of work. It is now realized that the air we breath, the food we eat and the water and milk we drink are all polluted to some degree, and there is now so much toxic waste being produced (one metric ton per person each year in the United States) that escaping from it is becoming impossible. Every year more than 3,000 new chemical products enter the environment, and of the 48,000 chemicals listed by the EPA, next to nothing is known about the toxic effects of 38,000 of these. Fewer than 1,000 have been tested for acute effects, and only about 500 for their cancer causing, reproductive or mutagenic effects. Occupational pollution has been recognized as a problem since the Romans realized that people who worked in lead mines suffered from general ill health. However such problems were seldom taken seriously by those in power because the health of workers were not held to be of any great significance. This is still largely true, especially in Third World countries, despite the increasing occupational pollution workers are now being subjected to.

Research has revealed associations between pollution (and additives to foods, pesticides etc.) and birth defects, mutations, cancer and heart disease, and various chronic illnesses, particularly respiratory diseases such as asthma. The most carefully analysed relationships have been those between pollutants and cancer, the great majority of which are now recognized to be due to environmental factors. However heart disease has also been shown to be closely associated with exposure to some pollutants. Both cancer and heart disease have been shown to be strongly correlated with the levels of sulphur dioxide, oxides of nitrogen, carbon monoxide and particulate sulphate. There is a positive correlation coefficient of cancer with sulphur dioxide of 0.56 and with nitrogen dioxide of 0.48; and of arteriosclerotic heart disease with sulphur dioxide of 0.48 and with nitrogen dioxide of 0.31.[89]

Cancer is the most striking illness caused by pollution. As Samuel Epstein wrote of this disease in the US in his monumental *The Politics of Cancer*:

> If one thousand people died every day of cholera, swine flu, or food poisoning, an epidemic of major proportions would be at hand and the entire country would be mobilized against it. Yet cancer claims that many lives daily, often in prolonged and agonizing pain... Cancer is now a killing and disabling disease of epidemic proportions. More than 53 million people in the United States (over a quarter of the population) will develop some form of cancer, from which approximately 20 percent of the US population will die.[90]

Between 1900 and 1960 the number of deaths from cancer in the USA increased more than six fold. Half of this increase, about 44% of the total, was due to factors other than increased population and

---

87. Otto Friedrich 'And if Mexico City Seems Bad...', *Time*, August 6, 1984, p.22f.

88. The extent to which people are being exposed to pollution has been revealed by James Bellini in *High Tech Holocaust*, Richmond: Greenhouse Publications, 1987.

89. Richard J. Hickey, 'Air Pollution' in *Environment: Resources, Pollution and Society*, William W. Murdoch ed., 1st ed., Stamford: Sinauer, 1971, pp.189-212, p.201.

90. Samuel S. Epstein, *The Politics of Cancer*, San Francisco: Sierre Club Books, 1978, p's.1,8.

life expectancy; that is, pollution.[91] Epidemiological studies have led to the conclusion that environmental factors cause between 70 and 90 percent of all cancers.[92] As Epstein put it, 'Just as germs cause infections, so do certain chemical and physical agents, carcinogens, cause cancer.'[93] More people are dying of cancer because they are more exposed to carcinogens. The prospect for the future is even worse than it appears from this because pollutants act synergistically, that is, they reinforce each other. For instance while people who smoke are 15 times more likely to die from lung cancer than those who do not, and people who work in uranium mines are 4 times more likely to die from lung cancer, those who both smoke and work in uranium mines are 120 times more likely to die from lung cancer than those who do neither.

While the main focus of attention has been the effects of general pollution on people in the wealthy countries of the world, the most deleterious pollution is that associated with the workplace, with the worst affected being in the poorer countries of the world. Pollution which affects the affluent in the cities of the wealthy nations is being attacked with some vigour while little effort is made to combat worker exposure to toxic substances. In the US there are official standards for less than 500 of the tens of thousands of toxic substances to be found in the workplace, and there is a wide disparity between what is considered acceptable for the general public and what is acceptable for workers. It is consequently hardly surprising that the life expectancy of blue-collar workers is 15 to 20 years less than teachers who are not exposed to occupational pollutants.

Over recent years wealthy countries have begun to 'export' their pollution to poorer countries by exporting toxic chemicals or by transferring to them their most polluting industries, the most well known instance being Union Carbide's factory in India which in December, 1984 leaked chemicals which killed more than 3,100 people and injured more than 200,000 others, 86,000 permanently. The extent of deleterious pollution in the Third World has only just begun to be investigated. One of the worst places is Brazil where preliminary figures reveal pollution to be largely responsible for its high infant mortality. Cubatao in particular is famous for its pollution, and associated with this, for the number of babies born without brains (anencephaly).[94] 32.4 babies in every ten thousand live births in Cubatoa die from congenital defects, with 52% of these deaths being due to congenital anomalies of the central nervous system, as compared to 29% elsewhere in Brazil. 6.5 in ten thousand are anencephalic.

Even the affluent of the wealthy nations have not been able to fully escape the deleterious effects of environmental degradation. And it is not only the diseases which cause death which are important. Degenerative diseases are occurring faster than can be explained by the longer expectancy of life.[95] Chronic disabilities which people have come to accept as a normal part of life or of growing old: asthma, myopia, headaches, allergies, ulcers, depression, neuroses, dental caries, high blood pressure and senility are in fact the products of the environment we live in. The Mebans in the Sudan have been found to be completely free of blood pressure, obesity, coronary thrombosis, duodenal ulcers, ulcerative colitis, allergies, bronchial asthma and dental caries.[96] Like a number of African tribes who accepted as normal the hideous sores of primary and secondary yaws, we have become accustomed to chronic dis-ease. One of the symptoms of this is the rapid growth of the field of medicine and the amounts spent on health care by the affluent. As André Gorz has noted: 'People are medicating themselves more because they are more morbid, and the

---

91. Samuel S. Epstein and Dale Hattis 'Pollution and Human Health' in Murdoch, *Environment*, 2nd ed. pp.195-221, p.208.
92. Epstein *The Politics of Cancer*, p.23.
93. Ibid. p.2.
94. Roque Monteleone-Neto et.al. 'Birth Defects and Environmental Pollution: the Cubatao Example', in *Prevention of Physical and Mental Congenital Defects, Part B: Epidemiology, Early Detection and Therapy, and Environmental Factors*, Alan R. Liss, Inc., 1985, pp.65-68, p.66.
95. André Gorz, *Ecology as Politics*, tr. Patsy Vigderman and Jonathon Cloud, Boston: South End Press, 1980, p.159.
96. René Dubos, *Man Adapting*, New Haven and London: Y.U.P., 2nd ed. 1980, p.240.

very rapid increase in their medical consumption doesn't at all keep their morbidity from increasing right along with it.'[97]

The rise of the medical profession, including psychiatry, amounts to an effort to refabricate people by excisions, admixtures of chemicals and psychotherapy to live in environments which are increasingly pathogenic and soul destroying. But while the development of medicine is celebrated as one of the major achievements of our society, close investigation has revealed most of its claims to success to be false.[98] The great scourges: cholera, typhoid and tuberculosis for instance, had practically disappeared before treatment of them had been developed. It is the environment, living conditions, lifestyles and hygiene which have been the true causes of changes in incidence of disease, and the environment is now beginning to generate new diseases. Some of these diseases can be effectively prevented or treated by medicine, but the major effect of medicine is to prolong the length of illness before death and to reduce infant mortality, resulting in the survival of large numbers of people who limp through life with permanent disabilities. Life expectancy of Americans once they have reached the age of 45 has increased scarcely at all since the nineteenth century, and what increases there have been are due to the prolonging of illness rather than to increasing the healthy years of life.[99]

## *The Degradation of Humanity*

There are more subtle consequences to living in cities, and these affect the affluent as well as the poor. Living in a world where everything has been formed for human purposes detached from the natural world changes the nature of human life. Previously the products of human activity were always seen in a context of a dynamic nature as achievements fulfilling human purposes. In large cities the environment is almost entirely a human production and is impregnated with sedimented human purposes which impose their logic on its individual members. As Winston Churchill asserted: 'We shape our buildings, and afterwards they shape us.'[100] In modern cities today this means reducing people from members of a community to exchangeable units without individuality, mere instruments at best, superfluous rubbish at worst. The effects of this have been pointed out by Samir Amin:

> Why ... is it that we love the old cities, we even love Manhattan, but no one, not even the city planners who conceived it, dares defend the perfect functionalism of the latest 'achievements' of post-war capitalism?... Perfect functionalism is necessarily compartmentalized and linear. It is always functionalism in relation to some one thing, not in relation to the whole. Add it up; the fastest possible means of transportation...(to go to work), the quickest possible places to rest (to regenerate labor power), the closest possible places to shop.[101]

Under the superficial glitter and diversity of cities, the life-worlds of most people are being progressively impoverished. Irrespective of climate, topography or language, from Washington to Tokyo, from Anchorage to Melbourne, the same pattern is unfolding.

In modern cities the space of offices and office buildings, shops and shopping centres, of factories, of schoolrooms and schools, streets and suburbs, is organized to ensure that people

---

97. Gorz, *Ecology as Politics*, p.168.

98. See Ivan Illich, *Medical Nemesis: The Expropriation of Health*, New York: Pantheon, 1976.

99. Dubos, *Man Adapting*, p.230.

100. Cited by René Dubos in 'The Crisis of Man in His Environment' in Bell and Tyrwhitt, *Human Identity in the Urban Environment* pp.178-184, p.183.

101. Samir Amin, 'In Praise of Socialism', *Monthly Review*, September 1974, pp.12-13.

conform to social norms, that they perform the tasks expected of them and that they do not use it in any other way than is prescribed for them. Foucault contrasted this with the cities of the past:

> Antiquity had been a civilization of the spectacle. 'To render accessible to a multitude of men the inspection of a small number of objects': this was the problem to which the architecture of temples, theatres and circuses responded. With spectacle, there was a preponderance of public life, the intensity of festivals, sensual proximity. In these rituals in which blood flowed, society found new vigour and formed for a moment a single great body. The modern age poses the opposite problem: 'To procure for a small number, or even for a single individual, the instantaneous view of a great multitude.'[102]

Private space is then defined as rigidly distinct from this public space, but it also serves the requirements of the economy. It is essentially the space which neutralizes people's political power and where labour power is recuperated, where, as Samir Amin wrote: 'men sink into the necessary state of stupor ... where they make a feeble effort to withdraw into themselves ... where they are bored.'[103] The space within which people are free to determine their goals and relationships, where they can act with spontaneity, is rapidly being defined out of existence.

The effect of these developments is to have destroyed the human community. As Martin Pawley argued in *The Private Future*, there is now nothing but a vacant, terrorized space between the government - which controls and maintains production - and the isolated consumer, who increases his consumption in proportion to his isolation.[104] The life of dialogue has been replaced by passive consumption of the products of the mass media in which information is decontextualized, and irrelevancies invested with a quasi-relevance, creating 'a neighbourhood of strangers and pointless quantity; a world of fragments and discontinuities.'[105] Modern cities isolate people from each other and 'serialize' them. People are constituted as separate individuals related to each other and for each other by their exterior conditions as a contingent gathering. Consequently, as Sartre put it: '...isolation becomes, for and through everyone, for him and for others, the real social product of cities.'[106] This is the condition described by Miroslav Holub in his poem *Subway Station*:[107]

> This evening    Mr Howard T. Lewis,
> of unknown address,    gloomy and tired,
> wearing a grey overcoat and brown hat,
> having decided to take    the B.M.T., Carnarsie Line,
> met    at the last station on 8th Ave.
> a man    in a grey overcoat and brown hat
> whose face,    gloomy and tired, was
> the face of Mr    Howard T.Lewis,
> while by the barrier    at the end of the empty platform
> stood a man    in a grey overcoat, of gloomy appearance
> whose face    was also the face
> of Howard T.Lewis    and    gazed dumbly
> at the bottom of the dirty steps    down which came
> a man in a brown hat,    gloomy and tired,
> with a face that was the face of Howard T.Lewis.

---

102. Michel Foucault, *Discipline and Punish*, tr. Alan Sheridan, Harmondsworth: Perigrene, 1979, p.216.
103. Amin, 'In Praise of Socialism', p.13.
104. Martin Pawley, *The Private Future*, London: Thames and Hudson, 1973.
105. As Neil Postman put it in *Amusing Ourselves to Death*, [1985] London: Methuen, 1987, p.72.
106. Jean-Paul Sartre, *Critique of Dialectical Reason*, [1960], tr. Alan Sheridan-Smith, London: NLB, 1976, p.257.
107. Miroslav Holub, *Although*, London: Cape, p.16f.

And then    through the worn wooden spokes
of the turnstile    came a woman, tired and gloomy,
of unknown address    with a handbag and in a brown
hat    whose face was the face
of all men    and therefore also of Howard T.Lewis and
the steps in the distance    and the nervously muffled steps
near by, steps    of figures bowed by the murkiness
and pale from the light were the steps
of Howard T.Lewis,  steps    from an unknown address
to    an unknown address,    now and then
the turnstile turned again    with a snap like a head
dropping in the basket,    or behind the barrier
could be seen    a figure without sex and of no
address,    but otherwise completely like
Howard T.Lewis,    steps    were heard,
heads,    spokes,    distances,    lights    and    tunnels
sucked in the sign    8th Ave.    8th Ave.    8th Ave.
in droning crescendo.

When the train left    a stray wind
scattered the pages of a paper    in which there was
a report on the unknown
address,    fate    and identification
of a man    in a grey overcoat and brown hat,
gloomy and tired.

The privileged, usually people of European origin, live highly regulated lives moving between their comfortable houses to their hermetically sealed office blocks, shopping malls, fast food outlets, international airports and so on according to fixed schedules in accordance with increasingly accurate clocks and watches. Membership of this privileged strata is dependent upon permanent conformity, with individuals being required to be the right colour, account for each year of their lives, speak with the right accents and display the correct mannerisms to be acceptable for each successive stage of advancement. Any deviation can derail a person's career. With the exception of a small and ruthless minority, the only power even successful individuals can aspire to is to become bigger cogs with pre-defined functions. Power is formally in the hands of experts, but each expert is an exchangeable cog. People who dream of a better world, who step outside the prescribed roles to take up causes or to attempt to realize ideals, soon discover how costly this can be. For the upwardly mobile without such dreams, living has become mechanized for the instant gratification of needs at the expense of a complete loss of power to shape their lives or to contribute to the direction of society. Many accept these pre-packaged lives from school to university to work career to retirement home to morgue with equanimity; others live in quiet desperation. Some superficial non-conformity is allowed to members of the working class, but at the expense of low status, permanent insecurity, and even greater powerlessness. Below these are the increasingly large number of people for whom there is no place for in the economy, defined as parasites, deprived of all sense of self-worth and dignity and utterly powerless to protest against their position.

## *Defence of the Existing Order*

So far the general degrading of nature, the threat to the world's ecosystem, the destruction of resources and imminent exhaustion of reserves, the pollution and the degradation of human life associated with the agglomeration of the world's population in built-up environments and increasing impoverishment, oppression and organized violence which is exacerbating all the other problems and preventing their solution, have been described. However there are also the inter-relationships between side effects of these problems: attempts to address each of these major problems will lead to increasing demands for raw materials and will cause more pollution. To increase food supply will require the use of more energy and resources to produce fertilizers and pesticides, which in turn will further destroy the soil and cause more pollution. The impoverishment of people will generate increased population and this in turn will lead to greater concentrations of people in urban centres. Any breakthrough in the production of energy will cause more heat pollution, which itself could destroy the conditions for life on earth. But perhaps the most important problem is that increased exploitation of reserves and resources will increase the intensity of struggle between people and nations, forcing countries into greater competition. This will not only use up more materials in armaments and military activity, but will intensify the competition for economic growth to support larger military forces, create more resource shortages and intensify the arms race in a vicious circle. Greater military competition and societies more fully mobilized to maximize economic growth will result in greater propensity for war, including nuclear war. At the same time fewer people will be left with the means to even think about the long-term problems facing humanity or their root causes. The problem of such unintended side-effects was summed up by Richard Adams in his argument against those who assume that it is running out of energy which is the main problem:

> ...the problem is both larger and more strategic: it is not where will energy come from, but where it is leading us. If we were to obtain ... a reliable and endlessly expandable source of energy, the real problem would be precisely that it would lead to ever greater complexity and indeterminacy, producing nonlinearities beyond the coping ability of human intelligence.[108]

Yet there are many who believe that environmental problems are an insignificant by-product of world economic progress. A closer examination of these thinkers will reveal more clearly the blinding effect of prevailing assumptions.

Kahn and his colleagues argued that 200 years ago there were few people in the world, they were poor and at the mercy of the forces of nature, while 200 years from now there will be large numbers of people, they will be rich and in control of the forces of nature. Such optimism about the future is almost always associated with a fixation on the promise of technological development and economic growth. Human progress is seen as developing mastery over nature, and it is argued that technology has been continuously improving the conditions for humanity. Each technological challenge has resulted in innovations which have not only solved the original problems, but have been a real improvement on the original conditions. There is no reason to think that such technological advances will not continue to occur, provided enough money is spent on technology. Associated with this technological optimism is the belief in economic growth, which is supposed to have lifted people out of their poverty and put them in a position where they are already, or soon will be, free of the tyranny of nature.

It is this more than anything else which reveals the simplistic thinking of the optimists and their blindness to context. It is this more than anything else which exposes the illusions of people who have come to see the world in terms of a linear notion of progress. Not only does the notion of economic growth as it has generally been formulated ignore the destructive effects of humans on

---

108. Richard Newbold Adams, *The Eighth Day*, Austin: University of Texas Press, 1988, p.241.

their environments, but it also ignores the actual changes which have taken place in the quality of people's lives. Since growth is seen in terms of activity in an exchange economy, the transformation of a subsistence economy into a market economy is always seen as economic growth, no matter how worse off the lot of the general population. An economy which destroys people's health and thereby generates spending on treatment, is seen as more economically advanced. The inadequacy of using the exchange of commodities as a measure for economic well-being should be immediately evident from the Japanese 'economic miracle'. Nominally the wealthiest people in the world, the conditions of life for most Japanese is if anything, now deteriorating. The main component of increased wealth is simply higher prices for resources, in particular, land. What was previously free or cheap is now a major cost of living, while entirely new needs requiring more expenditure have also emerged. And one third of Japanese workers now suffer from nervous and mental disorders caused by stress.[109]

In actual fact, while technology has continually improved throughout history, the amount of work people have had to do has generally increased, while the conditions of life for the majority of the population have generally deteriorated. It has been pointed out by Marshall Sahlins that people in traditional societies, notably the Australian Aboriginals in Arnhem land and the Bushmen of the Kalahari desert, live in a state of affluence in which all their requirements can be met with a minimum amount of effort. Of the two thirds of Bushmen who work at all, the work week is approximately 15 hours not counting cooking and preparation of implements, while the Australian Aboriginals work for only four or five hours a day, including cooking and preparation of implements.[110] In each case, the tribesmen are completely confident of their ability to obtain a livelihood.

As documented by Eric Wolf in *Europe and the People Without History* and John H. Bodley in *Victims of Progress*, it is these original affluent societies which have suffered most over the last few hundred years.[111] When the English arrived in Australia in 1788 there were between 850,000 and 1,250,000 aboriginals.[112] By 1950 there were only 40,000 full blood aboriginals left, mostly living as despised, impoverished outcasts in a land they had inhabited for at least 40,000 years.[113] American Indians suffered a similar fate, both in North and South America. In Africa, European colonists inflicted massive destruction on traditional societies well into the twentieth century. When the Belgians took over the Congo basin and set about exploiting it to obtain rubber latex for the manufacture of tyres, they also set about exterminating all those who would not work for them, paying a bounty for the ears and hands of such people. In this way they reduced the population of the Belgian Congo from 25 million to 15 million people in twenty years.[114]

The conditions for people in non-European civilizations were in many cases better before the era of European imperialism than they have been since, and before the nineteenth century non-Europeans, the Chinese in particular, were certainly not at the mercy of nature.[115] The state of affluence of these people was such that Europeans could only obtain goods from these countries by virtue of their military prowess, since in most places non-Europeans were not in need of anything that Europe could offer them. While in 1700 people in the Middle East were suffering as a result of

---

109. Jacabo Schatan, *World Debt*, London: Zed Books, 1987, p.66.

110. Marshall Sahlins, 'The Original Affluent Society' in *Stone Age Economics*, London: Tavistock, 1974, p.1-39.

111. Eric R. Wolf, *Europe and the People Without History*, Berkeley: University of California Press, 1982; and John H. Bodley, *Victims of Progress*, Menlo Park, California, 1982. See also papers in A. Ugaldo, ed., *Studies in Third World Society*, Austin, Texas, 1986.

112. Noel Butlin, *Our Original Aggression*, Sydney: George Allen & Unwin, 1983.

113. The subjugation of the Australian continent by Europeans has been poignantly described by William J. Lines, *Taming of the Great South Land*, Sydney: Allen & Unwin, 1991.

114. Mark Twain, *King Leopold's Soliloquy*, New York: International Publishers, 1961, p.52.

115. For a comparative study of societies when Western Europe gained ascendency and developed capitalism see E.L. Jones, *The European Miracle*, Cambridge: Cambridge University Press, 1981.

the corruption and decline of the Ottoman empire, and in northern India under the yoke of the Mughals, other areas of the world were prospering before Europeans began to intrude. An historian of Indonesia noted that: 'when the first Dutch merchants and sailors had come to the island world of the Indies, they had been amazed by the variety of its nature and civilization, and the more observant among them had recognized that southern and eastern Asia were far ahead of western Europe in riches as well as in commercial ability and mercantile skill.' In the eighteenth century the Chinese emperor pointed out to George III that 'our celestial empire possesses all things in prolific abundance' and therefore had little need of English goods.[116] To overcome this problem the British introduced opium into China which led to the Opium War when the Chinese tried to put a stop to this trade. The British won. Trade with India was developed by destroying its textile industry and forcing Indians to buy British cotton goods. By the twentieth century European imperialism had totally destroyed the civilizations of the Americas, and contributed to the impoverishment of Egypt, the Middle East, India, the East Indies and China. India and China were left subject to devastating famines.[117] And the effect of 'economic growth' in the twentieth century has often been associated with a reduction in production per head of population. Thus it was possible to declare on the basis of figures published in 1955 that 'the economic well-being of the average person in the world outside the USSR was in 1956 less than in 1913 and perhaps less than in 1900.'[118] With the incredible growth of poverty in Africa, in parts of Asia and Central and South America, it is evident that this trend is continuing. The legacy of 200 years of European progress has been the destruction or impoverishment of most other societies and civilizations, culminating in a world economic system in which the majority of the world's population live in permanent poverty with little hope for the future.

The most significant effect of improvements in Western technology has been to facilitate the greater exploitation by some groups of people by others. One of the greatest areas of technological progress has been in the means of oppression, ranging from military hardware and the means to coordinate its use to computerized spying technology and the technology of mind control. However even the standard technology of industrial production is intimately associated with exploitation. It is often a way of reducing the labour time expended by some people by forcing others to spend at least equal time producing the means for them to do so. This is not to say that some technology does not have the potential to reduce people's workload, but this potential is probably overestimated, and is unlikely to be realized without radically changing the economic organization of the world.

One example illustrating this is the vast technological advances in transport. Ivan Illich calculated that the amount of time put in by the average American for travelling and paying for travel amounts to one hour for every five miles, the same as for those who walk.[119] However this calculation is based on averages of expenditure, income, time worked and time spent in travelling by Americans and does not take into account the different amounts of time required by different groups to pay for their travel, nor does it take into account the cheap labour of Third World countries involved in mining, refining, manufacturing and shipping necessary for the production of the means for this travel. If this is taken into account, the labour time facilitating each five miles travelled in USA is even greater than one hour. The importance of the technology is that it has made it possible for some people to save time travelling at the expense of others.

---

116. Keith Griffin, 'Underdevelopment in History' in *The Political Economy of Development and Underdevelopment*, ed. Charles K. Wilber, 2nd ed. New York: Random House, 1979, pp.77-90, p.79. See also A.J.S. Reid, 'The Origins of Poverty in Indonesia' in James J. Fox et. al. *Indonesia: Australian Perspectives*, Canberra: A.N.U. Press: 1980, pp. 441-454.

117. For the history of this see Andre Gunder Frank, *Dependent Accumulation and Underdevelopment*, New York and London: Monthly Review Press, 1979.

118. Arghiri Emmanuel, *Unequal Exchange*, New York: Monthly Review Press, 1972, p.262.

119. I.D. Illich, *Energy and Equity*, New York: Harper and Row, 1974, p.19.

The improvement of life for people in the West since the nineteenth century has been largely the consequence of the increase in workload and further impoverishment of people in the economic peripheries who have supplies cheap commodities to the West. Before the Second World War this was achieved for the most part by direct colonial rule by European nations of the rest of the world. Since then it has been achieved by direct and indirect intervention in Third World nations to saddle them with oppressive dictatorships willing to act in the interests of the wealthy nations, particularly the USA, by orienting their economies away from home consumption towards the world market. Andre Gunder Frank described the effect of this in 1984:

> In the case of Brazil ... since the military coup in 1964, wages were reduced by over 40%. In Argentina, since the military coup in 1976, wages were reduced by over 50%. But already before the coup real wages were going down as a result of the economic policy of the right wing of the Peronist government in 1974-75. In Chile, real wages since the coup were reduced by two thirds, that is to say, from an index of 100 almost to an index of 30, and unemployment increased from 4% to 20%, fell to 12% and rose to 30%. To be able to do this it was necessary first to destroy or to control the unions, to eliminate - often physically - the leadership, to repress all political opposition, and to throw people in jail, torture them, murder them, exile them, and so forth.[120]

The International Labour Organization reported in 1987 that real incomes of labourers had fallen by up to 40% in South America and Africa south of the Sahara.

So-called economic progress is also proving increasingly costly to the members of affluent countries. The new international division of labour forces workers in Western societies to compete with the oppressed workers of the Third World while at the same time Western governments are financing their military budgets by cutting down on social services, welfare, health and education expenditure, thereby simultaneously reducing the size of the middle class.[121] It has been argued that capitalism has now transcended all national boundaries, and that the notion of a separate Third World no longer makes sense.[122] But with very few exceptions the conditions of life in the Third World are not improving or actually deteriorating, while the loss, or abandonment, of State sovereignty over their economies of First World countries has led to a growing deterioration of conditions for the majority of the population. Throughout the 1980's some 30 million people in the industrialized (OECD) nations were jobless. A study undertaken by physicians headed by Dr J.L. Brown entitled *Hunger in America: The Growing Epidemic* found that 20 million Americans were suffering from chronic hunger, and that the problem was worsening.[123] Three million Americans were homeless. Looking at the effects of agribusiness on North American agriculture, Jon Bennett pointed out:

> Parallels with the poorer South are increasingly apparent: fewer and fewer people control greater amounts of land; absentee landowners - a landed 'aristocracy' - is beginning to emerge; and the cost is measured in terms of joblessness as rural communities are gutted. Here, as in the developing world, wages are inappropriate to the needs of the struggling families, and a reckless disregard for conservation bodes ill for future generations.[124]

---

120. Andre Gunder Frank, *Critique and Anti-Critique*, N.Y.: Praeger, 1984, p.213f.

121. Folker Frobel, Jürgen Heinrichs and Otto Kreye, *The New International Division of Labour,* [1977] tr. Peter Burgess, Cambridge: Cambridge University Press, 1980.

122. Nigel Harris, *The End of the Third World*, Harmondsworth: Penguin, 1986.

123. *Harvard University Gazette,* Vol. LXXX, No.24, March 1, 1985, p.1.

124. Jon Bennett with Susan George, *The Hunger Machine*, Cambridge: The Polity Press, 1987, p.188.

Then in 1990 the world went into a deep recession. The current economic crisis heralds not the end of the Third World, but the beginning of the end of the First World.

Where there are better conditions associated with improvements in the efficiency of production, this is almost always offset by additional costs elsewhere, manifest in the continual production of new necessities which individuals must pay for to participate in the economy. Ivan Illich pointed out that not only do people in America spend the same amount of time travelling each mile as do people in societies deprived of traffic industries, but they are forced to spend a vastly greater proportion of their lives travelling and paying for travel. While people in traditional societies spend 3% to 8% of their time travelling, people in USA spend 28% of their time travelling or working to pay for their travel. And this is fairly typical of the apparent advances in Western societies.[125] In his essay on 'The Growth of Affluence and the Decline of Welfare', E.J. Mishan described similar cases in which what is represented as improvements in the quality of life are actually part of its deterioration. Books, journals, education, newspapers, vacations to recover from stress, employment agencies, marriage bureaus and the like were simply not needed in more traditional societies. The growth in modern education is not an advantage to the individual. Individuals have to spend more of their lives studying in highly competitive environments just to be employable. As Mishan pointed out, 'This sort of education is not education in the classical sense. It is not education in the humanities. It has no affinities with art or culture of civilized living.... [T]he universities, the centres today of what cynics call the knowledge industry, are, in the nature of things, no longer able to produce educated men, men of cultivated intelligence. They are geared to produce specialists...'[126] More must be spent on medicine to combat increasing disease. The increased mobility in modern life, whether associated with moving between workplaces or exchanging friendships and spouses, is not a liberation from old constraints. It is the imposition of lifestyles which demand of people that they give themselves totally to their careers to succeed, that they acknowledge no other constraints.

Social life has deteriorated accordingly.[127] There has been a breakdown of communities, the family, all the institutions which gave people's lives meaning and direction, while the increasing stress of meaningless work and intensified competition has produced an increasingly instrumental orientation in interpersonal relations. Mishan has described the effects of this:

> The unavoidable frustration resulting from this act of social vandalism has, alas, only aggravated man's lust for power and sparked his hopes with technological fantasies that can only remove him further from human fulfilment. The resulting despair has begot a craving to pierce more wantonly the seeming repressive integument of the social order, a craving expressed in the feverish search for novelty and excitement which, while admirably serving the forces of economic expansion, leads unerringly to the totalitarian state.[128]

---

125. On the low correlation between economic growth and social welfare, particularly when ability to sustain such welfare is taken into account, see Herman E. Daly and John B. Cobb, Jr., 'Misplaced Concreteness: Measuring Economic Success', *For the Common Good: Redirecting the Economy toward Community, the Environment, and a Sustainable Future*, Boston: Beacon Press, 1989, Ch.3.

126. E.J. Mishan, 'The Growth of Affluence and the Decline of Welfare' in Daly ed. *Economics, Ecology, Ethics*, pp.267-281, p.272.

127. As described by E.J. Mishan in *The Costs of Economic Growth*, [1967], Harmondsworth: Penguin, 1969.

128. Mishan, 'The Growth of Affluence and the Decline of Welfare', p.279f.

## *Conclusion*

Optimism about the future is due to the failure to consider the full range of environmental problems, and to indifference to the suffering of others. These problems cannot be understood in isolation from each other, or from other political, social and economic problems, either within nations or between nations; and they are forcing an awareness of the blindness of the modes of thought which try to do so. Such optimism therefore manifests the general problem which has been revealed by this chapter: the domination of society by a linear, abstracting form of thinking which blinds people to the side-effects of their efforts to control the world: the destabilization of the world ecosystem, the exhaustion of reserves and the destruction of resources, the growing conflicts between people and the concomitant militarization of societies, pollution, and the impoverishment of social and cultural life. This form of thinking has also blinded people to the significance of what is being affected or degraded: other life forms, ecosystems, people in the economic peripheries, future generations, and ultimately, their own lives. The environmental crisis reveals the need for developing modes of thought capable of analysing the inter-relationships between such diverse problems and for appreciating the significance and meaning of life. And it raises fundamental ethical and social issues. Why should people concern themselves with non-human life, future generations, oppression in the economic peripheries or the plight of the poor? And why in fact do so few people do so? In raising these issues environmental problems reveal the nihilism pervading modern civilization. Finally, the question presents itself of how can those who do accept responsibility for these problems act effectively. In the following chapter the efforts to address these problems will be examined, and it will be shown that even the ideas of those who have confronted environmental issues are vitiated by the prevailing forms of thought.

# 2

# RESPONSES TO ENVIRONMENTAL PROBLEMS

While there has been spasmodic concern about the environment since the nineteenth century, it was only in the late 1960's and early 1970's that it became a major international issue. Then after 1975, interest in environmental problems subsided, except in West Germany.[1] The economic crisis which began in 1973 and continued to worsen until 1982 forced environmental problems out of the limelight. Only after 1986 with unemployment temporarily in decline, after the Chernobyl nuclear reactor melt-down, with evidence of the greenhouse effect, with a hole in the ozone layer, and with increasing levels of pollution, did the environment become a major issue again. However by mid-1991, interest in the environment was again waning and up until 1996 has continued to wane. Under these circumstances, although green movements have gained some power and a number of green parties have been able to establish themselves, only superficial environmental problems have been addressed to any effect.[2] There have been a few efforts in the more affluent nations to do something about environmental problems of immediate economic significance or which have captured the public imagination - economising on the use of oil, sending some food to the latest victims of famine, reducing the visible forms of pollution or the pollution which is obviously killing people in the immediate vicinity, saving a few wilderness areas and preserving a few species of animals. More recently there has been some international co-operation to consider controlling the production of chlorofluorocarbons and international condemnation of Brazil's destructive economic policies towards the Amazon. A United Nations commission, headed by the Norwegian Prime Minister Gro Brundtland, has also investigated the more general relationship between environmental problems and economic development.[3] And there has been a major international conference on the environment. But, for the most part, these efforts have been merely reactive, dealing with problems in isolation without any effort to address their basic causes. Action has often been cosmetic, designed for electoral success or public relations rather than practical effect. When seen against the background of all the problems discussed in the previous chapter it should be immediately evident just how inadequate these responses have been.[4]

It is clear from the way the environmental movement lost momentum in the 1970's and again in the 90's that it is almost impossible to mobilize people to the extent necessary to deal effectively with environmental problems while they are economically insecure; and people will continue to be insecure for as long as the world economy continues to develop along its present path. However current economic instability is itself largely a surface manifestation of the far more serious environmental crisis:

---

1. On the decline of the early environmentalist movement see Francis Sandbach, *Environment, Ideology and Policy*, Oxford: Blackwell, 1980, pp.1-20.

2. For the history of the global environmental movement see John McCormick, *Reclaiming Paradise*, Bloomginton: Indiana University Press, 1989.

3. The findings of this commission were published as *Our Common Future* by Oxford University Press in 1987. For a critique of this, see Thijs De La Court, *Beyond Brundtland: Green Development in the 1990s*, [1980] tr. Ed Bayens and Nigel Harle, London and New Jersey: Zed Books, 1990. See also William E. Rees, 'Sustainable Development: Economic Development and Ecological Realities', *The Trumpeter*, Vol.5, No.4, 1988, pp.133-138.

4. For a good overview of the successes and failures of the environmentalists see Jonathon Porritt & David Winner, *The Coming of the Greens*, London: Fontana, 1988, and John McCormick, *Reclaiming Paradise: The Global Environmental Movement*, Bloomington: Indiana University Press, 1989.

> We are entering a period in which resource limits can no longer be ignored, nor can the interests of different sections of the world be assumed compatible: to solve one country's problems may well be to aggravate those of another... The [economic] crisis is not just a cyclical downturn nor even ... the slack phase in a hypothetical Kondratieff cycle... [It is] the culmination of a period of increasing strains on the world's productive structures, natural resources, and political systems. Thus a swift rise in world output would soon reveal shortages in oil, various minerals, and food, and increased international tension...[5]

The soaring prices for minerals, for wheat, wool and other primary products after 1987 bore out Seers' predictions.

The failure to address the full extent of environmental problems and their causes, and the failure of green political parties after some initial electoral successes, suggests an intellectual failure and a failure of imagination to face up to the magnitude and complexity of the crisis confronting the world. In this chapter the intellectual responses to the environmental crisis will be examined - with the exception of the work of Marxists and those ecological economists who are so alien to the spirit of European civilization that, until very recently, they have been unable to gain formal recognition within any Western country.[6] Marxists and ecological economists will be examined in later chapters.

Because of the diversity of approaches taken only the major trends in environmentalism will be presented, although inevitably this will not do justice to all those involved. But the major concern will not be with particular limitations, but to show that there is something fundamentally wrong with our culture - with the way people think about problems, particularly ethical and political problems. Dominant cultural forms have made it almost impossible to get the fundamental problems of society into perspective. To confront environmental problems it will be necessary to radically transcend the modes of thought which at present pervade society, to develop fundamentally new ways of thinking about the world and our place within it. Ethical ideas have come to be isolated from general discourse. Unless they reinforce tendencies which manifest more deeply rooted conceptions people have about themselves and their place in the world, they have become virtually irrelevant to the way people live.

One of the most important effects of prevailing metaphysical assumptions has been the prevalence of the dogma that interpretation of the world is separate from ethics and political philosophy, that scientific knowledge is separate from evaluation. Yet those scientists who attempt to explain environmental problems almost invariably make proposals for dealing with them on the basis of their explanations, and these proposals are taken far more seriously than the arguments of philosophers. The chapter will therefore begin by examining the proposed explanations of the problems and the associated proposals for solutions, showing the relationship between the nature of explanations and the proposals before looking at the work of philosophers and political theorists. Efforts have been made to account for environmental problems in terms of attitudes to nature, population growth, the nature of technology, and economic growth. Each of these will be examined in turn, showing how a defective notion of causation has led to simplistic diagnoses and to proposals which are not only inadequate, but which are oppressive and dangerous.

Attention will be turned to what philosophers and some philosophically oriented political scientists have said about environmental problems and three trends in environmental philosophy -

5. Dudley Seers, *The Political Economy of Nationalism*, Oxford: Oxford University Press, 1983, p.1.

6. Juan Martinez-Alier pointed out in the concluding chapter of *Ecological Economics* (Oxford: Blackwell, 1987) that: 'There have never been university chairs in ecological economics. Our writers lacked an academic forum (because of the separation and professionalization of the sciences) where their arguments could be presented. There was also a crucial absence of plausible political groups, outside academia, which could adopt them as ideologues.' (p.234)

representing the major stances in modern ethico/political thinking - will be identified.[7] The first follows the mainstream of ethical thought and attempts to extend rights theory and utilitarianism to encompass our relationship to future generations, animals and plants. It is characterized by a belief in moral progress in which irrelevant forms of moral discrimination are being overcome. Rights theory, utilitarianism and the notion of moral progress are the ethico/political doctrines which have been formulated on the foundation of the 'scientific' view of the world. The second trend is associated with Christianity. It is characterized by efforts to draw on the Christian tradition of thought and to defend Western thought and institutions against those who have argued for a radically new ethics and radical political action. The third trend associated with the 'deep ecologists' belongs to the tradition of 'romantic expressivism'. It is characterized by efforts to develop new ways of seeing people in relation to each other and to nature as a foundation for a new ethical and political philosophy and a new social and economic order. Arguing that we are part of nature, that nature has intrinsic value, and that the goal of life is to realize our potential, they come nearest to developing their ideas in terms of a new metaphysics.

## *Attitudes to the World*

The effort to explain the environmental crisis in terms of the attitudes to nature dominating Western society has produced more controversy and resulted in more research than any other proposed explanation. While the foremost contributors to this debate are Lynn White, Clarence Glacken,[8] Robin Attfield, Hans Jonas, John Passmore and Yi-Fu Tuan, there are also a great number of lesser known contributors. The thinker who has argued the strongest thesis along these lines, and who has consequently aroused the greatest controversy, is Lynn White.[9]

White argues that: 'What people do about their ecology depends on what they think about themselves in relation to things around them,'[10] and that the forms of thinking which have led to environmental problems in Western society have their roots in Christianity. According to White, Christianity accepted the Judaic conception of linear time and developed a view of both God and man as transcending nature. This was a radical break from pagan notions of a cyclical time and a world inhabited by spirits. However Eastern and Western Churches developed these ideas in different directions. While the Greek Churches of the East believed that sin was intellectual blindness and salvation was to be found in illumination, orthodoxy, or clear thinking, the Latin West believed sin was a moral evil and salvation was to be found in right conduct. As White wrote: 'The Greek saint contemplates; the Western saint acts.'[11] Finally, while both Greeks and Latins saw understanding nature as a means of attaining a better understanding of God, the Greeks saw nature as a symbolic system through which God speaks to us while the Latins attempted to understand God

7. These three trends have been identified by Charles Taylor in *Sources of the Self*, Cambridge: Harvard University Press, 1989. See especially p.495ff. for a summary of these.

8. Clarence Glacken's *magnum opus* is *Traces on the Rhodian Shore: Nature and Culture in Western Thought from Ancient Times to the End of the Eighteenth Century*, Berkeley, LA: University of California Press, 1967. However Glacken has contributed a number of papers to this problem including 'Changing Ideas of the Habitable World' in William L. Thomas (ed.) *Man's Role in Changing the Face of the Earth*, Chicago: University of Chicago Press, 1956, pp.70-92 and 'Man Against Nature: An Outmoded Concept' in Harold W. Helfrich Jr. *The Environmental Crisis*, New Haven and London: Yale University Press, 1970. The works of the other thinkers referred to here will be footnoted later.

9. Lynn White Jr. 'The Historical Roots of our Ecological Crisis', first published in *Science* Vol.155, (10 March, 1967) pp.1203-1207, republished in Ian G. Barbour (ed.) *Western Man and Environmental Ethics: Attitudes Towards Nature and Technology*, Reading: Addison-Wesley, 1973, pp.18-30. These arguments have been further developed in *Medieval Religion and Technology: Collected Essays*, Berkeley: University of California Press, 1978.

10. White in Barbour, p.24.

11. Ibid. p.26.

by discovering how his creation operates. The effect of these ideas of the Latin West in medieval society was to promote the development of technology and the domination of nature, conceived of as part of the perpetual progress of humanity. And White has provided overwhelming evidence that there was a far greater emphasis on developing technology in Western Europe than in any other civilization right from the beginning of the feudal era. This provided the impetus which led to the global imperialism of Western Europe, to the development of mechanistic science, to the emergence of industrial capitalism, and finally to the global ecological crisis.

Most of the criticisms of White's arguments are misdirected. They usually centre on White's interpretation of the Bible, and it is argued that the Bible justifies a much more positive view of nature than White allows. However White is not concerned with explicitly stated positions but with discovering the values which actually move people, and with showing how these were formed. Thus in a reply to his critics, he wrote:

> The artifacts of a society, including its political, social and economic patterns, are shaped primarily by what the mass of individuals in that society believe, at the sub-verbal level, about who they are, about their relation to other people and to the natural environment, and about their destiny.[12]

What White is concerned to point out is that the sub-verbal beliefs of Western society have their roots in Christianity. He acknowledged that no one publicly advocates pollution but also points out that the structure of values embedded in these sub-verbal beliefs gives priority to other goals than achieving a viable ecology. Unless we face up to this, he argues, the environmental crisis will not be overcome.

White's approach to explaining the formation and role of attitudes in the environmental crisis is more subtle than most other thinkers who have entered the debate. He rejects the idea that religious beliefs could be regarded as 'the cause' of environmental problems. He points out that the notion of cause is generally not used by professional historians since there is always more than one cause, and the search for causes always leads to deeper causes. He writes, 'It is this sense of pluralism, and the various strata of historical "causation," that led me to refer the metaphor of *roots*.'[13]

The weakness in White's argument lies elsewhere: in his analysis of the way beliefs attain and retain their dominance. Despite his claims to the contrary, White has not completely emancipated himself from the notion of cause as an event or factor which can be abstracted out to explain things. This is evident in his arguments against opponents, particularly Marxists, his failure to account for why these beliefs have persisted, and his argument that since the roots of the environmental crisis are religious, so must be its solution.[14] In attacking Marxists, White counterposes religious values and the economic-social-political component of human relationships as candidates for the basic explanatory factor in society.[15] These religious values are conceived extrinsically to society, as is evident from his reference to artifacts of society being 'shaped' by beliefs. Thus White's proposed solution amounts to the assumption that once the basic cause of the problem has been identified, then the problem can be solved by removing the cause; as though society were a machine to be repaired. What this reveals is that White is himself caught up within those sub-verbal beliefs which he is trying to reveal - that the world is a mechanical order to be controlled through causal intervention.

But the situation is more complex than this. If fundamental attitudes or beliefs account for people's relation to their environments one would have expected no environmental problems to have emerged in classical China. Yi-Fu Tuan has pointed out that the Chinese traditionally saw

---

12. Lynn White, Jr. 'Continuing the Conversation' in Barbour, *Western Man and Environmental Ethics*, pp.55-64, p.57.

13. Loc.cit.

14. White, 'The Historical Roots of our Ecological Crisis' p.30.

15. White, 'Continuing the Conversation', p.57.

themselves as part of nature and had a deep respect for it.[16] They also recognized the importance of vegetation for preserving stream flows, soil, and keeping out invaders. Yet they still managed to destroy most of their forests. On the other hand Communist China adopted the Western idea that nature is to be subordinated, yet until recently it has had a far better record than traditional China in relation to environmental problems. The Communists, at least when Mao Ze Dong ruled, did much to reafforest the country, to conserve resources and to improve the environment in other ways.[17] Conversely, the Japanese still profess a special appreciation of nature, yet have subordinated their environment to their drive for industrial development.

White needs to go beyond simply showing that there are beliefs about the world deriving from Christianity which underlie the environmental destructiveness of Western European civilization - important as this is. He needs to consider why these beliefs have prevailed and how they influence social action, how they have become institutionalised and how institutions embodying them have gained ascendancy. He needs to account for the stability of these institutions and for other institutions and social processes which they have engendered in turn. Finally, he needs to consider the relationship between the dynamics of these institutions and social processes and the beliefs which were the original cause of their emergence. By failing to extricate himself from the culture which he set out to explain, by failing to develop an adequate alternative to the mechanistic notion of causation, White seems to have been blinded to these problems.

Anticipating subsequent argument, I want to suggest that the beliefs about the world which actually affect the way people live are those which are articulated into the basic concepts or categories by which the world is defined in the practices of individuals and institutions, which underlie the organization of discourse and of the way people live. In the remainder of this chapter I will show how the categories which orient people towards the domination of the world vitiate in one way or another the work of environmentalists.

## *Population*

The mechanistic notion of causation pervades the thinking of those who have argued that growth in population is the real cause of environmental problems. The foremost exponent of this view was Paul Ehrlich, although he now appears to have modified his views. His place as the foremost proponent of this explanation has been taken by Garrett Hardin. Ehrlich pointed out that population is growing exponentially. This is clearly evident from the figures cited from Ehrlich's books in the previous chapter. Efforts to reduce population growth have had little effect. In 1990, with a world population well over 5 billion, the time for population to double has been slowed from 33 years to just 39 years.[18] Ehrlich used such figures to support the Malthusian view that an increase in food together with a reduction in disease will result in an increase in population, which in turn will involve increasing demands for resources.[19] He further argued that as the limits are approached, the struggle for resources will become more intense and humans will destroy much of their environment. What we are now seeing is the beginning of this destruction.

The solution proposed by Ehrlich in his early work and in a more extreme form by Garrett Hardin, is encapsulated in what they called 'lifeboat ethics'. As the name suggests, the principle of this ethics is that people in the wealthy nations should recognize that there is a limit to the number

---

16. Yi-Fu Tuan, 'Discrepancies Between Environmental Attitude Behaviour: Examples from Europe and China' originally published in the *Canadian Geographer*, Vol.12 (1968), 176-91; reprinted in Paul Ward English and Robert C. Mayfield (eds) *Man, Space, and Environment* New York, London, Toronto: Oxford University Press, 1972, pp.68-81.

17. See Leo A. Orleans and Richard P. Suttmeier, 'The Mao Ethic and Environmental Quality' in *Science* Vol.170, 1970, 1173-76 and Qu Geping & Woyen Lee eds, *Managing the Environment in China*, Dublin: Tycooly, 1984.

18. See Paul R. Ehrlich and Anne H. Ehrlich, *The Population Explosion*, N.Y.: Simon and Schuster, 1990.

19. Paul R. Ehrlich, *The Population Bomb*, New York: Ballantine, 1968 Ch.1.

of people who can be saved, and their primary concern should be to ensure that they themselves survive. It involves ceasing any attempt to aid the most impoverished of the Third World and concentrating on saving the First World and the wealthier Third World countries from destruction by imposing strict birth control.[20]

This analysis is based on the identification of population growth as an independent factor, as a mechanical process governed by its own laws. It is a simple consequence of evolution that people will struggle to reproduce to the fullest extent, since people so oriented have reproduced most successfully in the past.[21] The part played by society and culture is nothing more than to have improved the food supply and reduced the mortality rate, thus releasing the natural constraints to the operation of the natural tendency of populations to grow. The increasing disruption of the environment is simply the effect of this cause. When the 'lifeboat ethic' calls for an abandonment of efforts to help the impoverished it assumes this analysis. It also implies that such measures will either force poor nations to develop birth control or reduce their populations through starvation.

But while overpopulation of the world is exerting enormous pressure on the world environment, the situation is far more complex than represented by Ehrlich and Hardin. To begin with it is not the impoverished who are responsible for the most important environmental problems. A very small proportion of the world's population is responsible both for most of the environmental destruction and most of the resource depletion. As Ivan Illich pointed out in 1974, Americans use more fuel for transporting people than is used for all purposes by all the Indians and Chinese, that is, nearly half the world's population, use for all purposes.[22] The nature of the technology used by the wealthy, including military technology, tends to be far more destructive of the environment than that used by the poor. Commoner has pointed out that the greatest environmental pollution has occurred since the Second World War and has argued that this can be largely explained by the use of newer production processes. For example, detergents, aluminium, plastic and inorganic fertilizers have been replacing soap powder, steel, wood and manure, lead has been added to petrol and there has been a rapid increase in the use of chlorine for the production of organic chemicals.[23]

The 'lifeboat ethic' conjures up the image of a benevolent group of people deciding against their inclinations who not to help. But the wealthy nations have become wealthy by subjugating the rest of the world, and as is becoming increasingly clear, the impoverishment of the Third World has been the result of Western European imperialism.[24] Moreover, the wealthy continue to bleed the impoverished of the world. As Roy Hattersly pointed out in 1986: 'In 1985, the net transfer of resources from the Third to the developed world was $22 billion. The truth is that the poorest parts of the world are still subsidizing the richest.'[25] In 1988 capital was flowing from the Third World to the rich at a rate of $43 billion a year.[26] The situation is far worse than it appears since the terms of trade are such as to allow the affluent nations to import artificially cheap goods from the Third World while exporting to them artificially expensive goods. The extent of the exploitation involved in this is hidden by the apparently small part trade plays between the industrialized nations and the Third World when measured in monetary terms. It appears to be only about 2½% of the gross domestic product of the industrialized nations. But if workers in the Third World were paid at the same rate as the workers in the industrialized nations, this trade would be nearer 25% of their gross

---

20. Ehrlich *The Population Bomb*; Garrett Hardin, *Exploring New Ethics for Survival*, N.Y.: Viking, 1972; 'Living on a Lifeboat' *Bioscience* 24, (October, 1974): pp.561-68; and *The Limits of Altruism*, Bloomington, Indiana University Press, 1977.

21. Ehrlich, *The Population Bomb*, p.34.

22. Ivan Illich, *Energy and Equity,* New York: Harper and Row, 1974, p.11.

23. Barry Commoner, *The Closing Circle: Confronting the Environmental Crisis*, London: Jonathon Cape, 1972, Ch.8.

24. See for instance Andre Gunder Frank, *World Accumulation: 1492-1789,* New York: Monthly Review Press.

25. Quoted by Nick Rowling in *Commodities: How The World Was Taken To Market*, London: Free Association Books, 1987, p.174.

26. Cited from the World Bank in *The Guardian*, Jan.1, 1989.

domestic product.[27] And this does not take into account the nature of the products of trade, the enormous flow of energy rich products, non-renewable materials and high quality food from the Third World to the industrialized nations, an exchange in which the terms of trade have moved continuously against the producers of these products for over a hundred years. A global leech is a more appropriate image for the affluent nations.

As for population growth, the most important generating condition of rapid growth is precisely this impoverishment of people which the 'lifeboat ethic' is promoting as a cure.[28] The rapid growth in population in non-European nations other than those in which the original inhabitants were exterminated and replaced by Europeans can be directly attributed to the social disorganization and consequent impoverishment produced by European imperialism. In the case of Indonesia, the Dutch had a deliberate policy of promoting population growth to facilitate further exploitation.[29] In recent decades, most instances of rapid population growth have occurred where people are insecure and impoverished, and where people, especially women, are uneducated.[30] When infant mortality is high and people are dependent upon their children to support them in their old age, people have as many children as possible. Variations in birth rates between poor countries result primarily from the distribution of income. Egalitarian societies have low birth rates.[31]

## *Technology*

Some of those who have attributed environmental problems to modern technology have been aware of the limitations of prevailing notions of causation. This is true of both Barry Commoner[32] and E.F. Schumacher.[33] However while Commoner and Schumacher are at pains to emphasise the complexity of environmental problems, others, including some members of the appropriate technology movement, have a tendency to view technology as the causal factor responsible.[34] The questions asked by Dr Raúl Prebisch, director-general of the Latin American Institute for Economic and Social Planning, illustrates this way of thinking:

> If [the underdeveloped countries] have been unable to keep up with the major world trade flows, is not technology to blame? ... Can it be said that the higher capital-intensive level, especially in industry, is not the product of technology? Is not the population explosion likewise attributable to the advance of science and technology?[35]

This oversimplifies the problem and turns attention away from crucial considerations about social and economic relations within and between nations - relations which are important in their own right and which impact on technological development. In capitalist societies technology has to make a profit. There is a vested interest in developing the sort of technology which will give manufacturers power over both their workers and a competitive edge in the international market. Ecological considerations, considerations about the requirements of the impoverished of the world

---

27. This has been noted by Arghiri Emmanuel in *Unequal Exchange*, New York: Monthly Review Press, 1972, p.367f.

28. The best study of this is Susan George's, *How the Other Half Dies*, Harmondsworth, Penguin, 1977, ch.2, 'The Population Myth'.

29. See Clifford Geertz, *Agricultural Involution*, Berkeley: University of California Press, 1968.

30. Nathan Keyfitz 'Population Growth: Causes and Consequences' in Murdoch *Environment* 2nd ed. pp.39-64, esp. p.58.

31. George, *How the Other Half Dies*, p.63. Kerala in India as compared to Bangladesh is a good example of this.

32. Barry Commoner, *The Closing Circle: Confronting the Environmental Crisis*, London: Jonathon Cape, 1972.

33. E.F.Schumacher, *Small is Beautiful*, London: Abacus, 1974.

34. See David Dickson, *Alternative Technology and the Politics of Technical Change*, Glasgow: Fontana, 1974, p.159f. for a critique of this.

35. Cited by Dickson, p.159.

and the effects on the quality of life of workers are irrelevant from the point of view of companies struggling against competitors to maintain their viability. In these circumstances the part played by technology is to entrench environmentally destructive economic organization rather than produce it. And frequently environmental problems are not related to technology at all. In view of the analyses of the reasons for population growth referred to above, it is difficult to justify Dr Prebisch's suggestion that it is science and technology that has caused population growth.

The failure to take into account the complexity of the relation of technology to society when technology is treated as a causal factor is also evident when it comes to action. The members of the appropriate technology movement tend to assume that having identified the cause of the problem in technology, all they have to do is change the technology. However their work in Third World countries has illustrated that when technology is introduced to a society it does not determine the nature of that society.[36] For instance when methane plants designed to use cow dung were introduced into villages in many Third World countries, the effect was often to enrich the wealthy villagers and to take away the fuel supply of the poor. Along with further impoverishment of the poor, this put greater pressure on other fuel sources, in particular, trees. What was seen as 'appropriate technology' further entrenched the existing, environmentally destructive economic order.

The problem of the relation of interpretation to action engendered by deficient notions of causation is evident in the work of Barry Commoner. Commoner is aware of the importance of context, and has criticised explanations which over-emphasize population growth and growing affluence on the grounds that they fail to take into account the context within which such environmental problems occur. Accordingly, in his investigation into the role of technology, he has examined in detail examples of environmental destruction. In fact he has devoted a whole chapter of *The Closing Circle* to this problem. To deal with it he works with two different concepts of causation. A holistic form of causation is used when considering ecosystems and life systems, but when technology is described as disrupting these systems, it is treated as an extrinsic 'independent' causal factor. A holistic form of causation is again assumed when considering the society which produces such technology. So while the type of technology is held to be largely responsible for environmental problems, technology is not taken to be an independent cause but a facet of more complex processes which can only be understood holistically. This leaves Commoner unable to offer any solutions and he concludes his book on a negative note:

> ...the world is being carried to the brink of ecological disaster not by a singular fault, which some clever scheme can correct, but by the phalanx of powerful economic, political, and social forces that constitute the march of history. Anyone who proposes to cure the environmental crisis undertakes thereby to change the course of history. But this is a competence reserved to history itself... That we must act is now clear. The question which we face is how.[37]

## *Economic Growth*

Some of the most important efforts to understand the environmental crisis are those initiated by The Club of Rome in the early 1970's. With variations, the members of the Club of Rome explain environmental destruction by economic growth. Their approach is to construct mathematical models of this growth by identifying a number of causal factors and examining the relationships between them. The best known of these are the models constructed by Jay Forrester (1971) and

---

36. On this see Nigel Pollard, 'Appropriate Technology: Really Appropriate or just a Misfit?' *The Ecologist*, Vol.13, No.1, 1983, pp.27-34.

37. Commoner, *The Closing Circle*, p.300.

Dennis Meadows (1972).[38] In considering economic growth they have isolated population, industrialization, food production, pollution and consumption of non-renewable resources as factors causing environmental problems, and then shown how the growth of each of these affects the growth of the others. By projecting various possible rates of growth of these factors, they came to the conclusion that there will be a disaster within the next hundred years if growth continues at anything like the present rates.

While it can hardly be denied that the Club of Rome are correct to point out that exponential economic growth must end in massive environmental destruction, these early models manifest most clearly the limitations of conceiving the world in terms of interacting factors. Admitting a number of factors allows for a greater degree of complexity than analyses which focus on a single factor and consequently gives the appearance of a greater degree of realism. But this appearance is spurious. All these factors were considered in abstraction from their contexts, aggregated into single world indexes without any concern for the differences between regions, even between the rich and the poor countries, and with no effort to examine situations where the changes were occurring, where the changes were affecting the environment, or why the changes were taking place. Allowing for interactions between abstractions does not reconstitute this context, and at best only allows for a single level of interactions; a causal level at which a certain rate of exploitation of natural resources leads inevitably to a certain level of pollution, and that again has a certain effect on birth-rates, death rates and so on in the actual world. Most importantly, political and economic organizations and power relations within the world were totally ignored. Not surprisingly, the scenarios predicted have already been found to be grossly at variance with reality. Exponents of this approach have been unable to offer any guidance for action other than vague injunctions to slow the rate of growth which, without taking into account regional variations, implies freezing the world economic order at a state in which a billion people are suffering from malnutrition. The analyses bewildered people rather than oriented them for effective action.

Later studies produced by the Club of Rome successively struggled to come to better grips with political and economic contexts.[39] They divided the world into regions and considered some of the political and economic problems engendering environmental destruction. These studies are important for identifying the magnitude of the problems, where some of the greatest economic irrationalities lie, and just how radical are the changes which need to be made in order to deal with them. But (with the possible exception of *Reshaping the International Order* coordinated by Jan Tinbergen) these have not succeeded in confronting the power relations within nations or in world politics, nor have they adequately confronted the cultural diversity of the world. To the extent that they have dealt with the issues of power they have been reduced to atheoretical descriptions.[40] They have not begun to identify those tendencies within the world and within particular societies which need to be subverted and those which could be fostered to overcome their problems. And they are

---

38. Jay W. Forrester, *World Dynamics*, Cambridge, Mass.: Wright-Allen Press, 1971; and Dennis L. Meadows et.al., *The Limits to Growth: A report for The Club of Rome's Project on the Predicament of Mankind*, [1972] London and Sydney: Pan Books, 1974.

39. See M. Mesarovic and E. Pestel, *Mankind at the Turning Point: The Second Report to the Club of Rome*, N.Y.: Dutton, 1974; Y. Kaya and Y. Suzuki, *Global Constraints and New Vision for Development: Japan Work Team of the Club of Rome*, Tokyo: COR Technical Symposium, 1973; Jan Tinbergen (Co-ordinator) et.al. *Reshaping the International Order: A Report to the Club of Rome*, N.Y.: Dutton, 1976; D. Gabor et. al. *Beyond the Age of Waste*, Oxford: Pegamon Press, 1978; and Auerlio Peccei and Daisaku Ikeda, *Before It is Too Late*, Tokyo: Kodansha International, 1984. The last of these, written as a sequel to *The Limits to Growth*, is Donella H. Meadows et.al,, *Beyond the Limits: Global Collapse or a Sustainable Future*, London: Earthscan Publications, 1992. For critiques of the work this group see *The Global 2000 Report to the President*, pp.607-681, and D.M. Gvishiani, 'Methodological Problems of Modelling Global Development' in A.D. Ursal ed. *Philosophy and the Ecological Problems of Civilisation*, Moscow: Progress Publishers, 1983.

40. For example Aurelio Peccei, *The Human Quality*, Oxford: Pergamon Press, 1977; and Bohdan Hawrylyshsyn *Road Maps to the Future. Towards More Effective Societies. A Report to the Club of Rome*, Oxford: Pergamon Press, 1980.

unlikely to do so without a theoretical framework which allows for many levels and types of causation, including human adaptive behaviour.

## *The Ethics and Politics of Modernity: Rights Theory and Utilitarianism*

A variety of thinkers has attempted to extend modern ethical and political doctrines, that is, doctrines which have been developed since the seventeenth century, to deal with environmental issues. Principal amongst these doctrines are 'rights theory' according to which certain beings have natural rights discoverable by reason on the assumption that society is based on a contract, and 'utilitarianism' according to which actions, principles and the organization of society can be judged in terms of whether they maximise happiness and minimize pain for the greatest number. They are correlated with efforts modify mainstream economic theory to take into account the environment and to facilitate the management of the market. Some environmentalists have attempted to extend rights theory to include future generations, animals and trees, while others have been more concerned to argue that people do not have the rights they thought they had in order to justify constraining people's behaviour. Utilitarianism has been invoked as a basis for condemning our present treatment of animals, and it is implicit in the work of those who have attempted to develop cost-benefit analysis to include the negative value of environmental destruction and pollution in policy decisions.

The issue of whether future generations can be held to have rights has been considered by a number of philosophers.[41] K.S. Shrader-Frechette typifies the way these philosophers have approached the issue.[42] Following M.P. Golding she holds that a moral community does not have to be based on an explicit contract between its members, but can be based on 'a social arrangement in which each member derives benefits from the efforts of the other members.'[43] Against Golding, she argues that such a moral community can be intergenerational. The possibility of defending the existence of rights for future generations on this basis is seen to revolve around three questions: Is intergenerational reciprocity possible? Is explicit reciprocity a necessary condition for all social contracts based on self-interest? and Is it plausible to reason that we cannot know what notion of the good life future generations will hold and that therefore we cannot determine our obligations to them?

Two main answers to support the notion of rights have been proposed to the first question. The first, argued by Wagner, is that there is a reciprocity since in recognizing the rights of future generations we attain greater happiness and self-actualization. The second, argued by Faulkner, is that there is a reciprocity based on each generation paying past generations by taking responsibility for the conditions of life of future generations. However even if these arguments are not accepted, the notion that future generations have rights can be sustained by a negative answer to the question whether explicit reciprocity is a necessary condition of social contracts. Shrader-Frechette points

---

41. Daniel Callahan, 'What Obligations Do We Have to Future Generations?' *The American Ecclesiastical Review*, Vol.164, No.4 1971, pp. 256-68; W.C. Wagner, 'Futurity Morality' *The Futurist* 5, No.5 1971: 179-199; M.P. Golding, 'Obligations to Future Generations' *The Monist*, Vol.56, No.1 1972, pp.85-99; E. Delattre, 'Rights, Responsibilities and Future Generations' *Ethics*, Vol.82, 1972, pp. 254-258; J.B. Stearns 'Ecology and the Indefinite Unborn' *The Monist*, Vol.56, No.4 1972: 612-625; J. Feinberg 'The Rights of Animals and Unborn Generations' in *Philosophy and Environmental Crisis* William T. Blackstone ed. Athens, Georgia: University of Georgia Press, 1974, pp. 43-68; R.M. Green 'Intergenerational Distributive Justice and Environmental Responsibility' *Bio-Science*, Vo.27, 1977, pp.260-265; P. Faulkner, 'Protection for Future Generations' *Frontiers*, Vol.42, 1978, pp.35-37; and Ernest Partridge ed. *Responsibilities to Future Generations: Environmental Ethics*, Buffalo, N.Y.: Prometheus, 1981.

42. K.S. Shrader-Frechette, 'Technology, The Environment, and Intergenerational Equity' in *Environmental Ethics*, K.S. Shrader-Frechette ed. Pacific Grove: Boxwood, 1981, pp.67-81.

43. Golding, 'Obligations to Future Generations', p.91.

out that Rawls' concept of justice based on choosing principles to govern society from 'the original position' (that is, behind a veil of ignorance from which one cannot know what one's position in society will be) does not involve reciprocity and can be applied between generations. Another argument is presented by Callahan that reciprocity is not a necessary condition for there being rights because it is possible for one party to choose to accept an obligation, as occurs when parents accept an obligation to their children. This could be seen to provide a prototype for relationships between present and future generations. However if we cannot know what future generations will want, then such arguments have no weight. Against this contention Feinberg has argued that we can know that future generations will have an interest in living space, fertile soil, fresh air and the conditions which sustain life and health more generally. It does not follow that because we are ignorant of specific aspects of what future generations want that we can assume that their interests will not be the same as ours.

One of the most vigorously debated issues in environmental philosophy centres on the question of whether animals have rights.[44] Because the idea of extending rights to animals is more radical than the idea of extending rights to future generations, this debate has led to more fundamental questions. Firstly it has led to a basic inquiry into the nature of rights. Two answers have been offered: that to have a right is to have a claim or an entitlement. Neither of these would rule out the possibility of ascribing rights to animals, although if a right is a claim then animals would need someone to claim for them. At this point the question arises of what kind of rights there are. It might be conceded that animals could be ascribed rights - as for instance when money is left to animals in a will - but the important issue is whether they can have natural rights. Can they have rights by simply being the kinds of beings they are? This then raises the further question of what sort of beings can have natural rights. Proposed answers to this are that they must be rational, that they must have free will, that they must have interests or that they must be sentient. Obviously the specified condition will affect the question whether animals have rights, and also whether species, plants and ecosystems have rights. Those inclined to rule out animal rights on the grounds that the precondition for such rights is rationality or free will are left with difficult questions concerning young children and the mentally feeble.

The question of whether animals can be said to have interests has also been the subject of much argument.[45] R.G. Frey, for instance, has argued that animals cannot be ascribed interests because they lack a proper language and self-consciousness.[46] If sentience is the criterion for being able to ascribe natural rights to anything, then there would be no problems with ascribing rights to animals, though it would seem to exclude species, plants and ecosystems. This has led some philosophers to argue for panpsychism. However even if we conclude animals do have rights, we still have to face the question of what rights they have. Among the proposals are that animals have a right not to be treated cruelly, a right to life, and a right to property and liberty. Numerous arguments have centred on these issues.

Ascribing rights to species, plants or ecosystems is even more problematic. Those who wish to support this contention have generally extended the arguments that the basis for anything having rights is that they have interests, and they have then gone on to argue that species, plants and

---

44. A number of papers on this subject representing the different positions argued for have been printed in Tom Regan and Peter Singer (eds), *Animal Rights and Human Obligations* Englewood Cliffs, N.J.: Prentice-Hall, 1976. Regan has produced a book arguing for animal rights: *The Case for Animal Rights,* Berkeley and Los Angeles: University of California Press, 1983. See also Peter Miller, 'Do Animals Have Interests Worthy of Our Moral Interest?' *Environmental Ethics*, Vol.5, No.4. 1983, pp.319-334.

45. For opposition to this see Leonard Nelson, *A System of Ethics* trans. N. Gutman, New Haven: Yale University Press, 1956; and H.J. McCloskey, 'Rights' *Philosophical Quarterly* Vol.15, 1965, pp.115-27.

46. R.G. Frey, *Interests and Rights: The Case Against Animals*, Oxford: Clarendon Press, 1980.

ecosystems do have interests.[47] However the major arguments in relation to rights for these entities have centred around whether it is possible to ascribe any kind of rights to them. The major defender of this possibility is Christopher Stone.[48] The basis of his argument is that in the Anglo-American legal system we already ascribe rights to non-people: corporations, municipalities, ships and some animals, and that the difficulty in thinking of natural objects as having rights is simply the novelty of it. The extensions of rights to slaves, women, children, and aliens had appeared unthinkable when they were first proposed. However according to Stone, legal systems create persons, property and rights; so his arguments are predicated on the assumption that there are no such things as natural rights. His real concern is to make the destruction of natural objects a cost by incorporating them into the legal system. As he puts it:

> Wherever it carves out 'property' rights, the legal system is engaged in the process of *creating* monetary worth. ... I am proposing that we do the same with eagles and wilderness areas as we do with copyrighted works, patented inventions, and privacy: *make* the violation of rights in them to be a cost by declaring the 'pirating' of them to be the invasion of a property interest.[49]

This simply does not answer the ethical problem.

Where the political problems associated with environmental preservation have been focussed upon, a different emphasis is placed on rights theory. Here the concern is to show that what people have come to accept as basic rights are in fact without foundation. The two most significant thinkers to argue along these lines are Garrett Hardin and William Ophuls. Hardin argued in his now famous paper 'The Tragedy of the Commons' that the underlying cause of environmental problems is that where common property is concerned, individual interest does not coincide with the common interest.[50] Individuals will tend to exploit the commons to the maximum extent, the ultimate effect of which will be to destroy the commons for everyone. Those people who do not exploit the commons in this way through conscience might be publicly lauded, but privately will be regarded as simpletons, and such people will tend to be eliminated in the evolutionary struggle. The only solution to this, Hardin argued, is the coercion of people where the commons are concerned; that is, 'mutual coercion, mutually agreed upon by the majority of people affected.'[51] The exercise of this coercion is defended in a wide variety of instances where people had previously thought they had an inalienable right to do as they chose; for instance to have as many children as they wanted. But as Hardin wrote:

> When men mutually agreed to pass laws against robbing, mankind became more free, not less so. Individuals locked into the logic of the commons are free only to bring on universal ruin; once they see the necessity of mutual coercion, they become free to pursue other goals.[52]

William Ophuls recognized the 'logic of the commons' described by Hardin as a special version of Hobbes' 'state of nature' in which there is a war of all against all, and Hardin's solution of

---

47. See K.S. Schrader-Frechette, 'Ethics and the Rights of Natural Objects' in Schrader-Frechette ed. *Environmental Ethics*, pp.89-99. For opposition to this see Scott Lehmann, 'Do Wilderness Areas Have Rights?' *Environmental Ethics*, Vol.3, No.2, 1981, pp.129-146.

48. Christopher Stone, *Should Trees Have Standing?: Towards Legal Rights for Natural Objects* Los Altos, Cal.: William Kaufman, 1974.

49. Ibid. p.29.

50. Garrett Hardin 'The Tragedy of the Commons' originally published in *Science* Vol.162, 13th Dec. 1968, pp.1243-1246; reprinted in Herman E. Daly ed. *Economics, Ecology, Ethics: Essays Toward a Steady-State Economy*, San Francisco: Freeman, 1980, pp.100-114.

51. Ibid. p.111.

52. Ibid. p.112.

'mutual coercion, mutually agreed upon' as equivalent to Hobbes' sovereign power, erected by the majority to constrain all men to be reasonable and peaceful.[53] While Ophuls' work provides a more thorough analysis of the political issues associated with environmental problems, essentially he has concurred with Hardin's diagnosis of the situation. Ophuls pointed out that traditional rights theory deriving from Locke has failed to draw Hobbes' conclusions because it was formulated as the New World presented apparently unbounded wealth to be opened up.[54] Modern political institutions have been based on this view of the world as a Great Frontier. However we are now in a 'closed' world similar to that existing when Hobbes was writing. When offering solution to the problem, Ophuls tried to exploit a number of ideas, but in essentials agrees with Hardin:

> The only solution is a sufficient measure of coercion. Following Hobbes, a certain minimum level of ecological order or peace must be established; following Rousseau, a certain minimum level of ecological virtue must be imposed by our political institutions.[55]

Utilitarians base their arguments on a different foundation. Although they frequently speak of rights, what is meant by 'rights' is fundamentally different. For rights theorists, rights exist whether they are recognized or not, while for utilitarians rights are accorded by society. The most well known effort to use utilitarianism as a basis for defending environmental concerns is that of Peter Singer. In his book *Animal Liberation*, Singer argued on utilitarian grounds that it is wrong to kill or inflict suffering on animals.[56] He bolstered his utilitarianism with the idea that moral progress consists in applying utilitarian principles to an expanding circle of beings. Thus 'difference of species' is seen as the last spurious grounds for moral discrimination. Singer's argument is not original. It simply, as Singer acknowledges, elaborates upon an argument of Bentham:

> The day *may* come when the rest of the animal creation may acquire those rights which never could have been withholden from them but by the hand of tyranny. The French have already discovered that the blackness of the skin is no reason why a human being should be abandoned without redress to the caprice of a tormentor. It may one day come to be recognized that the number of the legs, the villosity of the skin, or the termination of the *os sacrum* are reasons equally insufficient for abandoning a sensitive being to the same fate. What else is it that should trace the insuperable line? ... The question is not, Can they *reason*? nor Can they *talk*? but, *Can they suffer*?[57]

Where more general environmental problems are taken into account by utilitarians, these are usually considered in terms of cost-benefit analyses - which is essentially an updated version of utilitarianism and of Bentham's felicific calculus.[58] In general those committed to cost-benefit analyses accept the market as the best means of distributing resources and consumer goods, but acknowledge that there are some imperfections in the market which require government intervention. Cost-benefit analyses enable these imperfections to be identified and evaluated in monetary terms commensurable with market evaluations, on the basis of which appropriate compensations or public investments can be made, taxes imposed, and protective laws enacted. The

---

53. William Ophuls, 'Leviathan or Oblivian', in Herman E. Daly ed. *Toward a Steady-State Economy*, San Francisco: Freeman, 1973; and *Ecology and the Politics of Scarcity*, op.cit.

54. Ophuls, 'Leviathan or Oblivian', p.222.

55. Ophuls, *Ecology and the Politics of Scarcity*, p.151f.

56. Peter Singer, *Animal Liberation: A New Ethics For Our Treatment of Animals*, New York: Avon, 1977.

57. Jeremy Bentham, *The Principles of Morals and Legislation*, [1789], Ch.XVII, Section 1; cited ibid. p.7f.

58. See Alan Coddington, '"Cost-Benefit" as the New Utilitarianism', *Political Quarterly*, Vol.42, 1971, pp.320-25. For its relevance to environmental problems see Robert E. Goodin, *The Politics of Rational Man*, London: Wiley, 1976, pt. 4.

development of cost-benefit analysis as a means of taking into account environmental destruction received a major boost with the National Environmental Policy Act in USA in 1970 which called for environmental impact assessments to consider the deterioration of the bio-physical environment.[59] They are most often used to evaluate what costs and risk levels can be justified for the benefits accruing from various environmentally degrading enterprises. However they also underlie efforts to develop environmental economics.[60]

While a number of different versions of cost-benefit analysis have been developed, the basic principles are all present in the Bayesian or estimated utility model. According to this model decision makers are confronted with a set of mutually exclusive and exhaustive possible future states of affairs and courses of action. Each combination of a state of affairs and an action yields an event to which a particular value or utility can be ascribed. Desired events are ascribed positive values and shunned events are ascribed negative ones. These are rendered commensurable by being measured in monetary terms by asking people what they would be willing to pay to bring about or prevent an event or state of affairs. This renders choices commensurable with market evaluations. By multiplying the values ascribed by the probability that the events will occur and summing the products of every event associated with each action, a quantitative value for performing each action can be obtained. The action with the highest value is the rational one to perform. While as many as 50 differing methodologies for studying environmental impact have been identified, these are all variations of this approach.

## *Problems with Rights Theory and Utilitarianism*

The most basic difficulty in the effort to extend rights theory and utilitarianism to environmental issues is the weakness of these doctrines in the first place. The modern theory of rights and utilitarianism have their origins in the efforts to replace the ethical and political thought of the disintegrating medieval society and world-orientation. They have been intimately associated with the rise of capitalism and mechanistic science, and assume a conception of nature as devoid of meaning and of society as nothing but a collection of egoistic individuals motivated by appetites and aversions. The secular doctrine of rights developed in the seventeenth century was formulated as an attempt to determine rights and obligations by the same resolutive compositive method (the method of analysis and synthesis) which had proved so successful in discovering the laws of nature.[61] The form in which this doctrine proved most successful was based on the idea that societies and political institutions are founded on social contracts or compacts expressing the self-interests of the contracting parties. Utilitarianism, developed in the eighteenth century by Helvetius and Bentham, is characterized by the attempt to found ethics and political philosophy entirely on subjective experience, recognizing only pleasure and pain as good and bad.

The immediate impetus for the development of rights theory was the need to find a basis for reconciling opponents within a disintegrating society characterized by violent religious, political and economic conflicts. Its major proponents were the rising capitalist classes who were attempting to justify their claims to political power on the basis of their growing economic power. Its major

---

59. For a brief overview of the principles of environmental impact assessment see Michael Carley, *Rational Techniques in Policy Analysis*, London, Heinemann Educational Books, 1980, Ch.8.

60. See for instance David Pearce et.al,. *Blueprint for a Green Economy*, London: Earthscan Publications, 1989. See especially Ch.3 - 'Valuing the Environment.'

61. A crucial transformation in the concept of rights had already occurred by the seventeenth century. Rights became something one *had*. See John Finnis, *Natural Law and Natural Rights*, Oxford: Calrendon Press, p.207. However it was only with Hobbes that rights were moved outside the juridical relationship altogether. (See ibid. p.208)

purpose was to provide a basis for distributing political and economic power in their favour.[62] The nature of claims to rights reflects this origin of the doctrine. As Simone Weil has pointed out:

> The notion of rights is linked with the notion of sharing out, of exchange, of measured quantity. It has a commercial flavour, essentially evocative of legal claims and arguments. Rights are always asserted in a tone of contention; and when this tone is adopted, it must rely upon force in the background, or else it will be laughed at... If you say to someone who has ears to hear: 'What you are doing to me is not just', you may touch and awaken at its source the spirit of attention and love. But it is not the same with words like 'I have the right...' They evoke a latent war and awaken the spirit of contention. To place the notion of rights at the centre of social conflicts is to inhibit any possible charity on both sides.[63]

Utilitarianism was developed within a society which had been reduced by commercialism to individuals and classes struggling for money. It reduced all nobility, all values, all meaning to the one quantifiable level of pleasure and pain, leaving the only reason to concern oneself with anything other than one's own pleasure: that it gives one pleasure to do so. This completely inverted the traditional way of thinking of the relationship between good and pleasure. Instead of pleasure being derived from what is good, what is good is defined as what is pleasurable. Thus the pleasure of the torturer and the satisfaction gained from achieving justice are equated. Since pleasure and pain are the only principles motivating people, there are no moral virtues or vices for which people can be held responsible, but only more or less efficient means of manipulating people - and from the very beginning utilitarianism was directed towards the efficient control of people.[64] The founder of utilitarianism, Helvetius (1715-1771), postulated the principle of the greatest happiness for the majority as a principle to be followed to achieve social stability.[65] Since humans are animated solely by 'a sentiment of love for pleasure, and of hatred for pain', then 'under a good legislation only fools would be vicious.'[66] It does not matter whether people are born virtuous since their sentiments are such that they can be made virtuous by skilful management. Bentham (1748-1832), who established utilitarianism as an influential doctrine in Great Britain, was primarily concerned to develop Helvetius' doctrine in relation to law and penal reform. His model for a prison, the *Panopticon*, was designed as a means to control people by keeping them under universal surveillance and as a laboratory which could be used to carry out experiments to alter behaviour, to train and correct people. Bentham took penal discipline as a model which could be applied to all the institutions of society. It was 'a new mode of obtaining power of mind over mind, in a quantity hitherto without example... Its great excellence consists in the great strength it is capable of giving to any institution it may be thought proper to apply it to.'[67]

In the form in which they were originally developed, rights theory and utilitarianism have been powerful forces in society. Where societies face disintegration the notion of forming a contract to prevent a war of all against all makes eminently good sense, especially when such an idea seems to

---

62. See C.B. McPherson, *The Political Theory of Possessive Individualism: Hobbes to Locke* [1962], Oxford: Oxford University Press, 1964.

63. Simone Weil, 'Human Personality' in *Simone Weil: An Anthology*, ed. Sian Miles, London: Virago, 1986, pp.69-98, p.'s 81 & 83.

64. The context and implications of Helvetius' doctrine have been analysed by Eric Voeglin in 'Helvetius and the Genealogy of Passions' in *From Enlightenment to Revolution*, edited by John H. Hallowell, Durham, N.C.: Duke University Press, 1975, pp.35-52.

65. Helvetius, *De l'Esprit*, sec.2, ch.24, *Oeuvres*, London, 1776, Vol.1, p. 300f.

66. Helvetius, *De l'Homme*, sec. 4, ch.22, *Oeuvres*, Paris, 1795, Vol,4, p.384; and *De l'Esprit*, sec.3 ch.17.

67. Jeremy Bentham, *Panopticon: or, The Inspection House, Works*, ed. Bowring, [1843], New York, Russell & Russell, 1962, Vol.4, p.39 & 66. The significance of Bentham's model has been examined in Michael Foucault, *Discipline and Punish*, tr. Alan Sheridan, Harmondsworth: Penguin, 1977, esp. ch.3.

accord with the atomistic, law governed view of the world purveyed by the science of nature. Utilitarianism is also plausible as a doctrine for maintaining a stable society and exercising social control. Being suitable for capitalist societies based on the selfish pursuit of individual interests, both these doctrines were adopted and have come to be important constituents of the economic organization and of the legal and political institutions of Western societies. But by becoming the reference point for ethical and political disputes, these doctrines have been extended beyond their original scope. Rights have been represented as universal and independent of contractual bargaining between people with power, while utilitarianism has been upheld as defining the highest good rather than as merely a principle for maintaining social order. These extensions have destroyed the coherence of the doctrines and have revealed their limitations.

Where individuals are enjoined to acknowledge the claims of others who have no other basis for n... this claim than the supposed contract on which society is based, the nature of this contract b... In fact society cannot be based on a contract, since forming a contract is ... pre-existent society. People only develop the capacity to enter ... a language, and so on. It is impossible to conceive ... precedes not only contracts, but ... rights established in terms of it as ... ing between opposing interests. ... d is that individuals are motivated by ... formulation of rights theory, that of ... ghts, Rawls posits a situation in which ... e's natural endowment and position in ... ent about these principles is held to be ... n society and assumes that they will base ... taken rule utilitarianism, formulated this ... rights as a claim for justice. In doing so ... ly elevated the principle of egoism. In ... nly be expected to be egoists, and therefore ... a contract when it is in their interests to do so. But ... when individuals have the power to threaten each other, and in fact ... rights theory has been accepted as a basis for adjudicating between people. Rights ... om been acknowledged in practice far beyond the community of those with the power to at least attempt to enforce their claims.

But even in these circumstances rights theory is inadequate since a condition for its success as a basis for adjudication is that it can provide an unequivocal answer as to what is right and wrong. But the notion of contract has been construed in different ways in order to defend various ethical and political claims, and different construals give different results. This has been clearly revealed by Alasdair MacIntyre who compared the conclusions of Rawls with those of Robert Nozick. Each of these thinkers had accepted as a starting point the idea that entry into social life is the voluntary act of rational individuals with prior interests who must ask the question 'What kind of social contract is it reasonable for me to enter into?'[69] On the question of whether taxes should be increased to pay social services, Rawls' version of rights theory based on the idea that what is right is what rational agents would choose if they were to choose from behind a veil of ignorance would not only support such a move, but would justify the claim that the underprivileged have a right to social services.[70] However Robert Nozick who argues that in a just society the only people entitled to appropriate anything for their use would be those who had justly acquired what they had by some original act of acquisition or by some just act of transfer, would claim that individuals have

68. John Rawls, *A Theory of Justice*, Oxford: Oxford University Press, 1972.

69. Alasdair MacIntyre, *After Virtue: a Study in Moral Theory*, London: Duckworth, 1981, p.227ff.

70. Rawls, *A Theory of Justice*, p.136 & 303.

an inalienable right to their income and no government has a right to appropriate this and give it to someone else.[71] Despite the common starting point, the two positions lead to different conclusions and there is no way of choosing between them. So not only is the notion of rights based on contract a fiction, but it is a useless fiction.

The extension of utilitarianism was initially effective as a basis for reform because it placed the suffering of the poor and oppressed on the same level as the suffering of the wealthy and exalted. But the notion of human motivation assumed by utilitarianism is the same as that of rights theory, and undermines any basis for entreating concern for other people or life forms. If people are only moved by what is painful or pleasurable, and there is no other basis for judging actions, then there can be no reason to expect people to concern themselves with ensuring that society is ordered for the greatest happiness for the greatest number other than it will be of less pain and greater pleasure to oneself. While this was implicitly assumed by Helvetius and Bentham, it loses its force when utilitarianism leads to the diversion of goods and activities to the powerless. Utilitarianism provides no basis for choosing between pleasures, in particular between pleasures associated with satisfying one's appetites and those which have been taken to be nobler ends such as concerning oneself with the welfare of others.

In an attempt to overcome this problem, John Stuart Mill rejected the notion that happiness could be reduced to pleasure and introduced qualitative distinctions between types of happiness, allowing that some types are higher than others. But this separated utilitarian principles from Bentham's hedonistic psychology, leaving Mill without any basis for demonstrating the superiority of one type of happiness over another. This difficulty in utilitarianism became increasingly evident towards the end of the nineteenth century and eventually Sidgewick concluded that it was only intuition which enjoins us to acknowledge the superiority of following the moral injunctions of utilitarianism rather than precepts enjoining us to the pursuit of our own happiness. As MacIntyre has pointed out, it was from this starting point that G.E. Moore developed his intuitionism.[72] According to this doctrine 'good' is a simple, non-natural quality which can be intuited in the same way as a basic experience such as 'yellow' can be experienced without being defined in terms of anything else. As it soon became evident that different people's intuitions are different, this paved the way for emotivism and prescriptivism in which moral injunctions were taken to be nothing more than expressions of feelings and attitudes, or the efforts to change the feelings and attitudes of others. Thus the attempts to develop utilitarianism beyond its original limited form led straight to nihilism.

The situation consequent upon the failure of both rights theory and utilitarianism as ethical doctrines has been well described by MacIntyre:

> The most striking feature of contemporary moral utterance is that so much of it is used to express disagreements; and the most striking feature of the debates in which these disagreements are expressed is their interminable character... There seems no rational way of securing moral agreement in our culture... [T]he culture of bureaucratic individualism results in their characteristic overt political debates being between an individualism which makes its claims in terms of *rights* and forms of bureaucratic organisation which make their claims in terms of *utility*. But if the concept of rights and that of utility are a matching pair of incommensurable fictions, it will be the case that the moral idiom employed can at best provide a semblance of rationality for the modern political process, but not its reality. The mock rationality of the debate conceals the arbitrariness of the will and power at work in its resolution... [P]rotest is now almost entirely that negative phenomenon which characteristically occurs as a reaction to the alleged invasion of someone's rights in the name of someone else's utility. The self-assertive shrillness of protest arises because the facts of incommensurability ensure that protesters can never win an argument; the indignant self-righteousness of protests

71. Robert Nozick, *Anarchy, State and Society*, Oxford: Blackwell, 1974, p.151 & 153.

> arises because the facts of incommensurability ensure equally that the protesters can never lose an argument either. Hence the *utterance* of protest is characteristically addressed to those who already *share* the protesters' premises.[73]

The debates on environmental ethics formulated in terms of rights theory are a graphic illustration of the debased state and irrelevance of prevailing ethical thought. The underlying egoism of the participants is evident from the way the issue of the rights of members of Third World nations to an equal share of the mineral reserves being consumed is never raised. Apart from this, it is clear from a reading of the arguments surrounding the questions of the rights of future generations, animals and wilderness areas that the proponents of different positions simply start from different definitions of rights and talk past each other. The degenerate nature of the ethical ideas is also made evident by the trivial nature of the arguments. The failures in the ethical doctrines drawn upon are ignored, and rather than addressing the major issues and deriving their conclusions from clearly established and defended positions, environmental philosophers argue by invoking those particular reformulations of the notion of rights which just happen to support their commitments, or by analogy with what extensions to rights theory have been made in the past. But the notion of 'rights' is extended in this way so far beyond the context of the original formulation of rights theory that it loses virtually all meaning. And in the case of animals the extension is achieved at the expense of reducing them to the status of severely defective humans. Generally the discussions are characterized by philosophers addressing themselves almost entirely to one another. There is little concern to work out what ideas would be convincing to anti-environmentalists.

Where utilitarianism has been invoked to oppose cruelty to sentient organisms the situation is somewhat different. It is after all in the calls to relieve suffering that utilitarianism has been most effective as a doctrine. However animals are in no position to rebel against their lot, and without the threat of this, the motivation for implementing the conclusions of the doctrine is lacking - except for those people who have been discomfited by awareness of animal suffering. This reaction has been minimised, however, by keeping such suffering out of sight, which is entirely in accordance with utilitarian principles of eliminating anything which could interfere with people's enjoyment of life. In fact, on utilitarian principles Singer's work can be condemned for upsetting people. But to invoke Benthamite utilitarianism is to commit oneself to all its implications. This leaves no grounds to justify concern with unique species of life or ecosystems. Since at most all that matters is that the animals do not suffer this would justify the replacement of wilderness areas with domesticated animals providing these were killed humanely. Such well looked after animals would suffer less than wild animals. While Singer condemns the treatment of animals as machines for converting low priced fodder into high priced meat, there is no basis in utilitarianism for opposing this. If we are to concern ourselves with animal suffering, then all that matters is that they do not suffer. This problem could be overcome by giving animals pain-killers and valium, or by putting electrodes into their brains to stimulate their 'happy' centres. In this way animals could be made to feel ecstatically happy even as they were being slaughtered. Or better still they could be genetically engineered to be ecstatic while being killed, while people could be genetically engineered to be ecstatic while killing them.

Where rights theory and utilitarianism have been promoted as a basis for political action the implications are more significant. There are several reasons for this. Firstly, the notions of rights theory and utilitarianism on which political proposals are based are not generally extensions of these doctrines but accord with the original reasons for their formulation. Secondly, these proposals fit in with the existing trends in late capitalist societies of placing of decision making in the hands of 'experts'. And thirdly, the idea of furthering such control by experts accords with the mechanistic way of thinking about the world. The development of these doctrines to deal with

---

72. MacIntyre, *After Virtue*, p.63.

73. Ibid. p's.6, 68, & 69.

environmental problems began in USA with the progressive conservation movement during the presidency of Theodore Roosevelt under the leadership of Gifford Pinchot, the Chief of the Bureau of Forestry.[74] The members of this movement were concerned with efficient use of resources, and believed that this could only be achieved in a society organized and controlled by experts applying scientific knowledge. They had an unbounded faith in science and the virtues of large scale organization. They believed that efficient business is big business, and wanted the nation to be managed by its elite. Roosevelt lamented to a friend that 'all modern legislative bodies tend to show their incapacity to meet the new and complex needs of the times' and Joseph N. Teal, chairman of the Oregon Conservation Commission asserted that 'The great difficulty in this country, and I presume in all democracies, lies in the fact that ... the views of experts are of little value ...'[75] Under the direction of Pinchot (who was also a prominent member of the eugenics society) this movement attempted to over-ride sectional interests and democratic procedures, and run the nation as an efficient business organization on the principle of the greatest good for the greatest number for the greatest length of time. Their utilitarianism eventually led to a clash with the preservationists led by John Muir, the founder of the Sierre Club, whose concern with saving wilderness was not just a matter of obtaining a sustained yield of timber. Garrett Hardin's and William Ophuls' accounts of environmental problems as being due to people acting in their own interests at the expense of the common interest extends the same tradition. Their proffered solution - 'mutual coercion, mutually agreed upon' for the interests of the society as a whole - also echoes the progressive conservationist's diagnosis of the problem and its anti-democratic solutions. Finally, the use of cost-benefit analysis as part of a scientific approach to environmental management is a refinement of their utilitarian criterion for administering society.

But this scientific politics is founded on misconceptions. The tragedy of the commons as a description of the problem in capitalist societies is undoubtedly correct. But its universalization as an analysis of the way resources are used by people is mistaken. To take one example, conservation has been practiced successfully in the commons of the villages of Japan with coercion playing a minimum role.[76] This was possible because egoism was constrained rather than promoted by the culture of these villages. What Hardin's, and to a lesser extent Ophuls', proposed solutions amount to is the acceptance of a society which engenders egoism, while offsetting this by giving more power to a central government. This leaves the problem of ensuring that governments will use their increased power to preserve the environment. Where societies are based on the selfish pursuit of private interests and are characterised by very unequal distributions of power, however, this is a problem. In societies of this kind, governments tend to be dominated by short-term problems of keeping the economic system going and by the most powerful pressure groups - the business community. Up until the early 1970's governments continued to pursue a range of wider public and social objectives. However with the development of transnational financial, agricultural and industrial organizations, this capacity has been eroded, and governments in most countries are less able to consider long term issues - even if they were willing to do so - as they struggle with a succession of crises engendered by this globalization of the economy. As a result, the increased power of State institutions has been increasingly directed towards managing crises, suppressing dissent and extracting more work for less both at home and abroad.

The history of environmentalism in USA follows this kind of trajectory. The progressive conservation movement did have some achievements to its credit, although only where conservation was defined in utilitarian terms. Pinchot and his engineering colleagues were actually hostile to the aims of the preservationists. But this movement began the process of transforming a capitalist

74. This movement has been described by Samuel P. Hays in *Conservation and The Gospel of Efficiency: The Progressive Conservation Movement, 1890-1920*, N.Y.: Athenium, 1972.

75. Cited ibid. p.133.

76. See Margaret A. McKean 'The Japanese Experience with Scarcity: Management of Traditional Common Lands' in Kendall E. Bailes ed. *Environmental History*, Lanham, N.Y., London, University Press of America, 1985, pp.334-373.

society based on competition in which the State was dominated by democratic institutions into a military-industrial complex governed by a managerial class of corporate executives, government bureaucrats, military officers and intelligence chiefs. This class has in fact continued its concern for the conservation of resources, especially where these resources are of military significance, but with the notable exception of New Dealers, and in particular Franklin Roosevelt's Secretary of the Interior, Harold Ickes, they have been as indifferent to more general environmental as to human degradation. Where wilderness areas and unique species have been preserved, this has been mostly due to the success of pressure groups over the military-industrial complex associated with the continued functioning of democratic processes. President Nixon even embraced environmental issues when the environmenalists were having most influence during the early 1970's, establishing the Environmental Protection Agency which significantly furthered the conservation cause. However, when President Reagan came to power, the environmentalists lost almost all the ground they had gained, and the Environmental Protection Agency was emasculated. As the power of the US government weakened in the face of the dynamics of the international economy, the environmental concerns of preservationists were sacrificed along with social welfare programs.

Where environmental problems have been taken seriously, in relation to the conservation of strategic resources, the greater part of the effort of the US military-industrial complex has been directed towards maintaining control of the resources of other nations. This, and the effects of it - the oppression of the Third World nations and the arms race - have been described in Chapter I. The ensuing political-military control of the world has meant that what success there has been in preserving forests and species within USA and other affluent nations has increased pressure on ecologically more important tropical forests in the Third World and on the populations which depend on them. While the USA and other affluent nations have maintained access to resources, this has been disastrous for the world as a whole, and particularly for those Third World countries unfortunate enough to possess strategically important resources.

Cost-benefit analysis is the instrument of the new managerial class of the military-industrial complex. It is vexed by all the traditional problems associated with utilitarianism.[77] To begin with, it is based on the subjective preferences of individuals. Since people generally have a very poor idea of what they want until they fail to get it, and even then are unlikely to appreciate precisely what it is they are missing, this provides a very weak foundation for decision-making. Cost-benefit analysis also ignores the problems of quantification such as the impossibility of comparing and weighing the desires and preferences of different individuals or comparing different types of ends.[78] For instance it is impossible to put a value in quantitative terms on an individuals' life, or to work out how much pleasure is equivalent to how much dignity or justice. Cost-benefit analysis is insensitive to distributions of happiness and generates problems by accepting a mechanistic view of the world which ignores the interdependence of things. It is an incremental approach, and it is the micro costs and benefits determined by aggregating the inputs and outputs of individuals (persons, firms or government corporations) rather than the dynamics of the global system which form the basis of decisions. The piecemeal nature of cost-benefit analysis takes the existing order of things as its starting point, and ignores the benefits of switching to different paths of development. The difficulties of switching paths, for example from centralised power production to decentralised power production or from private transport to efficient public transport, are counted as costs against making such switches. This always rules them out as unrealistic. By assuming an essentially inert world in which actions give rise to events and states of affairs rather than contributing to the

77. The difficulties in cost-benefit analyses are briefly summed up in Robert E. Goodin, 'Ethical Principles for Environmental Protection', Robert Elliot and Arran Gare (eds), *Environmental Philosophy: A Book of Readings*, St. Lucia: University of Queensland Press, 1983, pp.3-20.

78. This is the essence of Mark Sagoff's critique of environmental economics which assumes that all values can be made commensurable by examining what people are willing to pay. See Mark Sagoff, *The Economy of the Earth: Philosophy, Law and the Environment*, Cambridge and N.Y.: Cambridge University Press, 1988.

creation and destruction of dynamic processes, cost-benefit analyses overlook the significance of such factors as keeping society's options open, the sustainability of benefits, and the significance of the difference between causing harm and foregoing benefits. Langdon Winner has described the implementation of cost/benefit analyses to evaluating risks to health in the following terms:

> Rather than eliminate from human consumption any substance shown to cause cancer or birth defects in laboratory animals, we are asked to substitute 'risk/benefit analysis.' In that rapidly developing, highly quantified moral science, people are (in effect) asked to acknowledge cancer and birth defects as among the exhilarating risks - often compared to flying or mountain climbing! - we run in order to live in such a materially abundant society. Like our factories that need a dose of 'reindustrialization' to bring them back to life, our nihilism is now being completely retooled, becoming at long last a truly rigorous discipline.[79]

Finally, while environmental management based on cost-benefit analyses might ameliorate some of the worst environmental destruction, it does not face up to the fundamental problem that it is impossible for the planet to support exponential expansion in economic activity for very long. Those who have faith in the computations of cost-benefit analyses might ponder the words of the Chairperson of the Tennessee Valley Authority, David Freeman. After doing his sums he announced that 'on a discounted cash-flow basis, the earth is simply not worth saving.'[80] The cost-benefit approach with its utilitarianism and incrementalism is blind to the prospects of the world as a whole. and the arrogation of decision-making by 'policy experts' using this pseudo-scientific quantification procedure has served to blind the general population to what is happening.[81]

To sum up then, rights theory, utilitarianism and their modern reformulations have failed to provide decision-making procedures for attaining a consensus. Worsening environmental problems have highlighted this failure. But this is only the beginning of the story. Rights theory and utilitarianism have a significance beyond this local failure. They belong to the foundations on which modern Western societies rest, and their invocation reinforces these foundations. By framing arguments in terms of rights theory and utilitarianism, ethical philosophers are enhancing the credibility of the assumptions of these doctrines: that people are basically egoists, that beyond self-interest there are no reasons for considering others; that nature is devoid of intrinsic significance; that the only rationality is the rationality of efficient control; and finally that the human sciences incorporating these assumptions present a hard-headed view of what the world is really like. These are the very beliefs which underlie the institutions and social dynamics responsible for environmental problems but they are also now associated with a more sinister development in society. Capitalism is being transformed into a vast system of global corporations and governments are dedicating themselves increasingly to providing the necessary infrastructure of social control: policing, training and marshalling finance. The claims of conservative politicians to stand for small government means only small government where human welfare is concerned. Where big business, weapons and global military intervention are concerned, the conservatives stand for very big government, as record deficit budgets have revealed. This development is associated with the elimination of the public sphere based on open discussion to arrive at a consensus, and its replacement by central direction and a technocracy in which experts use cost-benefit analyses and implement decisions through social engineering. Another critic of cost-benefit analysis has noted:

> Persecutions, massacres, and wars have been coolly justified by calculations of the long range benefit to mankind; and political pragmatists, in the advanced countries, using cost/benefit analyses reared for them by gifted professors continue to burn and destroy. The utilitarian

---

79. Langdon Winner 'Techne and Politeia: The Technical Constitution of Society' in Paul T. Durbin and Friedrich Rapp (eds), *Philosophy and Technology*, Dordrecht: Reidel, 1983, pp.97-111, p.104.

80. Cited by Geoffrey Lawrence, *Capitalism and the Countryside*, Sydney and London: Pluto Press, 1987, p.66.

> habit of mind has brought with it a new abstract cruelty in politics, a dull, destructive political righteousness: mechanical, quantitative thinking, leaden academic minds setting out their moral calculations in leaden abstract prose, and more civilised and more superstitious people destroyed because of enlightened calculations that have proved wrong.[82]

The environmentalists who call for more coercion of the general population and government by cost-benefit analyses are significant because they are perversely contributing to this development. They are contributing to a social order which is blindly devoted to the expansion of the economy, of military power and of social control and which is blind to the more fundamental environmental problems and to the absolute limits of the world ecosystem.

## *The Christian Tradition*

Yet another attempt to respond to the environmental problems of modern industrial societies involves appeal to tradition, and traditionalists are those who wish to fall back on traditional ideas and institutions, and in particular to Christian ethics, to resolve environmental problems. The most well known of these is John Passmore who set out his position in *Man's Responsibility for Nature*.[83] Robin Attfield as the foremost defender of the Christian tradition of thought in relation to environmental problems also belongs to this group.[84] However while Passmore is a conservative, Attfield is closer to the 'deep ecologists' - despite his criticism of their efforts to develop a new foundation for ethics. Here I will be more concerned with Passmore.

At first sight Passmore appears to be wholly concerned with environmental problems. In the preface to the second edition of his book he writes:

> Let me then try to be more explicit about my intentions. I set out, first of all, to discover whether there are any sound arguments for the following conclusions:
> (a) the present level of pollution ought to be reduced
> (b) resources ought to be conserved for the use of future generations
> (c) the rate of population growth ought to be reduced
> (d) not only animal species but areas of wilderness ought to be preserved.[85]

However the introductory 'Note to the Reader' reveals a different motivation for his interest in environmental problems. He writes:

> The scientific literature fully convinces me, so far as personal observation does not suffice, that men cannot go on living as they have been living, as predators on the biosphere. But I find no less alarming the suggestion in so much of that literature that the West can solve its problems

---

81. On this see Francis Sandbach, *Environment, Ideology & Policy*, Oxford: Blackwell, 1980, esp. Ch.'s 2 & 3.

82. S. Hampshire, 'Morality and Pessimism', in S. Hampshire (ed.), *Public and Private Morality,* Cambridge: C.U.P., 1978, pp.1-22, p.4.

83. John Passmore, *Man's Responsibility for Nature: Ecological Problems and Western Traditions*, 2nd ed. London: Duckworth, 1980.

84. Robin Attfield's views are developed in *The Ethics of Environmental Concern* 1st ed., Oxford: Blackwell, 1983; - 2nd ed., Athens, Georgia: University of Georgia Press, 1991; 'Christian Attitudes to Nature', *Journal of the History of Ideas* Vol.44, 1983, pp.369-386, and 'Western Traditions and Environmental Ethics' in Elliot and Gare (eds), *Environmental Ethics: A Book of Readings*, pp.201-230.

85. Passmore *Man's Responsibility for Nature p.viii.*

> only by forgetting what it has so gradually learnt, only by reverting to attitudes and modes of thought which it painfully shook off.[86]

In other words his concern is also to defend the Western tradition of thought: he defends the resources of Western thought in order to pre-empt the efforts of those who would use the environmental crisis to justify a radical revision in our way of thinking or radical political action.[87] Thus he begins the first chapter by describing and criticizing the calls by Aldo Leopold and Lynn White for a rejection of traditional Western ethics, and later attacks Herbert Marcuse's call for 'a radical transformation of the very institutions and enterprises which waste our resources and pollute the earth.'[88]

In his examination of the Western tradition of thought Passmore begins by supporting the claims of environmentalists that it has helped to produce a despotic attitude towards nature. He cites the biblical report of God's instructions to Noah to:

> Be fruitful, and multiply, and replenish the earth. And the fear of you and the dread of you shall be upon every beast of the earth, and upon every foul of the air, upon all that moveth on the earth and upon all the fishes of the sea: into your hands are they delivered. Every moving thing that liveth shall be meat for you...[89]

and traces the development of such attitudes through to the nineteenth century when the radical American economist, H. C. Carey told his readers that 'the earth is a great machine, given to man to be fashioned to his purpose'.[90] However Passmore identifies two minor traditions within Western thought, one that sees man as a 'steward', as God's deputy actively responsible for the care of the world, and one that sees him as co-operating with nature in order to perfect it.[91] He then argues that it is because of the presence of these 'seeds' in the Western tradition that reformers have some hope for the future. Most of the rest of the book is devoted to cultivating these seeds.

Attfield is more defensive of Christianity than is Passmore. In fact he is very critical of Passmore's interpretation of Western traditions and all who have been influenced by this interpretation.[92] He argues that neither the Old Testament nor the New Testament can be interpreted to justify a despotic attitude towards nature. The Bible generally is characterized by injunctions against cruelty to animals and does not support the view that everything has been made for humankind's use. Furthermore, subsequent Christian history has been characterized by far more varied attitudes to nature than Passmore allows. Attfield also argues that Christianity provides the basis for a more positive attitude towards nature than Passmore himself defends. In general, he argues for the idea that we are stewards of nature, but interprets this in a less human centred way than does Passmore. On this basis he argues that views expressed by the 'deep ecologists' such as Routley and Naess can be supported within the tradition of Christian thought.[93] It is for this reason, Attfield holds, that exposure to new facts about the treatment of animals in factory farms and ecological findings about the interdependence between species have changed our attitudes.

---

86. Ibid. p.xiii.
87. In considering this tradition, Passmore even refers to the ideas of Heraclitus which have been eclipsed since Parmenides (ibid. p.183). As I will argue towards the end of this work, there would be a radical revision of ethics if it were founded on the thought of Heraclitus. However Heraclitus is only mentioned near the end of Passmore's book, and Passmore's work is devoted to defending more traditionally acceptable ideas.
88. Ibid. p.61.
89. Cited Ibid. from 'Genesis', ch.9, verses 1-3.
90. Cited Ibid. p.21.
91. Ibid. p.39.
92. Attfield, 'Christian Attitudes to Nature' pp.369-386; and 'Western Traditions and Environmental Ethics' p.201ff.
93. Attfield, *The Ethics of Environmental Concern,* 1st ed. p.5.

However Attfield's argument goes further: not only is the Western tradition of thought capable of being developed adequately to deal with environmental problems: 'a new ethic is impossible; the most that is possible is a revised normative theory accommodating and enlarging upon accepted judgements.'[94] Even if we could devise a new ethic, he argues, it would be impossible to 'establish its credibility unless it were not a new departure but an extension, analogical or otherwise, of existing patterns of moral thought.'[95]

## *Deficiencies in the Arguments of the Traditionalists*

One of the most admirable features of traditionalists such as Passmore and Attfield is that they do consider the issue of how ideas come to be effective. Passmore's approach to ethics is largely one of searching for what could motivate people to act in an environmentally responsible manner, and Attfield regards the development of a sound ethical theory as essential if education, the broadening of the imagination, pressure groups and political parties are to play their parts in confronting environmental problems. The reasons offered by Passmore and Attfield for developing the tradition of ethical thought rather than developing a new ethics are based on their concern for what ideas would be likely to influence people. However, they have probably failed to identify such ideas.

Ethical ideas are acceptable in the short run because of their familiarity. But in the long run their acceptability is determined by their coherence with what is taken to be the most defensible conception of the world. The Judeo-Christian tradition from which Passmore and Attfield derive their Stewardship Ethics was accepted in Ancient Rome largely because of the efforts of the Church Fathers to interpret and defend it in terms of the dominant metaphysics of the age: Neoplatonism. And the ethical ideas associated with this only became a significant influence on how people behaved after this system had been incorporated into the institutions of society to form the feudal order in the Middle Ages. The ideas dominating feudal society only retained their plausibility through the continued defence of the metaphysics on which they were based. At least part of the reason for the continued success of Christianity in the Middle Ages was the work of theologians such as Thomas Aquinas in assimilating to it the metaphysics of Aristotle, thus nullifying the challenge from the Averroists. It was the tradition of a whole world-view or world-design embodied in the institutions of society which gave the ethical thought of the Judeo-Christian tradition its force. It is this world-view rather than the tradition of ethical thought as such which is needed to provide the grounds for the acceptability of ethical proscriptions or prescriptions of Christianity. However the medieval world-view which justified Christianity has broken down and been replaced by the mechanistic world-view vouchsafed by the achievements of science, and this new world-view has been incorporated into the institutions of capitalism which have replaced the feudal order.

Passmore and Attfield do not acknowledge this and have obfuscated the issue by their approach to ethics and the history of ideas. Searching for the source of attitudes dominating the present in past traditions of thought, as did Lynn White, is a valid and important enterprise in the effort to understand the present situation. But to assume that there could be any significance for ethics in showing that there were alternative ways of thinking contained within the Christian tradition of thought or that present attitudes to nature have their origins in misinterpretations of the Bible or Christian theologians, is to fail to face the fact that it is science which is now the ultimate arbiter in matters of belief, not religion. It is in terms of science that ethical doctrines must now establish themselves if they are to carry any force, and mechanistic science does not lend itself to the support of any ethics associated with Christianity. Attfield's arguments to the effect that there can be no

94. Attfield, 'Christian Attitudes to Nature' p.225.

95. Attfield, *The Ethics of Environmental Concern*, 1st ed. p.4.

new ethical ideas assumes the highly implausible view that ethical ideas stand without any outside support.

The pitfalls of separating ethics from their foundations in metaphysics, and thereby from the natural and social sciences are evident in Passmore's concrete analyses of problems. He sees science as providing knowledge which can be used by technology which in turn can be used by the economy. Only then do ethics enter in association with politics and administration. For instance, in relation to pollution, after science shows 'how a particular form of pollution arises and in what its danger consists, the next problem is technological: to discover a method of reducing its incidence.'[96] Since a number of technological solutions can be offered, it will then be necessary 'to make fuller use of rational Western-type methods, cost-benefit analyses or decision procedures. Then the economist enters the picture as a specialist in such methods.'[97] At this point the proposed solution is examined for its moral acceptability, political feasibility and administrative consequences. At all points Passmore places his faith in specialists, writing, 'Close communication between specialists is a more effective procedure than the attempt to rely on "generalists."'[98] This piecemeal pragmatic approach in which the important judgements are left to 'experts', problems are treated in isolation from each other and economic control is seen in terms of specific bureaucratic interventions, is simply an endorsement of the status quo.[99]

As with most modern ethical pronouncements formulated in terms of Christianity, Passmore's ideas amount to nothing more than a gloss over the existing reality. In the 1950's Alasdair MacIntyre responded to Marxists in Eastern Europe, particularly Leszek Kolakowski, who were attempting to invoke principles of liberalism to ameliorate the effects of Stalinist Marxism. He wrote: 'One cannot revive the moral content within Marxism by simply taking a Stalinist view of historical development and adding liberal morality to it.'[100] Similarly one cannot ameliorate the effects of a capitalist society embodying a mechanistic view of the world, rights theory and utilitarianism, and as I will argue later, Social Darwinism, by grafting on to it elements of Christian thought. The attempt to do so changes nothing. This, moreover, appears to be consonant with the deeper aims of Passmore's stand.

## *The Deep Ecologists*

This brings us to the most radical environmentalists, the proponents of a new way of thinking, the 'deep' ecologists. Predictably, those who propose new ways of thinking differ from each other more than do those arguing for developments of conventional positions. There is however, some important common ground. All deep ecologists agree that prevailing modes of thought are inadequate. They are also united by their rejection of an ethic centred on humanity. Nature is seen to be intrinsically valuable, and humanity is always seen as part of nature.

Their views have their origins in the Hermetic philosophers of the sixteenth and early seventeenth centuries and in the Romantics of the late eighteenth and early nineteenth centuries, particularly in Germany under the influence of Herder and Schelling. The ideas spread from Germany to Russia, and to Britain via Coleridge and Wordsworth, then to America where they were taken up by the transcendentalists: Emerson, Thoreau and finally John Muir.[101] However

---

96. Passmore *Man's Responsibility for Nature*, p.48.

97. Ibid. p.50.

98. Ibid. p.48n.

99. The limitations of this approach have been pointed out by C.A. Hooker, 'On Deep Versus Shallow Theories of Environmental Pollution', in Elliot and Gare *Environmental Philosophy*, pp.58-84.

100. Alasdair MacIntyre, *New Reasoner* Vol.7, p.100; cited by himself in *After Virtue*, p.ixf.

101. The American tradition has been described by Roderick Nash in *Wilderness and the American Mind*, New Haven: Yale University Press, 1967.

while such thinkers extolled the value of nature for other than utilitarian ends, they still conceived of it anthropocentrically as means to a higher form of experience. It was only in the twentieth century that Western thinkers have clearly argued that nature has a value in its own right completely independent of humanity. Albert Schweitzer argued for a reverence for life as such, and backed up his views by appealing to Eastern philosophy. Then an American, Olaus Murie, formulated a totally non-anthropomorphic ethics of nature. Beginning in the 1920's Murie argued for the value of pests which are of no conceivable benefit to mankind, simply because they are living beings.[102] Murie had little immediate influence, but won over a leading exponent of the Pinchot school, Aldo Leopold who presented the position with great eloquence, if less consistency, in his posthumously published *Sand County Almanac* printed in 1949.[103] The *Sand County Almanac* made a case for a 'land ethic'. It argued that all ethics are based upon the premise that the individual is a member of a community of interdependent parts. A land ethic would enlarge the boundaries of the ethical community to include 'soils, waters, plants, and animals, or collectively: the land.'[104] On this basis Leopold concluded: 'A thing is right when it tends to preserve the integrity, stability, and beauty of the biotic community. It is wrong when it does otherwise.'[105]

Leopold's sentiments were taken up in the early 1970's by a number of philosophers, largely in reaction to the homeocentric and elitist attitudes of environmentalists such as Ehrlich who were exclusively concerned with population growth, resource depletion, pollution and the affluent of the world. The basic position of these new environmentalists was summed up by Arne Naess in 1973 when he drew a distinction between 'the shallow' and 'the deep, long-range ecology movement.'[106] Naess defined the deep ecologists as being characterized by their rejection of the man-in-environment image in favour of the relational, total-field image in which all organisms are seen as knots in the biospheric field of intrinsic relations, and by biospherical egalitarianism in which all forms of life are accorded a deep respect. Deep ecology was opposed to elitism, either within or between nations, and supported complexity of ecosystems, economies and ways of life, the decentralization of power and local autonomy.

A number of thinkers have since elaborated and defended these themes, including John Rodman,[107] Holmes Rolston III,[108] Richard Sylvan and Val Plumwood (formerly Richard and Val Routley),[109] Charles Birch and John Cobb,[110] Bill Devall,[111] George Sessions, [112] Henryk Skolimowski,[113] Warwick Fox,[114] and Freya Mathews.[115] And there has been some parallel or

---

102. See Donald Worster *Nature's Economy: The Roots of Ecology*, Garden City, N.Y.: Doubleday, 1979, p.282ff.

103. Aldo Leopold, 'The Land Ethic' in *A Sand County Almanac* [1949] Oxford: O.U.P., 1968 pp.201-26.

104. Ibid. p.204.

105. Ibid. p.224f.

106. Arne Naess, 'The Shallow and the Deep, Long-Range Ecology Movement. A Summary' *Inquiry*, Vol.16, 1973, pp.95-100.

107. John Rodman, 'The Liberation of Nature' *Inquiry*, Vol.20, 1977, 83-145.

108. Holmes Rolston III, *Philosophy Gone Wild: Essays in Environmental Ethics*, Buffalo: Prometheus, 1986 and *Environmental Ethics: Duties to and Values in the Natural World*, Philadelphia: Temple University Press, 1988.

109. In particular Richard and Val Routley, 'Human Chauvinism and Environmental Ethics', in Don Mannison, Michael McRobbie and Richard Routley (eds) *Environmental Philosophy*, Canberra: Australian National University, 1980, pp.96-189. For Richard Routley's more recent work see R. Sylvan, *Three Essays Upon Deeper Environmental Ethics*, Democracy Books, Canberra: A.N.U. Press, 1987 (both Routleys have changed their names).

110. Charles Birch and John B. Cobb, Jr. *The Liberation of Life: From the Cell to the Community*, Cambridge: Cambridge University Press, 1981.

111. Bill Devall, 'The Deep Ecology Movement' *Natural Resources Journal*, Vol.20, 299-322 and with George Sessions, *Deep Ecology - Living As If Nature Mattered*, Salt Lake City: Peregrine Smith Books, 1985.

112. George Sessions, 'Spinoza and Jeffers on Man in Nature', *Inquiry*, Vol.20, 481-528.

113. Henryk Skolimowski, *Eco-Philosophy: Designing New Tactics for Living*, Boston: Marion Boyars, 1981.

114. Warwick Fox, *Towards a Transpersonal Ecology*, Boston: Shambala, 1990.

derivative movements, differentiating themselves in some way from the deep ecologists.[116] Thus Murray Bookchin has attacked the anti-human bias of many deep ecologists and has formulated an alternative - 'social ecology', Richard Sylvan has redefined his position in opposition to 'deep ecology' as 'deep green',[117] and ecofemininists have differentiated themselves from deep ecologists, identifying the cause of human destructiveness in patriarchal society.[118]

In general, deep ecologists and derivative movements tend to concur with Hans Jonas' view that:

> Only an ethic which is grounded in the breadth of being, not merely in the singularity or oddness of man, can have significance in the scheme of things... an ethics no longer founded on divine authority must be founded on a principle discoverable in the nature of things.[119]

Rather than attempting to derive an ethics from the relationships between people or from the experience of individuals, deep ecologists have been attempting to derive an ethics from the nature of the world and the place of humanity within it.

One of the most forceful early expressions of this effort is the paper by John Rodman, 'The Liberation of Nature', a paper free of the faults which other radical environmentalists, including ecofeminists, have found in deep ecology. Rodman criticised the tendency of environmentalists to treat animals as defective humans and began developing a new ethic by pointing out that it is not the utilitarian arguments in Peter Singer's *Animal Liberation* which provide the most convincing condemnation of how animals are treated, but the description of them being treated as though they are not animals but machines converting low cost fodder to high priced meat, or as breeding machines.[120] Rodman argued instead that the 'non-human world is full of what Mill called "inward forces", potentialities striving to actualize themselves' and suggested 'we can ally ourselves with these tendencies and resist the efforts of other human beings to obstruct them.'[121] Further, Rodman argued that the repression by humans of these potentialities in nature has the effect of repressing their own potentialities. Just as it is necessary for males to stop oppressing females in order to free the feminine sides of their own personalities, so it is necessary for people to liberate nature from their domination of it in order to fully liberate themselves.

Other deep ecologists have been attempting to elaborate a political philosophy to accord with the new ethics.[122] Societies, they argue, should be judged by the quality of life they facilitate, generally defined in terms of 'self-realization'. In most cases deep-ecologists have argued for 'bio-regionalism' - the identification by people with their local environment, for a decentralization of

---

115. Freya Mathews, *The Ecological Self*, London: Routledge, 1991.

116. Some of the conflicts within this movement have been aired in *The Ecologist* Vol.18, Nos. 4/5, 1988.

117. This opposition is defined in a submission to UNESCO: *Australian Perspectives on Environmental Ethics*, 1989.

118. For early statements of ecofeminism see Ariel Kay Salleh, 'Deeper than Deeper Ecology: The Eco-Feminist Connection', *Environmental Ethics*, Vol.6, Winter 1984, pp.339-346; and Charlene Spretnak, 'Ecofeminism: Our Roots and Flowering', *Ecospirit*, Vol.III, No.2, 1987, pp.2-9. For more recent work, giving some account of the diversity of feminist positions on environmentalism, and analysing the relationship between ecofeminism and deep ecology, see also Irene Diamond and Gloria Feman Orenstein, *Reweaving the World*, San Francisco: Sierre Club Books, 1990. The most rigorously formulated defence of eco-feminism is Val Plumwood, *Feminism and the Mastery of Nature*, London: Routledge, 1993. For a review of the major positions within radical environmentalism, see Michael E. Zimmerman, *Contesting Earth's Future*, Berkeley: University of California Press, 1994.

119. Hans Jonas, *The Phenomenon of Life: Toward a Philosophical Biology* N.Y.: Dell, 1966, p.284; referred to by Rodman, 'The Liberation of Nature', p.95.

120. Rodman, 'The Liberation of Nature', p.100.

121. Ibid. p.117.

122. For example Val and Richard Routley, 'Social Theories, Self Management and Environmental Problems', Mannison et. al. *Environmental Philosophy* pp.217-332.

society and for the creation of self-sufficient communities.[123] The most influential proponents of such an organization of society are Murray Bookchin and Rudolph Bahro.[124]

To develop and defend such ideas environmental philosophers have had to go beyond ethics and social philosophy and engage in the fundamental issues of cosmology, metaphysics, epistemology and logic. The atomistic or mechanistic view of nature has been rejected in favour of a conception of the world which emphasizes inter-dependence. Ecological theory in particular has been pressed into service for this task, supplemented by the metaphysical ideas of such philosophers as Spinoza, Schopenhauer, Nietzsche, Whitehead and Heidegger and by Eastern religions and Amerindian cultures.[125] Efforts to justify both metaphysics and the rejection of the fact-value dichotomy have led to a questioning of prevailing epistemological doctrines and of the extensional logic on which many of these are based.[126]

## *Limitations of the Deep Ecologists*

Given the inadequacy of prevailing modes of thought, it is of prime importance to develop new ways of thinking. The question is whether the new modes of thought which have been proposed so far are adequate. From one point of view it can be expected that they will not be. Educational institutions have been transformed from communities within which students could appropriate and develop their cultural heritage into organizations for the efficient production of marketable skills and the way humanist oriented intellectual dissenters have been replaced by experts, specialists and generalist-integrators. It has therefore become extremely difficult for those critical of society or of the modes of thought underlying it to gain academic positions or funding for research. Consequently there have been far too few people involved in the field to develop and inter-relate ideas. However beyond this some of the deep ecologists can be criticized for their inconsistencies and the direction of their work,[127] and it can be argued that the deep ecologists need to develop their ideas in ways that they have hitherto not considered. The deficiency of these ideas lies not just in their details but in their failure to fully transcend the disciplinary divisions which embody the prevailing metaphysical assumptions, and to provide an overall perspective which would justify and reveal the relevance of their ideas (although some recent works, notably Arne Naess's *Ecology, Community and Lifestyle*, have sort to redress this deficiency.)[128]

What is most disturbing about the deep ecologists is that it is hard to imagine their ideas having anything more than a marginal impact on society. The question which needs to answered is: Why?

---

123. For a study of environmental political philosophy see Robyn Eckersley, *Environmentalism and Political Theory*, London: University College London Press, 1992.

124. See Murray Bookchin, *Towards an Ecological Society*, Montreal: Black Rose Books, 1980 and *The Ecology of Freedom*, Palo Alto, Calif.: Cheshire Books, 1982; and R. Bahro, *From Red to Green*, London: Verso, 1984 and *Building the Green Movement*, tr. Mary Tyler, London: Heretic Books, 1986.

125. See for example *Nature in Asian Traditions of Thought: Essays in Environmental Philosophy*, ed. J.Baird Callicot and Roger T. Ames, N.Y.: S.U.N.Y. Press, 1989 and Richard Sylvan and David Bennett, 'Tao and deep (Ecological) Theory: A Preliminary Investigation' in Peter Hay and Robyn Eckersley ed. *Ecopolitical Theory*, Hobart: University of Tasmania, 1992.

126. Val and Richard Routley, 'Social Theories...' pp.239-250

127. Major attacks on the deep ecologists exposing their inconsistencies have been made by Richard Sylvan (formerly Richard Routley) and Janna Thompson. See Richard Sylvan, 'A Critique of Deep Ecology' Part 1 and 2 in *Radical Philosophy*, Summer and Autumn, 1985, and Janna Thompson, 'A Refutation of Environmental Ethics,' *Environmental Ethics*, Vol.12, Summer, 1990. Another attack has been made from a more political standpoint by Tim Luke in 'The Dreams of Deep Ecology', *Telos*, No.76, Summer, 1988.

128. Arne Naess, *Ecology, Community and Lifestyle*, tr. David Rothenberg, Cambridge: Cambridge University Press, 1989.

There are, I think, two basic problems. Most of the deep ecologists seem to be addressing themselves to environmentalists in order to justify their intuitions; but they are not even fully successful at this task because with only a few exceptions they do not attempt to refute the assumptions of opposing positions. Ideas are simply thrown into the arena to exist beside prevailing ethical ideas. They do not contest or transcend them. Secondly, deep ecologists have failed to relate their ideas to practice because they have failed to situate them in a perspective which relates them to the rest of culture and society. So while they may succeed in producing a changed attitude to nature, this by itself is unlikely to have any more effect on the rate of environmental destruction than the Chinese reverence for nature had on the destruction of their forests. In each case these failures can be traced back to the tacit acceptance of metaphysical assumptions manifest in the acceptance of existing disciplinary boundaries.

All this is illustrated in the work of John Rodman. His seminal paper stands as an inspiration, a new way of thinking about the world, breaking significantly with the prevailing world-orientation. But his position can easily be dismissed from the standpoint of prevailing assumptions. His central concept is that of potentialities and the value of their realization. This contradicts the prevailing assumption that value pertains only to subjective experience. While the relationship between subjective experience of value and the 'objective' world has been looked at by other environmentalists,[129] the more important issue is the status of potentialities. As this notion has come to be understood on the basis of the prevailing metaphysical assumptions, potentialities are what could be the case. They range from potentialities which will be realized in the future without any unique form of intervention (such as for instance adulthood being the potentiality of a young organism), to potentialities which require contrived circumstances for their realization. For instance most children have the potentiality to become psychopaths if they are placed in certain environments, and the USA has the potentiality to obliterate all the cities in Russia if the order is given. But so conceived, what sense can be made of the notion that it is better to have any particular potentialities realized? How can one differentiate between potentialities on this basis? Rodman implies that potentialities which are not the product of contrived circumstances are superior, but this would mean that it is better to allow cancerous tumours to develop their full potentiality than to remove them surgically. Why not 'liberate' cancerous tumours? Why not liberate plagues? Without replacing the scheme of ideas which led to the present conception of potentialities, Rodman's ideas lack any real substance.

To put environmental ideas in the perspective necessary to achieve practical relevance would also involve rejecting the boundaries between ethical, political and economic theory. While deep ecologists are able to expose the radical deficiencies in economic theory, they generally fail to address economic issues in a way that is plausible,[130] and fail to show how their ethical views should or could be incorporated into economic thought or be built into the economic organization of society. The acceptance of these disciplinary boundaries is closely related to the failure of deep ecologists to transcend one of the most important features of the dominant metaphysics. Despite their concern to demonstrate that humans are part of nature, they have failed to overcome the dualism between thought and being in relation to ethics. This is manifest in the separation of ethical judgements and political ideals from interpretation of the present state of the world. The ethical thought of deep ecologists amounts to nothing more than injunctions to fight to preserve wilderness areas without any analysis of why such destruction is taking place. Similarly in relation to politics, the deep ecologists have generally proposed that power in society be decentralized. But little effort

---

129. Holmes Rolston III, 'Are Values Subjective or Objective' *Environmental Philosophy*, Elliot and Gare eds, pp.135-165; and Richard Sylvan, 'On the Value Core of Deep-Green Theory.' Unpublished.

130. For a critique of Bahro and others in this regard, see Boris Frankel, *The Post-Industrial Utopians,* Cambridge: Polity Press, 1987, esp. Ch's 1 & 2. I do not include the work of Herman Daly and John Cobb here. Their ideas, inspired by process philosophy, are part of the solution.

is made to explain why existing societies are badly organized or how to begin reorganizing them.[131] There are no comprehensive and plausible plans for changing the present destructive relationship of society to its environment.[132] By failing to situate their ideals within a critical understanding of the dynamics of existing societies they offer food for fantasy rather than direction for action.

The accusation that deep ecologists have not properly confronted prevailing assumptions might appear paradoxical. It is the deep ecologists who have argued that we need a new metaphysics and have proposed alternative metaphysical theories. However they have generally failed to acknowledge the significance of metaphysical thought, and consequently what they have proposed tends to be little more than a useful intellectual vision which it would be nice for people to believe. This is illustrated by George Sessions' effort to co-opt the metaphysics of Spinoza - the favourite philosopher of the deep ecologists. Sessions writes:

> The West is clearly in need of a 'perennial philosophy' with which to pick up the pieces of the shattered dream, the wreckage of both Nature and our own psyches, and begin the process of healing and integration... It seems that, short of 'pure mysticism', if this perennial philosophy is to be expressible at all on a conceptual plane, it would need to be a version of what Russell once called 'neutral monism'... I think ... that there is such a system in the West which has been largely overlooked in the scramble, and which, when properly interpreted and understood, would serve our needs admirably: namely Spinozism.[133]

But this is an odd place for environmentalists to seek metaphysical support. For Spinoza humans are inescapably egoistic, and he declares that 'it is plain that the law against the slaughtering of animals is founded rather on vain superstition and womanish pity than on sound reason. The rational quest for what is useful to us further teaches us the necessity of associating ourselves with our fellow-men, but not with beasts, or things, whose nature is different from our own...'[134] More fundamentally, however, a metaphysical system is not something to be shopped around for to serve our needs.

A metaphysical system, that is, a theory of being serving as the foundation for a cosmology, must be a claim to be the true understanding, or at least to be the best alternative for understanding, the world. Whether it picks up the pieces of our shattered dream and the wreckage of our psyches is irrelevant to this claim except where it can be shown that our shattered dreams and wrecked psyches are a manifestation of a defective understanding of the world and our place within it. To argue for a metaphysical system as Sessions has done is to assume the framework of a consumer society in which commodities are promoted in terms of the subjective experience of satisfaction to be gained from them. A successful metaphysical system must also be more than an intellectual vision. To think in such terms involves a tacit acceptance of the prevailing dichotomy between ideas and reality. A metaphysical system should provide the basic concepts in terms of which the world and humanity can be understood such that if it were accepted, these concepts would come to be presupposed in practical and theoretical thinking. Mechanistic materialism provides such concepts at present in Western societies, and they are constitutive of social reality. Spinoza developed his system as a creative reformulation of the mechanistic conception of the world in such a way that it would overcome the difficulties in Descartes' philosophy and avoid the implications drawn by Hobbes, reappropriating the ethical orientation of Plato and the Stoics. His system certainly provided a starting point for important advances over mechanistic materialism, but the implausibility of the idea that extension and thought are but two of an infinite number of attributes

---

131. While Naess's *Ecology, Community and Lifestyle* must be recognized as a promising development in this regard, it could hardly form the basis for government by a green political party should such a party ever win political power.

132. As Frankel has argued, *The Post-Industrial Utopians*, Ch.5.

133. Sessions, 'Spinoza and Jeffers on Man in Nature' p.492.

134. Spinoza, *Ethics*, Prop. XXXVII, Note I.

of being, and the difficulty in accounting for individuation, make it difficult to take his metaphysics as a serious contender to replace mechanistic materialism as the foundation of modern society.

Such failings are symptomatic of the unsystematic and undialectical approach adopted by most deep ecologists. They tend to appropriate ideas in an eclectic way to legitimate their preconceptions. To take a dialectical approach requires a metaphysical system which can reveal and interpret both the strengths and limitations of opposing systems, or at least provide a research programme which promises to be able to achieve this. Only in this way can reasons be provided which could persuade people to accept the validity of the new metaphysics. New ideas only refute prevailing ideas by replacing them. But a metaphysical theory must do more than elicit intellectual approval. It must also be able to provide an understanding of the tendencies and weaknesses within the prevailing social order which will enable people to transform it to accord with the new ideas. And it must inspire people to do so. Most deep ecologists are utopians (literally, at no place). What has been offered is a new commodity in the marketplace of ideas, neatly packaged in accordance with the categories of the prevailing system, enabling their works to be efficiently sold in the corners of bookstores reserved for eccentrics.

Without challenging prevailing ethics and without putting their ideas in a credible practical perspective, the deep ecologists feed into a romanticism which complements the prevailing order rather than challenges it.[135] The affluent are served in their efforts to preserve wilderness parks in their neighbourhoods, but they are also allowed to continue living as before. By concentrating on wilderness areas in abstraction from humanity the deep ecologists further support the notion that humans are separate and distinct from nature. The effect of successes achieved on this basis is to put more pressure on other, often more ecologically important, regions. It is environmentalism of this sort which discredits it in the eyes of the less privileged of the world.

## *Conclusion*

The ideas and criticisms presented in this chapter provide some indication of the limitations of prevailing thought and the directions which need to be taken to come to terms with environmental problems. Lynn White's work should have convinced people that at least a major reason for the rise of Western civilization has been the assumptions its members have held 'about who they are, about their relation to other people and to the natural environment, and about their destiny.' But while White was justified in ascribing the source of these assumptions to Western Christianity, the issue is more complex. It is also necessary to consider why these values and beliefs were accepted in the first place, and what forms of social relations and further modes of thinking have been engendered by this Christianity. These social relations now maintain themselves in existence and reproduce the attitudes deriving from Christianity rather than vice-versa, and the intellectual status of these attitudes derives from there having been incorporated into the mechanistic view of the world which is legitimated by the achievements of science. This conception of the world is now so pervasive that it vitiates the efforts of those who try to explain environmental problems, including those of Lynn White.

The pervasiveness of this conception of the world is evident in the work of almost all those who have addressed themselves to environmental problems. It is evident in concepts used by those who have attempted to explain environmental problems and in the division which most ethical and political philosophers accept between efforts to understand the world and ethical and political philosophy. While it can hardly be denied that population growth, economic growth and

---

135. Karl Marx noted in relation to the quest for an original fulness which the deep ecologists exemplify that 'The bourgeois viewpoint has never advanced beyond this antithesis between itself and this romantic viewpoint, and therefore the latter will accompany it as legitimate antithesis up to its blessed end.' (*Grundrisse*, tr. Martin Nicolaus, Harmondsworth: Penguin, 1973, p.162.)

technology are implicated in environmental destruction, there is an incapacity to see these forces in perspective or to comprehend their significance. While ethical and political reflection has raised the issues of our relation to future generations and to other life forms, the problems in the work of philosophers reveal the extent to which their doctrines are based on the forms of thinking which engendered the environmental crisis. Ethical and political philosophers have not addressed themselves to the beliefs which are the real basis for people's decision-making.

What the limitations of all these thinkers suggest is the difficulty of getting into perspective environmental problems while living in a culture which is itself the source of these problems. But the crucial inadequacies of the arguments surveyed above suggest a far deeper problem: that these arguments and the forms of thinking to which they belong are integral parts of a self-reproducing cultural system. And by their very nature as parts of this system, their development can only serve to reinforce this culture. While it is easy to convey the impression that something is amiss, to fully reveal the nature of prevailing beliefs appears to be almost impossible because we are so completely encircled by them. They are presupposed in social practices, in organizations, in educational institutions and in the organization of academic disciplines. Truth has also come to be identified with efficiency and wealth creation so that philosophical ideas, ideas which question prevailing assumptions and propose radically new ways of conceiving the world, now appear as trivial and irrelevant to the real world. It has become difficult to believe that reality could be radically different. Even those who have thought of themselves as radicals have seldom been able to break out of the circle, to transcend their cultural heritage. Only a thorough analysis of the history and cultural dynamics of Western civilization can now provide any hope of freeing people from this culture.

# 3

# IDEOLOGY, METAPHYSICS AND SOCIETY: THE METAPHYSICAL ROOTS OF EUROPEAN CIVILIZATION

The poverty of environmentalist thought is not simply a failure to adequately orient people in relation to environmental problems; it is a failure to be anything more than a disguise for what really orients people for action. In practice, environmental concerns are always taken as peripheral to the really important issues of economic growth and defence. We have arrived at a peculiar situation where the impoverishment of hundreds of millions of people and environmental degradation are taken for granted as unavoidable aspects of progress, and the immediate steps which must be taken to avoid the eventual destruction of the conditions for humanity's continued existence have come to appear utopian. If the environmental movement is to become effective, it is necessary to come to grips with the ideas and forms of thinking which actually move people to action and which lead to such attitudes.

In the following chapters it will be argued that people's refusal to confront environmental problems is not simply a matter of 'human nature', but as Lynn White argued, is at least partly a reflection of the basic, non-verbalized assumptions most people hold about what they are and what is their place in the world. Through a history of the development of this civilization, the history of the basic assumptions of this culture in the philosophies, discourses and social practices which have developed and interacted throughout the evolution of Western society will be presented. This will reveal how the situation has arisen where people can no longer conceive of life having any other meaning than survival, the daily satisfaction of their appetites, entertaining distractions, social climbing, and exhilarating power games; where, as T.S. Elliot wrote in *The Waste Land*:

> You cannot say, or guess, for you know only
> A heap of broken images...

In other words, this work will provide a genealogy of nihilism.

What is most important in the dynamics of cultures is the concepts and images which are embodied in society's dominant institutions and in the way people live - their modes of being in the world. In Western civilization nihilism is incorporated in such a way that it has become extremely difficult to attain a clear insight into what it is and how destructive it is; or of there being viable alternatives to prevailing modes of being or prevailing forms of society. To expose this, the ideas of Nietzsche, the Frankfurt Institute philosophers (Adorno, Horkheimer and Marcuse), and Heidegger and those influenced by him, including Foucault and Derrida, will be drawn upon. However while the cultural analyses offered by all these philosophers are illuminating, they lead to impasses (which in the case of the Frankfurt Institute philosophers, Habermas has only partially transcended through his theory of communicative action). The analyses offered here, grounded in the philosophies of Heraclitus, Bergson and Whitehead, will point the way to a successful overcoming of these impasses. The questionability of the concepts on which nihilism is based will be exposed by examining them at their inception, before they became so embedded in social practices that their validity came to appear self-evident; and the possibility of transforming the social world by reconstituting it on the foundation of a radically different conception of the world will be revealed. At the same time, through this genealogy, a justification for a Heraclitean view of the world will be provided by showing how in terms of the perspective provided by it, the practices, concepts and forms of thinking which have dominated in the past, with all their achievements and limitations, can be presented in a coherent historical narrative. And by analysing the relationship between practices, culture and social dynamics, what is required to effect the sort of cultural revolution necessary to successfully overcome environmental problems will be revealed.

## *Culture and Traditional Societies*

Anthropologists who have attempted to understand traditional societies have frequently been astounded by statements and actions which appear to be irrational. For instance Von den Steinen reported in 1894 the statement of the Bororo of Central Brazil that 'We are red macaws';[1] and Evans-Pritchard reported that the Nuer hold 'that a twin is a bird as though it were an obvious fact, for Nuer are not saying that a twin is like bird but that it is a bird.'[2] Further examination of such beliefs indicates that the criteria for evaluating their validity are different from those which prevail in Western culture. This is most clearly evident when particular beliefs are challenged. For instance Evans-Pritchard's examination of the Azande's beliefs in witchcraft seemed to catch them in flagrant contradiction. The Azande simultaneously believe that witchcraft is always inherited and that the post-mortem examination of a suspect's intestines will conclusively reveal whether or not 'witchcraft substance' is present. It should follow from this that a few post-mortem examinations will reveal for all time which families or clans are witches. But the Azande do not draw this conclusion and go on treating the question as an open one. Furthermore they simply brushed aside Evans-Pritchard's objections. This suggests that beliefs and criteria are part of a cultural system, and that it is impossible to evaluate any aspect of this system in isolation from the total culture.

Such anomalies have led to an appreciation of the radical difference between the cultures of people in traditional societies and of our own culture, and how these cultures constitute the cognizable world of the members of these societies. As Evans-Pritchard wrote of Azande culture:

> In this web of belief every strand depends upon every other strand, and a Zande cannot get out of its meshes because it is the only world he knows. The web is not an external structure in which he is enclosed. It is the texture of his thought and he cannot think that his thought is wrong.[3]

In other words, the Azande do not think of themselves holding beliefs about the world. They take their beliefs as reality; and this 'reality' is very different from what we take to be reality.

To fully understand the role of such modes of thought in society it is necessary to see how they are developed. In the past this has been ignored by anthropologists who have either considered the dynamics of societies independently of human agency, or have focused on the way in which society and culture form the individual. But recently the focus has shifted to how society and culture are produced and reproduced through human intention and praxis.[4] It is in terms of praxis that it is possible to transcend cultural relativism. In the efforts to understand societies it is necessary to develop a conception of humanity able to explain all its variety and manifestations. By focusing on praxis, the issue becomes what are people's ultimate ends in life. In accordance with both the metaphysics being defended and with the achievements of the human sciences, a conception of humans is proposed which implies that they are striving to orient themselves in the world, to gain

---

1. See Dan Sperber, 'Apparently Irrational Beliefs' in Martin Hollis and Steven Lukes, eds, *Rationality and Relativism*, Oxford: Blackwell, 1982, pp.149-80, p.152.

2. E.E. Evans-Pritchard, *Nuer Religion*, Oxford: Oxford University Press, 1956, p.131.

3. E.E. Evans-Pritchard *Witchcraft, Oracles and Magic among the Azande*, Oxford: Clarendon, 1936, p.194.

4. See Marshall Sahlins, *Historical Metaphors and Mythical Realities*, Ann Arbor: University of Michigan Press, 1981 and Sherry B. Ortner, 'Theories in Anthropology since the Sixties' in *Society for Comparative Study of Society and History*, 1984, pp.126-165, esp. p.158.

recognition of their significance, and to gain power; strivings which are manifest to some degree in people's every action and interaction. In most societies people recognize power and recognition as their ultimate ends. For instance J.J. Maquet writes that when questioned on what the people of his group wish for above all else, a Matutsi of Ruanda will answer immediately 'children and cattle'. But a further question, Why?, 'discloses that these are not ultimate values sought for themselves, but intermediate ones, means to reach more abstract ends. The latter are power (amaboko) and reputation (ugukomera).'[5] However whether they acknowledge it or not, there is evidence from the way people behave that they are deeply concerned to be able to orient themselves in the world, in relation to each other, to society and to nature. These struggles generate semi-autonomous processes characterized by cognitive structures, structures of legitimation and structures of power which constrain people to reproduce these structures. What is culturally relative is how people strive for these ends. To understand the role of culture it is necessary to see it in relation to socially situated praxis directed towards these ends.

Piaget has demonstrated how in individuals the capacity to reason which enables them to interact intelligently with their world develops through their practical engagement in their environments. Individuals assimilate the environment to interpretive schemes while at the same time accommodating these schemes to the environment. In other words, cognitive schemes which facilitate intelligent interaction with the environment are developed through their analogical use from one situation to another. In this way children constitute their environments as an intelligible world. Pierre Bourdieu's work has revealed how within societies there is a similar analogical use of interpretive schemes which then constitute the world for their members, although in the case of such social schemes of interpretation the schemes define appropriate behaviour and partially constitute social relations.[6] Such schemes involve the generalization of forms of relationships between people to facilitate 'the interchangeability of reactions and enable the agent to master by a sort of practical generalization all similar problems likely to arise in new situations...'[7] People act in situations according to how these are assimilated to such schemes.

These schemes are first and foremost schemes associated with practices and are not necessarily understood at the symbolic level by the actors:

> Lacking symbolic mastery of the schemes and their products - schemes which they are, products which they do - the only way in which agents can adequately master the productive apparatus which enables them to generate correctly formed ritual practices is by making it operate. This is what the observer is likely to forget, because he cannot recapture the logic immanent in the recorded products of the apparatus except by constructing a model which is precisely the substitute required when one does not have (or no longer has) immediate mastery of the apparatus.[8]

Schemes are embodied as a *habitus*, which Bourdieu defined as 'a system of lasting, transposable dispositions which, integrating past experiences, functions at every moment as a *matrix of perceptions, apperceptions, and actions* and makes possible the achievement of infinitely diversified tasks, thanks to analogical transfers of schemes permitting the solution of similarly shaped problems, and thanks to the unceasing corrections of the results obtained...'[9]

---

5. J.J. Maquet, 'The Kingdom of Ruanda', *African Worlds: Studies in the Cosmological Ideas and Social Values of African Peoples*, Plymouth: Oxford University Press, 1954, pp.164-189, p.176.

6. Pierre Bourdieu, *Outline of a Theory of Practice* [1972] tr. Richard Nice, Cambridge: Cambridge University Press, 1977.

7. Ibid. p.119.

8. Ibid. p.123.

9. Ibid. p.82f.

Through generalization of schemes from one situation to another, and by forming the interpretive basis for transformations of nature, they also come to be embodied in the humanized environment, so that all kinds of human action and products of human action in traditional societies come to reflect each other, from the manner of economic production and the cooking of meals to ritual ceremonies, the manner of dressing and the layout of buildings and villages.[10] Such reflection is evident among the Maolan Islanders of Eastern Fiji studied by Marshall Sahlins. This society is divided into two groups: the Land people or Animal people who were the original settlers, and the Chiefs who arrived later by sea and conquered the Land people. This division reflects the division of labour in society where the Land people concentrate on growing food while the Chiefs concentrate on fishing. Sahlins described the way this division has become integrated and embodied within Maolan society:

> A difference of social groups corresponds to the distinction of land and sea on the geographical plane, itself an instance of a general spatial differentiation of interior and peripheral, correlated with oppositions of indigenous and foreign, earlier and later, even animal and cultural; the same groups again are inferior and superior politically, ritual and secular functionally.[11]

In their efforts to orient themselves, members of traditional societies develop interpretive schemes beyond the scope required for the mastery of immediate situations. Such interpretive schemes provide global orientations, relating individuals with all their individual and social practices to each other, to their community and to the world generally. These schemes are usually developed by using nature as an analogy for understanding society and society as an analogy for understanding nature. As the French Marxist anthropologist, Maurice Godelier, described thought in society:

> Spontaneously, by systematically covering all the possible analogous parallels between Nature and Culture, thought constructs a gigantic mirror effect, where the reciprocal image of man and the world is reflected *ad infinitum*, perpetually decomposing and recomposing in the prism of Nature-Culture relations... By analogy the whole world makes sense, everything is significant, everything can be explained within the symbolic order, where all the positive known facts...may take their place with all their rich abundance of detail.[12]

This usually results in all aspects of the world being related in terms of one dominant thematic motif which defines the basic nature of the world. This was revealed by Roy Willis in his comparative study of the Nuer, the Lele and the Fipa. Willis found that the Nuer sense of distance from and equality with surrounding nature contrasted markedly with the Lele sense of the village's moral inferiority to and dependence on the forest and the Fipa sense of the village's properly dominant relation to the surrounding bush. These cultural differences in the perceived structure of the universe were found to correlate with different ideas of time, of historical consciousness or lack of it, and in ideas of the self; and these differences in turn were found to be grounded in the meaning given to the man-animal relation which in each case was conceived in terms of a hierarchical pyramid with one animal at the apex: the ox in the case of the Nuer, the pangolin for the Lele and the python for the Fipa. Willis argued:

> [T]hese three beasts symbolize, for these three societies, the ultimate value - what we might call the 'meaning of life'... [W]hat these animals symbolize is, respectively: transcendence of

---

10. This reflection based on analogy or 'paradigmatic transformation' has been described by Edmund Leach in *Culture and Communication*, Cambridge: Cambridge University Press, 1976.

11. Marshall Sahlins, *Culture and Practical Reason*, Chicago: University of Chicago Press, 1976, p.24.

12. Maurice Godelier, *Perspectives in Marxist Anthropology* [1973] tr. Robert Brain, Cambridge: C.U.P., 1977, p.213.

> individual personality in pure, inner selfhood; transcendence of individual differentiation in pure communalism; and pure becoming, or developmental change, both social and personal.[13]

He concluded from his study that 'The cohesion and vitality of human cultures appears to rest on a kind of cognitive and affective reductionism by which a wide range of disparate concerns are subsumed under a single thematic motif.'[14]

Yet the achievement of such a subsumption does not require symbolic mastery of the way the world is conceived by the members of society. While some traditional societies such as the Dogon do develop such mastery, others such as the Shilluk appear to have attained a unity in their orientation to the world without any capacity to represent their conception of the world to themselves.[15] Thus the world-orientation of a culture cannot be conceived as a set of beliefs, and in fact the very notion of belief in such contexts has been questioned by Rodney Needham.[16] Making a similar point, Mary Douglas wrote: 'It is a mistake to think of people as being set somewhere below and apart from their cosmological ideas. People are living in the middle of their cosmology down in amongst it...'[17] The coherence of a world-orientation is experienced by people as a coherence in the world and in their lives.

It is in terms of this socially defined world that individuals transcend their biological centredness and define themselves from a social perspective as a self among other selves. The concepts of self which are developed depend upon the cultures of the particular societies involved, and vary immensely. For instance the Dinka have no concept of the mind as storing up experiences. What Westerners encapsulate within themselves as a 'memory' related to past experience and taken to be an interior psychic phenomenon, Dinka regard as a feature of a timeless external world with the power to act upon them.[18] Through such concepts of the self, individuals define their relation to the world, and consequently define their needs and goals. As Irving Hallowell concluded on the basis of his studies of the Ojibwa Indians: 'In so far as the needs and goals of the individual are at the level of self-awareness, they are structured with reference to the kind of self-image that is consonant with other basic orientations that prepare the self for action in a culturally constituted world.'[19] People's behaviour is constrained by the status accorded to different actions, goals and achievements in their society.

But such constraints are not sets of principles which define correct action. The culture of a society is part of the self, and the ethos of a society through which people gain respect is a mode of bodily engagement in the world. As Bourdieu argued:

> However close it may come to the logic of practices, the abstract diagram which has to be constructed in order to account for that logic is liable to obscure the fact that the driving force of the whole mechanism is not some principle ... still less the set of rules which can be derived from it, but the *sense of honour*, a disposition inculcated in the earliest years of life and constantly reinforced by calls to order from the group, that is to say, from the aggregate of the individuals endowed with the same dispositions, to whom each is linked by his dispositions and interests... [T]he point of honour is a permanent disposition, embedded in the agents' very bodies in the form of mental dispositions, schemes of perception and thought, extremely general in their application, such as those which divide up the world in accordance with the oppositions between male and female, east and west, future and past, top and bottom, right and

13. Roy Willis, *Man and Beast*, Frogmore: Paladin, 1974, p.8f.
14. Ibid. p.86.
15. See Marcel Griaule and Germaine Dieterlen, 'The Dogon' and G. Lienhardt, 'The Shilluk of the Upper Nile' in Daryll Forde ed., *African Worlds*, Plymouth, Oxford University Press, 1954, pp. 83-110 and pp.138-63, esp. p.162.
16. R. Needham, *Belief, Language and Experience*, Oxford: Blackwell, 1972, esp. p.151.
17. Mary Douglas, *Implicit Meanings: Essays in Anthropology*, London: Routledge & Kegan Paul, 1975, p.60.
18. From G. Lienhardt, described in Willis, *Man and Beast* p.81.

> left, etc., and also, at a deeper level, in the form of bodily postures and stances, ways of standing, sitting, looking, or walking. What is called the sense of honour is nothing other than the cultivated disposition, inscribed in the body schema and in the schemes of thought, which enables each agent to engender all the practices consistent with the logic of challenge and riposte, and only such practices...[20]

Despite the way cultures are embodied, they still have an historical dimension as their members struggle to overcome contingencies within the world and generated within societies themselves. It is through this history that the nature of cultural coherence is manifested. Confronting new contingencies involves the accommodation of old structures of power, legitimation and cognition to new situations, and this produces a coherence in societies over time. But even in traditional societies this coherence is limited. To begin with inadequate interpretive schemes often result in people's actions producing effects unintended by them, which once established, often take on a life of their own and subsequently change the structures of society. Such societies are also characterized by residues of earlier formations of the society. These are maintained as societies gradually change through the process of confronting contingencies, though they play no significant role in the later formation and can be eliminated without much effect on the society. In other instances development of new practices through accommodation can generate tensions between the developed practices and the old practices, and the power relationships between people can be upset. For example in the late nineteenth century the Maolan Sea People established the village of Nuku. To maintain their dualistic way of conceiving the world, they divided themselves so that some groups of people came to be seen as Land People, with all the ritual forms of behaviour and degradation this involved. The blatant contradiction between the recognition that Nuku had been established entirely by Sea People or Chiefs and that it was now organized into Sea and Land people was explained by saying that the Land People had arrived earlier. Sahlins wrote of this:

> [T]here is no escaping the contradiction of a village at once composed of Land People and Sea People, and yet of Sea People alone... [T]he opposition of structure and event is overcome, but at the cost of a social complication which denies the structure even as it is confirmed. One dualism negates the other, is placed across the other, and it seems reasonable to suppose that any system will discover limits to its ability to thus accumulate historical contradictions, or at least that it will become vulnerable to some transformation.[21]

Such historical studies of cultures reveals their unity to be more like that of an ecosystem than that of an organism, with individuals and groups struggling to find niches within which they can gain control over their destinies, gain respect, and orient themselves in the world. Cultures provide a reserve of schemes of interpretation which can be analogized to comprehend new situations. These are selected and preserved if they work in practice and can be legitimated in the general community. Such legitimacy will to some extent be an expression of the power (political and economic) and status of the individuals involved, which in turn will be largely determined by the dynamics of semi-autonomous processes within the society. But legitimacy will also depend on whether such definitions and actions are felt to accord with the nature of things as they have been socially defined in the past. The various practices then tend to support each other when they share the same basic ways of conceiving things. A coherent society will be one in which the various practices resonate with each other so that people moving from one social situation to another find their conception of the world continually reinforced. Practices necessary for the survival of the society are likely to carry the most weight in determining which ways of conceiving things will be

---

19. A. Irving Hallowell, *Culture and Experience*, Philadelphia: University of Pennsylvania Press, 1955, p.76.
20. Bourdieu, *Outline of a Theory of Practice*, p.14f.
21. Sahlins, *Culture and Practical Reason*, p.42.

felt to be legitimate in other social practices. However the dominant thematic motif or analogy of a society which defines the basic nature of being will also be highly significant in this regard, especially in societies which have articulated explicitly the forms of thinking associated with this motif. With the Maolans the dualistic conception of the world is preserved even though this involves falsifying history.

Cultures of different traditional societies vary in their potential to be developed by their members. The ultimate stress for most of these societies has been contact with Western civilization. The highly flexible world-orientation of the Fipa studied by Willis enabled them to critically assimilate Western culture without their own culture losing its integrity. Other cultures have fared less well. Marshall Sahlins has described the transformation of the culture of the Sandwich Islands Kingdom in response to Europeans.[22] To begin with, islanders interpreted Europeans according to their cultural categories, seeing the Europeans as gods and acting accordingly. This led to the ritual slaughter of Captain Cook. But further contact eventually extended this culture to breaking point, destroying its coherence and in so doing, undermining the traditional social order. It led the different strata within society to act in ways which gave new conceptual meanings to signs by placing them in novel relationships with objects in the referential process, and by placing them in novel relationships to other signs in the instrumental process. For instance for good traditional reasons the Hawaiian chiefs consistently used the power of tabu in an unprecedented manner to accumulate property in trade, displacing the received relationships of the concept away from the supernatural and ritual towards the material and the political. Such actions destroyed the meaning of these signs, leading to the development of a new structural state, resulting in the Christianization of the population.

Other cultures completely failed in such contacts, eventually leading to the total destruction of these societies. One particular instance of such failure was the response of the Australian Aboriginals in their confrontation with Western civilization.[23] Joseph Banks, the botanist who accompanied Captain Cook on his voyage to Australia recorded in his journal in 1770 the first contact with these people:

> Under the South head of it were four small canoes; in each of these was one man ... These people seemed to be totally engag'd in what they were about: the ship passd within a quarter of a mile of them and yet they scarce lifted their eyes from their employment...At 1 we came to an anchor abreast a small village consisting of about 6 or 8 houses. Soon after this an old woman followd by three children came out of the wood...She often looked at the ship but expressed neither surprise nor concern. Soon after this she lighted a fire and the four Canoes came in from fishing; the people landed, hauld up their boats and began to dress their dinner to all appearance totally unmovd at us, tho we were within a little more than 1/2 a mile of them.[24]

It appears that a sailing ship was so different from anything that people in this society had ever seen or dreamt about before that it could not be assimilated to their culturally constituted world, and consequently for practical purposes they were virtually blind to it.

Some societies were destroyed because they refused to adapt to Western civilization because it was unacceptable in terms of their own understanding of the world. For instance a conflict arose between the Columbia Basin Indian tribes and American whites when the United States government tried to force the Indians to become farmers. The Indians resisted on grounds which clearly reveal

---

22. Sahlins, *Historical Metaphors and Mythical Realities*.

23. For the history of the struggle of aboriginals against white invaders in Australia see Henry Reynolds, *The Other Side of the Frontier*, Harmondsworth: Penguin, 1982.

24. J.C. Beaglehole, ed., *The Endeavour Journal of Joseph Banks, 1768-1771*, 2 volumes, Volume II, Sydney: Angus and Robertson, 2nd ed. 1963, p.53f.

the role of analogical thinking in defining their place in the world. The reasons for the resistance were given by their leader, Smohalla:

> You ask me to plough the ground; shall I take a knife and tear my mother's bosom? Then when I die she will not take me to her bosom to rest. You ask me to dig for stones; shall I dig under her skin for her bones? Then when I die I cannot enter her body to be born again. You ask me to cut grass and make hay and sell it and be rich like white men; but how dare I cut off my mother's hair?[25]

Acting on this view of nature, the Nez Perce Indians revolted, and were crushed by United States' troops who killed Indians of both sexes, both adults and children.

## *Culture and Civilization*

While the relationship between modes of thought, action and society is essentially the same in civilizations as in traditional societies, there are a number of complicating factors. With civilization comes literacy, which, as Walter Ong has convincingly argued, itself radically transforms people's thinking.[26] The complexity of civilizations facilitates the development of dynamic processes more autonomous from people's intentions, ranging from those associated with a few individuals to socio-economic systems which characterize whole eras; and civilizations are characterized by at least one major division between its members: between those who live in cities and those who live in the country. Since people are involved in radically different activities, this makes for a greater complexity in the relationship between the modes of cognition of different people. At least some people in civilizations are likely to be aware that other people understand or have understood the world in radically different ways than they themselves do. This can come about through sustained contact with people from different regions or classes within a civilization, through the development of literacy and acquaintance with the writings of earlier members of their own civilization, or through contact with members of other societies.[27] Finally, the differentiation of society within civilizations tends to lead to the defence of forms of thinking as part of the power struggle between different groups. Awareness that ways of thinking are only opinion, *doxa*, generally leads to efforts to impose one opinion as *orthodox*.

The awareness of alternative beliefs and the resulting conflicts between their proponents also have tended to generate far greater efforts to represent beliefs and to elaborate them. Consequently civilizations are more likely to be characterized by traditions of critical thought and by deliberate efforts to assimilate ideas from other cultures than are traditional societies. However with greater differences in the power and status of groups and individuals, augmented by more powerful semi-autonomous social processes, civilizations are also more likely to be characterized by the systematic reproduction of manifestly defective conceptions of the world, especially where such conceptions favour the interests of particular groups or classes within society, by the deliberate imposition of these on the general population and the persecution of anyone who questions them. Such developments are likely to be associated with the rise of systematic education and elaboration of ideas, associated with a professional class of intelligentsia to some degree insulated from other aspects of life. Consequently there will be no simple relationships between the ideas espoused by thinkers and the concepts in terms of which people in everyday life define themselves and their place in the world.

---

25. Quoted by Brian Easlea in *Witchcraft, Magic & the New Philosophy*, Sussex: Harvester Press, 1980, p.140.

26. Walter Ong, *Orality and Literacy*, London: Methuen, 1983.

27. R. Horton, 'African Traditional Thought and Western Science', *Africa*, Vol.37, No's 1-2, 1967, pp.50-71 and 155-87. See also Robin Horton, 'Tradition and Modernity Revisited', in Hollis and Lukes *Rationality and Relativism*, pp.201-60.

Yet despite this complexity, the nature of people's thinking and the principles leading to cultural unity remain the same. Most people are not engaged in the explicit elaboration of ideas, and their thought is primarily practical thinking in specific contexts. The immediate unity of a culture derives from the generalization by analogy of modes of engaging in the world from one situation to another, thus favouring those practices which resonate with the prevailing modes of engagement. Theoretical thinking has its roots in the practical thinking through which people have to live, and involves the development of world-orientations, frequently through using social relations as an analogy for understanding nature, and nature as an analogy for understanding society.[28] This theoretical thinking then provides the basis for coordinating practical thinking in other areas of society. The difference between traditional societies and civilizations is basically one of complexity and the amount of time required for the unifying tendencies of a culture to take effect, and in particular, the time taken for theoretical ideas to be appropriated by society and to dominate the way most people think and act.

To illustrate this and to pave the way for an analysis of the way mechanistic materialism has come to dominate the world, the emergence of the basic ways of thinking which have come to characterize European civilization will be described first, contrasting these with the modes of thought which came to dominate Chinese civilization. How European society institutionalized these modes of thought in the Middle Ages will then be described. The nature of the revolution of culture in Western civilization in the seventeenth century and the nature of the development of culture and society since then will be analysed in the following chapter.

Characterizing European culture throughout its history in the short space of four chapters presents obvious problems. My assumption that it is even possible rests on the view of cultures being defended, namely that cultures are dominated by particular conceptions of the nature of being. If this is the case it is possible to identify and describe such conceptions of being and their developments. However in civilizations, cultures lack the coherence of those of traditional societies and it is therefore necessary to emphasise the limited nature of this dominance. Various versions and transformations of the dominant conception of being are likely to co-exist, there is only a loose coherence between the ideas and practices legitimated in terms of it, and there are always adherents to opposing ways of conceiving the world. It is only by considering very long periods of time and by contrasting different civilizations that such a coherence will be revealed. In accordance with the historians of the Annales school, I am presupposing the existence of a long-term history distinct from and irreducible to the short term history which most historians focus upon.[29] And as Braudel wrote: 'Culture is the oldest character in human history: economies succeed each other, political institutions crumble, societies replace each other, but civilizations continue along their way.'[30]

The subject to be investigated is in fact analogous to the archeological sites in Europe which were only discovered after the invention of the helicopter. Archeologists on the ground could not identify these sites because the order which was perfectly clear from the altitude of the helicopter simply did not exist within the range of their vision. Those scholastics who, in the name of professionalism, deny the possibility of characterizing broad historical sweeps are simply rejecting the possibility of there being order of long durations because it does not exist in the trivia amongst which they compulsively immerse themselves.

---

28. This is clearly evident in the study of the world-views of the civilizations of Egypt and Babylon by Henri Frankfurt et.al. in *Before Philosophy* [1946] Harmondsworth: Penguin, 1964.

29. The nature of the long term is described by F. Braudel, *The Mediterranean and the Mediterranean World in the Age of Philip II*, Vol.1, tr. S. Reynolds, London: Collins, 1972, p.16ff.

30. Fernand Braudel, *Civilization and Capitalism 15th-18th Century, Vol.III, The Perspective of the World*, [1979] tr. Sian Reynolds, London: Fontana, 1984, p.65.

## *The Origins of Western Culture*

Western civilization is founded on a fusion of Greek and Hebraic culture. Greek culture is the more important of these since Hebraic culture, as with all other aspects of European culture, only attained or retained its plausibility by being interpreted in terms of Greek thought. So it is primarily Greek culture with which I will be concerned.

The main stimuli for the development of Greek culture were sustained contact with other culturally vigorous societies, particularly Egypt, the social transformations which occurred with the invasion and destruction of the relatively peaceful palace based civilization of Mycenae, the instability of these transformations and the rivalry between the members of different classes, which then included a slave class, and of the proponents of different tendencies within this unstable social order. This instability manifested itself in the variety of political forms of the different city states in Ancient Greece. Associated with the more militaristic organization of the invaders, there was first a replacement of monarchical rule by the military aristocracy, then in some cases, associated with the rise to pre-eminence of disciplined hoplite troops over the aristocracy's chariots, of rule by a form of limited democracy (usually associated with a lower proportion of slaves in society). Along with these changes there developed a form of decision-making in which the settlement of disputes by gladiatorial combat was replaced by open debate, discussion and argument in the public square, the agora.

The development of democratic rule opened the entire spiritual world to the community (*demos*), so that knowledge, values and techniques were brought into public view to be submitted to criticism and controversy. Under these conditions literacy became virtually universal among its citizens, and laws were written down so they could be applied equally to all. Justice (*dike*) which had been a divinity remote from common people, was brought down to earth where it could be incarnated on a human level without ceasing to be regarded as an ideal value. It became common to all while remaining superior to all, a standard subject to discussion and modification while remaining sacred. Associated with this the priesthood which had claimed a special intimacy with the divinity was dissolved. All religious symbols were removed to the public temple so that the protection of the deity would extend to the entire community. All the members of the polis, whatever their origin and rank, conceived themselves to be fundamentally alike, as interchangeable units within a system whose law was the balance of power and whose norm was equality. Correspondingly the highest virtue came to be temperance, or self-control and good sense (*sophrosyne*). The polis was seen to form an organized whole whose harmony was dependent upon its constituent members maintaining their places while receiving the share of power due to them by virtue of their own qualities. It was organized so that sovereignty (*arche*) was no longer concentrated at an apex but passed on from one group to another, from one individual to another, in a regular cycle so that command and obedience became two aspects of a reversible relationship. The social realm thus had the form of a centred and circular cosmos in which each citizen had to cover the entire circuit, successively occupying and surrendering each of the symmetrical positions that made up civic space.

To begin with, despite the restriction of power of the new monarchy, the mythology of the Greeks continued to be formulated in accordance with the tradition deriving from Babylon. The universe was seen as a structural analogue of monarchical order with a hierarchy of powers reflecting differences in function, value and rank. In this scheme of things order did not emerge from the play of elements but was established through the dramatic efforts of an agent, in the case of Greek mythology, Zeus, who then maintained this order through his exceptional powers. This was associated with a dual notion of becoming. The early Greeks had no sense of time independent of occurrences; time was the occurrences themselves. But there were two types of occurrences: the sacred and the profane. Sacred time, the time of the gods, was seen to fold back on itself to be relived as the eternal present, as eternal recurrence. In relation to this time Kurt Hübner wrote: 'the past was still *there*, still *existed*, for him like something eternal, something which could be directly and immediately seen in nature, in the heavens, in his own action, and especially in his cultural festivals.'[31] Thus it was *the* spring whose return was joyously celebrated, not a new spring. Profane time on the other hand was the time in which nothing recurs and everything passes away. But sacred time was constantly woven into profane time, not only through festivals, but through the constant intervention of the gods in people's lives. Emotions, sudden changes in fortune, spurts of energy and so on were generally attributed to the intervention of the immortal gods, and this raised them to a different plane of significance as sacred events outside profane time.

The first ethico-political conflict was expressed in mythopoeic form with co-operative ideals embodied in the original feminine nature Gods: Helias the sun, Selene the moon, Gaia the earth, and Demeter the god of fertility, vying with combative ideals associated with militarism expressed by Homer and embodied in the predominantly male sky gods brought to Greece by its invaders.[32] However with further transformation of society this realm of mythical thought ceased to convey social realities or correspond to ritual practices, giving the Greeks a critical detachment from such culture and the values expressed by it. In the ensuing ideological battles, a number of Greek thinkers: Hekataios, Pherekydes, Hellanikos, Xenophanes and Ephoros among others effectively destroyed sacred time by translating Greek myths into profane time, transforming mythical figures by presenting them in genealogies and then developing a system of dating the different mythical events. This virtually reduced the mythical realm to fairytales, and destroyed the force of the ethical imperatives it embodied. Expressing this critical detachment from mythical thought, Xenophanes argued:

> The Ethiopians say that their gods are snub-nosed and black, the Thracians that theirs have light blue eyes and red hair. But if cattle and horses or lions had hands, or were able to draw with their hands and do the works that men can do, horses would draw the forms of gods like horses, and cattle like cattle, and they would make their bodies such as they each had themselves.[33]

The polis itself became the sole locus of the sacred. Distinguished from the realm of economic necessity and the household, it became the realm in which people strove to achieve eternal glory through their words and deeds.[34]

The ideas advanced by the first philosophers can be understood as the development of a new orientation to the world by using the new democratic social relations as an analogy for interpreting the world as a whole. As Jean-Pierre Vernant argued in *The Origins of Greek Thought*:

> In constructing their cosmologies, they made use of ideas elaborated by moral and political thought, projecting onto the world of nature that conception of order and law whose success in the city had made the world a cosmos.[35]

This is most clearly evident with Anaximander. Anaximander rejected the idea that the order in the world was imposed and maintained by an agent and argued that the principles which govern the cosmos are immanent in the world, corresponding to the self-ordering of the polis. Similarly he argued that the earth could stay at rest without external support or roots because it is equidistant

---

31. Kurt Hübner, *A Critique of Scientific Reason*, tr. Paul R. Dixon Jr., and Hollis M. Dixon, Chicago: University of Chicago Press, 1985, p.242f.

32. M.I. Finley, *The World of Odysseus* [1954], Harmondsworth: Penguin, 1983, p.136f. Feminists have made much of the different orientation to nature and to females of pre-Homeric Greeks. On this, see Charlene Spretnak, *Lost Goddesses of Early Greece*, Berkeley: Moon Books, 1978.

33. G.S. Kirk, J.E. Raven and M. Schofield, *The Presocratic Philosophers*, 2nd ed. Cambridge: C.U.P., 1983, p.169.

34. This is described by Hannah Arendt in *The Human Condition*, Chicago: University of Chicago Press, 1958, p.28ff.

35. Jean-Pierre Vernant, *The Origins of Greek Thought*, [1962], Ithaca, N.Y.: Cornell University Press, 1982, p.108.

from all points on the celestial circumference and so has no more reason to sink than to rise, or to move to one side rather than another. No longer was any portion of the world to be privileged at the expense of the rest, or a physical power to be in a dominant position. An equilibrium is maintained through a regular cycle in which each force alternately prevailing and then falling back in accordance with justice, linking together expansion and contraction, strength and weakness, birth and death. It was this equality and symmetry of powers that made up the cosmos that characterized the new conception of natural order, and supremacy belonged exclusively to the law of equilibrium and continuous reciprocity. As Vernant put it: '*Monarchia* was replaced, in nature as in the city, by the rule of *isonomia*.'[36] Heraclitus can be seen as continuing this tradition of thought.[37]

However social developments and problems in achieving agreement through debate led to a failure in the ethics on which Greek society was based, a failure dramatically manifest in the failures of Athens, and leading democratic polis, during the Peloponnesian War. While in the early stages of the development of the democratic polis the sacred status of the polis and the principles of justice on which it was based replaced and compensated for the desacralization of the mythical realm, the desacralization of the polis associated with the development of commerce resulted in a loss of any point of reference outside the flux of profane time by which people could orient themselves. The connection between virtues and rewards had been sundered, leading to the promotion of a new vocabulary of evaluation by the Sophists extolling the qualities and goods of effectiveness at the expense of the qualities and goods of excellence.[38] It was the corrosive effect of such thinking which led to a a reaction against the democratic temper of the early philosophers, particularly by the members of the old aristocratic families. Proponents of aristocratic forms of life struggled to resacralize life and to find an immutable foundation to orient themselves and to re-establish their ideals of excellence. Such philosophers turned against democracy and attempted to establish an elitehood, either forming themselves into esoteric groups and cutting themselves off from the rest of society, or struggling to attain political power. Pythagoras, Parmenides, Plato and to a lesser extent, Aristotle, are the main representatives of this elitist tradition.

In the ensuing intellectual struggle, the problem of knowledge emerged as a major issue. The paradigmatic form of knowledge for the Greeks was what one was actually perceiving. Greek philosophers failed or refused to develop the notion of propositions which were valid independently of when they were uttered. They could say: 'It is raining.' or 'It was raining yesterday.' but not 'It rained on such and such a date at such and such a place.' Thus an eternally true statement could only be one said about something which was unchanging and with which one could always be acquainted.[39] Therefore to reorient themselves the Greeks searched for something omni-temporal outside the flux of becoming by which they could get their bearings.

Pythagoras offered a solution to this problem. He was opposed to the values being generated by the commercial life of Greek society, and classified men and evaluated them in terms of whether they were lovers of gain, lovers of honour or lovers of wisdom.[40] But in opposing commercial values he developed a philosophy based on modes of thought developed in commercial society. Arithmetic, which had been developed with commerce as exchange value and had come to be measured quantitatively in terms of coinage, was developed by Pythagoras beyond the needs of

---

36. Ibid. p.122.

37. This is evident from Charles H. Kahn's study of Heraclitus in *The Art and Thought of Heraclitus*, Cambridge, C.U.P., 1981, pp.9-23.

38. This has been acutely analysed by Alasdair MacIntyre in *Whose Justice? Which Rationality*? Notre Dame: Notre Dame University Press, 1989, Ch.'s II-V.

39. The relationship between temporality and knowledge in Greek thought has been brilliantly explicated by Jaako Hintikka in *Knowledge and the Known: Historical Perspectives in Epistemology*, Dordrecht: Reidel, Ch.3.

40. The writings of the Pythagoreans have been collected and translated by Kenneth Sylvan Guthrie, *The Pythagorean Sourcebook*, Grand Rapids: Phanes Press, 1987.

commerce.[41] His success in describing geometrical figures in numerical terms and in finding simple numerical ratios between the intervals of a string producing consonant harmonies led him to conceive all things as number. These arithmetical units were thought to maintain their separateness through the inhalation of the 'boundless breath' of the 'unlimited'. These numbers, being omnitemporal or eternal, and identified with the harmony of music, performed the same structural role in culture as the sacred realm of previous times.

Developing this idea led Pythagoras to some fundamental innovations. He rejected the belief in the mortality of humans. Since we appear to have knowledge of numerical relations independent of sensory experience, he concluded that such knowledge must have been gained in a former life, and that souls transmigrate.[42] Pythagoras did not simply attempt to get his bearings in this changing world by reference to the numerical realm; he exalted it at the expense of the changing world. While number and soul were seen to be the forms of the world, they were also seen as beyond it, and Pythagoras was concerned with purification from contamination by the profane world. The Pythagoreans used medicine to purge the body and music and scientific and mathematical study to purge the soul. The numerical realm was seen as the realm of true value, just as in a commercial society money is taken to be the ultimate, quantifiable value transcending and providing a measure for the value of things in the world.

This tendency to reject the sensible world for an eternal realm grasped in purely intellectual terms was consolidated and developed to its most extreme form by Parmenides. Parmenides argued that only that which is, is knowable. Since the notion of motion implies that what is not has come to be what is, motion is unknowable. Parmenides therefore distinguished the realm of the uncreated, indestructible, unchangeable One, a plenum which he characterized as the true world, from the world in which things come into being and perish which he held to be an illusory world, the way of mere belief.[43] Pythagoras and Parmenides were major sources of inspiration for Plato.

## *Plato*

Plato (427-348 B.C.) is the most important thinker in Western civilization. Yet his influence rests on his dialogues which Plato disaffirmed as expressions of his own views.[44] The dialogues were exercises, they were exploratory and frequently inconsistent with each other. Plato thought that only through conversation could true knowledge be achieved. Consequently in referring to Plato, it is rather Plato as he has been understood by Neoplatonists and thereby as he has provided the foundations of Western civilization that will be of concern.[45]

To overcome the problem posed by the follower of Heraclitus, Cratylus, of how there could be knowledge in a changing world, Plato accepted Parmenides' point that the object of significant knowledge must be unchanging. However he followed the Pythagoreans rather than Parmenides, conceiving the real world as the eternal forms.[46] He was also influenced by Socrates who had been primarily concerned with ethics and had focussed his attention on the definitions of ethical

---

41. The invention of number was attributed by the Greeks to Palamedes who organized the army at Troy (see Plato, *The Republic*, 522d), but it was through commerce that the general population became acquainted with numbers. On this, see John Burnet, *Early Greek Philosophy*, 4th ed. [1930] London: Adam & Charles Black, 1975, p.99.

42. See John Burnet, *Greek Philosophy: Thales to Plato*, [1914] London: MacMillan, 1981, p.33f.

43. See Burnet, *Early Greek Philosophy*, p.169ff.

44. See Plato, *Letters: II*, 314c.

45. That Plato can be understood differently has been well argued by a number of interpreters. For a recent anti-Neoplatonist interpretation of Plato's concept of the forms and of the Good, see Gail Fine, 'Separation', *Oxford Studies in Ancient Philosophy*, Vol.II, Oxford: Clarendon Press, 1984, pp.31-87, and 'Knowledge and belief in *Republic V-VII*,' *Epistemology* ed. Stephen Everson, Cambridge: C.U.P., 1990.

46. See Aristotle, *Metaphysics*, 987a29-987b20.

concepts. So the ethical forms were not only included along with mathematical entities as part of the real world, Plato gave greater status to ethical forms than to mathematical entities. However he conceived these ethical forms in mathematical terms.

According to Plato, we can only know the forms. To justify the existence of a diversity of forms, Plato rejected Parmenides' contention that if all that is knowable is what is, therefore the object of knowledge must be a plenum. His argument against this was that 'not being' could be made sense of as not referring to something contrary to what exists, but as something which is different.[47] In other words, to say that something 'is not' is to say that it is not one kind of thing but one of those indefinitely numerous other kinds of things. Plato used this idea to argue for a relational theory of knowledge in which to determine what a thing is, also entails determining what it is not. This allows us to conceive of differentiation within being, and the task of the dialectician is to divide things according to kinds, and to distinguish 'Kind by Kind, in what ways the several Kinds are or are not able to combine.'[48] Knowledge is then achieved by the method of synthesis and division. A synthesis 'is that in which we bring a dispersed plurality under a single form, seeing it all together' while division is 'the reverse of the other, whereby we are enabled to divide into forms.'[49] In this way, the world of discourse and the world itself were seen to be correlated as differentiated unities, wholes with internal relational structures.

The sensible world was seen to be knowable omni-temporally only insofar as it is participating in the forms. Philosophers were defined by Plato as those with 'a constant passion for any knowledge that will reveal to them something of that reality which endures for ever and is not always passing into and out of existence.'[50] And he railed against those mathematicians 'who constantly talk of "operations" like "squaring," "applying," "adding," and so on, as if the object were to do something, whereas the true purpose of the whole subject is knowledge - knowledge, moreover, of what eternally exists, not of anything which comes to be this or that at some time and ceases to be.'[51] Along with the Pythagoreans, Plato regarded that which is eternal as of greater value than that which is mutable.

This laid the foundations for the later development of substantialism in Western culture, the notion that the world consists of enduring things the essential properties of which are atemporal; and correspondingly, as Heidegger has argued, the exclusion from awareness or concealment of the actual presencing of what is present in the world.[52] As such, it reflected the influence of the commercial world disdained by Plato. Forms define the potential to be used on the basis of which entities have exchange value, a value which endures while, and only as long as, this potential endures. By privileging forms over becoming, Plato was devaluing the creative activity of nature and of humans involved in forming the world into useful things and in maintaining these forms, taking the perspective of a member of a privileged class who could buy what they wanted, and who were only interested in the world insofar as it could be bought and was worth buying.

Plato saw everything in the world, including people, as striving to participate in the forms. He developed his conception of the relationship between forms in the world by assuming that forms are the goals of action, and interpreting the relationships metaphorically as mathematical relationships. In this way he came to postulate the 'Good' as central to both ethics and being. In action people do not have goals in isolation but have a hierarchy of goals, higher goals being more fundamental than the particular goals. Thus in making the leg of a table the carpenter must have an apprehension of both the form of the leg while at the same time the form of the whole table. But beyond the goal of

---

47. Plato, *Sophist*, 256d-257b.

48. Ibid., 253d.

49. Plato, *Phaedrus*, 266d-e.

50. Plato, *Republic*, Bk VI, 485b.

51. Plato, *Republic*, Bk VII, 527a.

52. Martin Heidegger, *Early Greek Thinking: The Dawn of Western Philosophy*, tr. David Farrell Krell and Frank A. Capuzzi, San Francisco: Harper & Row, 1984, esp. p.8.

making a table the carpenter is also concerned to make a good table, a good piece of furniture, a good house, and so on. Anyone doing anything is always striving for the good; and each particular good is only comprehensible as part of some higher good. So the ultimate form, more fundamental than all particular forms, is the form of the Good. Plato conceived the relationship between the Good and other forms simultaneously on the analogy of an organism in which each part is only comprehensible in relation to the whole organism, and on the relationship between basic definitions in mathematics and all other mathematical forms. A few such definitions together with some construction postulates imply the existence of a vast number of mathematical figures and relationships.[53] The Good is then seen as both the ultimate goal of everything, the intelligible structure of the cosmos, and the source and basis of all particular structures in the cosmos, the basis of both the unity of the world and its diversity.[54] In the *Timaeus* Plato also postulated the existence of a creator which, being good and therefore desiring that all things should be as like himself as they could be, fashioned a cosmos from the chaos.[55] This he endowed with soul and intelligence capable of apprehending the forms and maintaining order in the world, its body, in accordance with reason. This accounts for the continued orderly nature of movement, particularly in the heavens where movements are the moving image of the eternal form, the circle.

Plato developed his ethical and political doctrines on the basis of this framework. He saw people as having two types of knowledge: one of how to do something without having any intellectual apprehension of what is to be achieved, as with the poets, and a higher form typified by artisans who have an image of what they are trying to make, that is, an intellectual apprehension of the form to be achieved.[56] He presupposed a world in which things are in the process of becoming and he believed that to have attained such an intellectual apprehension of a form is to have begun the process of actualizing or participating in this form. Therefore the most important task confronting the philosopher is to define the true form of humanity, or justice.

Plato's ethics, while influenced by Socrates, were essentially a development of Pythagorean ethics. He assumed that all forms, as with mathematical concepts, can be defined unambiguously, and consequently no form can have contradictory characteristics. Each thing in the world therefore exists with its own features clearly defined from all other things. It is only because he held this assumption that Plato could believe that he had demonstrated that justice cannot be 'what is to the interest of the stronger party' as Thrasymachus had argued by showing that this definition leads to the contradictory conclusion that 'it will be right to do what is not to the interest of the stronger party, as well as what is so.'[57] In place of this Plato argued that justice is attending to all that is, in the fullest sense, a person's proper concern. After arguing this for the polis as a whole, he went on to argue that the justice should also prevail within the individual. He presented the human Soul as consisting of three parts: reason, spirit and appetites, on the grounds that only if this were the case could the conflicts within each individual be explained.[58] The appetites were denigrated, with sexual appetite being compared to a 'savage beast of a master' of which a person is best free.[59] For Plato the earthly body with its characteristics of sex and death are unworthy of the true nature of the human Soul. Justice in the individual is achieved when the higher rules the lower, where reason, the immortal part of the Soul through which the eternal forms are apprehended, rules the spirit, which is concerned with honour, and spirit rules the appetites. As Plato put it: 'The just man does not allow the several elements in his soul to usurp one another's functions; he is indeed one

53. Plato, *Republic*, Bk VI, 510b-511e; Bk VII, 517.

54. For a thorough analysis of Plato's concept of the Good, see Hans-Georg Gadamer, *The Idea of the Good in Platonic-Aristotelian Philosophy*, tr. P. Christopher Smith, New Haven: Yale University Press, 1986.

55. Plato, *Timaeus*, 29e ff.

56. See Plato, *Apologia*, 22c and *Geogias*, 503d-e.

57. Plato, *Republic*, Bk I, 339c.

58. Plato, *Republic*, Bk VI, 439b.

59. Plato, *Republic*, Bk I, 329c.

who sets his house in order, by self-mastery and discipline coming to be at peace with himself, and bringing into tune those three parts, like the terms in the proportion of a musical scale, the highest and the lowest notes and the mean between them...'[60]

Plato constructed his idea of the just society, *The Republic*, that is, the form which he believed all societies should strive to participate in as fully as possible and which must be apprehended by rulers as the condition of rational statesmanship, in accordance with this idea of justice. This ideal was based partly on the militaristic, slave based society of Sparta - a society in which the size of the population of the enslaved compared to the rulers had engendered an extraordinarily high level of discipline, and in which slaves did all the work. Individuals were conceived of in abstraction from the community then evaluated in terms of their function in society.[61] This involved both an affirmation of individualism and of severe political powers to control individuals. It involved an affirmation of the division of labour with a vengeance, with society divided between those people whose reason is dominant, those people whose spirit is dominant and those people whose appetites are dominant. A just society was held to be one based on the principle 'that everyone ought to perform the one function in the community for which his nature best suited him',[62] one in which the wise rule over those dominated by spirit, and in which those dominated by their appetites are subordinated and do all the physical work. Plato's *Republic* rejected the family, argued for the sharing of wives, and opposed art. The rulers were justified in systematically deceiving the ruled, specifically to deny the reality of people's kinship relations, so as to be able to more effectively maintain order. The major part of education was to be in mathematics, and the ultimate level to be achieved involved a turning away from the changing sensible world to the contemplation of the 'first principle', the Good. Intellectual contemplation was exalted, and manual labour held in contempt. Art and particularly drama, both tragic and comic, were to be rejected or heavily censored for fostering the emotions of pity and sympathy, or laughter and buffoonery.[63] In effect Plato wanted to eliminate those forms of communication which focus on the ambiguity of life rather than seeing the world in terms of clear, arithmomorphic concepts.[64] Denying such ambiguity, Plato argued for the immortality of the soul and argued that the soul must reap the consequences of being just or unjust in the afterlife.[65]

Basically Plato's philosophy, especially as it has since come to be understood, represents an effort to attain absolutes: an absolute orientation to the world, an absolute sense of one's own significance, and absolute power. Platonists turned their back on the changing, sensible world to focus on the eternal world in which all meaning and value are unambiguous, in terms of which they could define themselves as significant independently of other people, and in which since there is perfect rationality and no death, they could have complete power over their destiny. Plato's ethical and political philosophy proposed a struggle to make the lived world conform to this ideal world.

---

60. Plato, *Republic*, Bk IV, 443d-e.

61. MacIntyre (*Whose Justice? Which Rationality?* p.98) argues that in fact Plato presupposes the philosophical community as the condition of acquiring knowledge and becoming virtuous. However this has generally been ignored by interpreters of Plato.

62. Plato, *Republic*, Bk IV, 433a.

63. Plato, *Republic*, Bk X, 605b-608c.

64. The difference between Plato's thought and that of traditional societies in this respect is elucidated by Stanley Diamond, 'Plato and the Definition of the Primitive' in *In Search of the Primitive: A Critique of Civilization*, New Brunswick: Transaction Books, 1974, pp.178-202, esp. p.187ff.

65. Plato, *Republic*, Bk X, 608c - end.

## *The Origins of Chinese Culture*

All this contrasts sharply with the basic modes of thought which came to dominate in China. China was never a slave based society, and the Confucian and Taoist philosophies which have dominated its thinking reflect this. Confucionism involves an affirmation, generalization and development of kinship forms of relationship, while Taoism involves an affirmation of the value of nature, of the need to harmonize with it, and of the limitations of conceptual thought and human inventions. Both are an affirmation of different aspects of the sensible world and of the forms of life existing before civilization.

Confucius (551-479 B.C.) lived when China was still divided into numerous city-states. Its unification was achieved three centuries later.[66] The central feature of his thought was the emphasis on the family. He emphasised the importance of filial piety and reverence for the aged. This was associated with Confucius' belief in the importance of culture or tradition in the formation of people, which also underlay his stress on the importance of education and the development of personality. Personality development involved the harmonious combination of such qualities as morality, education and refinement with a judicious balance of inner virtues and external polish. Such a personality should possess *chih* (inner integrity), *i* (righteousness), *chung* (conscientiousness towards others, loyalty), *shu* (altruism or reciprocity which is summed up in the *Analects* as 'not doing to others what you do not like yourself'), and above all, *jen* (humanity). However he should also possess *wen* (culture) and *li* (ritual or etiquette). Reischauer and Fairbank contrast such ethical thought with Western ethical thought:

> Great philosophical and religious leaders in ... the West have commonly dealt in absolutes; or perhaps one might say that they have thought in logico-mathematical terms. Confucius was a relativist, thinking in socio-human terms.[67]

Confucius' views on the state were based on his conception of it as a large family. While this led to stress on the virtues of obedience, he also emphasised that the ruler should set an exemplary moral example to his subjects, and did not rule out opposition to an unjust ruler, provided such opposition was open. Thus in replying to the question of how to serve the sovereign, Confucius said, 'Never oppose him by subterfuge, but do so openly if need be.'[68] The principles binding a state were family respect and the development and cultivation of humanity rather than rule by law. As Vitaly Rubin wrote of Confucius, 'He believed that the law had no importance whatsoever for the improvement of society. It was important only that the state possess a good ruler who would instruct the people by his own example, and influence them with the help of virtue and the rules of decorum - *li*.'[69]

After Confucius' death, the states began an almighty struggle for power which culminated in the unification of China under the Ch'in. This struggle led to the development of anti-Confucian ideas, first of Mo Tzu who argued for an austere utilitarianism and for the development of a more mechanically ordered state, and then of Shang Yang who provided the ideology of Legalism under which the Ch'in functioned.[70] Shang Yang was concerned with how to obtain absolute power and argued that people should be kept ignorant, treated as simply means for the purposes of the ruler, and should be controlled by rigidly enforced laws, with severe punishments for transgressions. While Legalism was vigorously opposed by Confucian thinkers such as Meng Tzu and Hsun Tzu

---

66. See Vitaly A. Rubin, *Individual and State in Ancient China: Essays on Four Chinese Philosophers*, [1970] tr. Steven I. Levine, New York: Columbia University Press, 1976, Ch.1.

67. Edwin O. Reischauer and John K. Fairbank, *East Asia: The Great Tradition*, Boston: Houghton Mifflin, 1960, p.71f.

68. Confucius, *Analects*, XIV:23; trans. Rubin ibid. p.20.

69. Rubin, *Individual and State in Ancient China*, p.18. see also the *Analects*, II:3.

70. See ibid. Ch.2 and 3.

and was eventually replaced by Confucianism as the state ideology, later Confucianism incorporated many of the ideas of the Legalists. However the rise of Legalism and an oppressive state gave rise to another opposing philosophy, that of Taoism.

The main work of Taoism, the *Tao Te Ching* was supposed to have been written by Lao Tzu, an older contemporary of Confucius. However it is now thought to have been compiled around 300 B.C.[71] This philosophy was also developed by Chuang Tzu (369-286 B.C.). Essentially it involves a rejection of society and an exaltation of nature, seen as dynamic and active, consisting of a multiplicity of fields of force, contained in and subsidiary to the main field of force of the Tao.[72] The ideal life is to follow the road or the way, the Tao; that is, to flow with nature. Thus:

> The best of men is like water;
> Water benefits all things
> And does not compete with them.
> It dwells in (the lowly) places that all disdain, -
> Wherein it comes near to the Tao...
>
> It is because he does not contend
> That he is without reproach.[73]

The Taoists were concerned to point out that success could only be achieved by harmonizing with the world, while people who compete must eventually be defeated. A later chapter concludes:

> The sage does not enter into competition
> And therefore no one competes with him.[74]

The Taoists opposed any departures from simplicity and plainness, and were sceptical of scholastic learning. Their doctrine was frequently a rallying point for peasant revolutionaries opposing oppression, and it inspired much of Chinese science and art.[75]

While Taoism and Confucianism were obviously at odds to some extent, Confucian political philosophy and the Taoist conception of nature also reinforced each other. This accord is brought out in the Taoist Wang Pi's third century A.D. commentary on the *I Ching*:

> The general meaning of the Tao of 'Kuan' is that one should not govern by means of punishments and legal pressure, but by looking forth one should exert one's influence [by example] so as to change all things. Spiritual rule is without form and invisible. We do not see Heaven command the four seasons, and yet they never swerve from their course.[76]

In essence, universal harmony comes about through spontaneous co-operation.

---

71. See D.C. Lau, 'Introduction' to Lao Tzu, *Tao Te Ching*, tr. D.C. Lau, Harmondsworth: Penguin, 1963, pp.7-52, p.11f.

72. See Joseph Needham, *Science and Civilisation in China*, Cambridge: C.U.P., 1956, Vol.2, p.322.

73. Lao Tzu, *Tao Te Ching*, Book I, VIII; tr. Lin Yutang in *The Wisdom of China*, ed. Lin Yutang, 1963, p.34f.

74. Lao Tzu, *Tao Te Ching*, Book I, XXII; tr. Waley.

75. Rubin, *Individual and State in Ancient China*, p.112ff and Joseph Needham with the assistance of Wang Ling, *Science and Civilisation in China*, Vol.2, Ch.10.

76. Quoted by Needham, *Science and Civilisation in China*, Vol.2, p.555f.

## *The Christian Synthesis*

There were a number of intellectual movements which developed after Plato, including Aristotelianism, Stoicism, Epicureanism, Scepticism and Gnosticism, though each of these had some relationship to Platonic thought. Aristotle rejected the transcendental tendencies within the thought of Plato and his followers, but supported the division of labour in the form of the prevailing slave based economy, and exalted the ideal of a contemplative life oriented towards eternal truths. Even more than Plato, Aristotle exalted the unchanging at the expense of the changing, holding the heavens to be the realm of perfection while the earth, characterized by mutability, to be the repository of the grossest dregs of the universe.[77] Stoicism, which was in effect a form of field theory,[78] came to be a major force in Roman society, although it was strongly attacked by the Sceptics and was eventually eclipsed by Neoplatonism. Neoplatonism was the product of an attempt to develop Plato's ideas more systematically, and it incorporated much of the thought of Aristotle and of the Stoics. It was this movement which was to be the most important for the future of Western civilization.

The most significant of these Neoplatonists was Plotinus (204-70 A.D.).[79] In his system reality was seen to consist of a hierarchy of hypostases of so many stages of degradation. The source of all reality is the self-identical and eternal One, corresponding to the One of Plato's *Parmenides* and the Good of his *Republic*. In conceiving the One as the single, transcendent divine source of all there is, Plotinus broke with Plato (and Aristotle) for whom all existence required three distinct sources: the forms, the receptacle and the demiurge. According to Plotinus this One emanates the other hypostases as an essential consequence of its infinite power which necessitates an outflow of reality which cannot be terminated until all that could possibly come into existence has actually done so. Each hypostasis is characterized by a descent from unity to multiplicity, from immobility to motion, from eternity to time. The second hypostasis, the Intellect, is both thought and object of thought, involving self reflection and hence a kind of duality between that which apprehends and that which is apprehended. The object of thought is the realm of forms or Ideas, the multiplicity through which the intellect grasps its unity. While the intellect grasps the world in a single timeless vision, the next hypostasis, the Soul, is forced to contemplate objects successively and is confined to images or verbal formulae reflecting the forms rather than the forms themselves. But the Soul is still non-spatial, being everywhere and nowhere. Differentiation into separate bodies, including human souls which emanate from the World Soul, occurs in the sensible world in which the forms are reflected as in a mirror in matter, the point at which the outflow of Reality from the One fades away into utter darkness. The imperfect diverse, changing and impermanent world in which we live is therefore to be understood as the least substantial reality. However each lower level of being retains its links with its source, and every being is seen as trying to return to its source. The levels of being are not spatially separated but are intimately present in each part of the universe and in each one of us. Thus each human soul is an intelligible cosmos reflecting in itself the whole universe, and its return to its source is achieved by casting off ties first with the sensible world, then with the world of forms to attain a mystical union with the One.

Although there were differences between Eastern and Western Christianity, Christian philosophy was essentially an interpretation and justification of Hebraic thought in terms of Neoplatonist thought. According to the Hebraic vision as expressed in the Bible man is in a fallen state after having been expelled from Paradise, in which God had originally intended him to live, because he had sinned. However God promised to restore man to Paradise and sent his Son to earth to make this restoration possible. The One of Neoplatonism was identified with the personal,

77. Aristotle, *On the Universe*, 400a3-39.

78. This is argued by S. Sambursky in *The Physical World of the Greeks*, London: Routledge & Kegan Paul, 2nd ed. 1960.

79. Plotinus, *Enneads*, tr. S. McKenna, London: Faber, 1962.

creative God of the Hebraic vision, the sensible world with the fallen world, and the realm of forms with the restored world. Thus man's soul was seen to have originated in heaven as a creation of God, descended to earth where it must live in a perishable world of deteriorated, half destroyed value, and if salvation is to be attained by the grace of God, to be destined to return to the realm of eternal forms.

However this fusion of Hebraic thought with Greek metaphysics within the Roman Empire involved a radical transformation of Greek thought, infusing it with the engineering mentality of the Romans. This change occurred with the translation of Greek words into Latin whereby, as Martin Heidegger put it: '*Roman thought takes over the Greek words without a corresponding, equally authentic experience of what they say, without the Greek word.* The rootlessness of Western thought begins with this translation.'[80] For example, as the early Greeks understood it, *physis* is the event of self-emergence, as when a bud bursts forth into a flower, and the appearing, shining forth or presencing of an entity. The Roman's translated *physis* as *natura* which, although it has its etymological roots in the notion of birth, was understood in terms of what is produced, caused or created. While the Greeks began the process of conceiving of Being as an underlying and constantly present 'ground' of the presencing of things, it was the Romans who conceived this ground as that which produces things. Thus 'Being' itself was reduced to the status of a superior kind of entity which produces the world, and the world came to be seen as the totality of all created beings.

To begin with Christianity was essentially a religious movement of the Eastern Roman Empire, and here the mystical and contemplative tendencies of the religion were developed. Eastern Christians were little interested in questions of morality. Within Rome itself, Christianity was originally only adopted because it had won political power in the East, and then it was adapted to the prevailing mentality in an effort to shore up a disintegrating society. However there were two places where Christianity was appropriated and developed with great vigour: Roman Africa, or what is now Tunisia and Eastern Algeria, and northern Europe in Britain and Ireland. Such appropriation took place because Christianity was seen by members of these societies as providing the means to support their initial orientation to the world. This meant that Christianity was assimilated to already existing cultures, and was developed in accordance with these cultures. Roman Africa (Carthage) was an essentially militaristic society, an older civilization than Rome itself, and Christians here formulated the doctrine to accord with their militaristic orientation. Britain and Ireland, on the other hand, appropriated Christianity because it accorded with a pre-existing individualism, and Christianity was interpreted according to this individualism. North African and Celtic Christianity together formed the basis of the Western Christianity which has formed the foundation of Western Civilization.

The first notable Christian in North Africa was Tertullian, son of a military proconsul and later a wealthy lawyer in Carthage. Tertullian was concerned to formulate a binding rule of faith and to enforce rigourous discipline in the church accordingly. His concern with enforcing such discipline to ensure the cleanliness of the church led him to conclude that people are predestined to be damned or saved, and that most of those within the church, including all conventional members of the church, are damned by nature. He then revived the Pauline doctrine of original sin. Since the Catholic church aspired to create a church for everyone it rejected Tertullian's ideal of excluding all but the spiritual elite. But the Africans could not accept the moral laxity of the Eastern church as did most of the Western church until the tenth century. Refusing to ignore questions of discipline, they were compelled to accept that there were sinners in the Church. Tertullian's notion of original sin provided the concept to reconcile this contradiction.

With this development, another North African, St Augustine (354-430 A.D.), who had been a Manichean holding to the doctrine that there is in every soul an inborn struggle between good and evil, found a church congenial to his psychological orientation. He developed this disciplinarian

80. Martin Heidegger, 'The Origin of the Work of Art,' in *Poetry, Language Thought*, tr. Albert Hofstadter, New York: Harper and Row, 1971, p.23.

form of Christianity into a coherent system, formulated in the Latin language. In doing so he became the greatest exponent not only of African Christianity, but of Western Christianity as such. According to Augustine, the division within humans between the corporeal and the spiritual natures is equivalent to the division between privation and what truly is, between the bestial and the rational and between evil and good. Only that which is without change was regarded by him as that which truly is, and it is the immutable which defines what kind or sort of thing each thing is.[81] This meant that the category of quality, defining the kind of being an individual is, was privileged and the other categories reduced to an ancillary status, often to mere accidents.[82] Individuals came to be seen as composites of form and matter bearing various traits, and since each thing was seen as complete in its being, the other categories, including quantity and relation, were seen to inhere in the thing. That is, 'quantity' and 'relation' which in Aristotle were categories of the same status as 'quality' and were understood as referring beyond each individual, were redefined and almost reduced to quasi-qualities of individuals. Following Augustine it then became common to speak of substances and their attributes, which included quantities. The sensible world of changing accidents was to be disdained and to be treated purely as a means to gain salvation. As Augustine put it:

> ...among all these things only those are to be enjoyed which we have described as being eternal and immutable; others are to be used so that we may be able to enjoy these.[83]

The things of the sensible world were to be treated as 'ready at hand' to be used.

Similarly, other people were not to be loved for their own sake, but were only to be used for the love of God. Augustine believed that in the existing situation in which people had to live in a corrupt world there were nevertheless many Christians living for what is eternal and immutable. This is the City of God as opposed to the earthly City, the City of Man. The earthly world is condemned to eternal decay. Since the City of God is in heaven, its members on earth are strangers or pilgrims. But while Augustine spoke of the City of God, this did not involve a real community. While there are some gestures towards the unity of humanity bound together by family affection, Christian salvation was essentially a matter of the relationship between the individual and God. The individual is absolutely dependent upon God, Who alone can satisfy all his desires. All love should be directed towards God. To emphasise this point, Augustine quoted the Old Testament prophet Jeremiah: 'cursed be the man that trusteth in man.'[84] As a soul the individual rules a body which should be treated as an instrument for salvation. Accordingly, Augustine revived the Pauline doctrine that sexual relations were to be tolerated within marriage solely for procreation.

While we might live for what is eternal and immutable, this is only to be achieved in the future, and in the present we must actively engage in the corrupt world. In this respect Augustine made a radical break with Eastern Christianity. This is evident in Augustine's interpretation of the Mary-Martha episode described in the Bible, Luke 10.38 - 42 where Jesus upbraids Martha for her efforts to wait on him rather than sitting down with him. In the East this was generally taken to mean that the active life symbolized by Martha is inferior to the contemplative life symbolized by Mary. However in opposition to the literal meaning of the text Augustine argued that Martha and Mary represent two stages in the perfect life: Martha the soul in time and space; Mary, in eternity. But since we dwell in time and not in eternity, we must live as Marthas and not as Marys.[85]

---

81. See St. Augustine, *Confessions*, Bk. 7, Ch. 11.

82. On this see Ivor Leclerc, *The Philosophy of Nature*, Washington: CUA Press, 1986, p.145. Leclerc's analysis generally accords with Heidegger's.

83. Saint Augustine, *On Christian Doctrine*, Bk I, XXII, 20; tr. D.W. Robertson, Indianapolis: Bobbs-Merrill, 1958, p.18.

84. Loc.cit. from Jer. 17. 5.

85. Saint Augustine, *Sermo*, 104.4. See Lynn White, Jr., *Medieval Religion and Technology: Collected Essays*, Berkeley: University of California Press, 1978, p.240 for an analysis of the significance of this.

Augustine interpreted the struggle between the corporeal and the spiritual, between the temporal and the eternal, in historical terms in accordance with the Judaic element of Christianity. He saw the whole of humanity advancing in a linear progression from the corporeal to the spiritual. This advance was seen as being due to the creativity of humanity, but its end was seen to be away from the sensible world. As he wrote in *The City of God*:

> The education of the human race, represented by the people of God, has advanced, like that of an individual, through certain epochs, or, as it were, ages, so that it might gradually rise from earthly to heavenly things, and from the visible to the invisible.[86]

Everything that happens in history, including the fall of Rome which was sacked in 410 A.D. by Alaric, was seen to have a meaning as part of this education of humanity away from the temporal realm to the eternal realm.

The most serious challenge to Augustine's domination of the thought of Western Christianity came from Pelagius, a Celt from Ireland or Britain. Pelagius was not giving expression to a merely personal view, but was expressing the form in which Christianity had been appropriated in these Northern regions. Borkenau has shown how in Northern Europe, particularly among the Irish Celts and the Norse Vikings, there developed an individualism which found expression in the prominence given to the first person pronoun, 'I', in Celtic and old Norse.[87] By tracing the origin of this emphasis in old Norse in the fifth century and showing how this spread, Borkenau was able to conclude that this manifested the individualism engendered by sea-faring practices. He argued that such a development of language, and therefore of individualism, had taken place among the Celts at an earlier period. The embracing of Christianity by these Celts in an era before there was pressure to do so can be explained by the attractiveness to such individualists of the Neoplatonic Christian doctrine of the soul as a permanent substratum of inner experience. This enabled them to situate their individualism within a cosmic perspective. In adopting Christianity, it was this individualism which was therefore emphasised.

The distinctive emphasis on individualism is evident in Pelagius' only extant work, a letter of religious guidance written to a young woman named Demetrius. The second chapter of the letter begins:

> Whenever I have to speak about the foundations of morality, and about the maintenance of holiness in life, I start by making people see the strength of the aptitudes of human nature, and how much it can achieve; so that by this very start I incite my pupils to every virtue.... The more perfect ... the sort of life we choose as our goal, the more fully must we understand that [human] nature is good. Otherwise the soul will lack determination and be slow in her efforts...[88]

In contrast to Augustine, Pelagius affirmed free-will and espoused a doctrine totally inconsistent with the notion of original sin. This involves an asceticism, but not a rejection of the world. As Borkenau wrote: 'His asceticism is a rule of permanent struggle *in* the world, a struggle to conquer the world. It ... [points] towards unceasing missionary work, and unceasing struggle towards a conquest of the world for the principles of a higher morality.'[89] It is in this Northern Christianity that the active orientation of Western Christianity noted by Lynn White has its deepest roots.

---

86. Saint Augustine *The City of God*, tr. Marcus Dodds, New York: Random House: 1950, Bk 10, Ch.14.

87. Franz Borkenau, *End and Beginning*, ed. Richard Lowenthal, N.Y.: Columbia University Press, 1981, Part II, Ch.1, 'The Rise of the I-Form of Speech'.

88. Quoted by Borkenau, *End and Beginning*, p.290.

89. Ibid. p.308f.

# 4

# CHRISTIAN NEO-PLATONISM AND THE EMERGENCE OF FEUDAL SOCIETY

With the growth of the Roman Empire, belief in the gods on which Rome's civil religion was based collapsed. When the cosmopolitan philosophy of Stoicism failed to provide an alternative foundation for Roman civilization, the Emperor Constantine made Christianity into a State religion in an effort to fill the subsequent vacuum, to legitimate the rule of Rome's emperors over Rome's diverse population. However in such a role Christianity was a decoration rather than the foundation of Roman society. Despite the vigour with which Theodosius had prosecuted all the opponents of the religion, the Church had to adjust itself to traditional institutions and their associated politics. It was only when European society had been thoroughly disrupted during the Dark Ages and had been reformed that it can be said that the Christian vision played a constitutive role in the formation of society. Feudal society emerged from the chaos of the Dark Ages as the social formation able to support a military class of heavily armed cavalry capable of defending Europe, firstly from the Saracens, then from the Vikings and Magyars.[1] This order was only made possible by the ideological mobilization of the population through its Christianization, and with the emergence of feudal society, the people of Europe came to define themselves, their relationships to each other and to nature, and to legitimate their behaviour, in terms of Neoplatonic Christianity.

The Merovingian age preceding the Carolingian renaissance which lifted Europe out of the Dark Ages had been characterized by widespread ethical chaos, at least among the ruling classes. In the 'penitential books', among the lists of church penances for typical crimes applicable to society, incest and murder within the family were invested with an inordinate prominence, reflecting their common occurrence.[2] The poetry of the era, the Gothic songs preserved in the Edda, and in particular the Frankish Nibelungen saga, also reflected this state of affairs and suggest the existence of high levels of anxiety, guilt and disorientation among the population, a state to which the most common response was paranoia: the projection by people of their own aggressive impulses on to others. These people saw the outside world as persecutory, full of implacable enemies deserving to be destroyed, while seeing themselves as persecuted paragons of all the virtues. Throughout Europe most of the population were pagan, conforming externally to the Christian faith while preserving their popular religion virtually intact. At best Christianity was a set of rules of protective magic by which a paranoid population attempted to ensure themselves from the persecution and vengeance of those around them.

Efforts by various Christian reformers to change this state of affairs failed until the eighth century. It was then that St Boniface organized or reformed the bishoprics and monasteries of Germany, a success which was then transferred to France. The basis for this success was in providing a form of Christianity which could be assimilated to the heroic attitude of the pagan or recently converted Germans. This was a form with a strong Pelagian element. The Irish Pelagians had set up monasteries in Northumbria, from which they had later been expelled. However the monasteries had retained their Pelagian orientation, and with the incorporation of these monasteries into the main body of the Latin church, this orientation had also been reincorporated. And it was the influence of these Pelagian monasteries which inspired St Boniface to set out to reorganize the

---

1. See Marc Bloch, *Feudal Society*, 2 Vols, Vol.1, tr. L.A. Manyon and M.M. Postan, 2nd ed., Routledge & Kegan Paul, 1965, Part I.

2. See Franz Borkenau, *End and Beginning*, ed. Richard Lowenthal, N.Y.: Columbia University Press, 1981, Part II, Ch.10, 'Primal Crime and "Social Paranoia" in the Dark Ages' for an analysis of this era.

Christian church within Germany. The result was an amalgam of Augustinian and Pelagian Christianity with a strong emphasis on the moral transformation of both the clergy and the laity. It provided the heroic goal of moral perfection.

With this development of Christianity life was seen as a struggle to establish the presence of the eternal world within the world of change. The monasteries played a major part in this, being 'the symbol of stability and immutability in a world of flux; they were the gate to heaven; they were replicas of heaven on earth.'[3] A title-deed to a new monastery proclaimed:

> Amid the fleeting and transitory world, all visible things hasten to their end more quickly than the wind, but the things which are not seen remain fixed and immutable forever. Seeking therefore to use our transitory and temporal riches to procure eternal rewards and lasting joy ... I give to the bishop and monastery of Worcester this piece of land to remain free from all human service till the end of time.[4]

Monasteries symbolized a divide through all life between immutability and the world of flux, a division which was reinforced by other symbols, from buildings to ceremonies, the liturgy and holy relics. These were all designed to emphasize the smallness of humans in contrast to the impersonal majesty of the spiritual world, only attainable in this life in symbolic ritual, and in the peace of spirit which could be found in rigid discipline. Associated with the notion of the creation, the temptation and fall of humans, the incarnation of God in Christ and the possibility of redemption in the afterlife, this divide engendered an over-riding concern with sin and salvation.

The Carolingians were aware of the importance of developing such an ethical order within society. Government was especially subject to the curse of meaningless flux, and it was therefore felt to be necessary to seek a supernatural sanction to give them a right to rule as the Vicars of Christ. While Pippin the Short encouraged St Boniface and persuaded him to transfer his efforts from Germany to France, Charlemagne inspired a renaissance in learning to revivify the inheritance of classical antiquity.[5] This renaissance found its foremost representative in John Scotus Eriugena, a philosopher promoted by Charles the Bald, who not only translated the work of the Pseudo-Dionysius but also wrote an original work *The Division of Nature*. This renaissance at first had only superficial effects, and the Carolingian era seemed to be falling back into a second dark age, and it was the growing strength of the Benedictine monasteries rather than this renaissance of ideas which was important for averting a new era of chaos. However this renaissance began an intellectual development which provided a general conception of the cosmos as a divinely ordained order linking the lower levels of creation with the heavenly realms.[6] There were three dimensions to this order: it was seen as a great chain of being, as a series of corresponding planes and as a cosmic dance. Deriving from Plato's *Timaeus*, the notion of the great chain of being involved seeing the world as an immense number of links, ranging in hierarchical order from the meagrest kind of existents through every possible grade up to the most perfect or highest possible kind of creature at the foot of God's throne. The idea that the world is a hierarchy of corresponding planes was a development of the Neoplatonic notion that every part of the universe reflects every other part. The main planes recognized were the divine and angelic, the universe or macrocosmos, the commonwealth or body politic, man or the microcosmos, and the lower creation. Each of these was understood to have the same basic characteristics, with each plane being a hierarchical order analogous to the body with a head, soul, heart, arms and legs. Consequently it was built into the

---

3. R.W. Southern, *Western Society and the Church in the Middle Ages*, Harmondsworth, Penguin, 1970, p.28.

4. Cited ibid., p.29.

5. On this renaissance see John Marenbon, *Early Medieval Philosophy (480-1150)*, 2nd ed., London: Routledge & Kegan Paul, 1988.

6. For a description of this conception order and the way it endured until the renaissance see E.M.W. Tillyard *The Elizabethan World Picture* [1943], Harmondsworth: Penguin Books, 1972, pp.12-16.

medieval world-orientation that the order of nature must be seen as a metaphor for the order of humanity and vice-versa. The idea that the universe is a cosmic dance, which derived ultimately from the Pythagoreans, involved the notion that the planets' orbits produce music, the harmony of the spheres. The whole universe was thought to be kept in order by this celestial music, and people were exhorted to keep their souls in harmony with it. Anything not keeping its place in this vast order of being was thought to be a threat to the whole of creation. It was this general conception of the world which provided the world-orientation for the high Middle Ages and which continued to provide the ideological foundation for European society until the Reformation.

Christianity was first appropriated by the aristocracy, but it soon permeated through society with the establishment of parish churches, which spread first to the cities and towns, and then into the countryside.[7] This radically changed people's lives. To begin with, peasants attending church knelt to talk to the saints, but this changed with the spread of the Irish innovation of individual confession to priests. In this way the illiterate parishioners came to be trained in moral self-examination and spiritual introspection, opening to the common people the experience of a new kind of highly personal, interiorized, religion. Although the confession did not become compulsory until 1215, it had become almost a universal feature of life long before then. Through the confession individuals received moral instruction, and efforts to provide this led to the development of the Christian Court of Conscience. Determining correct action in particular concrete situations came to be known as casuistry. This was not simply a matter of applying general principles to particular situations, but involved reference to the Christian world-orientation. As Bentley put it:

> Casuistry cannot attain its end, thus conceived, simply by taking over 'conclusions' - moral generalizations and axioms - from general moral theology and relating them to circumstances. Often the right resolution of a case requires direct reference to the fundamental mysteries of the Christian faith or fresh consideration of the God-given natures and ends of created beings.[8]

All aspects of life were considered by the casuists:

> Solutions for the conduct and regulation of man's life and all his relations in the market place, in the battlefield, the court, the home and elsewhere, were ... developed in innumerable treatises on the cases of conscience. All the urgencies of life and the aims of men as they moved about in their daily lives were indeed grist for the mills of the casuists.[9]

In this way, the medieval world-view was articulated into daily life.

This does not mean that Christianity was a monolithic structure of ideas which everyone accepted. It was unlikely to have completely permeated the thought of all classes in all regions, and to the extent that it did, it was a vision which could be developed in different directions by emphasising different aspects.[10] A study of the aristocratic chronicles and the records of the clerisy by William Brandt revealed that different groups maintained different conceptions of the world and their place within it for several centuries.[11] Differences in outlook were also developed by different heretical groups. The Heretics of the Free Spirit, influenced by John Scotus Eriugena, emphasised the immanence of God in the world and argued that redemption would be attained by establishing a

---

7. For a description of this see Lynn White Jr, *Medieval Religion and Technology*, Berkeley: University of California Press, 1978, p.145ff.

8. G.B. Bentley 'Casuistry', *Encyclopaedia Britannica*, 1961, Vol.5, p.13.

9. Benjamin Nelson 'Conscience and the Making of Early Modern Cultures' in *Social Research*, 1969, Vol.36, p.13.

10. Some idea of the complexity and diversity of medieval culture can be gained from Jacques Le Goff's study, *The Medieval Imagination,* tr. Arthur Goldhammer, Chicago: Uni. of Chicago Press, 1985.

11. William Brandt, *The Shape of Medieval History: Studies in Modes of Perception*, New Haven: Yale University Press, 1966.

new order on earth,[12] while the Albigenses emphasised the Gnostic or Manichaean elements of Neoplatonism, seeing the sensible world as the creation of an evil demon and abjuring it to the extent of starving their children to death.[13] Yet these opposing positions were all recognizably variations of the same basic Christian Neoplatonist world-orientation.

A more fundamental opposition to Neoplatonic Christianity came from the peasants. The content of medieval ideology: asceticism, providentialism, sin, atonement and suffering associated with fear, religious awe and humility, all in the service of an oppressive and intimidating ruling class, was challenged by the tradition of laughter: the carnival, the parody, the buffoonery and the celebration of the grotesque, by which the peasants attacked the icy, petrified seriousness of their masters. Depicting very old women as pregnant, they glorified the association between decay and generation, extolling becoming over being and feminine fertility over masculine domination. As Bakhtin pointed out, through parody the people 'were freed from the oppression of such gloomy categories as "eternal," "immovable," "absolute," "unchangeable" and instead were exposed to the gay and free laughing aspect of the world, with its unfinished and open character, with the joy of change and renewal.'[14] While such parodies of the Church as the feast of fools and the feast of the ass were condemned from the early seventh century onwards, judicial prohibitions had little effect. It was not until the Reformation that the liberating potential of laughter was extinguished. But even this tradition of laughter which had its roots in pagan culture can be seen as constrained by the dominant ideology. It was the negation of Christianity, exalting the corporeal at the expense of the spiritual, the bestial at the expense of the rational, the changing sensible world at the expense of the eternal. But as a negation it came to be dependent upon what it negated, and consequently could only soften the effects of the reigning ideology.

In what follows, how the Neoplatonist Christian world-orientation came to constitute relationships, first between the rulers of society, and then between the general population, will be described. It will be shown how this led to a more and more pervasive individualism which eventually led to an almost complete dissolution of the holistic terrestrial community and how this was associated with the development of a domineering orientation towards both people and nature. The concern will be to describe how a culture emerged in which all practices, beliefs and transformations of the physical world came to resonate with each other, and thereby sustain a general orientation to the world. This orientation was characterized by a growing anxiety, and by alienation from and hostility to the world which culminated in frenetic efforts to subjugate people, nature and other societies.

## *The Church and the Aristocracy*

As the Neoplatonic Christian vision of the world was assumed by the aristocracy, salvation came to be seen as the goal of life, and aristocrats came to see their lives and the world around them as means to attaining this salvation. In accordance with traditional Neoplatonic Christianity, meditation was thought to be one way of achieving this. The concern with salvation through meditation was manifest in the lives of the individuals described by David Douglas in his book *The Norman Achievement 1050-1100*:

> Duke Robert I ... father of William the Conqueror was a young, lustful and ruthless prince who was successfully reducing his turbulent duchy to order when he suddenly determined to mend his soul by departing to Palestine on a pilgrimage from which in fact he was never to return. Again, Simon de Crepi, count of Vexin, consolidated his power by winning in

12. See Norman Cohn, *The Pursuit of the Millenium* [1957], London: Paladin, 1970, Chs 8 & 9.

13. See Emmanuel Le Roy Ladurie, *Montaillou*, [1978], Harmondsworth: Penguin, 1980.

> profitable marriage Judith the daughter of the count of Auvergne. But he chose the occasion of his wedding night in 1078 to vow himself and his wife to perpetual continence and departed forthwith to become a monk in the abbey of Saint Claude in the Juar. To men such as these a pilgrimage might be as important as a war, or a monastic vow as compelling as the establishment of order, and it may be recalled how many of the warrior lords of this age retired after their strenuous lives to spend the evening of their days in monasteries.[15]

However it was action which came to be most stressed by Western Christianity as the primary means to salvation, with appropriate action being defined from the Christian perspective.

With the development of Neoplatonist Christianity, all power in the world was seen to flow downwards. Correspondingly, the source of royal power which had previously been invested in the electing body, no matter how limited the franchise, came to be taken as God, thus freeing the king from responsibility to the people. As Ullmann described this new relationship:

> The king by the grace of God had effectively emancipated himself from the populus itself and on the other hand freely acknowledged God as the source of his royal power. The ascending conception of kingship had faded out: in the Middle Ages its place was taken by the descending or theocratic thesis.[16]

This conception of hierarchical order was reinforced by the notion of the universe as a hierarchy of corresponding planes, each with the same basic structure. The hierarchical order of the kingdom of God was seen to correspond to the order of the kingdom of man, to the kingdom of the beasts, to the kingdom of the fishes, and so on. The central analogy of this hierarchy was the body and the relationship between its parts, and it was this which became the dominant thematic motif unifying medieval culture. Society came to be conceived of as a body with individuals related to each other as parts of a body to one another. John of Salisbury gave expression to this view in the twelfth century:

> The place of the head in the body of the commonwealth is filled by the prince, who is subject only to God and to those who exercise His office and represent Him on earth, even as in the human body the head is quickened and governed by the soul. The place of the heart is filled by the senate, from which proceeds the initiation of good works and ill. The duties of the eyes, ears, and tongue are claimed by the judges and governors of provinces. Officials and soldiers correspond to the hands. Those who always attend upon princes are likened to the sides. Financial officers and keepers ... may be compared to the stomach and intestines... The husbandmen correspond to the feet...[17]

This then replaced Plato's idea of *The Republic* as the ideal form of society, and individuals were required to play their appointed parts to fully actualize this form.

This organic analogy did not in any way affirm human community. A body was not seen in the way that anti-reductionist biologists see organisms today. To the extent that there were communities in the Middle Ages, this was despite the ruling culture, not because of it. People in feudal societies, at least at the higher levels, related to each other according to their positions, and loyalties and obligations were to positions rather than to people. The unity was ideal rather than real. The Mafia represents a survival of this medieval form of social relationship. Furthermore a central feature of

---

14. Michael Bakhtin. *Rabelais and His World* [1965] tr. Helen Iswolsky, Bloomington: Indiana University Press, 1984, p.83.
15. David Douglas, *The Norman Achievement 1050-1100*, London: Collins, 1972, p.19.
16. Walter Ullmann, *Principles of Government and Politics in the Middle Ages*, London: Methuan 1961, p.117.
17. John of Salisbury, 'Policraticus' in *The Portable Medieval Reader*, Harmondsworth: Penguin, 1977, p.47

feudal relations was that vassalage had to take precedence over kinship, and all action had to be seen as ultimately for the glory of God. So while John of Salisbury was concerned to ameliorate the lot of the lower orders of society by pointing out how essential they were to the body politic, he framed his argument in terms which reinforced their oppression. The analogy of the body was far more effective as a basis for arguing that society should be composed of members unlike in goodness, dignity and wealth, just as a body should be composed of unlike members, and this argument was widely used up until the Renaissance.[18]

In the early Middle Ages, rulers saw their power as deriving directly from God, claiming supernatural attributes to sanctify their decrees. They wore ecclesiastical vestments on ceremonial occasions, they were anointed with holy oil used in the consecration of bishops, and the sword, sceptre, ring and crown they received were blessed in formulae appropriate to ecclesiastical ordination. But the Papacy was not satisfied with being ruled by Christian Kings, no matter how devout. The combined effect of Augustinian and Pelagian aspects of Christianity, the disciplinarian and the activist, formulated within the framework of the hierarchical Neoplatonist conception of the world in which all power was seen as flowing downward, finally inspired the Church to struggle for temporal power superordinate to that of emperors and Kings. Pope Leo IV drove the Saracens out of Italy and established the Leonine City, and his second successor, Nicolaus I (858-67) asserted Papal political supremacy over the Frankish court and exerted in a new way Papal supremacy over the French bishops. In the struggle which followed, the hierarchical conception of the world with power deriving ultimately from God played a major role in giving ultimate victory to the popes over the emperors and in establishing a hierarchical social order throughout Europe. This culminated in 1075 with the *Dictatus Papae* which declared the previous political and legal order abolished. Henceforth, kings were held to derive their legitimacy solely from the Church.

Beginning in 1095, following the ascendency of the Papacy over the aristocracy, a series of crusades were launched against the Holy Lands. The aristocracy were then called upon to undertake these crusades, but at the same time were required to act as Christians. To this end the Court of Conscience developed a code of conduct for the aristocracy to accord with their social position and the role they were called upon to fulfil. The code of chivalry was the outcome of this Christianization of military behaviour.

The ethics which subsequently came to be embodied in the institutions of chivalry were thoroughly grounded in Neoplatonic philosophy. Right and wrong conduct were understood in terms of participation in forms, and individuals were only seen to be significant in so far as they transcended their bestial natures by participating in the forms of virtue affirmed by chivalry. People were always defined in the aristocratic chronicles by six or eight adjectives and their contraries, and no other possibilities were allowed for. Men were valiant, courteous, prudent, and so on, or they were cowardly, discourteous and reckless. Women were beautiful, charming and discreet, or their opposites. The feudal code was directed towards fostering the achievement of honour which, in the case of males, was to be pursued positively and aggressively through military action performed in the prescribed manner. William Brandt who made a detailed study of the chronicles of the feudal aristocracy wrote of their ethical assumptions:

> The aristocrat found his *summum bonum* in a kind of public posture taken with regard to his own class; he was an actor inventing a script which he hoped would turn out to be heroic... To understand the Middle Ages, we much realize that a great many activities - the most important - were pursued for their own sake, with no other end in view beyond the public posture they permitted.[19]

---

18. See E.M.W. Tillyard, *The Elizabethan World Picture*, [1943] Harmondsworth: Penguin, 1978, esp. p.19.

19. Brandt, *The Shape of Medieval History: Studies in Modes of Perception*, p.114.

The pursuit of honour through warfare became the over-riding concern, whether the war was directed at infidels or other Christian kingdoms. According to these new values there could be no such thing as a good, peace loving aristocrat. As the medieval French writer, Philippe of Navarre asserted '...he who passes his youth without exploit may have cause for great shame and grief.'[20] In fact it was only in so far as people participated in the ideal forms defining aristocratic honour that they could be considered to be fully 'real'. People who stood outside the circle of values defining human significance had only a shadowy existence in the eyes of the aristocracy, and they were treated accordingly. This was illustrated by the way non-combatants in battles were generally slaughtered, not out of any real malice, but because they were considered so insignificant. At the famous battle of Limoges, three French knights who had especially distinguished themselves in individual action were seen by the Black Prince who 'looked on them with pleasure, and he repressed and softened his ill-will.'[21] These knights were allowed to surrender and were presumably spared, while 3,000 men, women and children were slaughtered. As William Brandt wrote of this incident: 'In the midst of incredible carnage to which the Black Prince was apparently totally indifferent, three knights by their honourable stance touched the Prince where mere suffering never could.'[22] Such behaviour on the part of aristocrats was not exceptional.

Warfare not only remained the central focus of the aristocracy of Europe, but was given such ideological support by Neoplatonic Christianity that the world came to be seen and valued entirely in terms of military performance. As such the rulers of medieval Europe contrasted radically with the rulers of the Sung Dynasty in China which existed at roughly the same time (960-1279). The ruling class of this society were the scholar gentry whose main road to power was through competitive examination in the civil service. The military classes and the merchants were kept in a firmly subordinate position below that of peasants.[23] The founder of the Sung Dynasty, T'ai Tsu was described by Reischauer and Fairbank as characterized by:

> ...his policy of leniency and benevolence toward his former rivals and recalcitrant subordinates. He set a standard of generosity toward his officers, deference towards his ministers, and modesty in his own scale of living that approximated the Confucian ideal and influenced his less able but often equally conscientious successors on the throne.[24]

While there was a frequent administrative bias in favour of the wealthy landowning class from whom the scholar gentry were mainly drawn, this was not always the case. The Chief Councillor Wang An-shih appointed in 1069 instituted cheap loans to the peasants, redistributed land to eliminate old inequalities, established a graduated scale of land taxes according to the productive capacity of the soil, commuted the remaining corvee services, which fell most heavily on the poor peasants, to taxes which fell most heavily on the rich, and attempted to regulate prices, assess all personal wealth and carry out much needed water-control.[25] There was no parallel to Wang An-shih in Europe. The virtues of T'ai Tsu and Wang An-shih, being unrelated to the pursuit of military honour would not have been comprehended by the European aristocracy.

---

20. Cited ibid. p.111.

21. Froissart's *Life of the Black Prince*; cited ibid. p.137.

22. Loc.cit.

23. Joseph Needham, *The Grand Titration: Science and Society in East and West*, London: George Allen & Unwin, 1969, p.197f.

24. Edwin O. Reischauer and John K. Fairbank, *East Asia: The Great Tradition*, Boston: Houghton Mifflin, 1960, p.197f.

25. Ibid. p.206.

## The Efforts to Control Nature

Corresponding to the insignificance accorded by aristocrats to people lower in the hierarchy of being except as beings to be used or subjugated, non-aristocrats viewed nature as nothing but material for the fashioning of the human realm. Advances in the history of the Middle Ages, particularly by Lynn White and Jean Gimpel, have revealed it to have been characterized by rapid advances in technology. These advances transformed Europe from a society under siege by Norsemen, Magyars and Saracens into the most dynamic society in the world by the last decade of the 15th century when it 'burst its oceanic limits and swept the earth - exploring, trading, looting, conquering, and colonizing.'[26]

The preoccupation with technology in Europe can be partly explained by the turmoil of the Dark Ages. The Dark Ages destroyed the institutions which had inhibited the development of technology. Roman society had been strongly opposed to technological innovation.[27] There was a story in Rome that when an inventor of unbreakable glass demonstrated it to the Emperor Tiberius expecting to be rewarded, he was beheaded. Whether this story was true or not, it typified the general attitude of Romans to technical invention. A farmer who invented an ox-powered mechanical reaper was ignored, and the water wheel was not exploited. These attitudes were institutionalized within the Roman Empire. Along with the destruction of these conservative institutions, the depopulation of Europe also opened the possibility of new forms of organization. It was in this way that the heavy plough which required long strips of land was established in Northern Europe.[28] Being ravaged on all sides also provided a strong incentive to develop new forms of armaments, and early appropriations or innovations in technology such as the stirrup were largely for military purposes. A picture dating from the early ninth century used by White to illustrate Western Europe's high regard for technology shows the iniquitous sharpening their swords with an old fashioned whetstone while the virtuous were sharpening their swords with a rotary grindstone.[29]

However only the domination of society by a culture which could orient people to such technological domination of nature and which could legitimate such developments could account for the speed with which technological innovations were embraced in Western Europe as compared with Byzantium or the Islamic world.[30] The individualistic activism of Augustinian/Pelagian Christianity was supported by the general vision of the place of humanity in the world. Western Christianity saw God in more personalized terms than Eastern Christianity, and partly because of the influence of Plato's most well known work in the early Middle Ages, the *Timaeus*, this personalized God was seen, despite St Augustine's objections,[31] as an artificer who had created the world as a planned abode for humanity. This image of God as an artificer dignified the creative efforts of humans, and as Clarence Glacken argued, if a dominant idea existed in the Middle Ages, 'it was that man, blessed with the faculty of work, assisted God and himself in the improvement of an earthly home even if the earth were, in Christian theology, only a sojourners' way station.'[32]

---

26. White, *Medieval Religion and Technology*, p.xxiv.

27. See M.I. Finley, *Economy and Society in Ancient Greece*, B.D. Shaw and R.P. Saller, eds, London: Chatto and Windus, 1981, Ch.11.

28. White, *Medieval Religion and Technology*, p.138.

29. Ibid. p.186 & 194.

30. This is not to say that the development of technology was nothing but a manifestation of ways of thinking. There were a multiplicity of reasons why Western Europeans developed their technology more rapidly than elsewhere, most importantly geographical conditions, the decentralization of power partly as a consequence of these conditions, and the social forms which developed in these circumstances. See E.L. Jones *The European Miracle*, Cambridge: Cambridge University Press, 1981 for an attempt to look at all the significant conditions which distinguished Western Europe from other societies.

31. City of God, XII, 24, p.504.

32. Clarence Glacken, *Traces on the Rhodian Shore*, Berkeley, LA: University of California Press, 1967, p.175.

Monks such as Bernard of Clairvaux (1091-1153) who had retired from the world celebrated the imposition of a human order on nature and their dominion over nature as helping God finish the creation. They fancied that by their work they were re-creating the earthly paradise, reasserting the complete dominion over all other life that existed before the Fall. The transforming of nature from wastelands suitable only for beasts into pleasant abodes for humans produced sentiments of exaltation and was frequently likened to the work of the creation itself.[33]

In the early Middle Ages the work of such monks was far more significant than that of the laity in transforming nature. However the monasteries themselves tended to lose their original motivating ideology and to take on a melancholy character of greed, worldliness, and corruption. But their achievements revealed what could be done and inspired others to emulate them. And while such people were more motivated by practical concerns, either to secure a food supply with minimum effort or to provide a basis for expanding social, political or military power, the basic conception of nature and of humanity's role in creation promoted by the Church provided background support to such work.

Among the laity Western Christianity was important to begin with for enabling them to overcome inhibitions about interfering with the course of nature. As Christianity began to permeate the society of commoners the localized spirits and *daemons* were ousted from nature by the cult of saints conceived to be abstracted from time and place. This abstraction of spirit from matter was reinforced by the interiorized, personal nature of religion cultivated by the Confession. In this way people came to see themselves as transcending nature, and it was this which enabled them to see nature in instrumental terms. At the same time Christianity guaranteed the continuing order and permanence of nature as part of the Lord's covenant (Gen. 8:21; esp.9:8-11), freeing people from concern about the destructive effects of their activity.

This orientation was supported by the absence of anything in the mode of thinking developed in the Middle Ages to stand in the way of its being seen in purely instrumental terms. Nature was seen as a multiplicity of separate, unrelated entities. Following the Christian Neoplatonist tradition the sensible world was seen as composed of formed individuals. While the forms were related ideally or logically, each sensible individual was seen in isolation from any context. William Brandt, who revealed the constancy in the medieval view of nature from the most popular encyclopaediast of the early Middle Ages, Isidore of Seville (c.560-636) to Albertus Magnus (d.1280), wrote of Isidore's views:

> Isidore's world was astonishingly static... Some movement there was, but it did nothing to blur the splendid isolation of its self-subsistent entities because it was never really a relationship between objects; it was habitually located within a particular object.[34]

In essence, Western thought was substantialist, with things and actions being conceived of in isolation, having various properties. This substantialism was adopted by the general population, and corresponded to the individualism within society. Without any conception of the inter-relationship between forms of life, there was no reason why nature should not be transformed indefinitely.

The significance of this substantialism stands out when contrasted with the relationist conception of nature of the Chinese. For contemporaneous Chinese not only was everything in nature interdependent, but it was always in the process of becoming or decaying. This relationist conception of nature was most fully developed by the Sung Neo-Confucian philosophers for whom it was impossible to conceive of anything in total abstraction from its context.[35] While the Chinese maintained their technological superiority over Europeans at least until the fourteenth century, and according to Needham, up until the sixteenth century, they were always wary of the possible side-

33. Ibid. p.312f.

34. Brandt, *The Shape of Medieval History*, p.7.

35. See Joseph Needham, *Science and Civilization in China*, Cambridge: C.U.P., 1956, Vol.2, pp.455-505, esp. p.478.

effects of technology.[36] This concern was expressed in the famous story of Chuang Tzu in which Tzu Kung tries to tell a farmer, who is drawing water from a well with a bucket, about the simple labour saving device, the counter-weighted swape. The farmer laughs and replies:

> I have heard from my master that those who have cunning devices use cunning in their affairs, and that those who use cunning in their affairs have cunning hearts. Such cunning means the loss of pure simplicity. Such a loss leads to restlessness of the spirit, and with such men the Tao will not dwell. I knew all about the swape, but I would be ashamed to use it.[37]

There was also strong opposition to attempts to force nature. The famous parable deriding the man who, discontented with the rate of growth of his plants, started to pull at them to help them come up, reveals a widespread concern to act in accordance with the immanent dynamics of the world (even if this concern had only a limited effect).[38]

As the development of Western European technology gathered pace during the economic expansion in the twelfth and thirteenth centuries, the medieval ethics of virtues provided support for its continued legitimation. Temperance rose from being considered the least important of the seven Christian virtues at the beginning of the era to the most noble virtue in the fifteenth century.[39] With this rise in its status, its character also changed. Temperance meant basically the duty to know and rule oneself. In the eleventh century Temperance was represented in icons as a woman pouring water into a cup of wine to reduce its potency. By 1359 however, Temperance had come to be represented in Florentine art as a woman holding a pair of compasses. From a measurer of space, Temperance was rapidly transformed into a measurer of time, and was soon shown holding a sand-glass. Temperance had become a virtue primarily associated with work and the effective use of one's time.

With the lack of concern with the dynamics of nature, the high value placed upon effective work, and an orientation towards the future led to a glorification of all technological advances, the way was paved for a technological explosion. The most significant outcome of this was the 'invention of invention' and an enthusiastic search for new solutions to solve problems. For instance technicians laboured from the 1260's to the 1330's to develop the first truly mechanical clock. The anticipation of invention was epitomized in Roger Bacon's confident prophesy of an age of cars, submarines and aeroplanes. This technological progress was seen in terms of the Christian vision of the world as a continual progression towards a higher plane of existence.[40]

## *The Medieval World-View and the Universities*

One of the most extraordinary features of the middle-ages was the development of the universities and the associated development of scholasticism. These further developed the Western European abstract form of thinking and detachment from the world.

To understand this development it is important to clear up a general misconception about this. Contrary to popular belief, these developments did not replace Platonism by Aristotelianism. Aristotle supplied the technicalities of logic and physics but not the direction of Christian thought, and the assimilation of Aristotelian thought consolidated the prevailing assumptions rather than replaced them. When Aristotelian ideas did begin to challenge the medieval world-view, they were

---

36. Needham, *The Grand Titration*, p.213.

37. Quoted by Needham, *Science & Civilisation in China*, Vol.2, p.124.

38. Needham, *The Grand Titration*, p.202.

39. White, *Medieval Religion and Technology*, pp.187-204.

40. See Robert A. Nisbet, *Social Change and History: Aspects of the Western Theory of Development*, London: Oxford University Press, 1969, esp. p.98f.

suppressed, this suppression culminating in 1277 with a blanket condemnation of 217 propositions, holding any one of which was punishable by excommunication.[41] After this, any speculation which might have brought the prevailing world-orientation into question was neutralized by denying the reality of the postulates of theory, and conceiving the purpose of theory as nothing but 'saving the appearances.' So people continued to see the world in terms of forms, as a great chain of being, as a hierarchy of corresponding planes and as a cosmic dance - virtually up until the Renaissance.[42] Aristotle's political ideas did have revolutionary implications, but these did not form part of the medieval world-orientation at all. They were important for beginning to undermine it.

The nature of the universities reflected their function.[43] They were designed to teach people what was thought to be a body of essentially complete knowledge. The largest faculties were the Arts faculties which prepared people for further education in law, medicine or theology. Only a few universities were permitted to have theology faculties, and while these were nominally the most prestigious centres of learning, they were relatively small. The centre of theology, Paris, had in 1362 25 professors of theology and 449 professors of arts.[44] Furthermore, in most cases the Arts faculties tried to avoid theological questions and channelled their interests away from the concerns of the theological faculties. The largest faculties apart from the Arts faculties were the faculties of law, since the legal profession was the most lucrative, and the Arts faculties were more than anything concerned to prepare people for entrance into the faculties of law. Consequently they focused on dialectic, logic and an adumbrated rhetoric - with some physics. Metaphysics and ethics were generally of little significance as they were too close to theology. The works of Aristotle which had the widest influence were therefore those associated with dialectics, logic and rhetoric, though the medical faculties were also interested in Aristotle's physics and biology.

It was in the theological faculty that the metaphysics of Aristotle, as distinct from his logic and physics, was assimilated, and this assimilation was designed to bolster the basic Neoplatonic world-orientation. Conversely there was no really successful effort by Aristotelians to free themselves from the Neoplatonist perspective before Pomponazzi (1462-1525).[45] The assimilation of Aristotle to Neoplatonism was facilitated by a number of factors. To begin with, Neoplatonism and Aristotelianism were not completely antithetical doctrines. Aristotle's philosophy was a development of the formist philosophy of Plato. The difference between these philosophies tends to be overestimated because Plato tends to be caricatured as a defender of a heavenly realm of forms, whereas, as I have argued, his real importance was to uphold the primacy of the omni-temporal. In this, though he was more concerned with the nature of becoming, Aristotle was ultimately at one with Plato. Furthermore Neoplatonists from Plotinus to Thierry of Chartres in the twelfth century had already assimilated much of Aristotle's thought to Neoplatonism.[46] The increasing status of Aristotelian ideas began with the translation of Avicenna's ideas into Latin in the twelfth century. But Avicenna's Aristotelianism was profoundly Neoplatonist, and most subsequent Aristotelian philosophy was coloured by this. As James Wesheipl wrote: 'The history of medieval thought is

41. On this suppression and its effects see Edward Grant, *Physical Science in the Middle Ages*, Cambridge: Cambridge University Press, 1977, Ch.III and onwards.

42. See E.M.W. Tillyard, *The Elizabethan World Picture*, [1943], Harmondsworth: Penguin, 1978.

43. See Gordon Leff, *Paris and Oxford Universities in the Thirteenth and Fourteenth Centuries: An Institutional and Intellectual History*, N.Y.: Wiley, 1968 for a study of the formation and characteristics of the universities.

44. Walter J. Ong, *Ramus, Method, and the Decay of Dialogue*, Cambridge, Mass.: Harvard University Press, [1958], 1983, p.133.

45. This has been argued by John Randall Jr. in 'Pomponazzi: Introduction' in *The Renaissance Philosophy of Man*, Ernst Cassirer, Paul Oskar Kristeller and John Herman Randall Jr., Chicago: University of Chicago Press, 1948, pp.257-59. The case for the continued popularity of Plato in the Middle Ages has been made out by Raymond Klibansky, *The Continuity of the Platonic Tradition during the Middle Ages*, 2nd ed. London, 1950.

46. See Stephen Gersh, 'Platonism-Neoplatonism-Aristotelianism', in Robert L. Benson and Giles Constable eds., *Renaissance and Renewal in the Twelfth Century*, Cambridge: Harvard University Press, 1982, pp.512-34.

full of commentators on Aristotle whose inspiration came from Avicenna and his neo-Platonism...'[47] The most significant Aristotelian of the thirteenth century, Thomas Aquinas (1225-1274) was concerned primarily with refuting those thinkers who were developing Aristotelian ideas in ways which undermined the important tenets of Christianity as it had been understood up until then, and he was supported by the Church precisely for this reason.[48] For instance one of the major notions he opposed was the Averroist doctrine that the intellect is one and the same when manifested in different persons, because this contradicted the individualistic emphasis of Augustinian Christianity. The development of the logic of Aristotle consolidated medieval Europe's substantialist modes of thought, while Aristotle's view of the world as the repository of the grossest dregs of the universe reinforced the low status given to the changing world. Aquinas took up Aristotelian ideas in ethics and political philosophy in a way which gave a far more exalted status to earthly life than Augustine had allowed, but his work in this area was of little political or social significance until his ideas were taken up during the counter-reformation in the sixteenth and seventeenth centuries.

Most revisions made to the Neoplatonist framework tended to consolidate the emphasis on eternal forms, and Aristotle's work on becoming was obscured by the less significant issue of the ontological status of universals. In a very un-Aristotelian way Avicenna conceived of forms as efficient causes of all movement in the world, thus introducing a pre-Cartesian dualism into philosophy. Aquinas understood motion as the product of the unity of forms and matter, but saw individuals as actualized forms rather than as forming activity.[49] As Heidegger has pointed out, this involves conceiving Being as something made, or as an act of production rather than as disclosed or revealed, although it is God rather than humans which is conceived to be the producer.[50] Motion was understood by Aquinas as imperfect actuality - the actuality of a being whose potentiality is actualized while still remaining in potency to further actualization. This paved the way for William of Ockham (1295-1349) to claim that motion is nothing more than the succession of forms acquired by a subject.[51] Though this provided the basis for the study of kinematics which led fourteenth century thinkers to develop ideas which anticipated much of, and paved the way for the scientific revolution in the seventeenth century, these ideas continued to be formulated in terms of forms. Change in velocity was described in terms of the intensification and remission of forms, or how forms became more or less intense.[52]

Universities influenced thought not only through the ideas purveyed, but also by the way learning was organized. As has been pointed out, the most important faculties in universities were the Arts faculties, and the most important studies in these faculties were of philosophy, particularly logic and dialectics. But the way these studies developed was largely a manifestation of the nature of the institutions. For the Greeks philosophy was 'love of wisdom', and dialectics was dialogue through which people attained wisdom. Socrates argued with his disciples and opponents and wrote nothing. Plato committed such dialogues to writing, but as we have seen, disclaimed these as his own views or as the source of wisdom. They were meant only as preparatory exercises. For Plato as

---

47. James A. Weisheipl, 'Aristotle's Conception of Nature: Avicenna and Aquinas', in Lawrence D. Roberts ed. *Approaches to Nature in the Middle Ages*, Birmingham: State University of New York, 1982, pp.136-160, p.154.

48. See Arthur Little, *The Platonic Heritage of Thomism*, Dublin: Standard House, 1949, Ch.1.

49. For the difference between Aristotle and medieval Aristotelians in this respect see Emerson Buchanan, *Aristotle's Theory of Being*, University, Mississippi; Cambridge, Mass, 1962, esp. p.7n.

50. M. Heidegger, *Basic Problems of Phenomenology*, tr. A. Hofstadter, Indiana U.P., 1982, p.100. Heidegger sees the influence of Roman culture here. For an illuminating commentary on Heidegger's relation to Aquinas, see J. Caputo, *Heidegger and Aquinas*, N.Y.: Fordham U.P., 1982, esp. pp.72-4 and p.88.

51. See W.A. Wallace, *Prelude to Galileo*, Boston and Dordrecht: Reidel, 1981, p.13.

52. On this see Edward Grant, *Physical Sciences in the Middle Ages*, p.55f. and Gordon Leff, *The Dissolution of the Medieval Outlook: An Essay on the Intellectual and Spiritual Change in the Fourteenth Century*, N.Y.: Harper and Row, p.107f.

for Socrates, only through personal relationships could such wisdom be achieved. Aristotle adopted a more abstract approach but still acknowledged a major role for and evinced respect for dialogue, as manifest in the role given to predication in his logic. Medieval universities fundamentally altered this conception of philosophy.[53]

In accordance with the hierarchical nature of society and the Christian notion of revelation wherein God teaches man, and where his teaching is passed on from generation to generation through the Church, the relationship between professors and students was seen as one in which the professor imparted the knowledge of his subject, a 'teaching' (*doctrina*) or conversely a 'learning' (*disciplina*), to his students. In the Arts faculties, young students were instructed by professors little older than themselves. These students were in competition with each other to obtain degrees for access into relatively lucrative professions. The way knowledge was conceived and the modes of thinking developed in these institutions were largely reflections of this pedagogical situation. In the context of medieval teaching dialogue was replaced by a classroom monologue which the professor produced on a schedule at fixed places and hours. Philosophy came to be seen as something which was taught and learnt. Knowledge had to be taught so that the student could remember it in a way which could be examined and measured. Consequently clarity rather than profundity became the central concern of the medieval professors, and knowledge was developed into a form in which it could be easily reproduced. This led to the hypostatization of knowledge. This culminated in the fifteenth century in Germany with the new University of Tübingen being conceived not as a collection of masters and students, but as a collection of universal knowledge.[54] In place of the idea of knowledge as wisdom transmissible only in the context of personal relationships it had come to be seen as a commodity.

The developments in medieval logic can be understood as a result of this pedagogical situation. The point of departure from Aristotle's logic was the logic of Peter of Spain, expounded in *Summulae logicales*, probably written before 1246. This work began with a definition which was the most repeated in all scholastic philosophy: 'Dialectic is the art of the arts and the science of the sciences, possessing the way to the principles of all curriculum subjects.'[55] No distinction was drawn between dialectics and logic, and dialectics was reduced to logic. At the same time it was presented as the foundation of pedagogy. Peter departed from Aristotle in virtually excluding from consideration those aspects of logic: assertion, or the act of predication, proposition as such and semantics, which could not be represented visually, that is, those aspects of logic associated with dialogue. His focus was on structure and terms treated in a quantified fashion - Peter originated what came to be known as terminism. The central feature of this logic was the theory of supposition according to which terms were taken to be 'supposing' or 'standing in' for physical existents. This led to the view of logic as a study of the reflection of the material world in the structure of the mind. In his concern with clarity which led him to conceive things in spatial terms, Peter, as with almost all other logicians who have followed him, tended to treat his terms as substances. As a consequence he tended to a corpuscularian psychology and outlook in which the real and the mental realms were seen as agglomerations of discrete items.

By transforming the meaning of education the medieval universities further dissolved human sociality, and by concentrating on what could be visualized, the conception of knowledge promoted accorded with the Platonic view that the object of knowledge must be quasi-timeless. In this way the distinction between the real, eternal world and the changing, sensible world was further consolidated. And by substantializing terms and seeing the world in terms of these, the developments in logic further promoted a view of nature as consisting of isolated, enduring substances. Since every subsequent medieval university student had to spend a year studying Peter's logic, this had a major influence on medieval thinking.

---

53. See Ong, *Ramus, Method, and the Decay of Dialogue*, on this.

54. Ibid. p.164.

55. Ibid. p.56.

## *The Embodiment of Medieval Culture*

The culture of medieval society cannot be understood as simply a translation of the Western Christian world-view into action. This world-orientation came to be embodied in the medieval world. Firstly it came to be embodied in social practices, including speaking, so that all forms of life within which people participated came to resonate with and reinforce each other as people generalized schemes of action and interpretation from situation to situation. At the most general level, the conception of nature as a hierarchical order reflected the organization of society. Then at a more basic level this world-orientation was embodied in the transformations of nature. The main features of culture which came to be embodied into the medieval world in this way were individualism, emphasis on action and detachment from the sensible world, usually associated with an orientation towards the future.

While the uniqueness of the emphasis on action of medieval European civilization is immediately evident from the rapidity of its technological advances and by its military aggressiveness, its other distinctive features are less commonly recognized. Because it has been so all pervasive up to the present, its individualism only becomes apparent when people have sustained contact with other cultures.[56] In most societies people do not perceive themselves as entirely separate individuals but as one of a group, the point of emergence from a collective reality. In some societies such as those of the Australian Aboriginals, there is also a sense of belonging to and sociality with the land. Similarly the distinctive nature of the detachment from the world is difficult to describe to people who have been socialized to take this completely for granted. Nevertheless I will try to give some idea of the nature of the embodiment and evolution of these features.

To begin with, an individualistic, detached, activism came to be reflected in, embodied in, and thereby reinforced by language.[57] It has already been suggested that the individualism engendered by the sea-faring of Northern Europeans was reflected in the emphasis placed on the personal pronoun. The individualism of medieval Europe expressed itself in the spreading of this mode of speech to the rest of Western Europe and in concomitant developments such as the formal mode of address and the composite future tense. However the nature of Western European speech can be seen more clearly against the background of classical language and when contrasted with the way speech evolved in Eastern Europe.

In ancient Greek and Latin the first person pronoun was not used with the verb except when it was necessary to emphasise some contrast. When the Roman said '*facio*' the verbal ending made it clear that it meant 'I do' and not 'he does'. When he said '*feci*' he conveyed in one word the idea of doing, that it was done in the past, and that it was 'I' who did it; what we must translate in our analytical language as 'I have done'. Greek and Latin did not separate the individual from his acts and gave unreserved expression to the outward act in its purity. Every event was reduced to a momentary pure action with no temporal duration and no relation to anything else. (It was this view of a constantly changing world which gave birth to Plato's epistemological problem of how enduring knowledge is possible, the problem he overcame with the postulation of the eternal forms.) This language made it difficult for Greeks to describe an enduring action with a definite conclusion such as 'I have finished sleeping'. They originally said something like 'slept-slept', later slurred into 'sle-slept'.

From the first century on, Greek and Latin began to lose their synthetic character. In the East, Greek and the Latin Roumanian language came under the influence of Semitic languages dominated by the Aorist tense which is used for expressing indifferently past, present and future. This reflected a conception of the world in which time and action played a very small part. The Greek

---

56. See Louis Dumont, 'The Modern Conception of the Individual: Notes on its Genesis and that of Concomitant Institutions' *Contributions to Indian Sociology*, Vol.8, 1965: 13-61.

57. This relation has been most thoroughly analysed by Borkenau, *End and Beginning*, Part II, Ch.1. from which much of the following is drawn.

language however did not follow the Semitic use of the pronoun. These developments of Greek and the other languages of Byzantium: Roumanian, Bulgarian and Southern Albanian reflected a growing inwardness and a rejection of the worldly self-assertion of the Greek individual in the face of the eternal. Correspondingly the Slavic languages developed turns of phrase without a subject to a far greater extent than Western languages. Tenses, relating to time were weakly developed, while 'aspects' which relate actions to permanence (whether it is permanent, passing or repeated etc.) were strongly developed. Old Russian had no future tense, and the future had to be expressed by derivative forms of expression.[58] It was not until the sixteenth century that we find unambiguous use of the imperfective future, and it is clear that this construction came from the West. Slavic languages still do not require the use of pronouns before the verb and tend to avoid representing the individual at the centre of things

By contrast, in the West the dominant feature of language was the spread of the 'I saying' habit, the earliest known case occurring in a runic inscription found on a golden horn made just after 400 A.D. which runs: 'I, Hlegstr from Holt made this Horn.' It involves the use of the first person pronoun immediately before the verb. The history of Western languages pivoted on this form of speech. To begin with, Latin, like Greek, lost its synthetic character. In particular it lost its simple future tense and the case endings of the noun. However, in striking contrast to the East there was a recreation in the early Middle Ages in Latin languages of the synthesis of root and endings for the future tense, a clear indication of the striving of Italians, Frenchmen and Spaniards towards a new synthetic expression of future action associated with a more activist orientation to the world. At the same time, to varying degrees the Latin languages amalgamated various speech habits of Northern Europeans. This took place most fully in certain Northern French dialects, in Rhaetian and in certain dialects of the Po valley where the characteristics of synthetic speech were shed and the new Northern European use of the pronoun adopted with the Northern form of composite tenses. Towards the end of the Middle Ages the composite tenses displaced the simple future tense in all Latin languages. To understand the significance of this it is necessary to examine in more detail the Northern European languages.

In Northern Europe the use of the first person pronoun became more pronounced with the evolution of medieval society. While in old Norse, Anglo-Saxon, old German, old Irish and old French the use of the pronoun had become widespread, it was nowhere obligatory before the eleventh century. It became obligatory in English from the beginning of the eleventh century, in German soon after the emergence of Middle German during the eleventh century, but did not become obligatory in French until the seventeenth century. In English and Scandinavian this use was associated with the composite future tense, with the verbs expressing obligation and decision, 'shall' and 'will', serving as auxiliaries. These verbs are associated with inner experience rather than outward action, and in particular with firm determination. They are related to forethought and are inseparable from the sphere of the 'I', the person who shall and who wills. And they are oriented towards the sphere of concentrated planning of a subject aiming to control outward life. These make the 'I' more than simply a series of events; it is the primordial source of action. This development was associated with the decay of the simple passive verbs, forcing people to use clumsy word combinations to say that something was being done to them. In English in particular these verbs make life appear as an incessant flow of directed activities. The 'I' which Western speech emphasises is first and foremost a centre of action. French and Italian did not go to the same extreme, using 'to go' as auxiliaries. To say 'I am going to do' reduces the emphasis on intention and decision and is closer to the Latin expression of pure action. Nevertheless it can be said of all Western Europeans that as distinct from Eastern Europeans their language inclines them to live in the outward world almost to the same extent as the classical age, but this world is not spoken of as

58. See Robert Auty, 'The Russian Language' in Robert Auty and Dimitri Oblonsky, *An Introduction to Russian Language and Literature*, Cambridge: Cambridge University Press, 1980, pp.1-38, p.27.

a world of momentary objective events of which people are part, but as a world of objects which is the substratum of will and duty, of planning and doing and fitting actions into a coherent whole.

To the self-assertion manifest in the use of the first person pronoun a reaction took place, first in the literature of the troubadours in the extreme south of Western Europe, Provence, and from there to the whole of Western Europe. This self-assertion was mitigated by exaggerated expressions of veneration for others, which at once turned into forms to express distance and reserve and to exclude the idea of intimacy. This involved the use of the second person plural to address other individuals in polite speech. This form of speech was fully developed in the 'Chanson de Roland' which was composed in Northern France in the eleventh century, but a German translation in the twelfth century still uses the familiar '*Du*' to translate '*vous*'. It was not until the thirteenth century that the 'you' came to be used with any frequency. In general the formal form of speech was adopted more readily in more western parts of Europe, and with the exception of a few pockets in Northern England, the 'you' form of address completely replaced the 'thou' form. The English, where 'I' is spelt with a capital letter and in which inner decision and the orientation to action are expressed most forcefully, have gone much further than any other group of people in developing a sense of distance and reserve. By contrast it was not until Peter the Great's revolution at the end of the seventeenth century that the familiar '*ty*' began to be replaced by the formal and officially endorsed '*vy*', and this impersonal form is still used only in official situations.[59]

Such linguistic transformations took place in the context of a developing socio-economic formation in which the orientation embodied and expressed in language was at the same time being embodied in a multiplicity of other ways. At the most basic level, the transcending activism of Western culture became embodied in the way people acted on their physical environment. When European peasants used the scratch plough, land was distributed in squarish fields designed for the support of one family. But with the introduction of the deep plough, teams of people were required to work them, and the strips ploughed were distributed in proportion to the contributions of the peasant to the team. Rather than distribution being of the amount of land required to supply people's needs, distribution came to be based on the power to till the soil. As White wrote of this: 'No more fundamental modification in a man's relation to his environment can be imagined: he ceased to be nature's child and became her exploiter.'[60]

Later the advances in human control of nature came to reflect the image of humanity as participating with God in the process of creation, with economic progress coming to be seen as paralleling the progress towards salvation. The capacity of humanity to transcend and dominate nature then came to be symbolized by human productions and the technology associated with production. By the eleventh century every peasant in Europe was living in the presence of at least one fairly complex, semi-automatic power machine: the mill. These revealed the potentiality of nature to be subjugated, and it is hardly surprising that these quickly came to be used for industrial processes other than milling.

The struggle by individuals to transcend the material world was expressed in Western art, which began to diverge from Eastern art from the beginning of the eighth century.[61] In contrast to Byzantine art which represented only passive emotions, solemn immutability beyond joy and grief in which any sign of feeling was tamed by the harmonious movement in which it expressed itself, Western art tried to represent the struggle of the will against temptation, moral conflict and the loneliness of the tortured soul amidst the indifference of the world. With the representation of the Crucifixion itself, Western works of art strove to represent action, energy and passion as something belonging not to the realm of the devil, but to the world of the divine. In place of Oriental

---

59. See James A. Billington, *The Icon and the Axe: An Interpretive History of Russian Culture*, N.Y.: Vintage, 1970, p.184.

60. White, *Medieval Religion and Technology*, p.145.

61. See Borkenau, *End and Beginning*, Part II, Ch.9 'Rome's Break with Byzantium as Shown in Christian Painting', pp.365-380.

impassivity and readiness for martyrdom, depth of inner feeling became the paradigmatic attitude, and the dynamic human personality was sanctified as a central religious factor. Correspondingly, while in Byzantine sculpture characters were only represented in relief works, Western sculpture produced complete statues, accentuating the separation of individuals from each other and their independence from their material context.

This orientation was also expressed in Western European architecture. As the feudal era advanced, Europe was covered in stone churches, castles and towers. Churches were placed in the centres of villages, towns and cities, and from the end of the tenth century their height was steadily increased. This took a quantum leap with the development of pointed arches and vaults at Saint-Denis between 1135 and 1144, the first true Gothic church.[62] The soaring spires, flying buttresses and elongated vaulted arches of Gothic architecture symbolized the movement upward to incarnate ultimate values, to rise above and dominate the surrounding world. While this development marked a radical divergence with the architecture of the Orthodox Church, its distinctive nature was even clearer by contrast with Chinese architecture. Dominated by the principle of *feng-shui*, Chinese buildings were designed to fit into the landscape.[63]

The tendency for Western Europeans to orient themselves to a transcendent order manifested itself in social life with the development of law. The hierarchical organization of society with power conceived of as flowing downwards had first led to the diminution of the significance of folk law based on populist consent in relation to law promulgated by the pope or king. This was consolidated by the recovery of Roman law in the late eleventh century. But Roman law had never been integrated into a system. When Cicero had proposed to the Romans that law be systematically organized, the jurists were not interested. As Harold Berman argued: 'They had no reason to transform the Roman genius for consistent adjudication into a philosophical system.'[64] But in the medieval universities of Europe the application of the twelfth century scholastic technique of reconciling contradictions and deriving general concepts by analysis and synthesis led to precisely such a development. A science of law was developed by Western jurists in this way at precisely the same time as Western theologians were creating what they conceived to be a science of theology. Beginning about 1140 with the great treatise *A Concordance of Discordant Canons* of the Bolognese monk, Gratian, law was coordinated and integrated into a body or system of law for the first time in history.[65] Law came to be conceived of as 'an organically developing system, an ongoing, growing body of principles and procedures, constructed - like the cathedrals - over generations and centuries'[66] developing towards the incarnation of ultimate values, that is, natural law and eternal law - the 'ideal of divine wisdom considered as directing all actions and movements.'[67] As such, its development became the prototype for Western science. This transcendent system of law was first developed in the Church as canon law, but this became the model for the secular, temporal society, which developed systems of royal law, manorial law, feudal law, urban law and mercantile law. Such law eventually became the basis for defining all relations between all orders of people so that these relations came to be seen as rights and obligations between legal subjects, defined in terms of an abstract set of principles existing independently of people.

---

62. White, *Medieval Religion and Technology*, p.233.

63. Yi-Fu Tuan, 'Discrepancies Between Environmental Attitude and Behavior: Examples from Europe and China', reprinted in *Man, Space, and Environment*, ed. Paul Ward English and Robert C. Mayfield, New York: Oxford University Press, pp.68-81, p.74.

64. Harold J. Berman, *Law and Revolution: The Formation of the Western Legal Tradition*, Cambridge: Harvard University Press, 1983, p.139.

65. This is argued by Berman, ibid. p.9.

66. Ibid. p.119.

67. Thomas Aquinas, *Summa Theologica*, Question 93, Art. 1, concl.

The development of this transcendent order of law facilitated the development of the monetary economy. While money was not unique to Western Europe, the individualism associated with the tendency to define the world from the perspective of a transcendent order, which as we have seen was originally formulated on the analogy of money's relationship to people, was uniquely propitious for the development of a monetary economy. To begin with, there was much resistance to the use of money. From the late eighth century the Carolingians had enacted laws to enforce the acceptance of coin as tender, prescribing fines or flogging for persons refusing payment in the government's pennies.[68] The repetition and amendment of these laws suggests that they were not entirely effective. It was not until 1100 that money was fully established in the centres of Europe and had begun the long process of penetrating all parts of economic life. But once established, the expansion of the monetary economy met with less resistance than in other civilizations, even taking into account Church objections to simony and usury. Law in particular provided the conditions for this expansion, as it provided a universalist, temporally transcendent framework to define relationships between people. This enabled the status of property and the rights and obligations of people involved in commercial transactions to be clearly and unambiguously defined and enforced.[69] This monetary economy then reinforced this universalist, temporally transcendent framework as the reference point for defining social relationships, further reinforcing the individualism, the fragmentation of relations between people and the domineering attitude towards nature.

This is evident even among those who were least involved in the monetary economy, the peasants. The individualism of Western Christianity nurtured by the growth of the monetary economy gradually undermined their bonds of kinship which were being assiduously preserved in most other parts of the world. Marc Bloch pointed out that in early feudal society the legality of individual possession was severely constrained by the solidarity of kinship relations which extended to a community of goods.[70] These communal households provided security for their individual members, and were held collectively responsible for the payment of dues or the behaviour of their individual members. Because of the constraints of kinship, landed property was very seldom sold prior to the twelfth century, and after this was only sold after being first offered to kinsmen, or by giving kinsmen the right to take the place of any buyer. But by the fifteenth century, for a large proportion of the peasant population, commerce was replacing kinship as the defining basis of human relationships, and many individuals found themselves struggling for a livelihood against the forces of the market. And land was beginning to become just a commodity to be evaluated according to its profitability.

In the later part of the Middle Ages, the Western orientation to the world attained a most significant expression in the development of the mechanical clock. The measurement of time had originally become important for the regulation of monastic life. The culmination of the efforts to measure time with the development of the mechanical clock enabled people to measure time independently of the natural rhythms of nature. When mechanical clocks were first developed this relationship to natural rhythms was not completely severed, and clocks were shown being adjusted to conform to the different lengths of hours in the day in different seasons. But in the late fourteenth and early fifteenth centuries Europe abandoned experiential time for abstract time, the cycle of night and day conceived as twenty-four hours of equal length, and organized their lives accordingly.[71] This manifested to the highest degree the European orientation away from the sensible world to an abstract, transcendent world, and then the organization of the sensible world to accord with this transcendent order.

This orientation to a transcendent order established the trajectory of European technological development which has continued to the present. The permeation and domination of life by abstract

68. Alexander Murray, *Reason and Society in the Middle Ages*, Oxford: Clarendon Press, 1978, Ch.2, 'Money', p.37.

69. See Berman op.cit. Ch.11, 'Mercantile Law', pp.333-356.

70. Bloch, *Feudal Society*, Vol.1, p.130ff.

71. Ibid. p.197f.

time has become so complete that it is difficult to realize just how extraordinary this is.[72] Its uniqueness stands out when contrasted with the Nuer described by Evans-Pritchard:

> ...the Nuer have no expression equivalent to 'time' in our language, and they cannot, therefore, as we can, speak of time as though it were something actual, which passes, can be wasted, can be saved, and so forth. I do not think that they ever experience the same feeling of fighting against time or of having to co-ordinate activities with an abstract passage of time because their points of reference are mainly the activities themselves, which are generally of a leisurely character. Events follow a logical order, but they are not controlled by an abstract system, there being no autonomous points of reference to which activities have to conform with precision.[73]

The clock subsequently became the primary symbol and metaphor for the Western orientation to the world. Its use for this purpose first occurred with the representation of the virtue of Temperance. When she was first conceived as a measurer of time, she was shown holding a sand-glass. But in 1400 she was shown holding a clock. In the same year Christine de Pisan wrote a treatise explaining this: 'Temperance should be called a goddess likewise. And because our human body is made up of many parts and should be regulated by reason, it may be represented as a clock in which there are several wheels and measures. And just as the clock is worth nothing unless it is regulated, so our human body does not work unless Temperance orders it.'[74] In 1450 Temperance was represented wearing a clock as a hat, a bit and bridle in her mouth, eyeglasses in her hand, rowel spurs on her heels, and standing on a windmill. White wrote of this, 'The clock indicates regularity, promptitude and reliability; the spurs, maturity; the windmill, steady industriousness.'[75]

The development of the abstract conception of time was followed by the development of an abstract conception of space through the development of maps for navigation.[76] The characteristics of European thinking in this regard are again revealed most clearly when contrasted with other societies; in this instance with the Polynesians whose ability at navigation at least equalled that of the Europeans. The Polynesians navigated by understanding the dynamic relations within nature. They judged both their direction and the presence of land masses by the wave patterns of the ocean. They could detect distant islands by their effects on clouds and light. They knew how far birds travelled from land and when they were heading away from or towards it, and used this knowledge accordingly. Where stars were used for navigation, the Polynesians oriented themselves by knowledge of star paths over the horizon, with changes of direction timed to moving locations. With this mode of reading stars, all constants were dynamic and temporally changing. In effect the Polynesians navigated by orienting themselves to the world they were engaged in.

By contrast, when Christopher Columbus set out in 1492 he took with him instruments such as the magnetic compass, the clock, the compass, the astrolabe for calculating angles to the sun and other heavenly bodies, maps and various measuring devices. He conceived the world to be round, with the surface as a network of intersecting lines upon whose surface one moved and in terms of which ones' position could be plotted by means of navigating instruments. He oriented himself to an abstract order of space, and his relation to the world around him was mediated through this abstract conception of the world and the instruments which served this mediation.

---

72. For this development see E.P. Thompson, *Past and Present*, Vol.38, 1967, pp.56-96.

73. E.E. Evans-Pritchard, *The Nuer,* Oxford: Clarendon Press, 1940, p.103.

74. Cited by White, *Medieval Religion and Technology*, p.193f.

75. Ibid. p.199.

76. See Don Ihde, 'The Historical-Ontological Priority of Technology over Science', *Philosophy and Technology*, eds Paul T. Durban and Freidrich Rapp, Dordrecht: Reidel, pp.235-52, p.248f. For the development of maps see Bailey W. Diffie and George D. Winius, *Foundations of the Portuguese Empire: 1415-1580*, Minneapolis, University of Minnesota Press, 1977, p.129ff.

## *Emergent Dynamics of Medieval Society*

The embodiment of the medieval world-orientation already gave this an inertia and dynamics not entirely intended by the actors who lived according to it. However there were other side effects totally unintended, and processes generated which once established, forced people to conform to, and so reproduce, these dynamics. It is these unintended aspects of medieval culture which transmuted the individualism, the detachment from the surrounding world and the activism of Western European culture into a grotesque struggle for domination of people and nature.

The most significant side effect of this culture was to produce a high level of anxiety which increasingly expressed itself in aggression. In the Dark Ages ninety per cent of the population were directly involved in agriculture. With the advances in agriculture in the tenth and eleventh century increasing proportions of the population left the soil to live in the strange environment of the expanding towns and cities. Unlike China where people on reaching maturity were granted land, where efforts were made to keep people on the land, and where towns and cities developed in a culture in which the importance of acknowledging people's significance (expressed in such practices as always allowing people to 'save face'), the towns and cities of Europe simply accumulated all surplus population, and in a culture stressing individualism, did little to replace the social relations of the rural community.[77] This urbanization broke the community bonds of peasant society, and this appears to have profoundly affected people. As Lynn White wrote, 'We are beginning to see that the eleventh century in the West was an age of dreadful anxiety, and consequently of aggression.'[78] This anxiety manifested itself towards the middle of the century in the sudden invention of Purgatory and indulgences, the evolution of the image of Satan from a fallen angel into a hideous monster, the realistic depiction of the tortures of Hell for the first time, and the killing of Jews and heretics. The distinctive general orientation towards the world produced by such anxiety was an aggressiveness which infected the orientation towards domination and had ramifications for the whole of European history.

The most important expression of this aggressiveness was to intensify concern with warfare. In the middle of the eleventh century, participation in slaughter, 'washing the sinner in the blood of unbelievers rather than of Christ,' came to be seen as a means of salvation.[79] This mutation of Christian values was a precondition for the crusades. The Crusaders' apocalyptic frenzy and the plethora of psychotic incidents: the episodes of Peter the Hermit, the Children's Crusade and the Crusade of the Shepherds among others attest the intensity of the efforts to regain the lost sense of community at this time. The crusades in turn reinforced the militarism of the aristocracy. While such military activity was justified by Christianity as the pursuit of honour for the greater glory of God, more than the greater glory of God was involved. War became a passion, and the brutality which had, according to Christianity, been justly directed against the Saracens in the Holy Land and against the Baltic nations whom the Teutonic knights set out to 'Convert or Kill' (they killed almost all the original inhabitants of Prussia), was turned inwards and directed at other Europeans.

Warfare associated both with the crusades and between kingdoms forced the further feudalization of social relationships. Feudalism had originally involved expropriating land from monasteries and granting it to individuals on condition that they equip themselves to serve as heavy cavalry in the army to defend Europe. Towards the end of the eleventh century this form of relationship between people, the relationship of lord to vassal, spread. Monarchs began to regard their kingdoms as their property, and all people who worked it, as their tenants. In this way they were able to raise more knights, more fiefs and more castles. Following William the Conqueror,

77. For a description of life in European cities see J.K. Hyde, *Society and Politics in Medieval Italy: The Evolution of the Civil Life, 1000-1350*, London: Macmillan, 1973; and John Hine Mundy, 'Urban Society and Culture: Toulouse and Its Regions' in Robert L. Benson and Giles Constable eds, *Renaissance and Renewal in the Twelfth Century* pp.229-47.

78. White, *Medieval Religion and Technology*, p.295.

79. Ibid. p.294.

regents systematized their taxation to exploit their subjects as efficiently as possible. There was also an increase in the number of levels of aristocracy, with power of jurisdiction, to try and to tax the subject population devolving downwards not only to counts, but to castellans and even to lords of one or two villages. At the lower levels of feudal society large scale technology was developed as a means of expropriating surpluses from the peasants. For instance the lords built wind and water mills and forced the peasants to use these by making handmills illegal.[80] Increasingly expropriation from peasants in the form of labour services were replaced by expropriation of money by a variety of different methods. Because these new forms of relationships increased the military power of the ruling class, regents who opposed these developments were eventually forced to go along with the trend in order to survive. Once inaugurated, the development of feudalism was inexorable.[81]

The dynamic which generated and perpetuated feudal relations was the struggle to find the means to wage warfare in a war prone society. The resulting hierarchical order of society was generally legitimated by the Neoplatonic Christian ideology, and the hierarchy having been formed the heavily armed ruling élite were in a position to maintain their privileges. However there was no inner dynamic generated by the feudal socio-economic system maintaining and developing it equivalent to that of capitalism. The aristocracy attempted to increase their power by expropriating as much as they could from the peasantry or by conquest, but they invested almost nothing in the improvement of the land. As the opportunities for conquest diminished, the life of the military aristocracy evolved into mere pageantry, a process graphically described by Huizinga in *The Waning of the Middle Ages*. The long term significance of feudalism was to force the pace of commercialisation of human relationships. Land came to be seen in terms of how much revenue could be produced from it, and peasants were forced to think in terms of how to raise the money to pay the various impositions of the ruling class. Monarchs also supported the development of commerce as a source of taxes, and universities as a source of trained manpower. This provided the conditions under which commercial and university towns and cities could develop and achieve considerable autonomy from the rest of feudal society, and absorbing the growing number of peasants escaping from their feudal overlords, could sustain a growing commercial economy.

With an association between piracy and commerce in early medieval Europe and with the general aggressiveness of European society, townspeople displayed a remarkable belligerence and propensity for waging war. It was in the towns and cities that people could rise in the world. However the situation of such people was extremely precarious. Those involved in trade frequently lost everything, while there was always an excess of university graduates over employment opportunities. The enormous gulf between the highest and the lowest levels of society meant that there was widespread insecurity. One manifestation of this was the popularity of the Wheel of Fortune in medieval illustrations.[82] This insecurity generated competition between those who were attempting to rise on the social scale. The vicious debates within universities was one sign of the ferocity of this competition.[83] Correspondingly, there took root and flourished a remarkable merger of market and military behaviour in the most active economic centres. William McNeill wrote of this: 'Commercialization of organized violence came vigorously to the fore in the fourteenth century when mercenary armies became standard in Italy. Thereafter, market forces and attitudes began to affect military action as seldom before. The art of war began to evolve among Europeans with a rapidity that soon raised it to unexampled heights.'[84] It was in the context that a systematic preoccupation with predicting and controlling the world emerged.

80. Merchant, *The Death of Nature*, p.46.

81. See R.H.C. Davis, *A History of Medieval Europe: From Constantine to Saint Louis*, Harlow: Longman, 2nd ed. 1970, p.295f.

82. Alexander Murray, *Reason and Society in the Middle Ages*, Oxford: Clarendon, 1978, p.98ff.

83. See ibid., Ch.10, 'The Intellectual Elite.'

84. William H. McNeill, *The Pursuit of Power*, Chicago: University of Chicago Press, 1984, p.69f. McNeill also wrote of how such a development had been prevented by the Confucian literati. See p.'s 40 & 50.

This preoccupation was not just a concern to relieve the burden of work, or even to dominate militarily. It became a passion to subjugate the world as an end in itself. The power of the Church came to rest more and more on its supposed power to mediate in people's fate in the afterlife. A very high proportion of the efforts to develop technology were directed towards the production of armaments. However those involved in the development of armaments to subjugate people also tended to be involved in efforts to develop astrology and magic in order to subjugate nature. Typically Konrad Kyeser who produced a major text in military engineering - in which among other things he recorded an instrument for slow castration - was also an astrologer and practising magician, employing for one incantation candles made of fat from a hanged man.[85] The orientation towards domination is most clearly evident in this art of magic. Books of magic became increasingly popular in the thirteenth century. These books claimed to give total power to their readers. As one claimed, when the principles expounded were comprehended, 'all the intelligences and all the compositions of the things of this world, all things will serve him and he will serve none of them.' And another widely copied book ended: 'so ends the book of the rational soul ... the book by which every creature can be subjected except the nine orders of angels.'[86]

## *The Final Stage of Feudalism*

What I have tried to demonstrate here is how the orientation towards individualism, detachment and domination engendered by medieval society amounted to a particular type of embodiment of the philosophy of Neoplatonic Christianity. It had become a mode of becoming in the world, characterized by an emotional detachment from the sensible and personal world of the present as people defined themselves in relation to a perfect, eternal order, and with the reduction of the surrounding world to a means to attain the perfection of this eternal order in some distant future. Both the conceptions of and the attitudes towards nature and society reflected and thereby reinforced each other. All the institutions of medieval society, including language, resonated with this basic orientation and with the general conception of the world of Neoplatonic Christianity which stood as the ultimate foundation for the legitimation of beliefs and practices and for the resolution of conflicts. However the side-effects of this institutionalisation, the anxiety produced by the disengagement from the surrounding world and from other people infected this mode of becoming with aggression.

This anxiety and aggression manifested itself in hostility both to the natural world and other people, especially insofar as such people could be identified with the natural world. The rejection of nature found expression in the efforts of people to distinguish themselves from animals and to transcend their affinities with them.[87] In the vituperative writings of intellectuals, ignorance was equated with brutishness and knowledge was seen to raise one above bestiality. Proponents of courtly love, and particularly homosexual love also defined themselves in opposition to animals. A devotee of homosexual love wrote that 'only rustici ... who can be called beasts, should properly ... filthy themselves with women.'[88]

This individualist detached orientation facilitated the emergence of commercial capitalism which, despite the proscriptions against usury, was to eventually displace the feudal organization of society. This was characterized by a passion for the accumulation of wealth, particularly in the immutable form of gold. Associated with its development, nationalism began to emerge in some regions as a dynamic force, particularly towards the end of the fifteenth century when central

85. White, *Medieval Religion and Technology*, p.306ff.

86. Murray, *Reason and Society in the Middle Ages*, p.113f.

87. The extension of this into the sixteenth and seventeenth centuries is described by Keith Thomas in *Man and the Natural World*, Harmondsworth: Allen Lane, Ch.1.

88. Cited Murray, *Reason and Society in the Middle Ages*, p.238.

governments subordinated the cities. This was fostered as a means for mobilizing people more effectively for warfare. Such nationalism replaced the frenzy of the crusades as a substitute means of attaining a sense of being a significant member of a community. Commercial capitalism and nationalism in turn became the foundations of the rise of Europe's intercontinental imperialism (although to begin with, such imperialism was formulated more in terms of religion).

It was this development in which modes of cognition and behaviour were applied in new situations involving people from other civilizations that most clearly revealed the distinctive nature of the European culture. In 1405 a Chinese fleet of 63 ocean going junks with almost 28,000 men aboard had set sail and had visited many parts of the south seas, including Ceylon.[89] Over the next twenty seven years, seven such expeditions set forth, reaching and bringing back animals from as far away as Africa. But the Chinese ruling elites were not interested in conquest. Typifying the attitude, a minister, Fan Chi, wrote to the emperor in 1426:

> Arms are the instruments of evil which the sage does not use unless he must. The noble rulers and wise ministers of old did not dissipate the strength of the people by deeds of arms. This was a far-sighted policy.... Your minister hopes that your majesty ... would not indulge in military pursuits nor glorify the sending of expeditions to distant countries. Abandon the barren lands abroad and give the people of China a respite so that they could devote themselves to husbandry and the schools.[90]

In 1436 a decree was issued forbidding the construction of new seagoing ships. The Pope on the other hand divided the non-European world between the Portuguese and the Spanish. The Portuguese set out to conquer the trade of the Indian Ocean and the Spanish to conquer the civilizations of America, beginning the era in which succeeding European colonialists set about conquering, subjugating and largely destroying the rest of humanity, and paving the way for the present world-order.

When the Portuguese found a way into the Indian Ocean via the Cape of Good Hope, there was a well organized and relatively peaceful system of trade dominated by the Muslims. These Muslims had traversed the seas from Arabia to Indonesia and China since the eighth century, their adventures forming the substance of Sindbad the Sailor's tales in the *Arabian Nights*. The Portuguese had little to trade, and simply wanted to divert trade to Portugal. Consequently they set about destroying what had been a competently organized native economy, routinely demolishing Arab, Egyptian and Gujarati vessels. Their manner of proceeding in this task of conquest is illustrated by the way Vasco da Gama dealt with the ruler of Calicut, the Samorin. Da Gama insisted not only that Portugal should have access to the port, but that all Muslims be excluded. Samorin was willing to admit the Portuguese, but not to exclude the Muslims. Winius, an historian of the foundations of the Portuguese empire wrote of da Gama's response:

> Da Gama's response must have taken the Semorin completely by surprise: the admiral suddenly gave the order to hoist anchor, swung his ships in close to shore, and opened fire, hurling stone and metal cannonballs into the city's streets for an entire day before sailing off towards Cochin. In addition to this barbarous act, which killed indiscriminately, he committed one even more appalling: he butchered or burned alive several hundred innocent fishermen who had assumed

89. Wang Gungwu, 'China and South-East Asia 1402-1424' in J. Ch'en and N. Starling eds. *Studies in the Social History of China and South East Asia*, Cambridge: Cambridge University Press, 1970.

90. Cited by McNeill, *The Pursuit of Power*, p.46. For more on the difference between Portuguese and Chinese attitudes to imperialism see Robert Finlay, 'Portuguese and Chinese Maritime Imperialism', *Comparative Studies in Society and History*, Vol.34, No.2, April 1992, pp.225-241.

that peace was in the offing and had sailed out beyond the Portuguese fleet to pursue their day's catch.[91]

Later the Portuguese sacked and destroyed every city along the Arabian coast which refused to put itself under Portuguese protection.[92]

In these military engagements the Portuguese were distinguished by their courage and ferocity. As Winius wrote in relation to their conquest of Malacca, 'As at Goa, the Portuguese soldiers in Malacca were experienced in fighting as a team; they cut throats with a joy no Asians could match (save perhaps the Bandanese), and their group psychology made it a point of pride to vie with one another in discounting their injuries and fighting on... Men, like the Portuguese, who tore into their enemies with obvious relish, were a novelty in Malacca and in most of the Indian states.'[93]

However it was the Spanish conquest of the American civilizations which revealed the full destructiveness of European civilization. In murdering the rulers of these civilizations who had accepted them as guests, enslaving their populations and destroying their cultures, the Spanish conquistadors displayed a seldom matched degree of treachery, brutality, and self-righteousness. It resulted in the population of Mexico being reduced from 25 to 30 million at the time of the conquest to three million in 1568, less than fifty years later, and to 1.6 million in 1620.[94] The 7 to 8 million population of Hispaniola (Santo Domingo) had been reduced to less than 300 in 50 years, the Spanish having produced a death rate of 40% a year. Ultimately the only Indian survivors were a few half-breeds.[95]

There has been a tendency to whitewash this by describing the destruction of life as simply due to the introduction of disease. To begin with it is not known to what extent steps which could have been taken to stop the spread of diseases were ignored, or even that the spread of disease was not deliberate. It is known that the English settlers in North America gave blankets taken from people who had died of smallpox to the Indians, and that scabs from people with smallpox had been taken to Australia by its first white settlers, following which the aboriginal population was devastated by smallpox. Apart from this, people's susceptibility to disease is largely a function of the state of mental and physical health. For instance it is now being argued that it was malnutrition which paved the way for the plagues in fourteenth century Europe, and that the plagues decreased with improvements in food.[96] The high susceptibility of the Indians was not entirely lack of resistance. It is necessary to take into account that an independent people had been reduced to slavery with many being forced to work in mines under atrocious conditions. Furthermore it was not simply death which caused the collapse of populations. Henry Reynolds has pointed out that Australian aboriginals, and Mark Twain that the Congolese, avoided having children because of the insecurity, the threats and the conditions they were forced to endure as a consequence of the European invasion.[97] And lest the propensity of Europeans for destruction be underestimated, it should be recalled that at the end of the nineteenth century the Belgians killed 10 million of the 25 million population of the Congo, while in 1904 the Germans in South-West Africa almost completely exterminated the Herero people; in both cases without the help of disease.

The conquest of the Inca civilization, described in the famous work by William H. Prescott, *History of the Conquest of Peru* illustrates the manner in which the Spanish proceeded. Prescott

91. Bailey W. Diffie and George D. Winius, *Foundations of the Portuguese Empire: 1415-1580*, Minneapolis: University of Minnesota Press, 1977, p.224.

92. Ibid. p.238.

93. Ibid. p.259.

94. William H. McNeill, *Plagues and Peoples*, N.Y.: Anchor Press, 1976, p.203f.

95. Emmanuel Le Roy Ladurie, *The Mind and Method of the Historian*, [1978] tr. Sian Reynolds and Ben Reynolds, p.79.

96. Thomas Cox, 'The Impact of Western Expansion on World Ecosystems: Comment' in *Environmental History* ed. Kendall E. Bailes, Lanham: University Press of America, 1985, pp.587-93; reporting the work of Andrew B. Appleby.

97. Henry Reynolds, *The Other Side of the Frontier*, Harmondsworth: Penguin, 1982., p.125ff.

described the land of the Inca empire through which Pizarro and his band of would be conquerors travelled, and the nature of their advance:

> The industry of the inhabitants ... had turned these streams to the best account, and canals and aqueducts were seen crossing the low lands in all directions, and spreading over the country, like a vast network, diffusing fertility and beauty around them. The air was scented with the sweet odours of flowers, and everywhere the eye was refreshed by the sight of orchards laden with unknown fruits, and of fields waving with yellow grain and rich in luscious vegetables of every description that teem in the sunny clime of the equator. The Spaniards were among people who had carried the refinements of husbandry to a greater extent than any yet found on the American continent... Everywhere, too, they were received with confiding hospitality by the simple people; for which they were no doubt indebted, in a great measure, to their own inoffensive deportment. Every Spaniard seemed to be aware, that his only chance of success lay in conciliating the good opinion of the inhabitants, among whom he had so recklessly cast his fortunes.[98]

Having successfully deceived the Inca people as to their intentions and having been accepted into the presence of their leader, Atahualpa, the Spaniards prepared to attack. Having completed the military preparations, 'mass was performed with great solemnity by the ecclesiastics who attended the expedition; the God of battles was invoked to spread his shield over the slaves who were fighting to extend the empire of the Cross; and all joined with enthusiasm in the chant, "*Exsurge, Domine*," "Rise O Lord! and judge thine own cause."...[W]hatever were the vices of the Castillian cavalier, hypocrisy was not among the number. He felt that he was battling for the Cross... With feelings thus kindled to a flame of religious ardour, the soldiers looked forward with renovated spirits to the coming conflict...'[99]

After having lured Atahualpa unarmed into the Spanish camp, and before the attack, Valverde, a Dominican friar and chaplain to Pizarro, approached him, expounded to him the doctrines of the true faith and asked him to acknowledge himself a tributary of the Emperor, Charles the Fifth. Atahualpa showed offence at this, describing the Pope as being 'crazy to talk of giving away countries which do not belong to him.'[100] He declined to accept Christianity and affirmed his faith in his own God, the Sun 'which lives in the heavens, and looks down on his children.'[101] The friar hastened back to Pizarro, informed him of Atahualpa's response and exclaimed: 'Do you not see, that, while we stand here wasting our breath in talking with this dog, full of pride as he is, the fields are filling with Indians? Set on, at once; I absolve you.'[102]

The attack began, the unarmed retinue of Atahualpa was slaughtered and Atahualpa captured. Prescott wrote of this: 'The number of slain is reported, as usual, with great discrepancy. Pizarro's secretary says two thousand natives fell. A descendant of the Incas - a safer authority than Garcilasso - swells the number to ten thousand.'[103]

Having captured Atahualpa, the Spaniards used him as a means of protection, and then as a means of attaining gold and silver from the empire. Pizarro agreed to release him if he would fill a room twenty-two feet by seventeen feet with gold to a depth of nine feet, and to fill a smaller adjoining room twice with silver. An immense amount of gold and silver arrived, but Atahualpa was not released. When the Spaniards had no further use for him he was sentenced to death by

---

98. *The Complete Works of William H. Prescott: History of the Conquest of Peru*, Vol.1, New York: Thomas Y. Crowell & Co., 1847, p.259f.

99. Ibid. p.290.

100. Ibid. p.296.

101. Loc.cit.

102. Ibid. p.297.

103. Ibid. p.300.

burning. A copy of the judgement was submitted to the friar, Valverde for his signature, which he gave without hesitation, declaring that, 'in his opinion, the Inca, at all events deserved death.'[104] This was commuted to death by garotting after Atahualpa had agreed to become a Christian. The execution was carried out on August 29, 1533. The following morning, his funeral obsequies were performed with great solemnity, Pizarro and the cavaliers going into mourning and the troops listening with devout attention to the service for the dead delivered by Father Valverde.

Following this the Spaniards continued their pillaging of the empire, burning its villages and cities and torturing its inhabitants in their rapacious search for gold, and any Indian leader who attempted to stand in their way was burnt alive. In this way an entire civilization was totally destroyed. The Inca empire originally had a population of seven or eight million (perhaps as many as ten million). By 1560 the figure was two and a half million, and by 1590 between 1.3 and 1.5 million.[105]

While behaviour of Europeans towards the rest of the world revealed the nature of Europeans most clearly, this was merely the generalization of a mode of behaviour and thought which had come to permeate European civilization. In the same century as Spain destroyed the civilizations of the Americas, the inquisition in Europe developed into an orgy of sadistic, mass torture and murder. Witch-hunting became a craze, revealing an unparalleled degree of misogyny. Efforts to subjugate nature were expanded as never before, and associated with this, the lower classes were exploited with a new level of intensity. This gave rise to widespread peasant riots and wars. All these developments resonated with each other as heretics, women, peasants and nature were identified with each other as evil, disorderly matter which must be brought under control and made to accord with a higher principle of masculine, rational order.[106] Increasing social tensions throughout Europe culminated in the seventeenth century with the Thirty Years War of 1618 to 1648 which decimated Europe's population.

However these developments inaugurated a new social order and produced forms of life and ideological struggles which destroyed the coherence of the medieval world-orientation. The new social order, capitalism, was nevertheless a continuation of the developments which had been taking place in feudal society. It institutionalized atomistic thinking, individualism and the degraded status of nature and of people. But it also introduced new problems. It is an order in which the struggle for economic security produces effects which deprive people of such security. It is associated with a form of science in which each advance in knowledge further disorients people. And it has developed in such a way that the efforts of people to attain recognition of their own significance undermines the institutions which accord such recognition. So while this order originated in and was based on the modes of being and thinking of feudal society, it led to radically new developments. It is the nature of these developments which has given rise to the modern world-orientation, and it is the relationship between this world-orientation and the rise of world capitalism which will be examined in the next chapter.

---

104. Ibid. p.342.

105. Le Roy Ladurie, *The Mind and Method of the Historian* p.77.

106. See Easlea, *Witchcraft, Magic & the New Philosophy* for a detailed analysis of this.

# 5

# MECHANISTIC MATERIALISM AND CAPITALISM: THE ORIGINS OF NIHILISM

One of the most significant manifestations of the dissolution of the feudal order was the demise of the values of the aristocracy. With the development of nominalism which denied reality to universals, the ideal forms which had defined all human significance for the aristocrat lost their ontological status. As Shakespeare has Falstaff say in *King Henry IV*:

> Honour pricks me on. Yea, but how if honour prick me off when I come on? how then? Can honour set-to a leg? No. Or an arm? No. Or take away the grief of a wound? No. Honour has no skill in surgery, then? No. What is honour? A word. What is that word, honour? Air.[1]

However the dualism between forms and the sensible world which had been argued for by Plato in his efforts to establish the reality of eternal values had come to structure the whole culture of Western civilization. While the rise of nominalism undermined some aspects of Platonism, Western culture remained essentially Platonistic in two ways.[2]

Firstly, the dualism between a real, eternal world and a changing, sensible world was not only retained, but reinforced. The real world came to be seen as a mechanical order of inert, immutable matter governed by the immutable laws of motion represented by timeless logico-mathematical relations, while all reality was denied to the qualitative diversity of the sensible world. This amounted to a complete triumph of being over becoming, with all immanent tendencies to realize potentialities being eliminated from the world. Only the totally formed were granted the status of reality, so that the world came to be seen as consisting of bodies occupying space and changing their positions over time - with an active soul or mind an incomprehensible and optional extra.

The assumption of this eternal reality of space and time, matter and motion and the laws of science, was completely taken for granted even by those whose extreme nominalism seemed to contradict it. For instance David Hume assumed a world of atomic events in which the only relation is the observed constant conjunction between similar, contiguous events. But, as Kant realized, this requires an eternal realm of time and space against which events can be mapped. Similarly, despite his arguments that the mind is nothing but a sequence of sense impressions and ideas, Hume had to assume an enduring subject who could *expect* sense impressions of events to be constantly conjoined. Only by assuming such an enduring background to events could Hume avoid facing the question confronted by Plato of how there could be knowledge over and above perception in a changing world.

Secondly, despite appearances to the contrary, Platonism was retained in a modified form in the way people defined value and significance by participation in forms. The forms of virtue such as justice, valour, courtesy and prudence lost their status, but were replaced by the purely numerical form of money and by the form of the machine. Only that which could be valued in terms of money

---

1. William Shakespeare, *King Henry IV,* Part I, V.i., 126-36, *The Complete Works of Shakespeare*, London: Octopus Books, p.456.

2. The Neoplatonism of the mechanical view of the world has been argued by a number of philosophers and historians of science from Burtt and Koyré to Ivor Leclerc. This continuity is most clearly evident in Spinoza who calls for love to be directed away from what is perishable to what is eternal and infinite, the deterministic order of the world. (See Benedict de Spinoza, *On the Improvement of the Understanding, The Ethics, Correspondence* tr. R.H.M. Elewes, N.Y.: Dover, 1955, 'On the Improvement of the Understanding', p.5 & p.37.)

was thenceforth seen to be significant. The most highly valued things came to be those which maintained their value, and these became the goal of economic activity. As the seventeenth century economist William Petty wrote: 'The great and ultimate effect of trade is not wealth as such, but preferably an over-abundance of silver, gold and jewels, which are not *perishable*, not as *fickle* as other commodities, but are wealth in all times and all places.'[3] The continued connection between this concern for immutable forms of wealth and Neoplatonic Christianity is most clearly evident in Protestantism. Luther generalized the notion of vocation to secular life, and following Calvin, Protestants regarded the notion of vocation as applicable to business. But the notion of a vocation did not lose its religious roots. As Marx wrote: 'The cult of money has its asceticism, its self-denial, its self-sacrifice - economy and frugality, contempt for mundane, temporal and fleeting pleasures; the chase after the *eternal* treasure. Hence the connection between English Puritanism, or also Dutch Protestantism, and money making.'[4] And in the pursuit of money, the ideal type of order to be adopted by individuals and by society, came to be that of a machine - a totally predictable system made up of unifunctional, replaceable parts.

With this development nature came to be regarded as significant only insofar as it could enter the realm of monetary relations as a resource, relationships between people came to be seen as significant insofar as they were monetary relationships, and people came to be defined in terms of their participation in the economic machine. As Marx pointed out:

> That which exists for me through the medium of money, that which I can pay for ... that am I, the possessor of the money. The stronger the power of my money, the stronger am I. The properties of money are my, the possessor's, properties and essential powers. Therefore what I am and what I can do is by no means determined by my individuality. I am ugly, but I can buy the most beautiful woman. Which means to say that I am not ugly, for the effect of ugliness, its repelling power, is destroyed by money... I am a wicked, dishonest, unscrupulous and stupid individual, but money is respected, and so also is its owner. Money is the highest good, and consequently its owner is also good.[5]

And as relations in the Middle Ages had been understood as secondary and derivative of the ideal forms of religious thought, in the new order relationships have come to be understood in monetary terms, that is, as commodities, whether in the form of resources saleable or prepared for use in production, labour power to be exploited, or the products of industry. Again as Marx, who rejected both modes of thinking, pointed out:

> A commodity appears, at first sight, a very trivial thing, and easily understood. Its analysis shows that it is, in reality a very queer thing, abounding in metaphysical subtleties and theological niceties... because the relation of the producers to the sum total of their own labour is presented to them as a social relation, existing not between themselves, but between the products of their labour... In order, therefore to find an analogy, we must have recourse to the mist-enveloped regions of the religious world. In that world the productions of the human brain appear as independent beings endowed with life, and entering into relation both with one another and the human race. So it is in the world of commodities...[6]

---

3. Petty, *Political Arithmetick*, p.178; cited by Karl Marx, *Grundrisse: Foundations of the Critique of Political Economy (Rough Draft)*, tr. Martin Nicolaus, Harmondsworth: Penguin, 1973, p.231.

4. Marx, *Grundrisse*, p.232.

5. Karl Marx, *The Economic and Philosophic Manuscripts of 1844*, tr. Martin Milligan, New York: International Publishers, 1964, p.167.

6. Karl Marx, *Capital*, Moscow: Progress Publishers, 1974, Vol.1, p.76f.

These two ways of conceiving the world: as a mechanical order and in terms of money have complemented each other in the development of capitalism to form the mechanistic world-orientation. The avarice which developed with the new capitalist socio-economic formation had a radically different character and relationship to society than in early commercial society. Aristotle had noted that people tended to regard money as wealth and to make its accumulation an end in itself.[7] But as Marx noted, such mania for wealth was destructive of the ancient communities since money was nothing but a means of exchange in these societies.[8] This was true also of late medieval society, and Spain was economically destroyed as a consequence of its discoveries of vast amounts of gold in the Americas.[9] What makes capitalism different is that in the quest for money, entrepreneurs employ labour power not only to produce consumer goods, but by transforming nature, to develop the means of production. For such a system to have been established a new way of thinking was required which could justify these developments, reveal how nature could be transformed, and provide an idea of what the world should be transformed into. Mechanistic materialism served all these functions, and in doing so, the notion of mechanism became more than a means for understanding the world. The idea of the machine came to complement money as the Platonic form to which nature, societies and individuals must be made to conform in order make money.

However mechanistic materialism did not just emerge as the logical outcome of medieval thought. It was developed in the process of an ideological struggle as a particular conception of the world among a number of such schemes of thought which vied for dominance as feudal society and the philosophies on which it was based lost their coherence. To begin with these circumstances provided the conditions for the flowering of a number of philosophies or world-views which had either previously been held in check by feudalism and the Catholic Church, or which emerged for the first time as the self-maintaining dynamics of feudalism weakened. Among the intellectual movements which developed were humanism, partly based on the recovery of Greek and Latin writers, particularly the Stoics, but also on the logic of Peter Ramus; and various forms of Protestantism. Humanism and Protestantism in turn led to vigorous efforts by the defenders of the old order to develop the heritage of feudal philosophy, particularly Thomism, to oppose these new developments. Another major intellectual movement which developed at this time was a radicalized Neoplatonism in the tradition of John Scotus Eriugena and the Heretics of the Free Spirit, but enriched by the then recently translated writings of Hermes Trismegistus and the Kabbala.[10] Proponents of this form of Neoplatonism, the Hermetics or 'nature enthusiasts', took God to be immanent in the world, and tended to emphasise the world's unity and dynamism. At the same time they tended to support the peasant movements and the communalistic forms of organization which some peasants had developed in late feudal society in their efforts to resist the oppression of the aristocracy. Here I will show how mechanistic materialism was developed as an effort to oppose such political movements and to defend the rising bourgeoisie; how the conception of the world as devoid of meaning was developed and promoted for political purposes. And my contention is that it was at least partly because of the success with which mechanistic materialism was synthesized into a coherent world-orientation, and its enormous success as the foundation for a new science of nature, that capitalism was able to prevail over the more radical political movements.[11]

---

7. Aristotle, *The Politics*, Bk I, Ch.ix.

8. Marx, *Grundrisse*, p.223.

9. Ibid. p.225.

10. See Brian Easlea, *Witchcraft, Magic and the New Philosophy*, Sussex: Harverster Press, 1980; Carolyn Merchant, *The Death of Nature: Women, Ecology, and the Scientific Revolution*, [1980], London: Wildwood House, 1982, esp. Ch's 3 & 4; and Frances A. Yates, *Giordano Bruno and the Hermetic Tradition*, London: Routledge & Kegan Paul, 1964.

11. The discussion of the ideological conditions preventing the emergence of capitalism in China by Max Weber in *The Religion of China*, tr. Hans H. Gerth, N.Y.: Free Press, 1951, Ch.8 provides support for this contention. See also William McNeill, *The Pursuit of Power*, Chicago: University of Chicago Press, 1984, p.'s 40 & 50.

In the remainder of this work I will describe the role of mechanistic materialism in promoting and legitimating the development of capitalism, showing its relationship not only to intellectual ideas which had been developed through the Middle Ages, but also to modes of conceiving the world which had evolved in social practices. The development of mechanistic materialism was a crystallization of a large number of such forms of thinking. Forms of relations within society provided the analogies for the development of this view of nature, which was then analogized to comprehend society. This comprehension reincorporated these forms of thinking into new social practices. The relationship between the formation of this world-orientation and the rise of capitalism will be analysed first, then the inter-relationship between the development of mechanistic materialism and the development of capitalism from the seventeenth century to the present will be examined. The final chapter will show how this world-orientation has become embodied within society and individuals.

## *The Complications of Capitalist Ideology*

Attempting to understand the development of ideology over the last few hundred years presents even more problems than attempting to understand the development of thought in the Middle Ages. There is in Europe and its colonies a great diversity of people separated by languages, regions and national boundaries engaged in a variety of forms of life, which themselves engender a diversity of modes of thought. But also the modern era has also been characterized by the emergence of 'discursive formations' and 'cultural fields' - to use the concepts of Michel Foucault and Pierre Bourdieu - with considerable autonomy from the rest of society, and there is far greater freedom of expression than in the Middle Ages. Consequently the modern world provides a picture of a vast range of viewpoints rather than of a society dominated by a unified culture. Mechanistic materialism itself has been formulated in a variety of different ways. However the situation in capitalist society is essentially the same as in feudal society. Too small a focus, both spatially and temporally, hides the existence of a real pattern in modes of thought. When Europeans are compared with traditional societies or with other civilizations, or if modern Europeans are compared with feudal Europeans, it immediately becomes apparent that the diverse points of view expressed have far more in common than at first appears, and what tolerance there is for alternative viewpoints in modern Western societies can be partially explained by the degree of entrenchment of the prevailing ideology and how ineffective is the opposition to it.

Nevertheless a diversity does exist, and to simplify this study I will focus mainly on the dominant economic powers in the world, first Britain and then later, USA. It was in England that after the decline of population in the fourteenth and fifteenth centuries the that struggle between peasants and landlords resulted in the consolidation of land-holdings, the dispossession of peasants, and the employment of people as wage-labourers, and which thereby paved the way for the establishment of industrial capitalism.[12] The struggles in France and Germany west of the Elbe led to the consolidation of a free peasantry who resisted the development of a capitalist mode of production, while east of the Elbe, feudalism established itself for the first time. Consequently it was in England and Scotland, and then later in the United States that the forms of thinking required to justify and defend the developments of capitalism were developed in their most coherent form. Germany on the other hand, as a late-comer to capitalism, provided the most critical analysis of the forms of thought deriving from Britain, while France tended to echo both English and German thinking without producing the coherence of thought of either English or German thinkers (it was a Frenchman, Cousins, who coined the term eclectic to describe his own philosophical position).

Further difficulties arise when attempting to understand the role and significance of mechanistic materialism in the formation and development of capitalism. To begin with, there is an asymmetry

---

12. For an analysis of this see Robert Brenner, 'Agrarian Class Structure and Economic Development in Pre-Industrial Europe', *Past and Present*, No.70, Feb. 1976, pp.30-75.

between those who have defended it and those whose behaviour has been influenced and legitimated by it. One of the greatest mechanists, Thomas Hobbes, was a defender of the old order and was unsympathetic to the rising bourgeoisie, while those more sympathetic to capitalism such as Newton and the Latitudinarians attacked the extreme mechanistic thought of Descartes and Hobbes and argued for a diluted version. The French revolutionaries were also opposed to the extreme version of mechanistic materialism, executing its greatest proponent of the time, Lavoisier.[13] In more recent times very few of the supporters of capitalism have espoused such materialism as a doctrine. On the other hand, many of those who have been opposed to the oppressive effects of capitalism, for instance the radical utilitarians in Britain and most of the early Marxists, attempted to develop and justify their views in terms of mechanistic materialism.

Another complication is the strong countervailing set of ideas to mechanistic materialism which have developed along with it. For instance while mechanistic materialism implies that nature is totally devoid of significance except insofar as it can be put to use for human ends, and humans are simply mechanisms moved by appetites and aversions, there has emerged since the seventeenth century a new sensibility to and appreciation of the natural world,[14] and an elevation in the status of individuals. This has continued to develop in the Western nations up to the present, and finds its fullest expression in literature. Such opposition to the mainstream of ideas has two sources. Firstly there has emerged a countervailing philosophical vision which has seen nature as active and divine, and humans as social and creative.[15] This originated with the Hermetic philosophers, was promoted by Spinoza and was then taken up by the Romantic movement which was most fully developed in Germany by the *Naturphilosophen*. This has had a continuing influence on Western culture. Secondly and more deeply rooted in Western culture there has been a strong development of the individual moral conscience which is difficult to reconcile with mechanistic materialism.

A third problem arises from the relationship of mechanistic materialism to science. It is generally assumed in our society that science in the seventeenth century overcame superstition and the tendencies towards anthropomorphism to discover the true nature of the world. It is a simple matter to deny this and to show the extent to which the form science has taken has been strongly influenced by the social context in which it has developed. But this leaves the problem that if our most reliable body of knowledge is relative to social context, how can any belief be validated, and in particular, how can the argument that social context affects scientific beliefs be then defended against charges that this argument is simply the product of a particular social context?

My contention is that these should not be taken as reasons for rejecting the claim that mechanistic materialism provides the ideological underpinnings of capitalism, but as indicating the complicated nature of the relationship between ideology, truth, social dynamics and individuals. An effective ideology is one in terms of which people define the world as they engage in it. In claiming that mechanistic materialism dominates the modern world it is being claimed that it is the concepts of mechanistic materialism which actually mediate people's most important interactions with nature and with each other. It is the mode of thinking which structures the major institutions of society. It is not necessary for those people whose whole mode of being in the world incorporates the mechanistic materialist world-orientation in this way to actually espouse the doctrine. In examining the history of the ideology of capitalism, as with feudalism, what is being considered is the history of the modes of thinking embodied in institutions and social practices which individuals must incorporate to make their way in life.

Given my claim that people are struggling to make sense of the world and to attain a sense of their own significance, it is hardly surprising that those who are gaining the most power and material benefits from a society based on mechanistic materialist modes of thought should fail to

13. See Charles Coulston Gillispie, 'Science in the French Revolution' in *Proc. N.A.S.*, Vol.45, pp.677-684, esp. p.678.

14. Keith Thomas, *Man and the Natural World: Changing Attitudes in England 1500-1800*, London: Allen Lane, 1983.

15. See Margaret C. Jacob, *The Radical Enlightenment: Pantheists, Freemasons and Republicans*, London: George Allen & Unwin, 1981.

acknowledge this, even to themselves, and to openly embrace radically opposed, though impractical, ideas about the world. Not only is it easier to exploit people if they can be indoctrinated with the notion that they should subordinate their egoism to the principles of morality, or to treat them as objects to be manipulated while disguising the fact, but it is difficult to think of the world as making any sense, or of oneself as being anything of intrinsic significance, in a purely mechanical world. Consequently those in positions of power are generally unlikely to espouse mechanistic materialism, though they will always take its implications to be the hard-headed appraisal of any situation. Those who are most oppressed on the other hand are less likely to have embodied mechanistic materialism, and with the upper classes representing a desirable ideal to be attained, they are less likely to be concerned that life is of no significance. Conversely they are likely to be impressed by the modes of thought which actually move those with power over them, and will consequently be likely to wish to emulate what will appear to them to be their realism. That is, because they are oppressed they are likely to believe that the modes of thinking of their oppressors come to grips with the world as it really is. By defending this doctrine explicitly, spelling out its implications and forcing social relations to accord with it, the oppressed, especially those who aspire to upward mobility rather than social transformation, further its hegemony in ways those born to power could not achieve by force or fraud.

For such reasons the development of mechanistic materialism cannot be explained psychologically, but is characterized by an objective dynamic transcending the individuals who participate consciously or unconsciously in its development. As with the development of mathematical ideas studied by Lakatos, the basic analogy dominating a society 'becomes a living, growing organism, that *acquires a certain autonomy* from the activity which has produced it; it develops its own autonomous laws of growth, its own dialectic.'[16] The efforts of those people articulating this analogy (who frequently are at the same time committed to contradictory ideas) are only a fumbling realization of this dialectic.

The countervailing ideas to mechanistic materialism are also for the most part an aspect of its domination. Exalting nature or people explicitly in such a way that this does not interfere with participation in the whole complex of life forms which are degrading and destroying both nature and people blurs and softens the appearance of this degradation. The development of the mechanistic materialist ideology involves at the same time the neutralization of potentially competing world-orientations by reformulating them as ineffectual polar oppositions to it, as Christian Neoplatonism deformed the tradition of laughter and parody by reducing it to the polar opposite of its own exalted conception of life, and Romanticism, which was originally presented as rationally superior to mechanistic materialism and an alternative foundation for science, has been reduced to such an opposition by rendering it as irrationalist and emotionalist and expelling it from science to the humanities. In this 'feminine' role it is allowed a little influence, but only on inessentials. Romantic utopianism has become a mere negation of mechanistic science, in fact its shadow, where what is required to replace it is a negation of the negation which transcends the dichotomy between rationality and emotion, the sciences and the humanities.

But the situation is still more complex. The development of the heroic moralism and the highly developed individual conscience has as deep roots in Western culture as mechanistic materialism, but has been directly and effectively in opposition to its implications. Not only this, but both these forms of thinking have their roots in the activist individualism which was seen in Chapter IV to have given rise to the Promethean efforts to gain power over the world. The individual conscience has been accentuated in Protestantism and in the more Protestant branches of Catholicism through the internalization of the Court of Conscience, the generalization of the notion of vocation to the secular realm, and, despite the notions of predestination, the ascription of total responsibility to individuals for not only their actions, but also their thoughts; while at the same time it has been in

16. Imre Lakatos, *Proofs and Refutations: The Logic of Mathematical Discovery*, ed. John Worrall and Elie Zahar, Cambridge University Press, 1981, p.146.

these branches of Christianity that the rejection of the material world has been accentuated and which were most responsible for the development of mechanistic materialism. While the ethical notions associated with the individual conscience are irrational from the point of view of mechanistic materialism and have been disregarded or assumed to be irrational by philosophers, they have engendered an extension and distortion of rights theory, utilitarianism and other ethical doctrines beyond their rational limits to express this conscience. Kantian ethical theory based on the categorical imperative to act only on universalizable principles can be seen as providing bad reasons for this heroic moralism, and his whole philosophy with its division between the noumenal and phenomenal realms can be seen as a struggle to reconcile this moralism with mechanistic materialism. Heroic moralism has been responsible for the egalitarianism and concern for the individual characteristic of Western civilization, culminating in the development of democratic socialism in the Scandinavian nations, particularly in Sweden. Much of the driving force (as opposed to the sentimentalism) of environmentalism is a modern expression of this heroic moralism. Its existence alongside mechanistic materialism represents a major fissure or contradiction in the culture of Western Europe.

The question of what status scientific views have if mechanistic materialism is so intimately a part of the ideological foundation of capitalism is a more complex problem. It will be considered more fully in Chapter XII where a particular epistemological theory. However the essential point can be made briefly: that if understanding is taken to be the goal of disciplined enquiry it is not a question of one theory about the world being true or false, but of the depth of understanding and what is revealed and what concealed by the different theories. This allows for the possibility of different paths taken in science being successful in different ways. To demonstrate that a particular path has been taken because of the nature of the society within which natural philosophers or scientists were living and because of their ideological motivations does not invalidate the achievements made by following such a path. On the other hand, focusing on the effects of the context and motives underlying such decisions should be beneficial to science by revealing to what extent past decisions to support or develop theories have been based on their coherence with particular ethical commitments rather than problems within science itself, by revealing distortions in the interpretations of the achievements of science, by revealing alternative paths which could be explored and by making clearer the limitations of the paths which have been taken. At the same time, allowing that the development of science is an aspect of the development of the ideology on which society is founded without science being entirely determined by society implies that the direction of society's development is itself affected by the immanent dynamics of science. The dynamics of society, ideology and science are conditional causes of each other while being irreducible to each other.[17]

## *The Background to the Seventeenth Century Cultural Crisis*

The development of mechanistic materialism came at the end of a long ideological struggle associated with the loss of legitimacy of the Roman Church, the development of commercial capitalism, the rapid development of military technology, the rise in power of the territorial kingdoms, particularly of France and Britain, and the dispossession and impoverishment of large sections of the population. Despite the defeat of the Crusades in the Holy Lands, the power of the Roman Church increased, peaking towards the end of the thirteenth century when Conradin, the last of the Hohenstaufen, grandson of Frederick II, the Emperor of the Holy Roman Empire, was executed at Naples in 1268. By this stage popes had come to regard themselves, in accordance with the hierarchical medieval world-orientation, as more than men, though still less than God, and the

17. For an analysis of the problems associated with viewing science in its social context see Michael Mulkay, *Science and the Sociology of Knowledge*, London: George Allen & Unwin, 1979.

Church had come to regard the laity as nothing but serfs, instruments whose function was to yield willing obedience.[18] However in achieving their temporal power, conducting their power struggles against Christians as crusades against apostates, levying taxes and selling indulgences to finance these, and attempting to gain total control over intellectual life, they aroused increasingly strong opposition from all levels of European society. This manifested itself in the defeat of Boniface VIII in his struggle with Philip the Fair of France at the beginning of the fourteenth century. Following this the papacy was shifted to Avignon where it remained until 1376. In 1378, the Great Schism began with the election of two popes, one at Rome and one at Avignon, producing a period of utter chaos in the Church. In 1409 there were three popes. These events took place against the background of the greatest crisis of the Middle Ages, the Black Death which between 1347 and 1349 killed a third of Europe's population. Then in the fifteenth century Europe suffered an economic decline.

These developments were associated with the emergence of mercantile capitalism, first in the major city states of northern Italy, then in northern Europe. The increase in wealth of the merchant class unmatched by political power had led to sharp class divisions and eventually the transformation of these societies. From the fifteenth century onwards northern Europe was characterized by a progressive political centralization leading to the emergence of powerful absolutist monarchs. This was associated with the rise of Protestantism in a form which legitimated the rule of those opposed to Rome and gave much greater scope for economic activity. These developments gave rise to increasingly bitter conflicts which were intensified by the rapidly evolving technology of warfare, culminating in the Thirty Years War of 1618-1648.[19]

Underlying these religious, political and military struggles, social life was characterized by an intensification of the struggle for survival within a disintegrating social order. There were severe food shortages and price inflation, dispossession of land from the poor and the pauperization of a large proportion of the population, numerous peasant revolts or wars, violent millenarian movements, and a rapid growth in the population of cities. Cities were characterized by widespread poverty, malnutrition, frequent plagues and fires, with scarcely any social organization to deal with such problems. The general population were helpless before such catastrophes.[20] By the end of the fifteenth century kinship and bondage had already been largely replaced by commercial contract as the basis for relationships between most people, with the majority of the population being free peasant proprietors. However in Britain especially, the smaller proprietors were gradually forced off their land during the enclosure movement as arable land was converted to sheep pasture. The dispossessed became a roving population of paupers who were seen as such a threat to society that Henry VIII (1491-1547) executed 2% of the total population of England, 'without producing any improvement in the morals of the nation.'[21] While tradesmen were initially buffered from commerce by their guilds, this respite was short-lived. During the sixteenth century there was a rapid acceleration in the process of capital accumulation, accompanied by a sharp decline in real wages to less then 50% of their level at the beginning of the century.[22] As John Maynard Keynes wrote of this period: 'Never in the annals of the modern world has there existed so prolonged and so rich an opportunity for the businessman, the speculator and the profiteer.'[23] The upswing in

1990's ?

---

18. Friedrich Heer, *The Medieval World*, tr. Janet Sondheimer, Cardinal, London, 1974, p.329ff.

19. On the arms races within Europe and their effects at this time, see William H. McNeill, *The Pursuit of Power*, Chicago: University of Chicago Press, 1982, Ch.'s 2 & 3.

20. For a description of life in Great Britain at this time, see Keith Thomas, *Religion and the Decline of Magic*, [1971] Harmondsworth: Penguin, 1980, Ch.1.

21. P. Ravenstone *A Few Doubts*, p.464, cited by Arghiri Emmanuel, *Profit and Crisis*, [1974], London: Heinemann, 1984, p.19.

22. Andre Gunder Frank, *World Accumulation 1492-1789*, N.Y. and London: Monthly Review Press, 1978, p.53.

23. Cited ibid. p.7 from John Maynard Keynes, *A Treatise on Money*, 2 vols, London: Macmillan, 1930; New York: St. Martin's, 1972.

economic activity was paid for by increased toil, hardship, impoverishment and dejection of the majority of the population. The power of the guilds in Britain to control wages, the labour force and the quality of goods had been totally destroyed by the end of the seventeenth century, and individuals were left in isolation struggling for a livelihood against the forces of the market. Some 20% of the total population of Britain were unemployed, that is, some 45% of the active population. Individuals had to dominate or be dominated, use other people or be used. In this social environment all principles governing the struggle for power, disintegrated.

All these religious, political, economic, military and social changes generated vigorous intellectual efforts to defend different social movements and new forms of organization, and these radically changed the intellectual environment of Europe.[24] It was in northern Italy that the most original break with medieval political thought occurred. Before 1250, with the reign of Augustine's thought denying the significance of temporal life, politics was not considered as a distinct branch of moral philosophy. But the Italian city republics which had established themselves in the twelfth century contravened the hierarchical form of organization of feudalism, and found themselves struggling to defend their integrity both against the feudal powers (the Holy Roman Empire and the Church) and against the rise of despotism from within. Augustinian thought provided no basis to legitimate this. The most important early defence, and the point of departure for all subsequent defences of these republics was made in response to the efforts by the Papacy to gain control of the city republics by exploiting the class divisions to support the rise to power of tyrants. In defence of liberty Marsiglio of Padua (c.1275-1342) argued in *The Defender of Peace*, that the Church can only be a congregation, a voluntary gathering of the faithful and therefore cannot claim any jurisdictional power, and that in fact the members of the Church must be subject to the highest secular legislator of each independent kingdom or city republic. Following Aristotle, Marsiglio argued that the goal of such legislation must be at all times the common benefit of the citizens.

The Italian defence of liberty culminated in the Florentine Renaissance of the early fifteenth century. It drew on various aspects of Scholasticism, Neoplatonism and particularly Stoicism. At this time liberty or '*libertas*' took on an almost technical meaning in diplomacy as 'independence and self-government'. However after the French invasions of Italy from 1494 onwards, despots were able to consolidate themselves. The most original political thought developed at this stage was that of Machiavelli (1469-1527) who set out to 'draw up an original set of rules' for the Prince or despot on the basis of 'things as they are in real truth, rather than as they are imagined.'[25] In his concern that city republics preserve their liberty, Machiavelli focussed his attention on how political leaders could gain, then maintain power in a society in which the medieval forms of relationships between people based on hierarchy and honour had effectively dissolved, and political life was dominated by military force. He rejected Cicero's contention that the crowning splendour of virtue is justice since he argued this would not guarantee liberty. He exalted courage, orderly behaviour, temperance, and most especially prudence, but also recommended cruelty, perfidy and deception to achieve and maintain power. In this way he originated the doctrine that reasons of State transcend ethics.

Most of the political thought of northern Europe either developed or challenged themes developed by the Italians. However nearly always this was influenced in one way or another by the attacks of the nominalists on Platonic realism. This attack began with William of Ockham (c. 1285-1347) who, on the basis of a reinterpretation of Aristotle's logic, rejected the reality of universal forms and argued that only individuals have reality. In rejecting universal forms, Ockham and his followers severely limited the role of reason in ethical debates and the possibility of knowing God by reason, and supported Marsiglio of Padua's notion that the Church is nothing but a congregation of faithful individuals. This was then used to justify a sharp distinction between ecclesiastical and

---

24. See Quentin Skinner, *The Foundations of Modern Political Thought*, 2 vols, Cambridge: Cambridge University Press, 1978, for the best and most comprehensive study of these ideas.

25. Niccolo Machiavelli, *The Prince*, tr. George Bull, Harmondsworth: Penguin, Ch.XV, p.90.

secular authority and to defend political authority as an independent, autonomous corporation with the fullest authority to regulate its own affairs. Ockham's ideas were most important for their later influence. The first major movement away from the hierarchical conception of power was made at the time of the Great Schism by the conciliarists who defended the authority of the General Council over the Church, leading to the defence of the Ockhamite doctrine by Gerson (1363-1429). While the conciliarists were defeated, Gerson's ideas were revived and further developed at the beginning of the sixteenth century by John Mair and Jacques Almain at the Sorbonne in their defence of the notion that the authority of a ruler derives from the people.

The northern Renaissance was inspired by the spread of Italian ideas into northern Europe. Northern humanists developed the ideas on law, further extended the study of Greek and Latin writers, particularly the Stoics, and vigorously opposed the doctrines of Machiavelli. However this movement was soon eclipsed by the rise of Protestantism, in particular by the rise of Luther (1483-1546) and Calvin (1509-1564). The Lutherans adopted the Ockhamite doctrine that the Church is simply a congregation of the faithful and can therefore lay no claim to temporal power, and in doing so played an important role in legitimating the emergence of unified and absolute monarchies independent of Rome. However in the face of persecution by Catholic rulers later Protestants questioned the absolute power of rulers, developing a justification for resistance by force against injustice and for religious freedom.

These ideas and the political movements associated with them led to a counter-attack by the Jesuits based mainly in Spain, beginning with Francisco de Vitoria (c.1485-1546) and culminating in the work of Suarez (1548-1617), on the basis of a revival of Thomism.[26] However in the process of developing their ideas these thinkers were compelled to defend the notion that political society was a human invention, and were led to the attempt to deduce society as a 'social contract' arrived at from an imagined 'state of nature', although these canonical phrases were rarely used.

During the sixteenth century the conflict between Protestants and Catholics increased and more radical efforts were made to justify the right of individuals to resist their rulers, while at the same time more radical efforts were made to justify the absolute rule of kings. These developments occurred particularly in France where the Calvinist Huegenots were being massacred with the support of the Catholic government. While such defence was mounted from a variety of intellectual positions, the most original Protestant defence of violent resistance was made by John Mair's student, George Buchanan (1506-82). To argue his position, Buchanan adopted the Stoic idea that humans had originally wandered in the fields like animals, and that therefore political society had to be seen as the outcome of a series of decisions. On this basis he argued that since the whole body of a people agree together to set up a lawful government, the entire populace, and not merely the elected representatives, have a right to resist the government. When a Huguenot seemed likely to inherit the French throne, similar ideas were defended by the Jesuit theologian Juan de Mariana in his *History of Spain* published in 1592. Thus the defence of individual liberty from tyrannical rulers emerged as a by-product of the struggle between implacably opposed religious groups. Cromwell justified his execution of Charles I in terms of the principles of Buchanan and Mariana.

By the late sixteenth century it had become evident that the protagonists of rival religious creeds were willing to fight each other to the death. Addressing this problem Jean Bodin (1530-1596) argued both against the right of people to resist the ruler and for the divorce of the powers of the State from the duty to uphold any religious faith. He argued that the sovereign has power to legislate for everyone without their consent, that sovereignty is perpetual and unconditional, and the sovereign himself is not subject to the laws he promulgates. He is bound by divine laws, natural law, the law of nations and the provisions of the natural constitution, but not by any ecclesiastical law. But at the same time, Bodin insisted that 'wars made for matters of religion' are not in fact 'grounded upon matters directly touching his estate'.[27] Thus in Bodin the modern notion of the

26. This is described in Skinner, *The Foundations of Modern Political Thought*, Vol. 2, p.154ff.

27. Jean Bodin *The Six Books of a Commonwealth* tr. Richard Knowles, Cambridge Mass., 1962, p. A14.

State as 'a form of public power separate from both the ruler and the ruled, and constituting the supreme political authority within a certain defined territory',[28] the ultimate reification of social relations, was formulated for the first time. Henceforth, the problems of political philosophy centred on the nature of the State and its power, and the rights and duties of individuals *vis-a-vis* the State. Western civilization had succeeded in constructing a realm of reified political relations to mediate its social relationships. This had a major effect on how social relations were conceived. While Roman legal theorists had defined *ius naturae* as "that which nature has taught all animals, Hugo Grotius (1583-1645) redefined natural law in a totally homocentric way, expelling nonhumans from membership in the realm of cosmic justice.

While such developments in religious and political philosophy eventually paved the way for modern political thought, their original promulgation had the effect of helping to discredit old forms of thinking without providing any solid foundation for adjudicating between opposing positions. The Thirty Years War was a manifestation of the failure to solve this ethical chaos.

## *Mechanistic Materialism as a New World-Orientation*

Mechanistic materialism can be seen as expressing the orientation to the world engendered by such circumstances, and as the product of efforts to come to terms with them. First, it was a development of those modes of thought associated with the struggle to dominate nature. It emerged as feudal society was dissolving and mercantile capitalism was becoming increasingly important. As Edgar Zilsel has pointed out, this period, characterized by the rise in power of the merchant class bent on making and selling things, radically increased the status of artisans among the educated classes and further intensified the struggle to develop technology.[29] The pre-occupation with dynamics reflected this as was shown by Leonardo Olschki in the case of Galileo and B. Hessen in the case of Newton.[30] Specific instances of ideas central to mechanistic science having been inspired by the struggle to develop technology have been noted by a number of historians. Edward W. Strong has revealed the importance of practical operational methods intimately connected with technology for the development of mathematics.[31] J. Delevsky has shown how the efforts to provide a mathematical basis for Mercator's projection developed for navigational purposes helped inspire mathematical ideas which were largely responsible for the development of the calculus. More recently David Bloor has pointed out that the acceptance of 'one' as a number, and thereby the development of the conception of negative numbers and functions, followed from the replacement of the counting of things, by measurement in engineering, as the prime practical application of mathematics.[32] However what is perhaps most important about this period is that astronomy, which had been the preserve of the elite of society, and mechanics which had for the most part been the preserve of the lower orders, were brought together, and the theoretical and

---

28. This is cogently argued by Skinner, *The Foundations of Modern Political Thought*, Vol.2, p.353.

29. Edgar Zilsel, 'The Origins of Gilbert's Scientific Method' in Philip P. Wiener and Aaron Noland eds, *Roots of Scientific Thought: A Cultural Perspective* (from the *Journal of the History of Ideas*), N.Y.: Basic Books, 1957, pp.219-250.

30. Leornardo Olschki, *Galileo und seine Zeit*, Halle, 1927; and B. Hessen, 'The Social and Economic Roots of Newton's *Principia*', in *Science at the Crossroads*, London, 1931. These and other such studies have been discussed by Lynn White Jr. in *Medieval Religion and Technology*, Berkeley: University of California Press, 1978, p.24n.

31. Edward W. Strong, *Procedures and Metaphysics: A Study in the Philosophy of Mathematical-physical Science in the Sixteenth and Seventeenth Centuries*, Berkeley, 1936.

32. David Bloor, *Knowledge and Social Imagery*, London: Routledge & Kegan Paul, 1976, Ch.6.

practical approaches to understanding the world were thereby integrated into a unified approach to the world.[33]

Mechanistic materialism was also an expression of the growing detachment from, abstract attitude towards, and general alienation from the world, especially by the commercial class.[34] While René Descartes' philosophy in which the world was portrayed as a geometrical, extended order, and minds as isolated self-subsistent substances, clearly expressed this attitude, that this was not merely an incidental aspect of Descartes' ideas is evident from it accorded with other cultural developments. Art exemplified this. Throughout the Middle Ages, artists had become steadily more concerned to accurately represent the diversity of the physical world, a trend which culminated in the geometrical orientation of the vanishing-point perspective developed by Brunelleschi in Florence in the fifteenth century. Brunelleschi, an architect, was attempting to accurately portray an octagonal building on a flat surface, but his discovery of how to do this was adopted by virtually all artists until the late nineteenth century.[35] This involved the unification of pictures by the total subordination of the composition to a rigid geometrical scheme which privileged one point, contrasting radically with the composition of Chinese paintings in which different scenes had some autonomy while contributing to the pattern of the whole. In the sixteenth century Pieter Breugel (1525-69) made secular themes the subject of his paintings rather than religious or mythical themes. But in these pictures the activities of people were interwoven with their physical settings which were largely the work of imagination rather than representations of the world in which Breugel lived. Then in the seventeenth century in Holland, at the same time that Descartes was living in Amsterdam writing his works, artists such as Johannes Vermeer who were selling their paintings to the rising bourgeoisie, brought this to fulfilment in a new objective and generalized approach consisting of representations of landscapes completely independent of human activity. Walter Ong has convincingly argued that the primary impetus for this transformation in people's thinking and way of relating to the world was the development of printing and a print culture.[36]

However, the most significant thing expressed by mechanistic materialism was the loss of any meaning in the world. This is comprehensible against the background of the social changes taking place at the time. It was first recognized in northern Italy (where it occurred first) that the old order had broken down and that neither society nor the cosmos could any longer be seen as reflecting the natural ethical order on which feudal society had been based. This is the real significance of Machiavelli (1469-1527) who noted that: 'contemporary experience shows that princes who have achieved great things have been those who have given their word lightly, who have known how to trick men with their cunning, and who, in the end, have overcome those abiding by honest principles'[37] and observed:

> One can make this generalization about men: they are ungrateful, fickle, liars, and deceivers, they shun danger and are greedy for profit; while you treat them well, they are yours. They

33. Edgar Zilsel, 'Copernicus and Mechanics' in *Roots of Scientific Thought*, pp.276-280. See also Norman Diamond, 'The Copernican Revolution', Les Levidov ed. *Science as Politics*, London: Free Association Books, 1986, pp.7-37 who emphasises more the relationship between new concepts and changing social relations.

34. The nature and development of this abstract attitude is best described by F. Borkenau, in *Der Übergang vom feudalen zum Burgerlische Weltbild, Studien zur Geschichte der Philophie der Manufakturperiode* Paris: Librairie F. Alcan, 1934. For a defence of Borkenau's work against criticisms, especially those of H. Grossmann, see Valeria E. Russo, 'Les Transformations De L'Image Du Monde: Notes Sur F. Borkenau' in *L'Esprit du Mécanisme: Science et Société Chez Franz Borkenau*, Cahiers S.T.S., Paris: *Centre National de la Recherche Scientifique*, 1985, pp.133-154.

35. The developments in medieval architecture and art which culminated in Brunelleschi's achievement is brilliantly described by Walter Horn in 'Survival, Revival, Transformation' in Robert L. Benson and Giles Constable eds. *Renaissance and Renewal in the Twelfth Century*, Cambridge: Harvard University Press, 1982, pp.711-758.

36. His research on this is summarized in Walter J. Ong, *Orality and Literacy: The Technologizing of the Word*, London: Methuen, 1982.

> would shed their blood for you, risk their property, their lives, their children, so long ... as danger is remote; but when you are in danger they turn against you... The bond of love is one which men, wretched creatures that they are, break when it is to their advantage to do so; but fear is strengthened by a dread of punishment which is always effective.[38]

The rise of the Protestants in Northern Europe was also a manifestation of this recognition. For Luther (1483-1546) people are inherently sinful, and it is inevitable that their inclinations, their will and their reason will be opposed to God's commandments. They have no control over their fate. Everything is predestined, and if people obey the commandments of God, they do so by His grace alone. For Luther, all we can do is 'hope for grace that we may be justified and forgiven for our inability to obey the fiats of a cosmic despot.'[39] Consequently Luther rejected the value of the sensible world even more completely than St Augustine, recognizing inner experience as the only true value, and rejected 'the medieval effort to establish an objective structure linking God and man, eternity and time, the other world and this world, spirit and flesh.'[40]

The disorientation associated with the disintegration of the old order and all its values was expressed in literature. Shakespeare (1564-1616) and Racine (1639-1699) in particular gave expression to what this meant in their tragedies. For instance Shakespeare, who represented the perspective of the aristocracy, had Macbeth, portrayed as an upstart who had murdered his king to attain the throne and then maintained it by murdering all who threatened his position, totally isolating himself from others, expostulate on hearing of the death of his queen and former confidant:

> To-morrow, and to-morrow, and to-morrow,
> Creeps in this petty pace from day to day,
> To the last syllable of recorded time;
> And all our yesterdays have lighted fools
> The way to dusty death. Out, out, brief candle!
> Life's but a walking shadow, a poor player,
> That frets and struts his hour upon the stage,
> And then is heard no more; it is a tale
> Told by an idiot, full of sound and fury,
> Signifying nothing.[41]

The mechanistic view of the world can be seen as a reflection of this state of mind, and the rise of this conception of the world and the disintegration of the old society was drawn at the time by John Donne (1572-1631) in his *Anatomy of the World*:

> And freely man confesse that this world's spent,
> When in the Planets, and the Firmament
> They seek so many new; then see that this
> Is crumbled out again to his Atomies.
> 'Tis all in Peeces, all coherence gone;
> All just supply, and all Relation:
> Prince, Subject, Father, Sonne, are things forgot,

---

37. Niccolo Machiavelli, *The Prince*, tr. George Bull, Harmondsworth: Penguin, 1961, Ch. XVIII, p.99.

38. Ibid., Ch.XVII. p.96f.

39. Alasdair MacIntyre, *A Short History of Ethics*, London: Routledge & Kegan Paul, 1967, p.123.

40. Charles Trinkaus, 'The Religious Foundations of Luther's Social Views' in *Essays in Medieval Life and Thought*, ed. John H. Mundy et.al., N.Y.: Biblo and Tanden, 1965, p.73.

41. Shakespeare, *Macbeth*, V. v. 16.

> For every man alone thinkes he hath got
> To be a Phoenix, and that then can bee
> None of that kinde, of which he is, but hee.
> This is the world's condition now.

Such a vision of the world was to some extent anticipated by the Gnostics.[42] Accepting the Platonic division of reality between the sensible world and the ideal world of eternal forms the Gnostics accentuated the lowly status of the sensible world in relation to the ideal world, representing the sensible world as the creation of an evil demon. But even for the Gnostics the world was not totally indifferent to humans, and behind the appearances, discoverable by knowledge, there was still the realm of ideal forms. In mechanistic materialism on the other hand all reality is denied to the non-quantifiable qualities of the sensible world and the real world behind these appearances is conceived of as meaningless, inert matter. The world is understood from a perspective outside the world, from that of the infinite universe. The subject is conceived of as essentially unrelated to his or her world, and therefore in abstraction from time and place. As Pascal (1623-62) expressed the view of the world from the perspective of mechanistic materialism:

> I see the terrifying immensity of the universe which surrounds me, and find myself limited to one corner of this vast expanse, without knowing why I am set down here rather than elsewhere, nor why the brief period appointed for my life is assigned to me at this moment rather than another in all the eternity that has gone before and will come after me. On all sides I behold nothing but infinity, in which I am a mere atom, a mere passing shadow that returns no more. All I know is that I must soon die, but what I understand least of all is this very death which cannot escape me.[43]

## *Mechanistic Materialism as Ideology*

However quite apart from the creative originality required to formulate the mechanistic conception of the world, it is an oversimplification to see new ways of conceiving things as nothing but expressions of a way of experiencing the world. While the conditions of life made a mechanical view of the world plausible, its development must be seen as an explicit affirmation of the meaningless of life, and this affirmation only makes sense in relation to an ideological struggle between different groups of people. In particular, the development of mechanistic materialism only makes sense as a struggle against the Hermetic philosophers, the 'nature enthusiasts' who interpreted the same conditions as a sign that a new age was about to dawn.

Hermetic philosophy was a revival of a radical version of Neoplatonism in which, in opposition to the more conventional Neoplatonic Christianity, it was held that God is immanent in the world, that nature is active and divine and that the end of history, the millenium which was evidently at hand, would not involve a transcendence of the world, but the establishment of a new order on earth based on brotherly love and a reunification of humanity with nature. They believed that the religious and political reform of the world could be effected by uniting dissidents in a new religion of nature, with the sun as a visible, unifying symbol of the deity; and they proposed to unite humanity with God through an understanding of nature's hidden forces, forces which could be manipulated magically for human benefit. By bringing God down to earth they collapsed the hierarchical structure of the cosmos and the ideological foundations for the hierarchical structure of

---

42. Hans Jonas has pointed out the similarities between Gnosticism and mechanistic materialism in this respect in 'Gnosticism, Existentialism and Nihilism' in *The Phenomenon of Life*, [1966] Chicago: Uni. of Chicago Press, 1982, Ninth Essay, pp.211-234.

43. Blaise Pascal, *Pensées*, Louis Lafuma ed., tr. John Warrington, London: Dent, 1973, p.6, fr.11.

society, and by presenting the world as active, self-moving and self-organizing, they provided the justification for people to be self-organizing and to create a new egalitarian social order in harmony with the world. Thus Giordano Bruno (1548-1600) aspired to 'bring men once again into communication with divine, living nature.'[44] and Campanella (1568-1639), who typified the Hermetics, revived the call for an egalitarian distribution of wealth based on an original harmony between people and nature. These ideas underlay the development of alchemy and the Rosicrucian Enlightenment.[45]

These Hermetics were then vigorously attacked. Campanella spent twenty-seven years in prison and Bruno was burnt by the Inquisition in 1600. The Rosicrucians were crushed as a major force in Europe when the Austro-Spanish armies invaded Bohemia in 1620, the first act of the Thirty Years War, but their notion that a new order was about to be established had a profound influence on the whole of Europe. Ideas deriving from this movement were taken up by the radical elements in the English civil war. The True Levellers, the Seekers, Ranters and Diggers who embraced these doctrines became a significant force after the civil war, refusing to pay church tithes, demanding a redistribution of land and property, instituting lay preaching and acting according to their own definitions of marriage and morality.[46] In doing so they provoked a violent reaction on the part of the establishment. But more importantly these social movements and their millenarian ideas both inspired and impelled the ruling elites of Europe into efforts to provide an opposing philosophy.[47]

This opposing philosophy was the mechanical philosophy. The originators of this philosophy were inspired by the Rosicrucians to believe that the present was a new age and that a new science of nature would bring greater power to humanity, but because of their opposition to their religious and social ideals, or for reasons of personal interest in an environment in which such ideals were anathema, they were concerned to oppose Hermetic ideas and to distance themselves from this intellectual movement. What was defended was the effort to re-establish humanity's rightful dominion over the lower orders of creation in accordance with orthodox interpretations of the Bible, rather than harmonizing with it.[48] Thus, Francis Bacon (1561-1626), who was a pivotal figure in legitimating the new philosophy, wrote: 'For man by the fall fell at the same time from his state of innocency and from his dominion over creation. Both of these losses however can even in this life be in some part repaired; the former by religion and faith, the latter by arts and sciences.'[49] Unlike the Hermetic philosophy, in this philosophy there was no program for the ethical and political advancement of humanity. Bacon's *New Atlantis*, unlike Thomas More's *Utopia*, was elitist rather than egalitarian, and the moral order was not considered to be an issue. Furthermore, the proponents of the mechanical philosophy were concerned to affirm the meaninglessness of temporal life and to reject the ascription of intrinsic value to nature in order to undermine the revolutionary implications of Hermeticism and to legitimate the emerging capitalist political and economic order.

The transformation of the Hermetic ideal of power through harmonizing with nature to the ideal of power through subjugating nature was effected by the mechanical philosophers by analogizing

---

44. Easlea, *Witch-Hunting, Magic and the New Philosophy*, p.104.

45. See Frances A. Yates, *The Rosicrucian Enlightenment*, London: Routledge & Kegan Paul, 1972.

46. For a study of these movements see Christopher Hill, *The World Turned Upside Down: Radical Ideas During the English Revolution*, Harmondsworth: Penguin, 1975.

47. See Yates, *The Rosicrucian Enlightenment*, Ch's 8 & 9 for an analysis of this relationship. For a detailed analysis of Francis Bacon's relationship to his Hermetic predecessors, see Paolo Rossi, *Francis Bacon: From Magic to Science*, tr. Sacha Rabinovitch, Chicago: Uni. of Chicago Press, 1968, esp. Ch.1.

48. On this, see William Leiss, *The Domination of Nature*, Boston: Beacon Press, 1974, esp. Ch.3.

49. Francis Bacon, *The New Organon*, in *The Works of Francis Bacon*, James Spedding, Robert Leslie Ellis, Douglas Devon Heath, eds. 14 vols, London: Longmans Green, 1870, Vol.IV, pp.247-248.

their aggressive orientation towards other people, and particularly, towards women.[50] Bacon more than anyone illustrated this. Taking the social philosophy of those who had described existing social reality as a model for the study of nature, he wrote: '...we are much beholden to Machiavel, and writers of that kind, who openly and unmasked declare what men do in fact, and not as they ought to do...'[51] For such people, females had come to symbolize the disorder of the sensible world of nature in opposition to the ideal, rational, masculine world. The aggressiveness towards women which gathered pace in this era culminated in the witch-hunts of the sixteenth and seventeenth centuries during which 100,000 people were tried for witchcraft, eighty-three per cent of them women.[52] In 1585, two villages in Germany were left with only one female inhabitant.[53] Such women were generally held to have had sexual intercourse with the devil to satisfy their uncontrollable lust. Bacon, who had been involved in the persecution of women as a judge in the witch trials, generalized the method he had used to interrogate witches to investigate nature.[54] He called for an unbridled inquisition of nature, for nature to be put to the wrack and tortured to reveal her secrets as witches had been tortured to confess. He argued: 'For like as a man's disposition is never well known or proved till he be crossed, nor Proteus ever changed shapes till he was straitened and held fast, so nature exhibits herself more clearly under the trials and vexations of art than when left to herself.'[55] He described nature in terms which, as Brian Easlea wrote, amounts to a call for its gang rape:

> Nor is mine a trumpet which summons and excites men to cut each other to pieces ... but rather to make peace between themselves, and turning with united forces against the Nature of Things, to storm and occupy her castles and strongholds, and extend the bounds of human empire, as far as God Almighty permit.[56]

Because of this aggressive orientation, Bacon dismissed concern with the purposes of things as a suitable object of investigation, and focused on that aspect of nature which could be reshaped, namely matter. Forms were dismissed as fictions of the human mind - unless identified with the laws of action. Continuing his use of females as an analogy, he wrote that, 'teleology is a barren thing, or as a virgin consecrated to God'[57] and repeatedly referred to matter as a 'common harlot'.[58] He then suggested that Democritus, who had conceived of nature as nothing but atoms and the void, that is, as matter without purpose, had a superior understanding of causation than did Plato and Aristotle.[59]

The motivation for this evaluation of the different ancient philosophers is clear. If one's sole interest in nature is to subjugate it, then the only relevant form of causation is the mechanical causation considered by the atomists, and the relevant form of understanding nature is that which sees things as decomposable into bits of matter which can then be rearranged. However while

---

50. Both Merchant in *The Death of Nature* and Easlea in *Witch-Hunting, Magic and the New Philosophy* have examined this.

51. Francis Bacon, 'Advancement of Learning' [1605] in *The Physical and Metaphysical Works of Lord Bacon*, tr. Joseph Devey, London: Henry G. Bohn, 1853, Bk III, Ch.ii, p.281.

52. Merchant, *The Death of Nature*, p.138.

53. Loc.cit.

54. The language of aggressive sexual domination used by Bacon has been most thoroughly analysed by Merchant, *The Death of Nature*, Ch.7.

55. Cited ibid. p.169, from Francis Bacon, '*De Dignatate*,' *Works*, Vol.4, p.298.

56. Francis Bacon, 'The Advancement of Learning', [1605] in *The Physical and Metaphysical Works of Lord Bacon*, tr. Joseph Devey, London: Henry Bohn, 1853, Book IV, Ch.i, p.150.

57. Ibid. Bk III, Ch.v, p.144.

58. As pointed out by Merchant in *The Death of Nature*, p.171.

59. Bacon, *The Advancement of Learning*, ed. William Aldis Wright, Oxford: Clarendon, Bk.II, Ch.vii, Sect. 7, p.119f.

Bacon flirted with Democritean ideas, he still took formal and final causes to have a place in the world. Descartes (1596-1650), who was the first to fully elaborate a mechanical view of the world (although he opposed atomism), eliminated potentiality entirely from the material world by defining matter as extension. With this conception of matter all motion, apart from thinking which was conceived to be radically disjoined from the material world, had to be conceived of as locomotion: change in position, which is the form of motion relevant for efforts to control the world. Motion which is not change of position, that is, generation and growth which can be fostered or destroyed, but which having their own dynamics cannot be completely moulded to human purposes, was defined out of existence. Descartes concluded that in place of the speculative philosophy of the Schoolmen, his physics provided a practical philosophy which could 'make ourselves, as it were, the lords and masters of nature.'[60]

However, more important than the facilitation of the growing exploitative attitude towards nature, the mechanistic conception of the world was designed to undermine the nature enthusiasm of the Hermetics on which their radical religious, social and political doctrines were based, and to legitimate the form of life of the emerging capitalist order. Since capitalism completed the reduction of nature and people to instruments, all meaning had to be seen to come from a transcendental source. By projecting the meaninglessness of life and the absence of freedom onto the physical world, the mechanistic world-orientation presented the meaninglessness of life and powerlessness of the vast majority of the population as the only possible state of affairs, justifying the social order which rendered people's lives meaningless. Since in terms of this world-orientation any significance which could be attributed to the world and to life could only come from a transcendent deity, it justified passivity in the face of the vicissitudes of life and legitimated the established church which mediated between individuals and this deity.

This was clearly evident in the work of the French monk, Marin Mersenne, who published his first onslaught on the Hermetic tradition in 1623, the year of the outbreak of the Rosicrucian scare in France.[61] Mersenne was a friend of Descartes, and his massive attack on the Hermetics cleared the way for the rise of Cartesian philosophy. The project of reconceiving the nature of the world to undermine the appeal of the nature enthusiasts and to legitimate a new form of life was then taken up in England, particularly by Boyle and Newton. These thinkers feared and detested the kind of social order that would be established if men came to believe that they, and not God and his appointed representatives, could master the course of history. Boyle in particular, who had lost most of his estates in the English civil war, hated the revolutionaries and detested the philosophy which legitimated their aspirations. Consequently they formulated a version of mechanical materialism which emphasised at every turn the providential role of the deity as the source of order and harmony, imposed through laws at work in nature, and capable of being imitated in society. This served as the ideology of the latitudinarians: the low Church of England Whigs who subsequently came to dominate political and social life in Britain.[62]

Robert Boyle (1627-91) began his attack on the Hermetics by criticising their view of nature. He wrote in his 'Free Enquiry into the Vulgarly Received Views of Nature':

---

60. René Descartes, 'Discourse on the Method' [1637] in *The Philosophical Writings of Descartes*, tr. John Cottingham et.al., Volume I, Cambridge: Cambridge University Press, 1985, p.142f.

61. See Francis Yates, *The Rosicrucian Enlightenment*, London: Routledge & Kegan Paul, 1972, pp.111-13.

62. See Margaret C. Jacob, *The Radical Enlightenment*, London: George Allen & Unwin, Ch.2, esp. p.66; and S. Shapin, 'The Social Uses of Science', in G. Rousseau and R. Porter, eds., *The Ferment of Knowledge. Studies in the Historiography of Eighteenth-Century Science*, Cambridge: Cambridge University Press, 1980, pp.93-139.

> ...there is lately sprung up a sect of men, as well professing Christianity, as pretending to Philosophy, who ... do very much symbolize with the antient Heathens, and talk much of God, but mean such a one, as is not really distinct from the animated and intelligent universe.[63]

In opposition to this he declared that matter is brute and inanimate. Motion is an accidental property of matter which must be imposed from outside according to the laws of motion laid down by God and sustained by His will, as civil society must be controlled by law and sustained by the Anglican Church which interprets God's will to people. While denying the possibility of the general population taking political control of society, he defended the Protestant ethic: an ascetic life of unremitting hard work devoted to sober self-interest.

This also conception of being was then used to attack the respect for nature, supported by the Hermetic philosophers, which was standing in the way of its subjugation. This respect was particularly manifest in the opposition to the development of mining, and there was considerable opposition to such projects as the draining of the Fens by those people whose livelihoods were being threatened. As Boyle put it (still characterizing nature as female):

> The veneration wherewith men are imbued for what they call nature, has been a discouraging impediment to the empire of man over the inferior creatures of God: for many have not only looked upon it, as an impossible thing to compass, but as something impious to attempt ... and whilst looking upon her as such a venerable thing, some make a scruple of conscience to endeavour to emulate any of her works as to excel them.[64]

Boyle's efforts were completed by Newton (1642-1727) and those of his followers who developed the implications of Newtonian physics for religion and society in the Boyle lectures.[65] While Newton conceived space as the sensorium of God and gravity as His activity in the world, thereby offering an alternative to the extreme mechanism of Descartes or Hobbes, he was, as E.A. Burtt wrote:

> ... squarely behind that view of the cosmos which saw in man a puny irrelevant spectator (so far as being wholly imprisoned in a dark room can be called such) of the vast mathematical system whose regular motions according to mechanical principles constituted the world of nature.... The world that people had thought themselves living in - a world rich with colour and sound, redolent with fragrance, filled with gladness, love and beauty, speaking everywhere of purposive harmony and creative ideals - was crowded now into minute corners in the brains of scattered organic beings. The really important world outside was a world hard, cold, colourless, silent and dead... [66]

Newton was concerned to justify the vision of nature as a world of 'brute and stupid' matter because such a world would require God to order it, and this view of the world would enhance the ruling oligarchies and the established churches and justify a life of hard working asceticism. Only God, mediated by the Church of England which promised an afterlife to those who conformed to its dictates, could give any significance to life in such a barren world. Accordingly, Newton wrote to

---

63. R. Boyle, *The Works of the Honourable Robert Boyle*, London: Millar, 1744, Vol.4, p.376.

64. Boyle, 'A Free Inquiry Into the Vulgarly Received Notion of Nature' in *The Works of the Honourable Robert Boyle*, Vol.4, p.363.

65. See Margaret C. Jacob, *The Newtonians and the English Revolution 1689-1720*, Ithaca: Cornell University Press, 1976.

66. Edwin Arthur Burtt, *The Metaphysical Foundations of Modern Physical Science*, 2nd ed., London: Routledge & Kegan Paul, 1932, p.236f.

Bentley who in his Boyle lectures was using Newton's ideas to justify an ethics of self-denial and obedience: 'when I wrote my treatise upon our system, I had an eye upon such principles.'[67]

So, the view of the world as devoid of potentialities or powers, as consisting of nothing but inert matter moving blindly, endlessly, meaninglessly, a view which had become plausible as people's lives were rendered increasingly meaningless, was developed and promoted for ideological reasons. While political and social theory became the site of later ideological conflicts, it has been this underlying mechanistic conception of the nature of the world which has been the foundation for the legitimacy of the prevailing order. The struggle to maintain the conception of the world as devoid of meaning and creative potential has been a persistent feature of the modern world. It was to support the prevailing conservative ideology that the eighteenth century biologists Bonnet and Spallanzani argued for the preformationist theory of generation and rejected epigenesis (the creative emergence of order in the development of the organism) as argued for by Buffon and Needham, despite the support Spallanzani's own experiments and observations gave to the epigenetic theory. Diderot on the other hand embraced the notion of epigenesis as evidence of nature's creativity to support his more radical political views. Similarly, the politically radical Joseph Priestly promoted the conception of matter as essentially active and creative.

## *The Concepts of Mechanistic Materialism*

While the general orientation to life engendered the breakdown of the feudal order and the ensuing ideological struggle between supporters and opponents of different political and social tendencies within Europe provided the primary impetus towards the transformation of the world-orientation dominating European society in the seventeenth century, the nature of society also influenced the way nature was conceived by providing a mathematical, abstracting type of thinking, a way of refining concepts, and many of the basic concepts which could be used in the construction of the mechanistic conception of the world.

It is hardly conceivable that mathematical physics would have been established outside a society dominated by monetary relationships. Monetary relationships had provided the forms of thinking in Greek society which were then developed independently of commerce by Pythagoras and his followers and then applied to understanding nature. But in the ninth century money played a small role in exchange.[68] Charlemagne had to force people to accept the coin he paid his soldiers as tender, and he had great difficulty finding people with enough knowledge of mathematics to control his financial affairs. The study of mathematics was encouraged, but those who studied Greek works on the subject had the greatest difficulty understanding its most simple aspects. The reason for the establishment of many of the early medieval learning institutions was the demand for people knowledgeable in mathematics to deal with an increasingly monetarised economy. Mathematics also became increasingly important in warfare as commanders of armies found it necessary to calculate the relationships involved in troop formations. While mathematics developed beyond the requirements of accounting and military organization, it was the quantification of human relationships, culminating with the development of commercial capitalism, which led to a general familiarity with mathematical relationships and sustained what mathematical advances were made.

Closely associated with this quantitative mode of comprehension of relationships, the exchange economy, along with the development of a print culture, produced an abstracting form of thinking which denied objects their qualitative diversity. As Marx pointed out: 'Every moment, in calculating, accounting etc., that we transform commodities into value symbols, we fix them as

67. Quoted Jacob, *The Newtonians and the English Revolution 1689-1720*, p.156.

68. The most thorough study of the relationship between the development of a monetary economy and the ultimate rise of mathematical physics was made by Alexander Murray, *Reason and Society in the Middle Ages*, Oxford: Clarendon, 1978, Part II.

mere exchange values, making abstraction from the matter they are composed of and all their natural qualities.'[69] This leads things to be seen not in their context, but abstractly in relation to the transcendent order of the money economy. For instance a coat is something produced to be worn, and is a coat only insofar as it is used as such. But as a commodity, its being used is irrelevant. It is seen as a cypher with a certain exchange value expressible in monetary terms. This is precisely the form of thinking which was developed by the mathematical physicists, particularly Galileo, Descartes and Newton, who came to see matter as inert, devoid of non-quantifiable properties, in purely quantifiable relationships to other matter.[70]

This form of abstraction is central to the notion of conservation which has been a basic principle of the development of the mechanistic view of the world.[71] The principle of conservation had been formulated by the Ancient Greeks, notably by the atomists and by Anaxagoras who held that nothing comes into being and nothing passes away, and it was also defended by Lucretius. But the principle took on a new impetus with the redevelopment of a monetary economy in which accounting, both in relation to business and in personal affairs, was based on the principle that money must come from somewhere, and spending cannot be greater than income without going into debt. This assumption of quantitative conservation over time subsequently formed the principle of the conservation of mass, of momentum, of energy, of electric charge, of spin, and so on.

As noted in the last chapter, Western Europe was unique in its development of law. Universities began as schools for lawyers, and the scholastic technique of analysis and synthesis deriving from Plato was used to refine legal concepts, producing a highly coherent body of law which came to regulate almost all human relationships. This same concern for precision of concepts was then taken over by those investigating the physical world, so concepts used by natural philosophers in Western Europe came to be defined with far greater rigour than the concepts used by natural philosophers in China. Without this rigour, the revolution in thought of the seventeenth century would have been impossible.

The most basic concepts of the new world-orientation were those of space and time. While these were only developed explicitly in the theoretical efforts to overcome technical problems in the efforts to comprehend motion, they appear to have been developed in practice before this theoretical development. Aristotle had conceived of place as the interior bounding surface of a body. This notion had prevailed until the sixteenth century when Scaliger, Telesio and Bruno developed the notion of places independent of body, which Bruno referred to as '*spatium*'.[72] This new notion of space was then developed by the Cambridge Neoplatonists and assimilated into Newton's physics to become the infinite, uniform 'container' of all that exists, and in terms of which motion could be defined. While Aristotle's notion of place is the notion which would be developed in the process of accommodating action to the immediate world, and is intimately associated with the sense that everything has its proper place, the notion of space is the concept which emerges when this immediate engagement in the world is mediated by representations of this world. In late feudal society, as we have seen, representational thinking became increasingly important. To begin with, people were using more maps, these were portraying distances by means of a Ptolemaic grid, leading people to define places in the world in terms of an abstract order of representations. At the same time the development of vanishing-point perspective in painting conveyed a new way of seeing the world in which each thing was seen as having a position within a geometrically ordered perspective. Representations of the globe on a flat surface enabled the whole

69. Marx, *Grundrisse*, p.142.

70. This was noted by Alfred Sohn-Rethel. See his *Intellectual and Manual Labour*, London: Macmillan, 1978; and 'Science as Alienated Consciousness' in Les Levidow ed. *Radical Science Essays*, London: Free Association Books, 1986, pp.104-139, esp. p.129.

71. The significance of this for science has been analysed by Emile Meyerson in *Identity and Reality* [1907], tr. Kate Loewenberg, New York: Dover, 1962.

72. See Ivor Leclerc, *The Nature of Physical Existence*, George Allen & Unwin, London, 1972, p.157ff.

world to be appropriated in imagination according to mathematical principles, and thereby to be seen as containable and conquerable for purposes of human occupancy and action.[73] Finally, the intellectual world was undergoing a revolution with the new approach to knowledge of Peter Ramus inspired by the development of the printing press. This involved the representation of knowledge of the world in diagrams and tables which the printing presses could reproduce. The central focus became the ordering of knowledge by manipulating it visually in such diagrams and tables for remembering and teaching.[74] Bruno developed his new concept of space after having spent a great deal of time designing and manipulating complex diagrams to be used as mnemonic devices. This spatialized conception of the world was consolidated by conceiving it in terms of the easily visualized and representable Cartesian coordinates of analytic geometry, which have dominated scientific understanding of the world up to the present.

The notion of time developed along different lines, but eventually coalesced with the concept of space. Despite the Hebraic notion that the history of the world is a progression, people in the early Middle Ages were supremely indifferent to time, taking little effort to record the birth dates of its rulers, and there was little concern for uniformity in the division of the day.[75] As was seen in the last chapter, the later Middle Ages were characterized by tremendous efforts to measure time more accurately. The development of the mechanical clock eventually led people to measure their activities and their lives in terms of the abstract order of clock time, developing in practice the concept of time as independent of change. Developments of clocks were combined with refinements in chronologies and with increasing use of calendars, producing for practical purposes a unified conception of time corresponding to the unified conception of space. As Pierre Bourdieu pointed out: '... just as a map replaces the discontinuous patchy space of practical paths by the homogeneous, continuous space of geometry, so a calendar substitutes a linear, homogeneous, continuous time for practical time, which is made up of incommensurable islands of duration each with its own rhythm...'[76] The new conception of time was appropriated by Galileo and developed theoretically by representing time spatially to describe acceleration. Prior to Galileo all efforts to understand motion had been in terms of distances traversed. The spatial concept of time was fully elaborated and synthesized into a total world-view by Newton, and was reduced to a mere dimension of space by Laplace. Space and time thenceforth became the ultimate, eternal reference framework for all cognition and knowledge.

Other concepts which came to make up the mechanistic materialist conception of nature were associated with concepts in terms of which people had come to understand themselves. To begin with, the individualistic detachment of people and the view of society as composed of separate individuals had been a basic feature of Western culture from the beginning of the feudal era. It was pointed out in the last chapter how one of the reasons why the Irish adopted Christianity was because it gave them a means of conceiving themselves as enduring individual entities. This in turn resonated with the conception of nature as enduring individual substantial forms, and both modes of thinking were supported by the development of terminist logic. The growing individualism in society associated with the rise of commercialism in which individuals came to be seen as legal subjects free to enter into contracts and responsible for all past transactions and bonds, reinforced the view of individuals as enduring substances essentially independent of their relations. So substantialist thinking was thoroughly entrenched in Europe. The development of the conception of

---

73. The emergence of this conception of the world and of space and time, pointing out the connection between social life and the concepts articulated within the new mechanical philosophy, is described by David Harvey, *The Condition of Postmodernity*, Oxford: Blackwell, 1989, Ch.15, 'The time and space of the Enlightenment project.'

74. See Walter J. Ong, *Ramus, Method and the Decay of Dialogue*, Cambridge, Mass.: Harvard University Press, [1958], 1983, p.151.

75. See Marc Bloch, *Feudal Society*, Vol.1, tr. L.A. Manyon and M.M. Postan, 2nd ed., Routledge & Kegan Paul, 1965, p.73f.

76. Pierre Bourdieu, *Outline of a Theory of Practice*, Cambridge: Cambridge University Press, 1977, p.105.

nature as composed of immutable bits of inert matter, the ultimate in substantialist thinking, can thus be seen as a culmination of a form of thinking which had been resonating within Western culture for over a thousand years.

Another instance of social relations being used as an analogy to understand nature was the development of the conception of nature as law governed.[77] In China, the Legalists had developed a mechanical approach to understanding people, expressing laws in standardized, quantitative terms. But the Confucians had attacked the Legalist idea that society could be ordered by the imposition of strict laws, and the idea that nature was law governed never developed. Corresponding to the emphasis on spontaneous co-operation in society, nature was seen as dynamic. The nearest to a concept of law developed by the Chinese was the concept of *Li* developed by the Neo-Confucians of the Sung dynasty. But *Li* meant not law but the active principle of order or dynamic pattern of various things in nature.[78] In Europe on the other hand law as a means of organizing society was far more important. With the rise of the great monarchies after the death of Alexander the Great organized on the basis of law, the Stoics developed the idea of a divinity as Universal Law governing the world, including nature. The principle of organization of Hebraic society was also based on laws, and the idea of a divine law-giver is correspondingly the central idea of Judaism. With this background there was a corresponding tendency for the early Christians to suggest that there is a natural law independent of people governing everything. However the idea of laws of nature was not developed in the Middle Ages, though it was proposed by Roger Bacon, despite the fact that there had been a tremendous development and systematization of law and the notion of natural law was developed as part of Christian morality. But in later medieval society the growing emphasis on God as free to will as He chooses associated with nominalism paved the way for the notion that nature is ordered by laws promulgated by God. Then there was a burst of development in legal theory with the breakdown of feudal society and the rise of the absolutist states associated with early capitalism. The most important feature of this was the elimination of reference to social position within law. Law became a set of principles applicable to all people without exception.[79] It was at this time, first sporadically, then systematically in the work of Descartes, that the idea of the universe being governed by the laws of God was fully developed. And as Zilsel has argued, it was no mere chance that the idea of God as a legislator of the universe developed only forty years after Jean Bodin had advocated the development of civil government by statute law which was then most thoroughly implemented in France, the homeland of both Bodin and Descartes.[80]

## *The Analogy of the Machine*

However what really integrated mechanistic materialism into a unified vision, subordinating the Neoplatonist, Aristotelian and atomist elements which had contributed to its formation, was the analogy of mechanisms, and in particular, of the clock itself. In 1370, soon after the invention of

---

77. For an analysis of the relationship between the development of human law and the concept of the laws of nature, and a comparison of Europe and China see Joseph Needham (with the research assistance of Wang Ling), *Science & Civilisation in China,* Volume 2, Cambridge at the University Press, 1956, Ch.18, Sections 8-18. See also E. Zilsel, 'The Genesis of the Concept of Physical Law', *Philosophical Review, Philosophical Review*, Vol.51, 1942, pp.245-79, and Francis Oakley, 'Christian Theology and the Newtonian Science: The Rise of the Concept of Laws of Nature', *Church History*, 1961, Vol.30, pp.433-57.

78. Needham, *Science and Civilisation in China*, Vol.2, p.565ff.

79. As noted by Borkenau, *Der Überbang*, Ch.II and elaborated by Russo, 'Les Transformations de L'Image Du Monde', p.148ff.

80. Zilsel, 'The Genesis of the Concept of Physical Law', p.245.

the clock, Nicole Oresme had used its ordered movements to characterize motion in the heavens.[81] The role of mechanical analogies in the representation of the virtue of temperance was pointed out in the last chapter. H. Grossmann has pointed out how from the fifteenth century philosophers developed their ideas about the world by direct and conscious analogy with machines: artillery, clocks, hoists, water-wheels, pumps, bellows and so on. However it was in the seventeenth century that the clock metaphor came into its own. Kepler wrote to a friend in 1605: 'I am now much engaged in investigating physical causes; my goal is to show that the celestial machine is not in the likeness of the divine being, but in the likeness of a clock...' and accordingly rejected his former view of nature as ensouled matter.[82] Then William Harvey (1578-1657) used the analogy in relation to physiology in describing the heart as 'a piece of machinery in which though one wheel gives motion to another, yet all the wheels seem to move simultaneously.'[83] Finally Descartes, Hobbes and Newton elaborated a total picture of the world on this basis. The analogy of the machine assimilated into a unified perspective all the separate ideas of the opponents of Aristotlelianism and Hermeticism. It implies that the world is composed of brute and stupid matter, that it is a configuration of parts having a specified location in space, that to understand it requires a specification of these parts in exact quantitative terms, and that these are governed by precise principles which can therefore be expressed as laws describing constancy and change in relation to an abstract space and an abstract, spatialized time. By the beginning of the eighteenth century nature had become for Western civilization, if not for all its members, an infinitely large mechanism composed of inert matter located within space and changing its position over time in accordance with immutable laws of motion. While the idea of what constitutes a machine has evolved from a clock in the seventeenth and eighteenth centuries to a steam engine running down in the nineteenth century to an information processing mechanism in the twentieth century, the machine has remained the dominant metaphor to understand nature up to the present.

mech. imag.

However the analogy of the machine became more than just a means to comprehend the physical world. It became the thematic motif symbolizing the ultimate value, the 'meaning of life' for people in Western civilization, just as the ox symbolized the ultimate value for the Nuer, the python for the Fipa and the pangolin for the Lele. In this function it abounds in ambiguities. The machine symbolizes the means for achieving the subordination of nature, while at the same time the already existing total subordination of everything to the functioning of the whole. The mechanistic analogy reveals those aspects of the world which are relevant to its subjugation, but at the same time implies a deterministic world without potentialities which could be actualized. It oscillates between representing humans as separate from the mechanical order of things and therefore totally free to act at will, and as machines totally determined by the laws of nature. The machine symbolizes power over the world and corresponds to Plato's form of the Good as the ultimate end of action, with the mechanical order having functioned as an ideal to be attained from when the virtue of temperance was represented as a clock in the fifteenth century to the Nazi ideal of the Thousand Year Reich and the late twentieth century ideal of the information society. But at the same time the mechanistic analogy implies that there is no feature of the world which is not totally subordinated to this form of mechanical order, and denies any reality to power. In short, the machine as a symbol of the meaning of life represents total power as the ultimate end while denying both intelligibility and meaning to this end. It is an orientation to the world in which heroic moralism has been transmuted into heroic nihilism, the culmination of Western Europe's individualist and activist Christian Neoplatonism, which fulfils itself by denying meaning to everything in the world, including heroism. These ambiguities, which were already present in germinal form in Plato, have underlain all subsequent history of Western civilization.

81. See Richard Olson ed., *Science as Metaphor*, Belmont, Calif.: Wadsworth, 1971, p.59.

82. Cited by Olson in Olson ed., *Science as Metaphor*, p.60.

83. Cited by Herbert Butterfield in *The Origins of Modern Science 1300-1800*, 2nd ed., London: Bell, 1957, p.50.

# 6

# MECHANICAL NATURE AND MECHANICAL HUMANS: THE TRIUMPH OF NIHILISM

While the conception of nature as mechanical did much to justify the increasing efforts to control it in the interests of the developing economy and to discredit the Hermetics, another important feature from the point of view of the emerging capitalist socio-economic formation was to provide an analogy for understanding humanity. This amounted to a dismissal of the higher, spiritual side of the Christian Neoplatonic dualism and the affirmation of its degraded side as the sole reality, both in theory and in practice. The only meaning this allowed in the world was the satisfaction of appetites and the struggle for the power to achieve such satisfaction. It undermined both what was left of feudal ideas of nobility and of the communalistic ideology of the Hermetics. It justified a life devoted to the self-interested pursuit of profit by arguing that there is no alternative, and it was able to be used to legitimate the overthrow of any political order which stood in the way of self-interest. Though he himself was actually opposed to the claims to legitimacy and power of the rising commercial classes, the most important philosopher involved in the development of a mechanistic view of humans was Thomas Hobbes (1588-1679). To understand the achievement and significance of Hobbes it is essential that he be seen as the figure who reconceptualized the nature of humans and their relationships on the basis of the mechanistic conception of the world, since nearly all the specific notions which he argued for had already been developed by earlier thinkers.

Hobbes, like Bodin, was primarily concerned with justifying the rule of an absolute monarch in order to ensure peace. He took as his starting point the view of humans as they had largely become at that stage of European civilization and as they had already been described by Machiavelli: egoistic, ruthless and manipulative; and he formulated his ideas as a 'civil science' to investigate the rights of states and the duties of subjects. But the distinctive feature of Hobbes' approach was that he formulated his ideas in terms of the 'resolutive-compositive' method of the physical sciences, conceiving of society and of people as mechanisms which could be explained by analysing them into their constituents, then logically deriving their properties as a consequence of the motion of these constituents.

To begin with, Hobbes analysed society into its constituent members, arguing:

> For as in a watch, or some such small engine, the matter, figure and motion of the wheels cannot be known, except it be taken in sunder, and viewed in parts; so to make a more curious search into the rights of states, and duties of subjects, it is necessary, (I say not to take them in sunder, but yet that) they be so considered, as if they were dissolved.[1]

Since the driving principles of the constituents of society have to be understood independently of society in order to explain it, individuals have to be seen as moved entirely by self-interest. The conception of individuals as self-interested was justified by seeing them as mechanisms, since as such, by the very fact of their existence they must be seen as arrangements of matter organized to maintain themselves and augment their power to do so. As Hobbes argued: 'every man ... shuns ... death; and this he doth, by a certain impulsion of nature, not less than that whereby a stone moves downward.'[2] All aversions were seen as impulsions to avoid anything impeding such motion, and

1. Thomas Hobbes, *De Cive*, N.Y.: S.P. Lamprecht, 1949, p.10f.

2. Ibid. Ch.1, sec.7, p.26.

all appetites as impulses towards anything which would assist this motion. Accordingly, he concluded that: 'men from their very birth, and naturally, scramble for everything they covet, and would have all the world, if they could, to fear and obey them.'[3] All the mental faculties were understood accordingly. Thought was seen to be internal motion deriving from sensations, motions produced from the effects of objects outside the body on the sense organs. Reason as the regulation of the contents of the mind by desire and design was thereby reduced to the process of calculation in the service of appetites and aversions, and voluntary action was seen as action brought about by this internal motion. Thus Hobbes inverted Plato's scheme in which the appetites are properly subordinated to spirit and spirit is properly subordinated to intellect. Appetites and aversions are the unchosen ultimate human ends. Striving for honour is only striving for signs of power as a means to satisfy appetites or avoid aversions, reason is simply an instrument in such striving and science is merely knowledge of how to bring about different effects.

Good and evil were then redefined as simply what is subjectively desired and what arouses aversion.[4] Laws of nature were referred to, but Hobbes defined this notion in a quite different sense from that associated with the natural law tradition. His is not a moral rule but a council of prudence, for, he says, 'A Law of Nature ... is a Precept, or generall Rule, found out by Reason, by which a man is forbidden to do, that, which is destructive of his life, or taketh away the means of preserving the same...'[5] Similarly the notion of 'right' is simply a description of how men do in fact act -: the Right of Nature, 'is the Liberty each man hath, to use his own power, as he will himselfe, for the preservation of his own Nature; that is to say, of his own Life...'[6], with liberty being understood as the absence of external impediments. Justice was then redefined as the performance of covenants in accordance with a civil power able to coerce people to accept them. These reformulations involved not only the abandonment of the notion of justice as fairness, but a reformulation of the notion of liberty so as to separate it from the power to shape one's destiny and thereby also to separate it from its relation to the liberty of society as a whole. People are not seen as choosing their own ends or requiring the means to do so. These ends are brute realities impelling people to act, and the task of political philosophy is to show how these ends can best be achieved. Hobbes argued on the basis of his presentation of individuals as engaged in an endless struggle for power, that without the covenants of society backed by force people are perpetually at war with one another so there is 'continuall feare, and danger of violent death; And the life of man, solitary, poore, nasty, brutish and short.'[7] Hobbes' argument for obeying the covenants of society, acknowledging the power of an absolute sovereign in all circumstances except in cases where one's life is threatened, is that it is the best way of achieving one's egoistic ends. Correspondingly, Hobbes' recommendation that sovereigns should concern themselves with the welfare of their subjects was based entirely on the argument that this is the best way to maintain their power.

Despite the obviously mechanistic language used by Hobbes, there has been a tendency in modern political philosophy to ignore the role of the mechanistic materialist metaphysics in his political writings. This began with Leo Strauss' argument of 1936 that Hobbes developed all his political ideas before becoming acquainted with the methods and ideas of the new science, and that the framework in which his ideas are presented is a mere cloak for these.[8] Strauss himself seems to have abandoned this view,[9] but recent scholarship continues to describe Hobbes as merely having

---

3. Thomas Hobbes, *The English Works of Thomas Hobbes*, ed. Sir William Molesworth, 11 volumes, London: John Bohn, 1839-45, Vol. VII, 'Decameron Physiologum: The Dialogues of Natural Philosophy', p.73.

4. Thomas Hobbes, *Leviathan*, [1651] ed. C.B. Macpherson, Harmondsworth: Penguin, 1980, Part I, Ch.6, p.129.

5. Ibid. Part I, Ch.14, p.189.

6. Loc.cit.

7. Ibid. Part I, Ch.13, p.186.

8. Leo Strauss, *The Political Philosophy of Hobbes: Its Basis and Its Genesis*, tr. Elsa M. Sinclair, [1936], Chicago, University of Chicago Press, 1952, esp. pp.1-5.

9. Leo Strauss, *Natural Rights and History*, Chicago: University of Chicago Press, 1953, pp.166-77.

revised the concepts of his predecessors.[10] These tendencies bear commenting on because they reflect the sterility of modern political philosophy.

Firstly they reflect the failure of political philosophers to appreciate the nature of the intellectual creativity involved in conceptual revolutions. To fail to recognize Hobbes' achievement is equivalent to failing to recognize the advances Galileo and Newton made in the study of kinematics over the fourteenth century proponents of 'impetus' theory. The point about Hobbes' achievement is that while most of Hobbes' particular views about people and society had already been proposed, these were part of a vast profusion of ideas with no criteria for choosing between them. The Medieval world-orientation was disintegrating and there was no solid foundation for legitimating either political or ethical views. In particular, few people accepted Machiavelli's ideas, but the basis for rejecting them was dissolving. The Thirty Years War which decimated Europe was largely a reflection of this state of affairs. What Hobbes did was to use the achievements of the new mechanical philosophy to provide such a foundation. In doing so, he provided a new way of conceiving humans and transformed the concepts of ethics and political philosophy.

The second way in which modern political philosophy reveals its sterility is in taking for granted that the abstruse trivialities of modern political thought, in which philosophers produce ideas with a multiplicity of minor variations, is the only possible form of this subject. This fails to recognize the extent to which Hobbes' succeeded in providing a reference point and a research program for understanding humanity which has been the foundation of Western thought, particularly Anglophone thought, in this area ever since. It has underlain the fields of politics, economics and psychology, and the modern doctrines of rights theory, utilitarianism, mainstream economic theory, Social Darwinism and behaviourist psychology are simply the working out of this research program.

## *The Evolution of Mechanistic Materialism and Capitalism*

Once framed, the mechanistic world-view not only served to legitimate the capitalist socio-economic formation and its supporters. It came to be embodied by this formation and its members so as to largely constitute the relationships between people and nature, individuals and society, and interpersonal relationships. In this way it has been reproduced as part of the self-production of the capitalist system and its social relations. Consequently the development of capitalism and the development of the mechanistic world-view have been intimately related, and has further exemplified the tendency for people to use nature as an analogy for understanding society and society as an analogy for understanding nature. To begin with the mechanistic conception of people was spelt out to interpret and legitimate the development of capitalism. This culminated with the rise of modern economics. This, together with the associated Malthusian theory of population, then served as an analogy for understanding the diversity and evolution of species in nature. Evolutionary theory in turn provided the basis for a new development in ideas about society, which are at present being reapplied to nature.

The first major figure to develop the mechanistic conception of humanity to legitimate capitalistic social relations was John Locke (1632-1704). Locke's ideas were not particularly consistent; his importance at the time rested with his willingness to justify the claims to power and the practices of the rising bourgeoisie. This ability to present the venal practices of this class in a favourable light can only be understood against the image of humans implied by the mechanistic view of the world. Locke developed Hobbes' notion that political power was based on a social contract, but argued that the state of nature was not a struggle of all against all, but a social order

---

10. See for example Richard Tuck, *Natural Rights Theories: Their Origin and Development*, Cambridge: Cambridge University Press, 1979. However for a contrary view see R. Ashcraft, 'Hobbes's Natural Man', *Journal of Politics*, Vol. XXXIII, 1971.

governed by natural law. However he represented this social order as essentially an exchange economy of rational, property owning egoists. He argued that in such a state of nature, 'every man has a *property* in his own *person*' which includes his labour, and extends to whatever he has 'mixed his labour with'.[11] It is labour, he argued, which contributes almost all the value to anything, that in most of what is useful, 'ninety-nine hundredths are wholly to be put on the account of labour.'[12] The acquisition of property from the commons by labour was justified by reference to God's command that we subdue the earth, and the reason for establishing political institutions and for creating civil society was to enforce the rules under which this economy functioned: 'The great and chief end therefore, of men's uniting into commonwealths, and putting themselves under government, is the preservation of their property'.[13] By conceiving of property, including labour, as independent of civil society, Locke was able to defend the unequal distribution of property and, once money had been introduced, the unlimited acquisition of wealth by the propertied class, to justify the sovereignty of the propertied class to maintain the conditions under which wealth could be appropriated, and at the same time to rule out any interference by government in their acquisition of property. Apart from the right to sell their labour for wages, the working class were to have no political rights, as their impoverished lives would not allow them to develop a sufficient degree of rationality.[14] The proper end of the economy of the nation was to acquire gold and silver in order to quicken and increase trade.

Locke used his philosophy to justify the enclosure movement by which the commons was appropriated by the wealthy farmers at the expense of the peasants. This reduced the yeomanry who had formed the backbone of Cromwell's strength to poverty, depriving them of access to the minimum of natural resources required to live, and forcing them to live solely by selling their labour power for wages. This was the first instance of capitalist appropriation of surplus value.[15] Locke argued that where such people were unable to obtain employment, they should be forced into workhouses which were to become sweated-labour manufacturing establishments. He believed that the children of the unemployed above the age of three, who had hitherto been a burden on the nation, could also be forced to earn more than their keep.[16]

Locke was also important for his development of Hobbes' mechanistic account of the mind. He conceived of thought as a mechanical association of ideas produced in the mind by the effect of matter on the body, and argued that knowledge is the representation of the primary qualities of the external world. This conception of mind and its associated empiricist epistemology undermined any effort to question or replace the mechanical view of the world and any rational ethics, leading Locke to claim that 'Good and evil are nothing but pleasure and pain, or that which occasions or procures pleasure or pain to us.'[17] These irrationalist implications of this conception of mind and its contents were fully spelt out by David Hume (1711-76) who argued that: 'Reason is the discovery of truth or falsehood. Truth or falsehood consists in an agreement or disagreement either to the real relations of ideas, or to real existence and matter of fact. Whatever, therefore, is not susceptible of this agreement or disagreement is incapable of being true or false, and can never be the object of

---

11. John Locke, 'An Essay Concerning the True Original, Extent and End of Civil Government' sect. 27; in *Social Contract*, ed. Sir Ernest Barker, London: Oxford University Press, 1971, p.17.

12. Ibid. sect. 40, p.25.

13. Ibid. sect. 124, p.73.

14. On Locke's view of the working class see C.B. Macpherson, *The Political Theory of Possessive Individualism: Hobbes to Locke*, Oxford: Oxford University Press, 1964, p.223ff.

15. Karl Marx, *Capital*, Moscow: Progress Publishers, 1974, Volume 1, p.676f.

16. See Macpherson, *The Political Theory of Possessive Individualism*, Ch.5, esp. p.222.

17. John Locke, *An Essay Concerning Human Understanding*, Bk 2, Ch.28, sect. 5. Similarly Spinoza wrote: 'By *good* I here mean every kind of pleasure, and all that conduces thereto ... By *evil*, I mean every kind of pain..." *Ethics*, Part III, Prop.XXXIX, Note in Spinoza, p.156.

reason.'[18] Hume concluded from this that passions, volitions and the ends of actions have nothing to do with reason, that, 'Reason is, and ought to be, the slave of the passions, and can never pretend to any other office than to serve and obey them.'[19] And he condemned metaphysics, calling for works on this subject to be committed to the flames.

With nothing of value in the world but the subjective experiences of the individual, there could no longer be any reason for individuals to concern themselves with anything but the satisfaction of their own appetites and obtaining the means thereto. This justified libertinism. As Diderot (1713-84) has his interlocutor in *Rameau's Nephew* say:

> So long live philosophy and long live the wisdom of Solomon - drink good wine, blow yourself out with luscious food, have a tumble with lovely women, lie on soft beds. Apart from that the rest is vanity... What does it matter whether you have a position or not so long as you are rich, since you only take up a position in order to get rich? Fulfilling your duties, where does that land you? Into jealousy, upsets, persecution. Is that the way to get on? Butter people up, good God, butter them up, watch the great, study their tastes, fall in with their whims, pander to their vices, approve their injustices. That's the secret.[20]

But libertinism assumes the conception of people as objects to be manipulated, and libertinism and attempts to gain control over people have always been closely associated. If all that is of significance is one's own subjective experiences, then everything in the world, including other people, are only of significance as instruments for achieving one's desired subjective states. And if people are moved only by the desire to achieve pleasurable subjective states, then controlling the conditions for achieving such pleasure provides the means for their total control. The idea of the machine at the same time provides an ideal of what such complete control would be.

So the real importance of conceiving of people as mechanisms animated by appetites and aversions was that it paved the way for a socially engineered society in which people were reduced to nothing but instruments. Michel Foucault described the emergence of this society of engineered people:

> The great book of Man-the-Machine was written simultaneously on two registers: the anatomico-metaphysical register, of which Descartes wrote the first pages and which the physicians and philosophers continued, and the technico-political register, which was constituted by a whole set of regulations and by empirical and calculated methods relating to the army, the school and the hospital, for controlling and correcting the operations of the body.... The human body was entering a machinery of power that explores it, breaks it down and rearranges it. A "political anatomy", which was also a "mechanics of power", was being born; it defined how one may have a hold over others' bodies, not only so that they may do what one wishes, but so that they may operate as one wishes, with the techniques, the speed and the efficiency that one determines. Thus discipline produces subjected and practiced bodies, "docile" bodies.[21]

The most important thinkers for the development of this orientation were the utilitarians, in particular Helvetius (1715-71) and Bentham (1748-1832). They proposed a mechanics of politics to maintain order in society by manipulating the mass of pleasure-pain mechanisms composing it for the good of society as a whole as they conceived it, that is, to produce the maximum amount of pleasure with the minimum amount of pain. Bentham's proposals for prison reform were

---

18. David Hume, *A Treatise of Human Nature*, London: Dent, in two volumes, Volume II, Bk III, Part 1, sect. 1. p.167.
19. Ibid. Volume II, Bk II, Part III, sect. 3., p.127.
20. Denis Diderot, *Rameau's Nephew and D'Alembert's Dream*, tr. L. Tancock, Harmondsworth: Penguin, 1976, p.65.
21. Michel Foucault, *Discipline and Punish*, p.138 & 140.

particularly important in facilitating the embodiment by society of the mechanistic conception of people as objects to be efficiently controlled. This reform was based on a new architectural model, the 'Panopticon', consisting of a central tower from which all prison cells could be observed without the prisoners knowing when they were being observed. Each prisoner was to be perfectly individualized and constantly visible. Bentham's influence began with prisons and reformatories, but the Panopticon was designed as a general model for the control of people and he proposed it explicitly for 'manufactories', 'mad-houses', 'hospitals' and 'schools'. In concluding his work, Bentham accurately prophesied that people 'should see a new scene of things spread itself over the face of civilization.... All by a simple idea in architecture.'[22] As Foucault noted: 'The panoptic arrangement provides the formula for ... generalization. It programmes, at the level of an elementary and easily transferable mechanism, the basic functioning of a society penetrated through and through with disciplinary mechanisms.'[23] By such generalization all the major institutions together with the buildings in which they were situated came to define people as objects to be manipulated and moulded to function as cogs within the social machine, creating a totally transparent society. These resonated with each other, allowing the same social schemes of perception and thought to be generalized from institution to institution.

## *Political Economy*

The development of political economy was also founded on a mechanistic conception of people, and furthered its domination. Traditionally economics had meant 'household management'. Xenophon's *Oikonomikos* written before the middle of the fourth century B.C. began with a long introduction on the good life and the proper use of wealth, and included sections on the leadership qualities necessary for a householder and on wifely virtues and the training of a wife. The longest section of all was on the practicalities of farming. There was nothing on economic analysis, the efficiency of production or on marketing. This way of understanding economics survived almost unaltered to the eighteenth century.[24] In his *Short Introduction to Moral Philosophy* published in Latin in 1742, Adam Smith's teacher, Francis Hutcheson, devoted the first chapters of Book III, entitled 'The Principles of Oeconomics and Politics,' to marriage and divorce, the duties of parents and children, and the relationship between masters and servants. Otherwise it was exclusively about politics. Various subjects associated with modern economics, such as money, interest rates and trade were investigated, but except for the finances of rulers which were treated as economic issues by analogy, these subjects were dealt with as part of the domain of ethics or political philosophy. The 'economy' had not become an object in its own right, and at least theoretically the ends to be achieved by society: enabling people to live the good life, to achieve salvation or whatever, had been defined independently of economics.

But Hobbes had begun a reorientation in thinking. In developing his conception of humans in *Leviathan*, he had written a chapter, 'Nutrition and Procreation of a Commonwealth', using Harvey's conception of the body as an analogy to describe society.[25] The nutrition of the social body was seen to depend upon the fruits of land either given freely or through the exchange of labour. The circulation of nourishment was seen to be facilitated by money, which Hobbes compared to blood. The driving force for the functioning of this process was seen to be the self-interest of society's component individuals. And the only end of value in this system was seen to be the satisfaction of appetites and avoidance of aversions of individuals and the reproduction of the

---

22. Bentham, 'Panopticon', *The Works of Jeremy Bentham*, Bowring ed. Edinburgh, Vol. IV. p.66.
23. Foucault, *Discipline and Punish*, p.209.
24. See M.I. Finley, *The Ancient Economy*, Berkeley and L.A.: University of California Press, 1974, p.18f.
25. Hobbes, *Leviathan*, Pt II, Ch.24.

system as a whole. While Hobbes still did not separate political economy from political philosophy, he provided the framework on which all later economic thought has been based.

Hobbes' ideas were first taken up by William Petty. Though Petty was primarily concerned with such issues as raising finance for the government and developed his ideas unsystematically, the influence of Hobbes led him to conceive money in terms of its role in the functioning of the system and to conceive wealth as the effect of present or past labour. Petty was the first to investigate the velocity of circulation of money and to attempt to measure national income, though he still recognized land as a major contributor to wealth. 'Labour', he wrote 'is the Father and active principle of Wealth, as Lands are the Mother.'[26]

Petty influenced the French thinker Cantillon and the Physiocrats, who then fully developed the conception of society as an organism requiring a flow of nutrition, and it was in relation to this effort that the term Political Economy was first coined. Cantillon defined land as the source of wealth, labour as the power which produces it, and all material products as its constituents, and was followed in this by the Physiocrats. The culminating work of the Physiocrats was Quesnay's *Tableau Economique* which presented in an abstract mathematical model the flow of commodities throughout the entire process of production and consumption in accordance with natural law, moral as well as physical, preordained by God. The system was divided into three classes: the farming class, the landowner class, and a 'sterile' class that included manufacturers. The farming class was seen as the source of all wealth. Only the land was seen as having a real potential to generate more than had been put into it by labour. The surpluses from farming flowed to the landowners and from there to the 'sterile' sectors of the economy. Thus the basis of wealth was rent on land, and the surpluses acquired by the landowners and spent by them on consumption flowed through the entire system.

The notion of the economy as an object of investigation in its own right was developed to its fullest extent in the eighteenth century by Adam Smith (1723-90) in his *The Wealth of Nations*. Smith was strongly influenced by the epistemology of his close friend, Hume. He took scientific theories to be means of organizing sense impressions. In his study of the development of astronomy he compared such theories to imaginary machines which are 'invented to connect together in the fancy those different movements and effects which are already in reality performed.'[27] His study of political economy was thus undertaken in the spirit of the physical sciences as a disinterested search for the simplest imaginary machine to account for the phenomena. He described the principles (the original formulation of which he attributed to Descartes) which he believed distinguished Newtonian astronomy as superior to its predecessor:

> [I]n Natural Philosophy, or any other science of that sort, we may either, like Aristotle, go over the different branches in the order they happen to [be] cast up to us, giving a principle, commonly a new one, for every phenomenon; or, in the manner of Sir Isaac Newton, we may lay down certain principles, primary [known?] or proved, in the beginning, from whence we account for the several phenomena, connecting all together by the same chain. This latter, which we may call the Newtonian method, is undoubtedly the most philosophical, and in every science, whether of Morals or Natural Philosophy, etc., is vastly more ingenious, and for that reason more engaging, than the other.[28]

Smith recast Political Economy to accord with this ideal.

---

26. *The Economic Writings of Sir William Petty*, ed. C.H. Hull, 2 vols, Cambridge: C.U.P., 1899, II, p.377.

27. Adam Smith, 'The History of Astronomy' in *Essays on Philosophical Subjects* ed. W.P.D. Wightman and J.C. Bryce, Oxford: Clarendon Press, 1980, IV.19, p.66.

28. Adam Smith, *Lectures on Rhetoric and Belles Lettres* ed. John M. Lothian, Edinburgh: Edinburgh University Press, 1963, p.139f. For a study of Smith's philosophy of science, see Andrew Skinner 'Science and the Role of Imagination' in Andrew S. Skinner, *A System of Social Science: Papers Relating to Adam Smith*, Oxford: Clarendon, 1979, Ch.2.

To achieve his task, Smith construed Political Economy in a narrower sense than the French political economists. Rejecting the physiocratic conception of it as dealing with the happiness and improvement of political society, Smith conceived of it as a branch of the science of a statesman or legislator having only two objects: to enable people to enrich themselves, and to provide public revenue.[29] While in *The Theory of Moral Sentiments* Smith had developed a conception of human motivation antithetical to that of Hobbes, in *The Wealth of Nations* Smith was compelled to conceive people in abstraction from their social relations and therefore as egoists in order to have independent constituents and a single principle in terms of which the economic mechanism could be explained. He assumed that all people have an innate disposition to struggle to better their condition which 'comes with us from the womb and never leaves us until we go to the grave',[30] and that there is 'a certain propensity in human nature... to truck, barter, and exchange one thing for another.'[31] This self-interest is the driving force of the economy. As Smith wrote: 'It is not from the benevolence of the butcher, the brewer, or the baker that we expect our dinner, but from their regard to their own interest.'[32]

This view of humans together with the Lockean notion that people have a right to consume or exchange the products of their labour (what they have mixed their labour with) provided Smith with the basis for an objective theory of value. He argued:

> The real price of everything, what everything really costs to the man who wants to acquire it, is the toil and trouble of acquiring it... What is bought with money or with goods is purchased by labour as much as what we acquire by the toil of our own body... They contain the value of a certain quantity of labour which we exchange for what is supposed at the time to contain the value of an equal quantity.[33]

Although he did not fully develop this labour theory of value (it was only fully developed in the nineteenth century by Ricardo and Marx), Smith's theory was sufficient to enable him to characterize and explain in accordance with his conception of science the economy as a mechanism, driven by the efforts of individuals to better their condition, in which commodities are produced and circulated through the exchange of money. The quantity of labour embodied in commodities functioned as the equivalent of matter in physics as the unchanging substance circulated through society. At the same time it provided the basis of a new conception of progress as the development of the economy's capacity to produce saleable commodities.

This reformulation of economics involved the rejection of the Physiocrats' analysis of the productivity of the different classes of commercial society. Smith argued that this society consists of landlords, wage-earners and capitalists receiving rents, wages and profits for their participation in the system. Since the pursuit of each of these is required for labour to produce and for its productivity to increase, he argued that all three classes are productive. This paved the way for the idea that it is the search for profit by capitalists which is the prime driving force for the accumulation of capital, and thereby for economic progress. In this way nature was relegated in status from the generator of all true wealth in society to a mere abstract limitation on economic progress, and all human ends apart from those exchangeable as commodities were excluded from consideration. This conception of nature and this exclusion of all but the most base human interests was later presupposed by Ricardo and Malthus.

---

29. Adam Smith, *An Inquiry into the Nature and Causes of The Wealth of Nations* (1776) ed. Edwin Cannan, N.Y.: The Modern Library, 1965, IV, Introd., p.397.

30. Ibid. p.324.

31. Ibid. p.13.

32. Ibid. p.14.

33. Ibid. p.30.

Smith's system was based on an image of the economic system as a machine and of humans as Puritans moved by sober self-interest. This image of the economy was presented as a model of reality and that of humans as a universal characterisation of human nature; but surreptitiously both these images function in his system as Platonic forms of how society should be and how individuals should act to realize the ideal form of society. This is no accident; it was clearly recognized as such by Smith who, strongly influenced by Plato, had argued in *The Theory of Moral Sentiments*:

> ...if you would implant public virtue in the breast of him who seems heedless of the interest of his country, it will often be to no purpose to tell him, what superior advantages the subjects of a well-governed state will enjoy... You will be more likely to pursuade, if you describe the great system of public police which procures these advantages, if you explain the connexions and dependencies of its several parts, their mutual subordination to one another... if you show how... all the several wheels of the machine of government be made to move with more harmony and smoothness, without grating upon one another, or mutually retarding one another's motions.[34]

A good machine is one which functions efficiently, and for the economic machine to function efficiently, individuals must accord with the image of humans presented by Smith. Thus idleness, profligacy and unproductive spending were seen as vices and attributed to the royal courts and their aristocratic retinues while industry and thrift were continually praised. This ideal of efficiency was synthesized by Smith into his theory along with the mechanistic conception of nature as devoid of significance except insofar as it could be moulded for human purposes and sold on the market, the atomistic view of society, rights theory and utilitarianism, and the notion of divine providence, 'the invisible hand' which ensures that individuals pursuing their own interests will increase the wealth of the nation to the benefit of all. As Marx put it:

> This sphere... within whose boundaries the sale and purchase of labour-power goes on, is in fact a very Eden of the innate rights of man. There alone rule Freedom, Equality, Property and Bentham. Freedom, because both buyer and seller of a commodity, say of labour-power, are constrained only by their own free will. They contract as free agents, and the agreement they come to, is but the form in which they give legal expression to their common will. Equality, because each enters into relation with the other, as with a simple owner of commodities, and they exchange equivalent for equivalent. Property, because each disposes only of what is his own. And Bentham, because each looks only to himself. The only force that brings them together and puts them in relation with each other, is the selfishness, the gain and the private interests of each. Each looks to himself only, and no one troubles himself about the rest, and just because they do so, do they all, in accordance with the pre-established harmony of things, or under the auspices of an all shrewd providence, work together to their mutual advantage, for the common weal and in the interest of all.[35]

Along with Millar, Lord Kames, Ferguson, Robertson and Dalrymple (and paralleling the work of the physiocrats in France, particularly Turgot), Smith placed this analysis of capitalism in a broader scheme of philosophical history, a secularized version of the Christian notion of Universal History centred on the notion of 'Progress'. According to this general theory, which was inspired by the work of Montesquieu, society naturally tends to progress over time through four more of less distinct modes of subsistence.[36] These are hunting, pasturage, agriculture and commerce. To

---

34. Adam Smith, *The Theory of Moral Sentiments*, Indianapolis: Liberty Press, 1982, Part IV, Ch. 1, Section 11, p.186.

35. Marx, *Capital*, Vol.l, p.172.

36. For an analysis of the development of this theory of history see Ronald Meek, 'The Scottish Contribution to Marxist Sociology', in Ronald L. Meek, *Economics and Ideology and Other Essays,* London: Chapman and Hall, 1967, and

each of these modes of subsistence there correspond different sets of ideas and institutions relating to law, property, and government, and also different sets of customs, manners, and morals. The driving force leading from one stage to the next is the self-interest of historical actors striving to better their condition. Commercial society represents the highest achievement in this effort.

Thus Smith both described, legitimated and reinforced the development of commercial society and the market form of relationship between people. While Smith conceived of his work on economics as part of a sociological history of society, and as subordinate to political philosophy which in turn was seen as part of moral philosophy, he paved the way for the creation of political economy as an independent discipline. Economics came to be the prime interpreter of society to its members, providing them with the concepts in terms of which they were able to define and legitimate their relationships to each other, to society and to nature. The effect of economics was to contribute to the growing autonomy of the market system from political and social control. With the development of this system, people became nothing but labour power to be sold on the market. The bulk of the population, including children, were forced to work in miserable conditions for long hours in factories and mines, while at the same time large numbers of people were deprived of any means of obtaining a livelihood. This led Malthus to develop his ideas on population according to which there is an inevitable process of immiseration of the working class, since any increase in income leads to an increase in population until population growth is again checked by starvation. Malthus argued that since economic growth could only occur arithmetically while population growth occurs geometrically, it is impossible for economic growth to ever get ahead of population growth for any length of time. God has designed the world so that the indolent are automatically punished. Malthus concluded:

> A man who is born into a world already possessed, if he cannot get subsistence from his parents on whom he has a just demand, and if the society does not want his labour, has no claim of right to the smallest portion of food, and, in fact, has no business to be where he is. At Nature's mighty feast there is no vacant cover for him. She tells him to be gone.[37]

## *Darwinism and Social Darwinism*

The picture of humanity inspired by capitalist society provided an analogy for the further development of the mechanistic view of nature.[38] Hitherto, it had been impossible to account for the ordered nature of matter composing the forms of life in terms of mechanistic science except by invoking the intervention of the Deity. Evolutionary theories had been proposed to account for this order using progress in society as an analogy, but there was no satisfactory mechanism proposed to account for such evolution. With the development of capitalist society, the emergence of the new political economy and the publication of Malthus's work, Darwin and Wallace were provided with the means for conceiving such a mechanism. Darwin, who avidly read the works of the political economists, and particularly Malthus, was familiar with the way cattle breeders improved their stock through breeding. Using nineteenth century English society as an analogy, Darwin was able to conceive the breeder within nature which led to the origin and diversity of species as population

---

'Smith, Turgot, and the "Four Stages" Theory', in Ronald L. Meek, *Smith, Marx and After*, London: Chapman and Hall, 1977, and Andrew Skinner, 'A Scottish Contribution to Marxist Sociology?' in *Classical and Marxian Political Economy*, ed. Ian Bradley and Michael Howard, London: Macmillan, 1982, pp.79-114; and 'Historical Theory' in Skinner, *A System of Social Science*, pp.68-103. See also Gladys Bryson, *Man and Society: The Scottish Inquiry of the Eighteenth Century*, Princeton: Princeton University Press, 1945.

37. Cited from the first edition of Thomas Malthus' *An Essay on the Principle of Population* by Eduard Heimann in *History of Economic Doctrine*, Oxford: Oxford University Press, 1964, p.86f.

38. On this, see Robert M. Young, *Darwin's Metaphor: Nature's Place in Victorian Culture*, Cambridge: C.U.P., 1985.

pressure generating a struggle of all against all, allowing only the fittest offspring to survive. So, as economic development came to be understood as the product of individuals pursuing their own interests in competition with everyone else, with the population being regulated through the starvation of the less fit, evolution came to be understood as the product of competition for survival in nature; and as economic progress came to be understood in terms of technological adaptation, the evolution of life came to be understood in terms of the development of new forms of adaptation. As Marx wrote to Engels, 'It is remarkable how Darwin recognizes among the beasts and plants his English society with its division of labour, competition, opening up of new markets, "inventions," and the Malthusian "struggle for existence."'[39]

Darwin made some attempt to apply his ideas about evolution to humanity in *The Descent of Man*. In this work he endorsed the oppression of non-white races on the grounds that the 'unfit' must inevitably make way for the more 'fit', and he praised the role of capitalism in facilitating this, asserting: 'the inheritance of property by itself is very far from an evil; for without the accumulation of capital the arts [technologies] could not progress; and it is chiefly through this power that the civilized races have extended, and are now everywhere extending their range, so as to take the place of the lower races.'[40] However Darwin himself did not develop evolutionary theory into a general theory of humanity and society, and into the general cosmology which came to be known as Social Darwinism.[41]

The most important figure in the development of Social Darwinism was Herbert Spencer - although it would be more appropriate to describe Darwin as a Spencerian than Spencer as a Darwinist. Spencer had argued in *Social Statics*, published in 1851, eight years before the publication of *The Origin of Species* for a reformulation of utilitarianism on evolutionary grounds. He argued that the greatest happiness for the greatest number could only be realized in society when each individual could 'claim the fullest liberty to exercise his faculties compatible with the possession of like liberty by every other man.' While this required a society of people who took pleasure in the pleasure of others, Spencer argued that there was an evolution of humanity towards this goal, beginning with the predatory instincts of aboriginal man 'clearing the earth of inferior races of men', followed by slavery which provided the 'stringent coercion ... required to make [aboriginal man] submit contentedly to the necessities of his new state', and finally by modern *laissez-faire* capitalism which provided the conditions for the individuation of and union between people, 'by the most elaborate subdivision of labour; that is, by the extremest mutual dependence...' In this process 'all desires inconsistent with the most perfect social organization are dying out, and other desires corresponding to such an organization are being developed.'[42] Correspondingly, Spencer argued against social reformers:

> That rigourous necessity which, when allowed to operate, becomes so sharp a spur to the lazy and so strong a bridle to the random, these pauper's friends would repeal... Blind to the fact that under the natural order of things society is constantly excreting its unhealthy, imbecile, slow, vacillating, faithless members, these unthinking, though well-meaning, men advocate an interference which not only stops the purifying process, but even increases the vitiation - absolutely encourages the multiplication of the reckless and incompetent by offering them an

---

39. Marx to Engels, June 18, 1862, in Saul K. Paddover ed. *The Letters of Karl Marx*, Englewood Cliffs, Prentice Hall, 1979, p.157.

40. C. Darwin, *The Descent of Man and Selection in Relation to Sex*, 2nd ed. London: Murray, 1874, p.207.

41. See Kenneth Bock, 'Theories of Progress, Development, Evolution' in Tom Bottomore and Robert Nisbet eds, *A History of Sociological Analysis*, London: Heinemann, 1978, pp.39-79.

42. From *Social Statics*, London: John Chapman, 1851, extracted in Herbert Spencer, *On Social Evolution*, ed. J.D.Y. Peel, Chicago: Uni. of Chicago Press, 1972, p.16 and pp.21-25.

> unfailing provision, and discourages the multiplication of the competent and provident by heightening the difficulty of maintaining a family.[43]

Thus Britain's *laissez-faire* capitalism was to realize a utilitarian heaven on earth.

However it was only after the publication of *The Origin of Species* that Spencer saw his way to developing a whole cosmology which would represent and legitimate the evolution of humanity and the development of capitalism as being part of a cosmic evolutionary scheme, and to developing a general science of humanity - an intellectual manoeuvre recognized by Frederick Engels who wrote of it:

> The whole Darwinian theory of the struggle for existence is simply a transference from society to organic nature of Hobbes' theory of *bellum omnium contra omnes* and of the bourgeois economic theory of competition, as well as the Malthusian theory of population. When once this feat had been accomplished, it is very easy to transfer these theories back again from natural history to the history of society, and altogether too naive to maintain that thereby these assertions have been proved as eternal natural laws.[44]

So in the sixth edition of *First Principles* published in 1862 Spencer argued that: 'Evolution is definable as a change from an incoherent homogeneity to a coherent heterogeneity, accompanying the dissipation of motion and integration of matter...',[45] an idea which he then devoted the rest of his life to elaborating into his 'System of Synthetic Philosophy', the theoretical framework for his massive *The Principles of Sociology*.

While grounded in evolutionary theory, Spencer's sociology can be seen as developing the philosophical history, the 'natural history of humanity', of the Scottish Enlightenment as represented by Adam Smith's *Wealth of Nations*, Millar's *Distinction of Ranks* and Ferguson's *History of Civil Society*, and as a continuation of the work of Saint-Simon and his followers towards developing a new intellectual synthesis in the service of the new industrial society. Comte, a former disciple of Saint-Simon, characterized and attempted to lay the foundations for this intellectual synthesis, representing it as 'positivism', the final triumph of science over theology and metaphysics, and as a positive philosophy to live by in place of the critical philosophy of the Enlightenment. It was Comte who coined the term sociology to characterize the study of society which was to be the highest development of positivism and the basis for a secularized ethics. Spencer reformulated Comte's project, and in taking over the concept of sociology he attempted to lay the principles for and to build a rigourous science which would also serve as an ethics.

While the notion of evolution through differentiation and integration served as the basic principle of Spencer's sociology, it was elaborated by comparing societies to organisms. Both were both seen as growing, generating increasingly complex structures with increasing dependence between parts or systems. While in accordance with a basically mechanistic view of the world Spencer defended methodological individualism, the view that society must ultimately be explained in terms of the behaviour of its component parts, he also argued that societies tend towards equilibrium. This then provided sociology with its basic concepts: structure, function, system and equilibrium which have dominated mainstream sociology ever since. In terms of these concepts Spencer examined the differentiation of societies into various institutions: ceremonial, political, religious and economic, including joint stock companies and unions.

With this Darwinian conception of humanity, the Hobbesian idea that reasoning is simply a mechanical process by which individuals calculate what is to their advantage in the struggle to increase their power, was reinforced. An organism, population or system maintains itself by

43. Herbert Spencer, *Social Statics*, abridged, together with *The Man Versus the State*, New York: Appleton, 1892, p.151.

44. Frederick Engels, *Dialectics of Nature*, Moscow: Progress Publishers, 1976, p.307f.

45. Herbert Spencer, *First Principles*, 6th ed. London: Williams and Norgate [1862], 1904, p.291.

demarcating itself from and adapting itself to a changeable, hypercomplex environment. This view of life undermines the assumption underlying rights theory that society is a rational construction based on principles which all rational people must acknowledge. As the American Social Darwinist sociologist, William Sumner wrote: 'There can be no rights against Nature except to get out of her whatever we can, which is only the fact of the struggle for existence stated over again.'[46] Similar sentiments were expressed by Austrian Social Darwinist sociologist Gumplowicz: 'The premises of "inalienable human rights" rest upon the most unreasonable self-deification of man and overestimation of the value of human life.'[47]

## *Further Refinements of the Mechanistic Image of the World*

The development of Social Darwinism coincided with the demise in economic theory of the labour theory of value with its implicit commitment to the rights of people to the products of their labour. In the 1870's classical economics began to be challenged by the neo-classical marginalist school, inspired by Jevons in England, Walras in France and Menger in Austria, which defined value entirely in terms of the subjective decisions or preferences of individuals. Neo-classical economists were committed to utilitarianism and to developing economics as 'the mechanism of utility and self-interest', as Stanley Jevons put it.[48] This facilitated the further elaboration of the mechanistic conception of society. Jevons noted:

> The Theory of Economy thus treated presents a close analogy to the science of Statical Mechanics, and the Laws of Exchange are found to resemble the Laws of Equilibrium of a lever as determined by the principle of virtual velocities. The nature of Wealth and Value is explained by the consideration of indefinitely small amounts of pleasure and pain, just as the Theory of Statics is made to rest upon the equality of indefinitely small amounts of energy.[49]

By using mathematical physics as a model, the neo-classical economists developed utilitarianism along new lines.[50] In the reformulation of physics by Lagrange and Hamilton, the total energy of a system was represented as dependent in a critical way upon the position of the mass-point. Position was defined in terms of a gravitational field, later identified as potential energy, which was described by partial differential equations with the sum of potential and kinetic energy being taken to be conserved within a closed system. This conservation law then served as the foundation for constrained maximization techniques to calculate the paths of mass-points under the influence of impressed forces. Adapting this scheme to economics, forces are redefined as prices, displacements as infinitesimal changes in the quantities of individual goods, gravitational potential energy as utility, and kinetic energy as expenditure. Constrained maximization or minimization of an imponderable quantity, 'utility', led directly to a conservative field, which in turn is seen to fix the permissible configurations of prices.

With this development of neo-classical economics, which still underlies mainstream economic thought, the scope of economics was further restricted, and in 1890, with the publication of Alfred Marshall's *Principles of Economics*, the term 'economics' as opposed to 'political economy' came into use for the first time. In the neo-classical scheme, agents are classified as consumers or

---

46. William Sumner, *What Social Classes Owe To Each Other*, New York: Harper and Rowe, 1883, p.135.

47. Ludwig Gumplowicz, *The Outlines of Sociology*, Philadelphia: Philadelphia Academy of Political Science, 1899, p.180.

48. W. Stanley Jevons, *The Theory of Political Economy*, 4th ed. London: Macmillan, 1911, p.21.

49. Ibid. p.viii. Walras defended the new approach to economics on the same grounds.

50. The nature of this development has been analysed by Philip Mirowski in 'Physics and the Marginalist Revolution', *Cambridge Journal of Economics*, Vol.8, 1984, pp.361-379 and in 'Mathematical Formalism and Economic Explanation' in *The Reconstruction of Economic Theory*, Boston: Kluwer-Nijhoff, 1986, pp.179-240.

producers and are simply assumed to have 'tastes' or 'goals' which, subject to certain constraints, they seek to satisfy to the maximum. There was no attempt to analyse the content of these tastes, goals or constraints. And while economics based on the labour theory of value had at least included nature as something to be worked on and had focussed attention on how much surplus labourers were capable of producing over their own needs, marginalism virtually excluded nature from consciousness. The economic process was represented as a circular diagram between production and consumption within a completely closed system. Any shortage of natural resources due to the destruction of nature was registered as increasing prices, and therefore as increasing national income. The jettisoning of the labour theory of value also excluded from consciousness all consideration of the real contribution to production of various participants in the economy. It justified the efforts of people to obtain whatever returns they could through the market, whether through the selling of their labour power, through exploiting labour power in manufacturing, or through speculative buying and selling of commodities and companies.

This fitted in with Social Darwinism, and by identifying evolutionary progress with the survival of the fittest and economic success with survival, those who were successful in business were provided with justification for their behaviour. Thus John D. Rockefeller declared in a Sunday-school address: 'The growth of a large business is simply the survival of the fittest ... This is not an evil tendency in business. It is merely the working out of a law of nature and a law of God.'[51] By winning out in the economic struggle ruthless capitalists were furthering evolutionary progress. The amount of money they made could then be regarded as both as a quantitative measure of their value to evolutionary progress, a reward for their contribution to this progress, and a means to enable them to perpetuate themselves. Another multimillionaire American capitalist, Andrew Carnegie, expressed in his autobiography how troubled he was at the collapse of Christian theology, until he read Darwin and Spencer. 'I remember that light came as in a flood and all was clear.' he wrote. 'Not only had I got rid of theology and the supernatural, but I had found the truth of evolution.'[52] Through competition humanity was continually evolving upwards in an endless march towards perfection. In another essay he wrote of this law of competition: 'It is here; we cannot evade it; no substitutes for it have been found; and while the law may sometimes be hard for the individual, it is best for the race, because it ensures the survival of the fittest in every department.'[53] As the evolutionary notion of progress provided the secular equivalent of the Christian notion of historical progress, being wealthy became the secular equivalent of being one of the elect, with participation in the form of money replacing participation in the forms of Christian virtue as the defining criterion of election.

However the development of capitalism outside Great Britain generated global competition between industrial powers for control of resources. Under these circumstances Herbert Spencer's optimistic identification of industrialism with the end of military societies lost its plausibility, and Social Darwinists were concerned not only with the competition between individuals, but also with the competition between nations and races. As such, it was used to justify both the destruction of races already conquered, and the destruction of people yet to be conquered. In Britain Sir Francis Galton wondered that 'there exists a sentiment, for the most part quite unreasonable, against the gradual extinction of an inferior race.'[54] In the 1870's the Argentine government decided in to apply the scientific principles of Social Darwinism by exterminating all the remaining Indians in the country. In 1885 in the USA the Rev. Josiah Strong published the best selling *Our Country: its Possible Future and Its Present Critics* in which he argued that the world was entering on a new

---

51. Cited by Richard Hofstadter, *Social Darwinism in American Thought*, Boston: Beacon Press, 2nd ed., 1955, p.45.

52. *Autobiography of Andrew Carnegie*, Boston, 1920, p.327; quoted by Hofstadter, p.45.

53. Ibid. p.45f.

54. F. Galton, *Inquiries into Human Faculty and Its Development* 2nd ed. New York: Dutton, 1883; cited by R.C. Lewontin, Steven Rose, and Leon J. Kamin, *Not In Our Genes: Biology, Ideology, and Human Nature*, N.Y.: Pantheon Books, 1984, p.22.

stage of history, 'the final competition of races for which the Anglo Saxon is being schooled.' Strong argued that this race would soon move down on Mexico, Central and South America and over Africa, and declaimed: '[C]an anyone doubt that the result of this competition of races will be the "survival of the fittest"?'[55] This sentiment swept the nation. Later the US President, Theodore Roosevelt, declared: 'If we stand idly by, if we seek merely swollen, slothful ease and ignoble peace, if we shrink from the hard contests where men must win at hazard of their lives and at the risk of all they hold dear, then the bolder and stronger peoples will pass us by, and will win for themselves the domination of the world.'[56]

With this more nationalist and racist form of Social Darwinism, a more activist orientation by business managers and governments was called for. This was associated with developments in the human sciences oriented towards predicting and controlling human behaviour. It was this, according to Hamilton Cravens, which effected a major shift in civilization.[57]

To begin with Social Darwinism inspired the eugenics movement which led to efforts by Galton, Cattell, Burt, Spearman and others to develop tests to measure mental ability in order to estimate people's genetic endowment of intelligence. These tests were subsequently used to stream students in education and to develop institutions to improve the genetic stock of society, either by screening immigrants or by segregating the intellectually feeble from the rest of the population to prevent contamination of the genetic stock. Justifying the use of intelligence testing in schools, the influential American educator Ellwood P. Cubberley wrote:

unis.

> Our schools are, in a sense, factories in which the raw products (children) are to be shaped and fashioned into products to meet the various demands of life. The specifications for manufacturing come from the demands of twentieth-century civilization, and it is the business of the school to build its pupils according to the specifications laid down. This demands good tools, specialized machinery, continuous measurement of production to see it is according to specifications, the elimination of waste manufacture, and a large variety in the outputs.[58]

The testing movement, financed by corporate foundations - particularly the Carnegie Foundation, helped meet the need for 'continuous measurement' and the 'elimination of waste manufacture', and it justified eugenic policies. As one of the most influential psychologists involved in the development of I.Q. testing, Edward L. Thorndike wrote in 1940:

> By selective breeding supported by a suitable environment we can have a world in which all men will equal the top ten per cent of present men. One sure service of the *able* and *good* is to beget and rear offspring. One sure service [about the only one] which the inferior and vicious can perform is to prevent their genes from survival.[59]

However the most important role of I.Q. testing was to justify the social divisions in society as being part of the natural order of things.

At the same time, F.W. Taylor developed his time and motion studies of work into a system of scientific management, dedicated to reducing people to 'trained gorillas' so totally submerged in the rationalized work process that pleasure, sensuality, and critical thinking would be almost totally

55. Cited by Hofstadter *Social Darwinism in American Thought*, p.179 from *Our Country*, p.175.

56. Theodore Roosevelt 'The Strenuous Life' [1899] in *The Works of Theodore Roosevelt*, XIII, p.331; cited ibid. p.180.

57. Hamilton Cravens, *The Triumph of Evolution*, Philadelphia: University of Pennsylvania Press, 1978, esp. p.7.

58. From *Public School Administration*, Boston: Houghton Mifflin Co. 1916, p.338. Cited by Clarence J. Karier, 'Testing for Order and Control in the Corporate Liberal State', in Ned Block and Gerald Dworkin eds, *The I.Q. Controversy*, London: Quartet Books, 1977, pp.339-369, p.344. This is a superb study of the proponents of I.Q. testing, their relation to eugenics and their influence.

59. E.L. Thorndike, *Human Nature and the Social Order*, N.Y.: Macmillan, 1940, p.957; cited ibid. p.347.

stifled, in order to maximize industrial output. This was associated with the development of the production line, exemplified in Ford's car factories.

The impetus to conceive life mechanistically combined with a concern to control people more efficiently then inspired the development of the behaviourist sciences of humanity in which humans were represented as nothing but stimulus-response mechanisms. Behaviourism can be seen as the continued development of the utilitarian program to control society according to scientific principles. It was a science of human behaviour designed to accord with Darwin's revolution in biology, modelled on the prevailing image of the physical sciences - in the hope that this would enable it to achieve a similar level of technological control over people as had been achieved over the physical world. People trained in behaviourist approaches to understanding people found employment in advertising agencies, in large business organizations and in government, particularly in education and military organizations. Hannah Arendt spelt out the significance of this:

> If economics is the science of society in its early stages, when it could impose its rules of behaviour only on sections of the population and on parts of their activities, the rise of the 'behavioural sciences' indicates clearly the final stage of this development, when mass society has devoured all strata of the nation and 'social behaviour' has become standard for all regions of life.[60]

Social Darwinism played a major part in the ideology of all participants in the First World War and it remained the dominant framework of social and political thought until the Second World War. In Germany, where Darwinism and Social Darwinism had been reformulated and popularized by Ernst Haeckel (1834-1919), Social Darwinism flowered with the rise of Nazism.[61] The Social Darwinist basis of Nazism was given typical expression by the Austrian Nobel Prize laureate Konrad Lorenz who wrote in 1940:

> The selection of toughness, heroism, social utility ... must be accomplished by some human institutions if mankind in default of selective factors, is not to be ruined by domestication induced degeneracy. The racial ideas as the basis of the state has already accomplished much in this respect.[62]

Himmler was an ardent exponent of Haeckel's ideas. Dr Mengler was an eminent geneticist. Hitler so fully embraced Social Darwinism and the idea that progress occurs through the struggle between nations and races that he accepted the right of the Russians to destroy Germany after they had demonstrated their military superiority. But the Nazis were only putting into practice and carrying out the logical implications of the most respectable ideas of twentieth century science.

## *The New World of Information and Cybernetics*

But Hitler's Germany was defeated, and the identification of Social Darwinist ideas with both Nazism and the economic ideas which had brought about the Great Depression temporarily weakened their influence. The Allies disowned their earlier Social Darwinian rhetoric, while the Germans were more or less compelled as losers to disavow Social Darwinism as part of their display of contrition for killing thirty odd million people. Consequently for the twenty five years after the Second World War the main goal in Western societies came to be the maintenance of full

---

60. Hannah Arendt, *The Human Condition*, Chicago: University of Chicago Press, 1958, p.45.

61. On this see Daniel Gasman, *The Scientific Origins of National Socialism: Social Darwinism in Ernst Haeckel and the German Monist League*, New York: Elsevier, 1971.

62. Quoted by Lewontin et.al. *Not in Our Genes*, p.30.

employment without inflation and the maintenance of political stability by providing for the needs of the entire population, with the exception of a few dark-coloured minorities.

The development of the welfare State which emerged from this was associated with the dominance of Keynesian economics (reformulated in USA to accord with neo-classical economics) in which macro-economics was separated from micro-economics, and macro-economics raised to pre-eminence. This tended to further reify the economy, shifting the object of concern away from satisfying the needs of individuals to the state of the economy as a whole. The aim of Keynesian economists was to eliminate economic fluctuations and inflation, and to maintain a sufficient rate of growth to maintain full employment. Since the economy must expand at 4% per annum to achieve this, the sale of the most wasteful, harmful goods and services was economically justified to keep the economy expanding. This involved the replacement of the utilitarian conception of humans as rational hedonists with definite wants to be satisfied, by a conception of humans as irrational hedonists whose wants had to be manipulated so they would act in a way appropriate to the requirements of the economy.[63] At the same time this form of economics continued to exclude from consciousness the environmental impact of economic growth and the ways in which the centres of the world-economy are draining off capital and non-renewable resources from the economic peripheries.[64]

With the growing complexity of government and business organizations associated with the more active role of the State and the expansion of capitalism after the Second World War, a new science of management emerged designed to enable managerial elites to control society in their interests. The 'socio-technical systems theory' which developed from this united information theory, cybernetics, operations research, games theory and cost-benefit analysis with systems theory into a generalized instrument for control.[65] In doing so, it has refined and reintegrated the mechanistic conception of humanity, and finally enabled Social Darwinism to be revived. As Mike Hales pointed out:

> Although developed in reaction to the mechanistic nature of Taylorism, the socio-technical approach contains its own form of mechanism, rather broader in conception and more subtle, but still mechanistic. The operator of an effectively designed automated system fits into the machinery of production in as calculated a way as does the assembly-line worker.... Management science which leans on systems theory tends to have a strong stream of cybernetics in it, and both system theorists and cyberneticians view organisms and organizations as organic machines in which the parts are significant only with respect to the functions they perform in the adaptation of the whole to the environment.[66]

The most important component of this new science of control has been information theory. Information theory was originally developed as a means to understand and develop the technology of communications and to provide a mathematical description of the laws governing systems designed to transmit and manipulate information. To do so quantitative measurements of information and of the capacity of various systems to transmit, store and otherwise process it were set up. Cybernetics, which emerged from the efforts to develop control systems for machines, was then formulated in the same terms. In information theory, information is treated as a statistical

63. The changing image of humans in economics is described in Walter A. Weisskopf, *The Psychology of Economics*, Chicago: University of Chicago Press, 1955; and 'The Image of Man in Economics' in *Social Research*, Vol. 40, 1973, pp.547-63.

64 On this, see Dieter Groh and Rolf-Peter Sieferle, 'Experience of Nature in Bourgeois Society and Economic Theory: Outlines of an Interdisciplinary Research Project,' *Social Research*, No.3, 1980, pp.557-581.

65. For an analysis of this see Robert Lilienfeld, *The Rise of Systems Theory: An Ideological Analysis*, N.Y.: Wiley, 1978.

66. Mike Hales, 'Management Science and "The Second Industrial Revolution"', *Radical Science*, ed. Les Levidow, London: Free Association Books, 1986, pp.62-87, p.72f.

quantity, a signal is considered as a particular choice from a statistical ensemble of possible signals, and the effectiveness of a control system in processing and measuring information is then measured by some kind of average for all the possible signals in the ensemble. This enables it to be related to the rest of physics through thermodynamics. Thermodynamics originated in nineteenth century efforts to measure the efficiency of machines, but was reformulated in terms of the principle of the conservation of energy and then systematized by Clausius, who introduced the concept of entropy to characterize the irreversibility of energy transformations. At this stage, thermodynamics was thought to be inconsistent with a mechanical view of the world, and 'energism' based on thermodynamics was promoted in opposition to it. But then Bolzmann sought to explain the macroscopic thermodynamic properties of systems in terms of the laws of motion of elementary particles - that is, in accordance with Newtonian physics, by conceiving entropy in terms of the statistical probability of occurrence of arrangements of atoms within systems. A low entropy system is a highly improbable arrangement of matter. A number of physicists who had speculated on the nature of entropy suggested that it could be described as loss of information. However it was not until C.E. Shannon working on the theory of information produced an equation equivalent to Clausius' equation defining entropy, that negative entropy could be formally defined as information (although thermodynamics and information theory remain largely independent disciplines). While systems theory originated in 'energism', and while its most eminent proponents, von Bertalanffy, Paul Weiss and Ervin Laszlo have been concerned to replace the mechanical view of the world, the development of information theory facilitated its incorporation back into the mechanistic world-view. The mechanistic version of systems was refined with the development of information processing technology, particularly computers, and all organizations came to be represented as information processing mechanisms, as cybernetic or self-regulating feed-back systems.[67]

With this development, mechanists were provided with new concepts to analyse the nature of life, society and the human mind, and a means to further specify the meaning of efficiency as the form to be aimed at by organizations. The ideas of systems theory were applied in ecology in the 1950's by Eugene Odum who developed his notions on the basis of flow charts of energy, using such concepts as 'productivity', 'efficiency', and 'yield' in accordance with the emphasis on efficient control. As Donald Worster commented on this: 'In this age of computer-run organizations and the carefully arbitrated resolution of all discords, it was probably inevitable that ecology too would come to emphasize the flow of goods and services - or of energy - in a kind of automated, robotized, pacified nature.'[68] With the reformulation of systems theory in terms of information theory, information theory was also incorporated into biology. In population biology forms of optimization theory have been taken over from management science as techniques for the prediction and explanation of evolution.[69] It is assumed that organisms are struggling for resources that are in short supply, that they must invest time and energy to acquire these resources and then reinvest the returns partly in acquiring fresh supplies of resources and partly in reproducing. That organism is most successful which acquires the greatest net surplus for investment in successful reproduction. Similarly, as noted by Lewontin and his colleagues, Crick's metaphor of DNA controlled protein production is based on the sophistication of modern economies in which considerations of production are becoming less significant than control and management:

> It was to this new world that information theory, with its control cycles, feedback and feed-forward loops, and regulatory mechanisms, was so appropriate; and it is in this new way that the molecular biologists conceive the cell - an assembly-line factory in which the DNA

---

67. For a representative sample of writings on early systems theory, see F.E. Emery ed. *Systems Thinking*, Harmondsworth: Penguin, 1969.

68. Don Worster, *Nature's Economy: The Roots of Ecology*, New York: Anchor Books, 1979, p.313.

69. See for example John Maynard Smith, 'Optimization Theory in Evolution' in *Conceptual Issues in Evolutionary Biology*, Elliot Sober ed., Cambridge Mass.: M.I.T. Press, 1984, pp.289-315.

> blueprints are interpreted and raw materials fabricated to produce the protein end products in response to a series of regulated requirements.[70]

Such imagery dominates the textbooks and teaching of the new biology, with drawings laid out in assembly-line style.

With this revised conception of nature, it could again serve as an analogy to understand and legitimate developments within society. Social Darwinism had been in the background as the hard-headed view of society even during its apparent eclipse after the Second World War. It was implicitly held by most of those people working on I.Q. testing and elaborating mechanistic views of humans and functionalist views of society. Typically, if somewhat more honestly than most behaviourist psychologists, the foremost exponent of the stimulus-response analysis of human behaviour B.F. Skinner wrote: 'A scientific analysis may lead us to resist the mere blandishments of freedom, justice, knowledge, or happiness in considering the long run consequences of survival.'[71] However until the mid-1960's this Social Darwinism was only defended surreptitiously. The resurfacing of fully fledged Social Darwinism incorporating the new information systems theory began in sociology.

Sociology in the United States had been dominated since the 1930's by Talcott Parsons. Parsons' main concern had been to legitimate and shore up liberal capitalism while it was felt to be under threat from Communism.[72] While formulating his ideas through a critical analysis of European thinkers, notably Weber, Durkheim and Pareto, and while overtly committing himself to the philosophy of Alfred North Whitehead,[73] Parsons inherited and worked within the basic framework of categories of Spencer's sociology - those of system, structure and function - but without Spencer's historical dimension. In the 1960's Parsons attempted to reformulate his theory to take history into account by using ideas from cybernetics, and in so doing, reconstructed the full Social Darwinian framework.[74] Society was represented as evolving by functional differentiation and then reintegration through a hierarchy of cybernetic systems, ranging from the economy which gets information feedback through the symbolic medium of money to regulate adaptation to the natural environment, the personality system which gains information feedback through the symbolic medium of power to serve the function of goal attainment: establishing priorities between goals and mobilizing system resources for their attainment, the social system getting feedback through the symbolic medium of influence which functions to integrate the acting units, to the cultural system which gets its feedback through the symbolic medium of commitment and functions to maintain and control tensions in the social pattern. Each of these four subsystems was seen as differentiating into four sub-subsystems, each subsystem with its own information feedback. Everything in society was then explained and evaluated in terms of its contribution to improving the capacity of society as a whole for survival and expansion in its competition with other such systems.

While in the United States Parsons' ideas were eclipsed by approaches devoted to providing knowledge for social engineering, this redirection of sociology was itself underpinned by Social Darwinism and was justified by Parsons' systems approach; and Parsons' ideas have recently been

---

70. Lewontin et.al. p.59.

71. B.F. Skinner, *Science and Human Behaviour*, New York: Macmillan, 1953, p.436.

72. On this, see Alvin Gouldner, *The Coming Crisis of Western Sociology*, N.Y.: Equinox, 1971.

73. On Parsons' relationship to Whitehead, see Andrew S. Dawson, *Parsons and Whitehead*, M.A. thesis, La Trobe University, 1991.

74. See Talcott Parsons, *Societies: Evolutionary and Comparative Perspectives*, Englewood Cliffs: Prentice Hall, 1966 and 'Some Problems of General Theory in Sociology' in J.C. MacKinney and E.A. Tyriakian eds, *Theoretical Sociology: Perspectives and Developments*, N.Y.: Appleton, Century & Crofts, 1970.

revived.[75] They have been taken up and developed by general systems theorists, most notably by the German sociologist and former student of Parsons, Niklas Luhmann whose ideas fully justify the reduction of sociology to social technology.

After the revival of Social Darwinism in sociology, the new field of sociobiology emerged. While this was ultimately inspired by Spencer, it was based on the synthesis of population biology, the new genetics of Watson and Crick, and ethology.[76] Genes, represented as repositories of information, were made the focus of evolutionary theory, and were represented as organizing and constructing bodies, including humans, as mechanisms for their survival.[77] Drawing on the cost-benefit analysis, investment opportunity cost, games theory and so on, sociobiologists again used social relations as analogies for understanding nature to explain away the appearance of altruistic behaviour of animals, representing it as nothing but the programming of bodies by genes to perpetuate their own kind. These ideas were then extrapolated back to society, slightly modified in some cases to take into account the self-replicating nature of culture, to account for human behaviour. To take culture into account, Richard Dawkins coined the term 'meme' to refer to the units of cultural transmission.[78] In this way socio-biologists have offered a new justification for seeing late capitalism as in accordance with nature, and Dawkins has accordingly criticised the welfare state as unnatural.

Complementing sociobiology both in terms of the concepts used and the implications of the ideas being promoted, philosophers and psychologists have been using information theory, cybernetics and analogies with computers in their renewed attempts to explain the mind, or rather 'intelligence', mechanistically, representing humans as cybernetic organisms or 'cyborgs'.[79] Anglo-American philosophy has come to be dominated by the philosophy of language, the major proponents of which are those dedicated to representing language as a rule-governed mechanism for encoding and conveying knowledge or information about the world. Cognitive psychology has replaced behaviourism as the dominant paradigm in psychology, but this merely allows for an inherited information processing mechanism to function between the stimulus and the response, and occasionally for reflexive feedback circuits. Closely associated with developments in the philosophy of language and cognitive psychology, evolutionary epistemologists have refurbished the instrumentalist theory of knowledge, in some cases representing scientific theories as nothing but ways of gaining, organizing and processing information about the world selected according to their contribution to systems in their struggle for survival.[80] And the doyen of computer science, Marvin Minsky, has argued accordingly that since computers are rapidly overtaking humans in their capacity to process information, they must be regarded as the next stage of evolution.

---

75. On this see Richard Münch, 'Parsonian Theory Today', in Anthony Giddens and Jonathon Turner, *Social Theory Today*, Cambridge: Polity Press, 1988, pp.116-155.

76. The founder of sociobiology, E.O. Wilson reportedly looks upon Spencer as 'one of his heroes' (Michael Ruse, *Taking Darwin Seriously: A Naturalistic Approach to Philosophy*, Blackwell: Oxford, 1986, p.97). E.O. Wilson's main work is *Sociobiology: the New Synthesis*, Cambridge: Harvard Univerity Press, 1975. The most popular exposition of sociobiology is Richard Dawkins *The Blind Watchmaker*, Harmdondsworth: Penguin, 1988.

77. Richard Dawkins, *The Selfish Gene,* Oxford: Oxford University Press, 1976, esp. p.21.

78. Ibid. Ch.11.

79. See D.C. Dennett *Brainstorms*, Hassocks: Harverster Press, 1978; J.A. Fodor, *The Language of Thought*, Hassocks: Harvester Press, 1976; and M.A. Boden, *Minds and Mechanisms,* Hassocks: Harverster Press, 1983 and *Computer Models of the Mind: Computational Approaches in Theoretical Psychology*, Cambridge: C.U.P.: 1988. Jeremy Campbell has described these developments in very general terms in *Grammatical Man: Information, Entropy, Language and Life*, Harmondsworth: Penguin, 1984.

80. On Darwinian epistemology see Ruse, *Taking Darwin Seriously*, pp.148-206. Not all evolutionary epistemology is mechanistic. See Kai Hahlweg and C.A. Hooker eds, *Issues in Evolutionary Epistemology*, N.Y.: S.U.N.Y. Press, 1989 on this.

At a practical level the resurfacing of Social Darwinism has been associated with renewed efforts to mechanize social control. It has been associated with the use of data from I.Q. testing by Jensen and Eysenck to justify discriminating between races and to justify less spending on the education of the underprivileged, in the explanations of rebellious behaviour in terms of people's defective brains, and in the rejection of the demand by women for equality on the grounds that male domination is built into our genes by generations of evolution. It underlay support for the new Cold War of the 1970's and 80's and the massive increases in expenditure on armaments. With increasing levels of unemployment in wealthy nations, with a massive redistribution of income from the poor to the rich both within nations and between nations and with increasing starvation in Third World countries, it has been important for enabling the sufferings of the poor to be dismissed as the inevitable by-product of progress as the weak and unintelligent are expelled from the system.

The new Social Darwinism provided the intellectual climate necessary for the demise of Keynesian economic theory in favour of monetarism, rational expectations theory, and supply-side economics, the doctrines associated with efforts by the New Right to dismantle social welfare provisions and institutions and to promote free markets to allow a struggle for survival between individuals and firms to generate economic progress.[81] And Social Darwinism is now filling a gap left by Keynesian economic theory and behavioural sciences as to what is the point of continued economic growth. Rather than seeing the purpose of economic growth as simply maintaining full employment and indefinitely increasing levels of consumption, and the development of the human sciences as simply means to manipulate people to ensure that they continue to consume more and feel good while they are doing so, economic growth and the control of people required to achieve it are increasingly understood as part of the struggle for power between nations. Comparisons are continually being made between economic growth rates and rates of technological advance (particularly of information technology) in different countries. It is assumed that those nations with the fastest growth rates will dominate those nations with slower growth rates in the future as Western societies with their more efficient economies have dominated other societies in the past. The next stage of this growth is seen as the integration of information technology into administration to achieve greater control over economic and social processes of society and economists have turned to systems analysis, games theory and cost-benefit analyses to interpret and regulate the functioning of the economic system. Anything which does not contribute to such economic advance and the development of technology (particularly information technology), whether this be social welfare or education which does not churn out business administrators and technocrats, is seen as an expendable luxury.

The political programme associated with this new Social Darwinism was developed and partly implemented by Zbigniew Brzezinski. Brzezinski argued that we are entering the 'technetronic' (technological + electronic) age in which humans will be remoulded by the new technologies and sciences associated with information processing, and he outlined the possibilities and dangers this posed for US power.[82] He then played a leading role in the establishment of the Trilateral Commission composed of leading ruling class figures from North America, Europe and Japan designed to guide a network of inter-imperialist co-ordination in accordance with this image of the future, and then as President Carter's National Security Advisor, steered US foreign policy into the new militarism which was brought to fruition under President Reagan.[83] The political agenda of

81. For an analysis of the new economics see Lester C. Thurow, *Dangerous Currents: The State of Economics*, Oxford; O.U.P., 1983; and Andrew Gamble, 'The Political Economy of Freedom', in *The Ideology of the New Right*, ed. Ruth Levitas, Cambridge: Polity Press, 1986, pp.25-54.

82. See Z. Brzezinski, *Between Two Ages: America's Role in the Technotronic Era*, Harmondsworth: Penguin, 1976.

83. For a critique of this, see Collettivo Strategie, 'The "Technetronic Society" According to Brzezinski', in Tony Solomonides and Les Levidow eds, *Compulsive Technology*, London: Free Association Books, 1985, pp.126-138. On Brzezinski's role on the foreign policy of the Carter administration, see Fred Halliday, *The Making of the Second Cold War*, 2nd ed., London: Verso, 1986, p.216, 218 and onwards.

Reagan was made clear by a Reagan advisor, Simon Ramo, who argued for 'the principal political need as being not more democracy but more systems analysis, since the more precisely a system can be modelled, the more easily it can be controlled from the top.'[84]

## *The Collapse into Nihilism*

Yet this is not a full description of the ideology of the modern era. Quite apart from the presence of its negation by various romantic idealists and the persistence of relics from earlier eras: concern with nobility by upper classes in Europe, concern with rights at least for people of European origin in USA, and so on, there is a dim recognition that Social Darwinism, or euphemistically, belief in 'progress', does not overcome the meaninglessness of a mechanistic vision of the world. If life is just an endless struggle for survival, then what is the point of anything? In fact Social Darwinism was the final product of a culture which has reduced the world to an instrument for a transcendental purpose, and foreshadowed the complete dissolution of all reference points which had given meaning to this instrumentalization.

It was Nietzsche who realized the impasse which had been reached by European civilization. As he summed this up:

> From time immemorial we have ascribed the value of an action, a character, an existence, to the *intention*, the *purpose* for the sake of which one has acted or lived: this age-old idiosyncrasy finally takes a dangerous turn ... there seems to be in preparation a universal disvaluation: 'Nothing has any meaning' - this melancholy sentence means 'All meaning lies in intention, and if intention is altogether lacking, then meaning is altogether lacking too.' In accordance with this valuation, one was constrained to transfer the value of life to a 'life after death,' or to the progressive development of ideas or of mankind or of the people or beyond mankind; but with that one has arrived at a *progressus in infinitum* of purposes: one was at last constrained to make a place for oneself in the 'world process' (perhaps with the dysdaemonistic perspective that it was a process into nothingness).[85]

Correspondingly, the heroic moralism of European culture has been gradually whittled down. As Max Weber, who regarded Nietzsche as one of the greatest thinkers of the nineteenth century, argued in the conclusion to his study of the Spirit of Capitalism: 'The rosy blush of its laughing heir, the Enlightenment, seems to be irretrievably fading, and the idea of duty in one's calling prowls about in our lives like the ghost of dead religious beliefs... [O]f the last stage of this cultural development, it might truly be said: "Specialists without spirit, sensualists without heart; this nullity imagines that it has attained a level of civilization never before achieved."'[86] This was written a few years before the First World War which was soon followed by the Great Depression, then by the Second World War. All that remains for this culture which now dominates the world is an endless quest for more efficient means to dominate nature and people, a quest which is foundering on the environmental crisis. We now live in a post-Christian, post-Enlightenment age.

---

84. David Dickson, *The New Politics of Science*, 2nd ed., Chicago: Chicago University Press, 1988, p.54.

85. Friedrich Nietzsche, *The Will to Power*, tr. Walter Kaufmann and R.J. Hollingdale, N.Y.: Vintage, 1968, §666.

86. Max Weber, *The Protestant Ethic and the Spirit of Capitalism*, tr. Talcott Parsons, London, George Allen & Unwin, 1930.

# 7

# NIHILISM INCORPORATED

The essential features of the world-orientation which has dominated European civilization have now been revealed. By comparison with virtually every other culture that has existed, European culture promotes an extreme individualism and an extreme detachment from and instrumentalisation of the world: both of nature and of people. Rather than experiencing themselves as participants in the stream of life, in the becoming of the world, people formed by European culture experience themselves as transcendent consciousnesses in a world of 'things' or 'objects' located in space and only externally related to each other. Progress, the ultimate concept of evaluation, is conceived as the increasing subordination of the world to this transcendent consciousness; as the transformation of the entire world into a vast machine serving human purposes (or rather, the purposes of the power élite). As Robert Jungk wrote of the most extreme development of European culture, USA:

> America is striving to win power over the sum total of things, complete and absolute mastery of nature in all its aspects... To occupy God's place, to repeat his deeds, to recreate and organize a man-made cosmos according to man-made laws of reason, foresight and efficiency: that is America's ultimate objective... It destroys whatever is primitive, whatever grows in disordered profusion or evolves through patient mutation.[1]

The rejection of life's spontaneity in favour of what is eternal has been most clearly manifest in the intellectual life of European culture. As Nietzsche noted, 'To impose upon becoming the character of being - that is the supreme will to power.'[2] Thus Nietzsche wrote of the idiosyncrasies of philosophers:

> There is ... their hatred of even the idea of becoming, their Egyptianism. They think they are doing a thing *honour* when they dehistoricise it, *sub specie aeterni* - when they make a mummy of it. All that philosophers have handled for millenia has been conceptual mummies; nothing actual has escaped their hands alive. They kill, they stuff, when they worship, these conceptual idolaters - they become a mortal danger to everything when they worship. Death, change, age, as well as procreation and growth, are for them objections - refutations even. What is, does not *become*; what becomes *is* not ... Now they all believe, even to the point of despair, in that which is.[3]

The mainstream of science is based on this philosophy, and is committed to explaining the entire universe in terms of identities, in terms of eternally valid, mathematical describable laws describing eternally self-identical elements (whether these be elementary particles or space-time points). It must deny genuine creativity, and thereby any significance to life. Again, Nietzsche identified the essential problem:

> Has not man's determination to belittle himself developed apace precisely since Copernicus? ... Ever since Copernicus man has been rolling down an incline, faster and faster, away from the

---

1. Robert Jungk, *Tomorrow is Already Here*, tr. M. Waldman, London, 1954, p.17.
2. Friedrich Nietzsche, *The Will to Power*, § 617, tr. Walter Kaufman and R.J. Hollingdale, N.Y.: Vintage, 1968, p.330.
3. Friedrich Nietzsche, *Twilight of the Idols* [1889], tr. R.J. Hollingdale, Penguin Books: Harmondsworth, 1968, p.35.

centre - whither? ... All science ... is now determined to talk man out of his former respect for himself, as though that respect had been nothing but a bizarre presumption.[4]

But the debasement of life goes much deeper than this. Here it has been shown how through the use of society as an analogy for nature and nature as an analogy for society, and the incorporation of these analogies into social practices, the whole of Western culture has incorporated this orientation.

Through Neoplatonic Christianity, the source of meaning in the world was projected onto an eternal, supersensible realm, leaving the changing sensible world to be seen as having only instrumental value in relation to this eternal order, and that after having evaporated off all meaning from the sensible world, the supersensible realm of forms itself lost its significance. After two thousand years the Platonic pursuit of Being and the rejection of becoming had been successful. To some extent this total elimination of becoming can be seen as a return to Democritus or Parmenides rather than a development of Platonic thought; but the eternal in the form of the economic machine still functions to some extent as the ideal to which everything must be made to conform. The effect of a thousand years of Christian Platonism had been to produce not simply an intellectual acceptance of these doctrines, but to have transformed and delivered the whole of Western society to this dead, meaningless, inert, Parmenidean One: the deterministic, totally predictable, totally controllable 'block' universe of mechanistic materialism. Darwinism, in which all living entities, including humans and human societies are seen as nothing but arrangements of matter instrumentalizing each other in an endless struggle for survival, the outcome of which is predetermined, not only is a major part of the scientific world-view, but has come to express the reality of everyday social life.

The pervasiveness of this world-orientation is evident everywhere. Piaget noted how very young children appropriate Newtonian concepts of space and time which took the work of men of genius to formulate in the seventeenth century. Such concepts have come to structure their entire world. The philosophy of linguistic analysts who attempted to clarify the conceptual schemes of everyday life, and the phenomenologists who attempted to apply a presuppositionless method to describe the world as it presents itself to consciousness, also revealed this mechanistic world-orientation. The linguistic analyst P.F. Strawson in *Individuals* defended a watered-down Newtonianism, while the phenomenologist Heidegger in *Being and Time* argued that the world is devoid of significance except insofar as it is being used as an instrument. He concluded that the surrounding world is essentially a world of things *zuhanden*, 'ready to use' and that it is only through their relation to our project of existence that we care for them. Anything seen as *vorhanden*, merely present, is stripped of our care and becomes a deficient mode of being, an object of mere curiosity.[5] (Heidegger rejected this mode of thinking in his later work.) Jean Paul Sartre's phenomenological investigations in *Being and Nothingness* (which were also partially transcended in Sartre's later works) revealed how this nihilism is extended to human relationships. He argued that individuals have no basis for choosing one course of action rather than another in a meaningless world, and described the relations between people as a struggle in which one individual either reduces the other to an object in relation to his or her own projects, or is reduced to an object by the other.

---

4. Friedrich Nietzsche, *The Genealogy of Morals*, tr. Francis Golffing, N.Y.: Doubleday, 1956, Third Essay, XXV, p.291f.

5. For an analysis of the roots of such attitudes see Hans Jonas, 'Gnosticism, Existentialism and Nihilism' in *The Phenomenon of Life*, Chicago: University of Chicago Press, 1966.

## *Neurotic Adaptations to Mechanistic Materialism: From Osiander to Lyotard and Habermas*

This attempt to reveal the coherence of the world-orientation underlying Western culture is in direct opposition to presently fashionable views about the modern world. For instance the 'post-modernist', Jean-François Lyotard, argues that there are no longer any grand narratives of legitimation, and that the attempt to reconstitute such a grand narrative is to be caught up in a type of thinking which is out of step with the most vital modes of knowledge in the postmodern world.[6] In the postmodern world it is necessary to recognize the heteromorphous nature of language games, none of which are epistemologically superior to any other, and to respond to the possibilities for creativity provided by these. But such claims can be best understood as the latest symptom of a neurotic adaptation to and denial of the domination of society by the mechanistic world-orientation.

Neurotic adaptation to the mechanistic world-orientation began with Osiander who wrote a preface to Copernicus' *De Revolutionibus* claiming that the theory of the sun-centred universe was only a device for simplifying the mathematics for making predictions about the motion of stars and should not be taken as a representation of reality. Then, attacking the philosophy of Descartes, Vico argued that knowledge of nature is defective and that we can only really know what humans have created: the social world. After Newton, Berkeley attempted to circumscribe the Newtonian conception of the world by claiming that the real world is the familiar world, and the world postulated by physics only deals with the 'grammar' of reality. Kant argued that the world as conceived by mechanistic science is merely experience ordered by the forms of intuition and the categories of the understanding, and is therefore not reality but the world of appearance, the phenomenal world. The real world, the noumenal world, could therefore still provide a ground for morality, though it could not be known in the same way as the world of appearance. Each of these strategies was elaborated throughout the nineteenth century and are still being elaborated. In the twentieth century, attempts to delimit the significance of scientific materialism have come from Cassirer's philosophy of symbolic forms, from the phenomenologists and from the linguistic analysts. Spelling out the implications of phenomenology, Husserl argued:

> In regard to nature and scientific truth concerning it ... the natural sciences give merely the appearance of having brought nature to a point where for itself it is rationally known. For true nature in its proper scientific sense is a product of the spirit that investigates nature, and thus the science of nature presupposes the science of the spirit... Spirit is not looked upon here as part of nature or parallel to it; rather nature belongs to the sphere of spirit... It was [transcendental phenomenology] which overcame naturalistic objectivism...[7]

And Wittgenstein argued that science is just one form of life or language game among others, without any privileged status:

> Our language can be seen as an ancient city: a maze of little streets and squares, of old and new houses, and of houses with additions from various periods; and thus surrounded by a multitude of new boroughs with straight streets and uniform houses.

---

6. Jean-François Lyotard, *The Postmodern Condition: A Report on Knowledge*, [1979] tr. Geoff Bennington and Brian Massumi, Minneapolis: University of Minnesota Press, 1984, p.14.

7. Edmund Husserl, 'Philosophy and the Crisis of European Man' in *Phenomenology and the Crisis of Philosophy*, tr. Quentin Lauer, N.Y.: Harper Torchbooks, 1965, pp.188-190.

The symbolism of chemistry and the notation of infinitesimal calculus were then described as 'suburbs of our language'.[8]

More recently Jürgen Habermas argued that there is a different rationality involved in communication from the rationality of domination which characterizes the natural sciences. In *Knowledge and Human Interests* he characterized knowledge as constituted by interests, with different interests generating different and equally valid forms of knowledge. The world as understood by the physical sciences is only the world conceived in relation to our efforts to control it, that is, to technical interests. An entirely different approach is required in the human sciences - one in which a 'practical' interest: achieving common understanding, and an emancipatory interest: freeing people from distorted forms of communication, should be the constitutive interests. Retreating from the concept of knowledge constitutive interests with its fixation on consciousness, Habermas then attempted to develop a theory of language which would reveal speech to involve commitments beyond instrumental rationality. He argued that all speech implies a commitment to an ideal speech situation free of external coercion and internal distortions in which participants would respond to the force of the better argument alone. In his *magnum opus*, *The Theory of Communicative Action*, he reformulated this, arguing that there are three validity claims implicitly raised and reciprocally recognized with the utterance of every speech-act - that the propositional content is true, that the performative component is correct, and that intentions are being expressed sincerely. These raise speech above the functionalist rationality of economic and administrative systems. The political problem of the modern world is to prevent functionalist rationality from invading the life-worlds of people and the norms established within it through relatively undistorted communication based on non-instrumental validity claims.

What makes all these efforts to bracket out or devalue the scientific conception of the world neurotic is that they not only deny the real problem, but they prevent its being properly addressed. They have insulated mechanistic materialism from more fundamental questioning. The implausibility of the attempt to circumscribe mechanistic materialism has become progressively greater with the advancement of its research programme into the life sciences and the human sciences. It is simply absurd to present evolutionary theory as anything but an account of the nature of reality, and it is hardly surprising that people take it as such and act accordingly. Husserl's philosophy did nothing to check the rise of Naziism based on Social Darwinism. At the same time such approaches pre-empt the efforts to get at the roots of the problem, to challenge mechanistic materialism on its own ground as the best metaphysical foundation for understanding nature. Thus the tradition which began with Leibniz of attempting to replace mechanistic materialism, the tradition of which process philosophy is the furthest development, has been pushed into the background and almost submerged. Postmodernism and similar intellectual movements which fail to acknowledge the existence of a dominant and coherent world-orientation dominating our whole civilization can be seen as the latest stage of this neurotic adaptation.

Illustrating this, Lyotard defines his position in opposition to the 'modern', in opposition to 'any science that legitimates itself with reference to a metadiscourse ... making an explicit appeal to some grand narrative, such as the dialectics of Spirit, the hermeneutics of meaning, the emancipation of the rational or working subject, or the creation of wealth.'[9] He sets out to oppose two legitimating discourses: the speculative narrative which from the time of Humboldt's university reforms in Germany in the early nineteenth century had been used to justify science as part of the Spirit's self-formation, and the emancipative narrative (of Habermas) which divides the domain of truth from the ethico-political and denies any necessary passage between the two, thereby placing both on the same level. But his real concern is not with these, since he claims that: 'The grand narrative has lost its credibility, regardless of what mode of unification it uses, regardless of

8. Ludwig Wittgenstein, *Philosophical Investigations*, tr. G.E.M. Anscombe, 3rd ed. Oxford: Basil Blackwell, 1968, §18, p.8.

9. Lyotard, *The Postmodern Condition*, p.xxiii.

whether it is a speculative narrative or a narrative of emancipation.'[10] His anger is directed against the performativity criterion of knowledge (described and defended by Niklas Luhmann), where scientific knowledge is seen as self-validating by virtue of the power it generates, the criterion which grand narratives fail to effectively challenge. With this new criterion, truth, efficiency and wealth come to be synonymous. Science becomes a force of production and scientists are purchased to augment power. Such power is absolute legitimation 'since performativity increases the power to produce proof, it also increases the ability to be right... It is self-legitimating, in the same way a system organized around performance maximization seems to be.'[11]

Lyotard's proposed solution to this problem is based on his argument that in fact science is not as the systems theorists such as Luhmann have presented it to be. He argues that modern science - postmodern science - has revealed through quantum theory, thermodynamics, catastrophe theory etc. the impossibility of attaining the form of total control and predictability aspired to by systems theorists. According to Lyotard, this science, a science 'concerning itself with such things as undecidables, the limits of precise control, conflicts characterized by incomplete information, "fracta", catastrophes and pragmatic paradoxes',[12] cannot be legitimized by grand narratives or by maximized performance criteria, but only by local narratives which inform its powers of imaginative invention. It is to Wittgenstein's philosophy of language games we should turn for intellectual salvation from nihilism.

By taking as his starting point German philosophy which, as the product of Germany's late involvement in the rise of the West, had been dominated by the attempt to encompass and supersede the mechanistic world-orientation through either a modified Neoplatonism deriving from Hegel or a free floating critical rationality deriving from Kant, Lyotard has failed to see that the systems approach and the performativity criterion he is attacking is the culmination of mechanistic materialism and Social Darwinism. It is the apparent lack of success of the project of German philosophy which has created the illusion that there is no super-ordinate framework of legitimation. But what this failure indicates is the victory of mechanistic materialism and Social Darwinism, a success so complete that no further narrative is required. The performativity criterion which Lyotard is so opposed to does not make science self-legitimating, nor is power self-legitimating. These are legitimated by the underlying theory of being which dominates the modern world, and by the Social Darwinism which is based upon it, and only an explicitly developed alternative theory of being articulated into a grand narrative could challenge this world-view. The developments in science which undermine the performativity criteria which Lyotard refers to are in fact the products of a struggle against the mechanistic world-view - and would not have been possible without the efforts by philosophers such as Bergson and Whitehead to create an alternative metaphysics, and associated with this, an alternative grand narrative to that which has dominated Western civilization. Lyotard seems blind to these broader intellectual struggles.

Lyotard's argument that grand narratives which put all particular narratives and discourses into perspective are neither possible nor desirable, echoes the ideas of Nietzsche who spelt out the implications of the collapse of the attempts by Kant and Hegel to transcend the mechanical view of the world.[13] However Nietzsche thought through the implications of the failure to discover a

---

10. Ibid. p.37.

11. Ibid. p.46f.

12. Ibid. p.60.

13. Lyotard as part of the postmodern movement in France is a manifestation of a dialectic of ideas which has been reproduced at least three times, first with the development in Germany from Kant's attempt to circumscribe the mechanical world-view to Hegel's attempt to subsume it under a broader metaphysical scheme, to the collapse of systematic philosophy culminating in the work of Nietzsche, and second in the development of German Marxism from Kautsky's mechanistic formulation of Marx's ideas to the efforts to graft a Kantian ethics onto this formulation by Vörlander and Bernstein, followed by a reformulation of Marx from the perspective of Hegel by Lukács and Korsch, followed by a rejection of the quest for a totalizing perspective by Horkheimer and Adorno. In the third instance, French philosophy

universally shareable framework of discourse - that relations between people can only be relations of power, that 'life is essentially appropriation, injury, overpowering of the strange and weaker, suppression, severity, imposition of one's own forms, incorporation and, at the least and mildest, exploitation - '.[14] That is, allowing perspectives to proliferate without making any effort to evaluate them must lead in practice to the adoption of the performative criterion that Lyotard is opposing, which in turn will lead to an acceptance of mechanistic science which at the same time legitimates the performativity criterion. This is in fact what happened when the Wittgensteinian philosophy championed by Lyotard came to dominate philosophy in Great Britain. Philosophy was smothered, leaving the way for the complete domination of intellectual life by the performativity criterion of a debased science. And the developments in science which Lyotard praised as a source of hope for the future are being severely hampered precisely because of the unchecked redirection of science into the development of technology which has resulted from the triumph of the performativity criterion. As Peter Dews argued in his critique of postmodernism (which included Foucault, Derrida and Lacan as well as Lyotard), the magical assumption that the fragmentation of knowledge will somehow break the grip of an oppressive social order is unsustainable. He concluded:

> ... the rejection of the claims of an integrated critical stand-point in post-structuralism, in the mistaken belief that such a stand-point implies repressive totalization, is far from providing a more decisive liberation from the illusions of philosophy, and a more powerful illumination of the contemporary world. The fate of post-structuralism makes clear that critique is not a question of the arbitrary and coercive espousal of premisses and precepts, but rather of commitment to that coherence of thought which alone ensures its emancipatory power.[15]

Dews himself had in mind the later work of Habermas as the coherent system of thought which could serve this purpose. But while one can admire Habermas' heroic effort to systematically defend rationality in a nihilistic age, and agree with his effort to replace a philosophy of consciousness by a philosophy of communicative action, his whole approach represents a failure of nerve in the face of the mechanistic world-view. As he wrote in the first page of *The Theory of Communicative Action:*

> Philosophy can no longer refer to the whole of the world, of nature, of history, of society, in the sense of a totalizing knowledge. Theoretical surrogates for worldviews have been devalued, not only by the factual advance of empirical science but even more by the reflective consciousness accompanying it. With this consciousness philosophical thought has withdrawn self-critically behind itself; in the question of what it can accomplish with its reflective competence within the framework of scientific conventions, it has become metaphilosophy.[16]

In deferring to present scientific conventions, Habermas has accepted the divisions between domains of rational discourse imposed by the mechanistic world-orientation - essentially those proposed by Kant (after having given up the effort to replace the categories of Newtonian physics): theoretical, practical and aesthetic discourse, with two more domains added on: therapeutic critique and explicative discourse.[17] So, in rejecting totalizing philosophy, he has formulated his own

---

developed from Leon Brunschvicg's Neo-Kantianism to the Hegelianism of Kojève, Hyppolite, Sartre and Merleau-Ponty, to the Nietzschean rejection of Hegelian dialectics and totalizing perspectives by the poststructuralists.

14. Friedrich Nietzsche, *Beyond Good and Evil*, Harmondsworth, Penguin, 1973, §259, p.175.

15. Peter Dews, *Logics of Disintegration*, London: Verso, 1987, p.242.

16. Jürgen Habermas, *The Theory of Communicative Action*, Volume 1, [1981] tr. Thomas McCarthy, Boston: Beacon Press, 1984. p.1f.

17. See ibid. p.23.

defence of rationality within the totalizing perspective of mechanistic materialism and closed off the possibility of displacing it.

With all facets of life permeated by the mechanistic world-orientation, Habermas's conservatism and his subsequent formulation of the main problem of the modern world as the colonization of the life-world by the functionalist rationality of systems - together with his solution: the promotion of values associated with the validity claims of communication as a basis for counter-attacking this colonization - must be rejected as inadequate. To begin with, this leaves the systems of purposive-rational action, notably the economies of affluent core zones of the world economy, to continue their destructive expansion at the expense people in the Third World and of the world ecosystem. Beyond this, this approach does not address, or addresses only very inadequately, the main point raised by Nietzsche - that in the modern world the highest values have devalued themselves, that even if it could be shown that there is an implicit commitment in all speech to non-instrumental notions of validity, the development of a view of the world on the basis of such validity claims has rendered such values meaningless. What is the point of acknowledging such commitments and creating situations of uncoerced communication when people no longer have any grounds for justifying any conviction beyond instrumental or functional efficiency and the need for uncoerced communication? Under these circumstances the provision of sites where communication is free from coercion can only be an instrument of social formations in the struggle for greater instrumental efficiency to win out in the struggle for survival. Social Darwinist systems theorists such as Niklas Luhmann can allow the validity claims associated with communicative action, but then deny their universalist claims, relativizing instances in which they are invoked to instruments in the power struggles of particular social systems.[18] Habermas' explanation for the lack of resistance to the colonization of the life-world by systems dedicated to instrumental control, that consciousness is now no longer false but fragmented, is only one side of the story. It involves a failure to acknowledge the coherence of the mechanistic world-orientation dominating society, including the life-worlds of people, underlying this fragmented consciousness. Consciousness is only fragmented among those who are excluded from power and where this world-orientation is failing as a means to understand the world.

Most commentators on and critics of modern culture have failed to come to grips with humanity's predicament because they have been blind to the coherence and power of the ideas which underpin it. This blindness results from an excessive preoccupation with the ideas of other humanistic intellectuals. But the coherence of this culture lies in the mainstream of science itself and in the forms of thinking embodied within the main institutions and organizations of society and within individuals. A culture must be seen as an ecosystem of practices and general orientations as well as explicitly developed ideas, with each constituting, conditioning, resonating with, disguising and providing the conditions for the development of each other. In European civilization, as in traditional societies, the general orientation to the world is incorporated as a habitus, 'a permanent disposition, embedded in the agents' very bodies in the form of mental dispositions, schemes of perception and thought, extremely general in their application ... the cultivated disposition inscribed in the body schema and in the schemes of thought...'[19] Underneath the fog of cliches and platitudes and polite gestures which take the place of thinking in the modern world, mechanistic materialism is lived as a mode of bodily engaging in the world; it is the orientation to the world which individuals must adopt if they are to make their way in life in modern capitalist societies and which is merely cloaked by neurotic adaptations to it. How is the domination of the mechanistic materialist conception of the world maintained and reproduced?

---

18. Habermas has recognized the relationship between systems theory, Social Darwinism and the Nietzschean undermining of reason in his confrontation with Luhmann, and has recognized the social significance of such thinking, but in my view he does not present a convincing refutation of Luhmann and all he stands for. See Jürgen Habermas, *The Philosophical Discourse of Modernity*, tr. Frederick Lawrence, Cambridge: M.I.T. Press, 1987, pp.336-385, esp.p336f.

19. Pierre Bourdieu, *Outline of a Theory of Practice* [1972] tr. Richard Nice, Cambridge: C.U.P., 1977, p.l4f.

## *Science*

To begin with, the officially sanctioned view of the world is the one seen through the objectivist categories of mechanistic science. Scientific experts have gained the status of a priesthood with a virtual monopoly of power to adjudicate on questions of truth and falsity on all but minor issues.[20] Scientific jargon, especially when it is interspersed with mathematical expressions, has attained the status Latin had in the Middle Ages, a superior language, accessible only to a higher order of beings; and scientific experts are those with the credentials to mediate between this higher discourse and the discourse of ordinary mortals. For the normal scientific expert, the world is a mechanical order of matter in which everything is entirely explicable in terms of the properties of its constituents.[21] While mechanistic materialism has been rejected in theoretical physics and in thermodynamics, this rejection is more than compensated for by the dogmatic adherence to a mechanistic framework for research in the mainstream of chemistry, biology, psychology and economics. Mechanistic materialism continues to be identified with science as being the objective, true account of the nature of the world.

While scientists are thus exalted as the custodians of truth, the conditions for the promotion of science guarantee that most scientists will continue to conceive the world mechanistically. To retain their financial support, institutions of education and research must produce and grade people to function within this society, to increase the market price of their labour power and produce knowledge which can be sold as a commodity; and it is knowledge developed on the basis of a mechanistic conception of the world which serves these functions. For instance genetic engineering which has a saleable technological payoff and purveys a mechanistic world-view is being provided with research funds, while funding for the study of epigenesis: the differentiation and genesis of form in organisms, and anti-reductionist ecology, which do not produce marketable knowledge or people with marketable skills and which undermine the mechanistic view of the world, is poor.

This bias is being accentuated with the deliberate reorganization of tertiary and research institutions in almost every Western country to shift research away from the humanities to the sciences, and away from pure science to applied science. As Lyotard has noted:

> Research funds are allocated by States, corporations, and nationalized companies in accordance with [the] logic of power growth. Research sectors that are unable to argue that they contribute even indirectly to the optimization of the system's performance are abandoned by the flow of capital and doomed to senescence. The criterion of performance is explicitly invoked by the authorities to justify their refusal to subsidize certain research centres.[22]

This process is most clearly evident in the United States, where, as David Dickson has shown: 'The notion of scientists as independent scholars, motivated soley by a thirst for knowledge and unconcerned about the eventual utility of their results, has been banished for good.'[23] Almost all studies which could reveal the limitations or lead to alternatives to mechanistic materialism are thus being eliminated as irrelevant to the development of technology, and it is almost impossible for most scientists to go against this trend. Of the 700,000 scientists in the United States, only 5,000 have any power to determine what research they do, and such power is usually only obtained

---

20. The parallels between science and religion have been explored by Paul Feyerabend in *Science in a Free Society*, London: Verso, 1978, pp.73-122. On the promotion of 'scientists' as an alternative clerisy in nineteenth century Britain, see Ruth Barton, *The X Club: Science, Religion, and Social Change in Victorian England* (Ph.D. diss., Uni. of Pennsylvania, 1976).

21. As pointed out by Ilya Prigogine and Isabelle Stengers, *Order Out of Chaos*, Toronto: Bantam, 1984, p.68.

22. Lyotard, *The Postmodern Condition: A Report on Knowledge*, p.47.

23. David Dickson, *The New Politics of Science*, 2nd ed., Chicago: Uni. of Chicago Press, 1988, p.46.

through conformity to existing powers in the first place.[24] An élite of 200 to 300 key people actually make all the important decisions about science, and have shown themselves prepared to discipline dissidents who step out of line to address the public.[25] Thus, truth, first of all identified with science, is now identified with technology which is saleable as a commodity.

This devaluation of truth is legitimated by empiricist philosophies of science. The empiricism which developed with mechanistic science was designed for the most part to justify its claims to knowledge and invalidate rival claims, while being consistent with the view of humans as complex machines. In the twentieth century logical empiricists attempted to buttress empiricism by a revived and greatly developed tradition of logic, continuing the effort to account for knowledge mechanistically. Fused with pragmatism, operationalism and more recently, mainstream North American philosophy of language and evolutionary epistemology, this has produced a one dimensional intellectual world.[26] Science, promoted by logical empiricists as the only worthwhile intellectual pursuit, was represented by them as the accumulation of objective knowledge verified by observation and experiment, formulated into mathematically describable laws enabling predictions to made from one observation to another. While subsequent work by historians and historically oriented philosophers of science showed this image of science to be totally at variance with what was involved in the great scientific achievements of the past,[27] logical empiricism has become increasingly institutionalized within science itself. To begin with this took place in a straightforward way, particularly in the human sciences where to demonstrate their scientific credentials, behaviourist theories promoting a mechanistic image of humans were modelled on the view of science purveyed by logical empiricists. But the real success of logical empiricists has occurred in a less direct way. By striving to conceive knowledge and rationality mechanistically, they have provided the basis for the manufacture of computers: machines, which as the logical empiricists understand the process, can think, and through the incorporation of computers into research these machines are now affecting the direction of science. Those areas of scientific enquiry to which computers can be applied to 'model' reality are gaining inordinate prestige and research funding. The effect of this is that in fields such as ecology and economics only those aspects of the world which can be comprehended in terms of bits of information processable by computers are being granted the status of science. This tendency is being reinforced through the incorporation of computers into education. In this way logical empiricism is coming to be the true account of science through its effects upon it.

This has made it difficult to question prevailing scientific ideas or to develop new lines of research. Science is understood and represented in a dogmatically realist way insofar as it is mechanistic (without ever revealing mechanistic materialism to be only a particular research programme which, as such, might be questioned), while those theories inconsistent with mechanistic materialism such as relativity theory, quantum theory and non-linear thermodynamics are presented in a highly abstract way which focuses attention almost exclusively on their capacity to make correct predictions.[28] As Levy-Lebond described the typical handbook on such theories:

---

24. On the state of modern science see David Dickson, *The New Politics of Science*. This updates Hilary Stephen Rose's, *Science and Society*, Harmondsworth: Penguin Books, 1970; and J.R. Ravetz's, *Scientific Knowledge and its Social Problems*, Penguin: Harmondsworth, 1973.

25. In particular those academics who have taken up the cause of the environment have had their careers jeopardised. On this see Brian Martin, 'The Scientific Strait-jacket' in *The Ecologist*, Vol.11, No.1, Jan.-Feb. 1981, pp.33-43.

26. As argued by Herbert Marcuse in *One Dimensional Man*, Boston: Beacon Press, 1964 and more recently by Allan Bloom in *The Closing of the American Mind*, N.Y.: Simon and Schuster, 1987.

27. The main actors in this were Whitehead, Burtt, Fleck, Koyré, Polanyi, Toulmin, Hanson, Kuhn, Lakatos, Feyerabend, Hesse, Shapere, Lauden and Harré. However there were many others.

28. This has close parallels with the medieval world in which the development of astronomy was inhibited because to make the heavens intelligible meant undermining the categories on which medieval social life was based. Astronomy was therefore reduced to making predictions, and even Copernicus was interpreted by Osiander, who wrote the preface to his

> It consists, in general, of purely theoretical, exaggeratedly formalistic accounts, from which references to real experiments steadily vanish. Not a single impression is left of the real procedures of scientific activity, of the dialectic between theory and practice, heuristic models and formalism, axioms and history. Modern physics appears as a collection of mathematical formulae, whose only justification is that 'they work'.[29]

Scientists who question the conceptual incoherencies within or between domains of science or attempt to develop new metaphysical foundations which could overcome these problems - such as Ilya Prigogine or David Bohm - immediately lose standing before the priesthood of scientific experts. As a consequence, science has splintered into a cacophony of sub-disciplines inhabited by ultra-specialized ignoramuses pouring out ever increasing quantities of unreadable and unread papers.[30]

This dogmatism is then reinforced by identifying the elaboration of a mechanical view of the world with masculinity. The resonance between gender relations and the development of mechanistic science has been revealed most clearly by Brian Easlea who has exposed the aggressively misogynistic sexual imagery of those espousing mechanistic, reductionist views of the world from Bacon and Newton to the present.[31] The identification of mechanistic science with masculinity is manifest in the division between the 'hard' sciences, most especially physics, and the 'soft' sciences, most especially the human sciences (with the exception of economics, modelled on physics); or as a former Professor of Philosophy at the Australian National University J.J.C. Smart described them, the boy and the girl sciences. This has led biologists and human scientists to struggle to make their works into hard, masculine sciences, and all those who have opposed the reductionist research programmes entailed by this, however convincing their arguments, are seen to have thereby revealed themselves to be 'soft-headed'.[32]

The effect of the identification of different scientific ideas with gender differences has been revealed by Evelyn Fox Keller.[33] Keller described how Barbara McClintock's work on the genetics of corn, revealing how transformations occurred within the genome by transposition of elements, was not given due attention at the time of the publication of her results. The way McClintock was subsequently marginalized revealed how the pervasiveness of a domineering attitude towards the world, symbolizing the Western ideal of masculinity, made it difficult for scientists to contemplate the breakdown of a form of explanation in which an immutable power centre controls in a

---

book, as though this were the sole end of his theory. See Norman Diamond, 'The Copernican Revolution: Social Foundations of Conceptualizations in Science' in Les Levidow ed. *Science as Politics*, London: Free Association Books, 1986, pp.7-37. In the modern world quantum theory has been reduced to making predictions and David Bohm's efforts to reconceptualize the nature of reality to make sense of quantum phenomena have been dogmatically dismissed by mainstream scientists.

29. Jean-Marc Levy-Lebond, 'Ideology of/in Contemporary Physics' in Hilary Rose and Steven Rose (eds), *The Radicalization of Science*, London: Macmillan, 1976: pp.136-75, p.160.

30. See the poignant memoirs of Erwin Chargaff *Heraclitean Fire* N.Y.: Rockefeller University Press, 1978 for a description of what it means to be a scientist in the modern world.

31. Brian Easlea, *Witchcraft, Magic and the New Philosophy*, Harvester Press, Sussex, 1980, *Science and Sexual Oppression: Patriarchy's Confrontation with Woman and Nature*, London: Weidenfeld and Nicholson, 1981, and *Fathering the Unthinkable: Masculinity, Scientists and the Nuclear Arms Race*, London: Pluto, 1983.

32. Thus T. A. McMullen in 'A Critique of Humanistic Psychology.' *Australian Journal of Psychology* 1982, Vol. 34, pp.221-229 equated humanistic psychology with a retreat from objectivity, belief in the paranormal, magic, prophecy and astrology, and as a symptom of tender mindedness. For a critique of this paper see Arran Gare and Leigh M. Smith 'The Philosophical Foundations of Humanistic Psychology: A Reply to McMullen' in *Australian Journal of Psychology*, Vol.36, No.1, 1984, pp.103-108.

33. Evelyn Fox Keller, *Reflections on Gender and Science*, New Haven, Yale University Press, 1985. Also *A Feeling for the Organism: The Life and Work of Barbara Mcclintock*, New York, Freeman, 1983.

mechanical way the organism's growth. Keller has pointed out how various possible types of explanatory theory have been similarly ignored in other branches of science, including quantum physics, because they break with this domineering orientation.[34]

## *Education as Indoctrination in Nihilism*

The mechanistic world-orientation, along with the status given to science and scientific experts, is explicitly and implicitly inculcated through education. People are now forced to endure up to twenty years of intense indoctrination, moulding and struggling against each other in educational institutions in order to obtain one of the small number of privileged positions within society. The most important way in which a conception of the world is inculcated by education is through the way disciplines are organized and the status each one is given. As Mary Douglas has argued:

> [The curriculum] is a scheme for fitting together bits of knowledge. As they are connected in the curriculum, so they enter into the minds of the pupils, and, though the details of the contents will fade, the connections are likely to guide their judgements and perpetuate the system of power which the curriculum represents. This feedback, which gives stability to educational systems, also stabilizes cosmologies. The cosmological scheme connects up the bits of experience and invests the whole with meaning; the people who accept it will only be ale to justify their treatment of one another in terms of these ultimate categories.[35]

With the rise of mechanistic materialism the medieval organization of universities, in principle based on the trivium and the quadrivium and with the pre-eminent place given to theology, was replaced by an organization in which the top place was given to science, and in particular to physics, although in recent years, physics has been displaced by economics, business studies and computer science. The humanities make up a rump where they are slowly withering away - along with humanist intellectuals.[36]

Along with this explicit curriculum, the mechanistic world-orientation is reinforced by a hidden curriculum.[37] Students are taught science (including economics and reductionist psychology) as though there are simple, pre-ordained conclusions to complex issues, and all that is required of students is the mastery of what is regarded as absolutely certain knowledge. Scarcely any time is devoted to investigating alternative hypotheses or to problems scientists have found insurmountable, and the assumptions of science are never looked at. By contrast humanities studies begin with simple phenomena and end by showing the complexities involved. They generally do not arrive at definite conclusions and thinkers studied in the humanities are often presented as having confronted insurmountable problems. The effect of this dichotomy is that even if students appear to be scientifically illiterate, they have come to accept implicitly and uncritically that science, including economics, is dealing with the truth, with objective facts, while the humanities are just playing with ideas, with subjective feelings and preferences. And while acquiring an image of the humanities as just a game, students acquire what Karl Popper called the 'bucket image' of science according to which scientific method is a quasi-mechanical process starting with

---

34. *Reflections on Gender and Science*, Ch.7.

35. Mary Douglas, *Natural Symbols*, Penguin: Harmondsworth, 1973, p.10. See also Richard Bernstein, 'On the classification and framing of educational knowledge' in *Class, Codes and Control,* St Albans: Paladin, 1973, pp.227-256.

36. For an account of the dismal state of higher education, see Kevin Robins and Frank Webster, 'Higher Education, High Tech, High Rhetoric', in Tony Solomonides and Les Levidow, *Compulsive Technology: Computers as Culture*, London: Free Association Books, 1985, pp.36-57.

37. See David Gordon, 'The Image of Science, Technological Consciousness and the Hidden Curriculum', *Curriculum Inquiry*, Vol.14, No.4, 1984, pp.367-400.

observation and experiment, then going on to inductive generalization, hypothesizing and verification, finally leading to an addition to the stockpile of objective knowledge. More generally, the hidden curriculum leads students to assume that nature is in the service of humanity, that what exists in nature must be quantified and purified of mysteries and spirits and that the extension of such scientific knowledge is the quintessence of progress. The human sciences conforming to this image: neo-classical economics and reductionist psychology which assume humans to be information processing mechanisms moved by appetites and aversions, are then blest with the mantle of science and accorded the status of objective knowledge, while humanistic social sciences and humanities assuming an image of people as creative, are invalidated. A modern education is an indoctrination in nihilism.

This state of affairs is now being cemented in the humanities departments of universities, the last refuge from the mechanist world-orientation. In concluding his recent book, *The Philosophy of Nature*, Ivor Leclerc argued:

> ... contemporary scientific development has thrown into question in an extremely fundamental way all our inherited philosophical concepts, categories, and basic presuppositions. Nothing like this has happened since Parmenides. Philosophy is being faced in our time with the necessity for a more thoroughgoing rethinking of the fundamental philosophical problems, concepts, and categories throughout its entire range, than philosophy has undertaken since the time of Plato and Aristotle. This rethinking will affect science no less deeply than it will philosophy itself. And the consequences for human life will be no less great than were those of the new science and philosophy of the seventeenth century.[38]

Philosophers have rejected this challenge. Leclerc belongs to a very minor tradition of philosophy and his works are little known. Mainstream philosophers have resolutely adhered to the prevailing world-view to the point of attacking and almost destroying philosophy as a discipline.[39] Anglophone philosophy has been reduced to epistemology, logic and philosophy of language designed to legitimate the claims of 'common-sense' and orthodox science to knowledge; and in the process it has become as sterile and trivial as late medieval scholasticism.[40]

Following the usurpation of Anglo-American philosophy by such anti-philosophy, students turned to the study of literature and Continental (usually French) philosophy for views about life to oppose the prevailing intellectual wasteland. But the post-structuralists under the influence of Barthes, Lacan, Foucault, Derrida and Baudrillard have usurped this domain. Following Derrida, the devotees of French philosophy are now 'deconstructing' philosophy, literary theory and literature.[41]

According to Derrida, the standard practice of metaphysicians is to conceive the world in terms of binary oppositions, one of which is assumed to be prior and superior to the other. The second

---

38. Ivor Leclerc, *The Philosophy of Nature*, Washington: CUA Press, 1986, p.208.

39. To the point at which an anthology embodying the work of the major living philosophers can be published entitled *After Philosophy: End or Transformation* (ed. Kenneth Baynes, James Bohman, and Thomas McCarthy, Cambridge, Mass.: MIT Press, 1987.) Neither Leclerc nor anyone else in the tradition of process philosophy is represented.

40. A good overview of mainstream Anglo-American philosophy is provided by Richard Rorty in *Philosophy and the Mirror of Nature*, Oxford: Blackwell, 1980.

41. For an account of Derrida's ideas and their relation to literary criticism in the United States, see Christopher Norris, *Deconstruction: Theory and Practice*, London: Methuen, 1982. The best overall critical studies of this school are Peter Dews, *The Logics of Disintegration: Poststructuralist Thought and the Claims of Critical Theory*, London: Verso, 1987 and Jürgen Habermas, *The Philosophical Discourse of Modernity*, tr. Frederick Lawrence, Massachusetts: M.I.T. Press, 1987. Most of the central ideas of the post-structuralists were anticipated by Michael Bakhtin who at the same time avoided their relativism. In particular see M.M. Bakhtin and P.M. Medvedev, *The Formal Method in Literary Scholarship: A Critical Introduction to Sociological Poetics*, [1928], Cambridge, Mass.: Harvard University Press, 1985.

term is made out to be external, derivative and accidental in relation to the first. The second term usually connotes something that endangers the value the first term assumes - thus difference is opposed to identity, absence to presence, undecidability to decidability, and so on. Through a strategy of opposition and prioritization, metaphysics represses everything that troubles its founding values. Deconstruction consists in overturning this system of oppositions and priorities by revealing how what is excluded as secondary and derivative in relation to an originary concept of foundation, ground or origin is in fact more primordial and more general than the metaphysical original. Difference is not derived from identity, but makes identity possible. Derrida exposes how all concepts of foundation, ground or origin must be similarly displaced, and shows that these concepts are points situated in relation to larger systems, chains, or movements.

As taken up in literature departments, particularly in the United States, deconstruction reveals the process of and conditions for the construction of meaning in texts from the available discourses to reveal the lack of unity in what is constructed, to undermine any claim to authoritative reading, and thereby to use the texts themselves against the intentions of their authors. Meaning is located not in the intentions of writers, but in texts and their relation to other texts. Meaning is seen as constantly deferred in the never-ending webs of intertextuality in which all texts are located. Deconstructionists purport to defend the status of the reader against the text and literature against philosophy, and claim thereby to be justifying the proliferation of different points of view, the play of discourse, or as Derrida put it: 'the Nietzschean *affirmation*, that is the joyous affirmation of the play of the world and of the innocence of becoming, the affirmation of a world of signs without fault, without truth, and without origin which is offered to an active interpretation'.[42] In so doing, these cultural critics are undermining any effort to develop any perspective or narrative by which different texts, or even different readings of texts, could even provisionally be evaluated. They are reducing all literature, all philosophy, all rhetoric and all interpretation to one level. As Tzvetan Todorov characterized the views of the deconstuctionists: 'The world itself is inaccessible; discourse alone exists, and discourse refers only to other discourse... Even so, we are not to believe that discourse is better endowed than the world: the latter may not exist, but the former is necessarily incoherent... As no discourse is exempt from these contradiction, there is no reason to prefer one sort over another, or to prefer one value over another. In fact, in the deconstructionist perspective, any value-oriented behaviour (criticism, the struggle against injustice, hope for a better world) becomes subject to ridicule.'[43]

So while the French neo-Nietzschians have been opposing the Platonist fixation on eternal forms and the Cartesian fixation on the executive consciousness and affirming the reality of becoming, they have steered philosophy back to the dead end of Cratylus, the Heraclitean philosopher who argued that since everything is in flux there is virtually no point in speaking - precisely the dead end which was Plato's point of departure. In so doing they have inspired an intellectual movement which has contributed significantly to the devaluation of literature and philosophy and created the illusion that any abandonment of the Platonist commitment to the eternal - which is now identified with scientific knowledge - must lead to total relativism.[44] And by purporting to reveal the underlying play of power in the construction of meaning in discourse they have reinforced the prevailing view that life is nothing but a struggle for power.[45]

---

42. Jacques Derrida, 'Structure, Sign and Play in the Discourse of the Human Sciences,' in *Writing and Difference*, London: Routledge and Kegan Paul, 1978, p.292.

43. Tzvetan Todorov, *Literature and Its Theorists*, tr. Catherine Porter, Ithaca: Cornell University Press, 1987, p.184.

44. This is not to say that Derrida's ideas should be rejected *in toto*. For an attempt to salvage Derrida's insights while avoiding epistemological nihilism see Michael Ryan, *Marxism and Deconstruction: Critical Articulation,* Baltimore: John Hopkins University Press, 1982.

45. For an overview of discourse analysis see Dianne Macdonell, *Theories of Discourse*, Oxford: Blackwell, 1986.

## *The Free Market-Place of Ideas*

But we appear to live in a time when more is published, when more ideas are canvassed, than ever before. We appear to have a free market-place of ideas. How then can the mechanistic world-orientation prevail under these circumstances?

With the fragmentation and atomisation of society, most people's consciousness of the world after they have finished their schooling derives almost entirely from newspapers, radio, and most importantly, television. In most countries the ownership and control of these are becoming ever more highly concentrated, and with the constraint on the media to satisfy the interests of owners and advertisers there is very little questioning of superficial aspects of the perspectives of those in power, let alone the basic assumptions on which these perspectives are based.[46] As John Pilger described the profession of journalism:

> It is censorship by subterfuge: the manipulation of thought and language, such as labels and clichés that deceive and polarise ('moderates' versus 'extremists', etc.) and a conditioned deference to authority and to the 'prevailing view' in the name of objectivity. This is journalism's most insidious restrictive practice. And here the absurdity is Orwellian; for to reject this bias is to be 'controversial' and 'committed' and to invite both direct censorship and the indignation of those whom Robert Louis Stevenson aptly described as 'your sham impartialists, wolves in sheep's clothing, simpering honestly as they suppress'.[47]

In the case of television, everything is transformed into entertainment, and the centralized control of television ensures that it does not aspire to be anything else.[48] As a writer observed in the American Spy magazine, television is not a window on the world, but on the minds of 20 Hollywood cokeheads. So, bombarded with decontexturalized and systematically biased information the individual loses all capacity to put things in perspective. He or she becomes a 'subjectless subject'; a character marked by a 'scattered, disconnected, interchangeable and ephemeral state of "informedness", which one can see will be erased the very next moment to be replaced by new information.'[49]

This information is then fused with advertising and fiction to create an image of the social world consistent with the prevailing world-orientation,[50] an image of glamorous, attractive, high-consuming members of respectable society (the 'winners' in life), very often involved at the forefront of advance into an exciting, high-tech future, fringed by affable, good humoured, usually comic, sometimes a bit roguish people from the lower orders; a nether world of terrorists, gangsters, communists, murderers, Arabs, militant trade union organizers, thieves, habitually unemployed, drug runners and so on bent on the destruction of respectable, glamorous people and the subversion of society; heroic, super-intelligent members of the security forces dedicated to protecting respectable, glamorous people and outwitting these villains; and on the far side of all this, a great mass of hopelessly impoverished, unattractive low-tech people in the Third World (or

---

46. On the concentration in media ownership and its effects see Bruce M. Owen, *Economics and the Freedom of Expression,* Cambridge, Mass., Balinger, 1975.

49. John Pilger, *Heroes*, London: Pan, 1987, p.xiiif. On this, see also Chomsky's *The Culture of Terrorism*, London: Pluto Press, 1988.

48. This is the argument of Neil Postman in *Amusing Ourselves to Death*, London: Methuen, 1987.

49. Theodor Adorno, 'Theorie der Halbbildung', in *Gesammelte Schriften,* Vol.8, p.115, cited and translated by Peter Dews in *The Logics of Disintegration: Poststructuralist Thought and the Claims of Critical Theory*, London: Verso, 1987, p.142.

50. The best study of advertising and its impact is Stewart Ewen, *All Consuming Images: The Politics of Style in Contemporary Culture*, London: Basic Books, 1988.

occasionally on the fringes of the economic centres) starving or killing each other or themselves, totally beyond redemption (the 'losers' in life).[51]

Books, even where they are still read, can no longer counter such images. Apart from a few exceptional cases which are usually unknown to the general public, books no longer provide a medium where new ideas can be presented and evaluated. There are enough examples of major critical thinkers having their works rejected, Noam Chomsky and Andre Gunder Frank, for example, to show the extent to which the major publishers, who now control almost all the publishing houses, censor their publications.[52] But this is not the only problem. There are more subtle ways in which books have been prevented from functioning as a means of communication.[53] Publishers concerned with guaranteed markets and constrained by the way bookshops operate either target specialist academic audiences (where libraries provide a guaranteed market) or mass markets (which means aiming at the lowest common denominator). This, combined with changing urban and suburban environments, has virtually destroyed the habitat of autonomous intellectuals, driving them to virtual extinction.[54] So as H. Stuart Hughs wrote: 'The meticulous scientists of words and the "terrible simplifiers" of Jacob Burchardt's nineteenth century nightmare, in their mutually incompatible endeavours,... have the field all to themselves.'[55]

The suppression of ideas consequent to this is disguised by the large numbers of publications which are critical of the prevailing order. But what people are being subjected to is a mass of simplistic social critiques from a multiplicity of different perspectives. What are conspicuously lacking are intellectually rigourous critical works of broad scope which are addressed to ordinary people about their most pressing concerns; works able to challenge the ruling culture by putting it in perspective and providing ideas which could actually displace it. A work equivalent to Marx's *Capital* - consisting of three large volumes and intermingling philosophy, economics, history, social critique, being both technical and polemical and highly critical of all preceding thinkers - would be virtually unpublishable today. Without such works, the dominance of society by the mechanistic, Social Darwinist world-orientation is obscured rather than undermined by this proliferation of opposing ideas. In fact this dominance is reinforced by being disguised. It provokes continual efforts to expose the egoistic motives behind people's moral utterances, and in doing so engenders the illusion that there are no other motives. The pervasiveness of moralistic rhetoric enables people to believe that they are uniquely hard-headed in having understood that there is nothing more to life than the struggle for the means to self-gratification. The market-place of ideas then allows these mass produced minds to maintain the illusion of their individuality by providing them with materials from which to concoct for display their own unique blends of religious, philosophical and scientific exotica. Since everyone involved in making such displays 'knows' that mechanical view of the world and logical empiricism are passé, and no self-respecting member of the power élite would be so gauche as to espouse Social Darwinism, there appears to be little point

---

51. This is not an entirely fair picture. Despite the main thrust of the mass media, television *has* served to reveal and thereby to help check oppression in the world, and through programs on nature to promote a greater awareness of the natural environment and its significance.

54. Noam Chomsky and Edward S. Herman described how the first version of their work *The Political Economy of Human Rights* was suppressed by their publisher in the prefatory note to Volume 1, *The Washington Connection and Third World Fascism*. Andre Gunder Frank was unable to publish his major theoretical work in English, and was unable to publish a major anthology on underdevelopment. See his *Critique and Anti-Critique* N.Y.: Praeger Press, 1983, p.vii and p.297ff. for an account of this. There are many other examples of such suppression of ideas.

53. On the book publishing business see Lewis Coser, Charles Kadushin and Walter W. Powell, *The Culture and Commerce of Publishing* Chicago: University of Chicago Press, 1982.

54. On this, see Russell Jacoby, *The Last Intellectuals*, N.Y.: Basic Books, 1988.

55. H. Stuart Hughs, *Consciousness and Society*, N.Y.: Random House, 1958, p.405. That this is the situation is now recognized by both conservative and radical thinkers. For the state of intellectual life see Allan Bloom, *The Closing of the American Mind*, N.Y.: Simon and Schuster, 1987.

in attacking these doctrines. And if attacks are made, they can be reduced to another commodity and smothered in the mass of mushy, woolly-minded romanticism which is already widely disseminated throughout society.

## *Mechanistic Materialism and Everyday Life*

However the mechanistic world-orientation is not just a set of ideas and attitudes. It is embodied in everyday life as a mode of being in the world and is reproduced in daily practice as part of the self-reproduction of capitalism. The development of capitalism has constrained people in almost every aspect of their everyday lives, from childhood to old age, to conceive the physical world as an order of things (or commodities) to be used or efficiently exploited, life as a struggle of all against all, and value as purely subjective, revealed only by what people are willing to pay. And the human subject has itself become fragmented into saleable parts. As Georg Lukács described the effect of capitalist social relations:

> ... time sheds its qualitative, variable, flowing nature; it freezes into an exactly delimited, quantifiable continuum filled with quantifiable 'things'... in short, it becomes space.... [This] transformation ... cannot ... content itself with the reduction of all objects for the gratification of human needs to commodities. It stamps its imprint upon the whole consciousness of man; his qualities and abilities are no longer an organic part of his personality, they are things which he can 'own' or 'dispose of' like the various objects of the external world. And there is no natural form in which human relations can be cast, no way in which man can bring his physical and psychic 'qualities' into play without being subjected increasingly to this reifying process.[56]

This is associated with the tacit acceptance of the view of reason as nothing more than an instrument of power, as nothing but a means for calculating what is in one's self-interest and for increasing the efficiency of obtaining given ends. As a consequence (and as the Frankfurt Institute philosophers have pointed out), instrumental efficiency has become the ultimate (explicit) reference point for the legitimation of institutions and courses of action.[57] The whole culture of society is now organized around the acceptance of this. As science is justified by its contribution to the control of nature and people, democracy is justified as a means of legitimating power, resolving conflicts and maintaining peace. Public opinion, which has replaced critical discourse as the foundation for democracy, has become an object of scientific manipulation. Outside the realm of objective technical control is only the realm of the subjective, irrational feelings belonging to the domain of private consumption. Individuals are left without grounds for justifying their ideals. They may exalt the dignity of humanity, but they have no rational grounds for doing so. Art has become a mere decoration, and works of art are no longer seen as communicating visions of the world but are reduced to commodities to be invested in and consumed as a series of haphazard emotions.

Subjection to this way of thinking begins at an early age and intensifies thereafter. Children are exposed to extraordinary pressure to adopt it and to behave like predictable mechanisms by parents and teachers, who, venting their own childhood and adulthood frustrations at being made to conform to an oppressive society, strait-jacket children to eliminate any spontaneity or creativity

---

56. Georg Lukács, *History and Class Consciousness*, tr. Rodney Livingstone, London: Merlin Press, 1971, p.'s 90 & 100.

57. This was the main point made by the Frankfurt Institute philosophers. For the clearest account of their analysis of the debasement of rationality see Max Horkheimer, *Eclipse of Reason*, New York: Continuum, 1974. However these philosophers focussed their analysis on positivist epistemology rather than the mechanistic materialism and Social Darwinism which is the real foundation for legitimating this instrumentalist conception of reason.

which might hinder them in the rat race they must enter as adults.[58] The competitive organization of society then ensures that those who have successfully embodied this mechanistic orientation to life, who have internalized the aggressive stand-point of their parents and teachers, succeed at the expense of those who are more sensitive, more compassionate, more creative, who think that there might be more noble ends to life than the grubby struggle for self-advancement. The functioning of the market, and the rat-races of business, political, educational and governmental organizations can be counted on to level such illusory ideals and to strip idealists of any influence. Those who succeed will be those who have mastered the arts of ingratiating themselves to their superiors, stabbing their rivals in the back, and treating their subordinates as expendable instruments.

The extension of this world-orientation is obscured, and thereby made more effective, by a division of labour in its advancement. The basic mechanistic framework and the translation of its implications into interpersonal relationships is mostly effected by people who identify themselves as politically leftist and who, albeit in a limp and cynical way, uphold anti-Social Darwinian ideals in relation to economics, politics and international relations,[59] Social Darwinian economic practices are defended by liberals or neo-liberals who are often opposed to reductionist conceptions of people,[60] while conservatives promote a Social Darwinian outlook in politics and international relations while at the same time being critical of science and evolutionary theory, upholding the family in terms transcending utilitarian principles, and defending the intrinsic value of life (providing it is a human foetus).[61]

However this still does not account for the extent to which the mechanistic world-orientation pervades people's thinking. The mechanistic world-orientation is reinforced in ways which escape people's consciousness. It is inculcated through a multiplicity of minor practices without any insight by individuals into how their conception of the world is being shaped. Pierre Bourdieu pointed out:

> If all societies and, significantly, all the 'totalitarian institutions'... that seek to produce a new man through a process of 'deculturation' and 'reculturation' set such store on the seemingly most insignificant details of *dress, bearing*, physical and verbal *manners*, the reason is that, treating the body as a memory, they entrust to it in abbreviated and practical, i.e. mnemonic, form the fundamental principles of the arbitrary content of culture. The principles em-bodied in this way are placed beyond the grasp of consciousness, and hence cannot even be made explicit; nothing seems more ineffable, more incommunicable, more inimitable, and, therefore, more precious, than the values given body, *made* body by the transubstantiation achieved by the hidden persuasion of an implicit pedagogy, capable of instilling a whole cosmology, an ethic, a metaphysic, a political philosophy, through injunctions as insignificant as 'stand up straight' or 'don't hold your knife in your left hand'.[62]

It is by such means that people are led to embody the mechanistic world-orientation as a habitus.

The extent to which the institutions of modern societies are devoted to the inculcation of the correct habitus has been brilliantly revealed in the work of Foucault.[63] With capitalism there

58. On the effects of childhood treatment on adulthood, and how this reproduces itself, see the works of Alice Miller.

59. Leftists who are not cynics are almost invariably involved in defending anti-mechanistic conceptions of people as creative agents.

60. Von Hayek being the exemplary case of this.

61. The peculiar, partly contradictory but essentially complementary relationship between neo-liberals and neo-conservatives is analysed in Ruth Levitas' anthology *The Ideology of the New Right*, Cambridge: Polity Press, 1986.

62. Bourdieu, *Outline of a Theory of Practice*, op.cit. p.94.

63. While Foucault is generally associated with the thinkers who were trying to undermine the effort to gain a totalizing perspective, his placing of the discursive formations he studied in relation to the development of capitalism, and his own description of his work as going beyond Marx's efforts to understand exploitation by attempting to comprehend the nature

emerged a multiplicity of new discursive formations: the asylum, the clinic, the prison, the factory, the school etc., all enforcing norms of behaviour. The archetypal example of this was the prison, but the principles developed with the prison reforms, particularly those proposed by Bentham in his *Panopticon* to 'rehabilitate' delinquents by keeping them under continuous surveillance, became paradigmatic for all these institutions. As Foucault wrote:

> With this new economy of power, the carceral system, which is its basic instrument, permitted the emergence of a new form of 'law': a mixture of legality and nature, prescription and constitution, the norm.... [T]he activity of judging has increased precisely to the extent that the normalizing power has spread. Borne along by the omnipresence of the mechanisms of discipline, basing itself on all the carceral apparatuses, it has become one of the major functions of our society. The judges of normality are present everywhere. We are in the society of the teacher-judge, the doctor-judge, the educator-judge, the 'social worker'-judge; it is on them that the universal reign of the normative is based; and each individual, wherever he may find himself, subjects to it his body, his gestures, his behaviour, his aptitudes, his achievements.[64]

The institutions of this control organize a field in which people are objectified and treated as things to be moulded into disciplined, predictable functionaries or cogs within the capitalist economy. At the same time, this objectification has created the individualized subject, and through delving into this subject ever more deeply, it has enmeshed people further in these new power relations. In particular sexuality has been promoted to the centre stage of society.[65] Behind the facade of sexual repression, sexuality has been continually aroused, providing the basis through the deployment of images of normality for an unparalleled control by society over the bodies of its members. Sex has been elevated over the soul, over love, to become the ultimate value, almost more important than life itself. Desire for it, to have access to it, to discover it and to liberate it has attached to each person the injunction to track along the paths laid out by society and to place oneself, one's body, in the grip of its power.

The mechanistic world-orientation is then encoded in a pattern of evaluative conceptual oppositions centred on the relationship between the sexes. Apart from the association of science with masculinity and the arts and humanities with femininity, the division between masculine and feminine is correlated with the divisions between hard and soft, strong and weak, enduring and changing, active and passive, dominant and subordinate, unsentimental and sentimental, dry and wet, rational and emotional, logical and intuitive, definite and indefinite, light and dark, objective and subjective, conscious and unconscious, straight and curved, right and left, good and bad.[66] Masculinity as hard and rational is exalted in opposition to soft and passive femininity. Being female is the only acceptable excuse for being feminine. Ideas are divided between those which present the hard-headed, rational, unemotional, objective, masculine view of the world; and romantic idealist notions which are identified with soft headedness, irrationality, sentimentality, subjectivity and femininity. To see the world as devoid of meaning in which life is a struggle of all against all is to have a tough, masculine view of the world, while to portray the world as having a meaning in itself wreaks of femininity, the sort of sentimental mush which could clog up and impair the efficient functioning of the economy if it were taken too seriously.

---

of power, belies this. (See Michel Foucault, *Language, Counter-Memory, Practice*, ed. Donald F. Bouchard, Ithaca, Cornell University Press, p.212f.)

64. Michel Foucault, *Discipline and Punish: The Birth of the Prison*, [1975] tr. Alan Sheridan, Harmondsworth, Peregrine, 1979, p.304.

65. Michel Foucault, *The History of Sexuality*, [1976] tr. Robert Hurley, Harmondsworth: Peregrine, 1984.

66. On this see Genevieve Lloyd, *Man of Reason*, London: Methuen, 1984.

Finally, the mechanistic world-orientation has been inscribed in social institutions and in nature so that people are continually confronted with it as the organizing principle of their world. As Hamilton Cravens concluded in his study of the influence of Darwinian evolutionary theory and its associated science of social control in America:

> ... the question of the influence of evolution and of the science of man probably resolves itself not into the number of American citizens who accept these ideas freely but to the extent to which these ideas are and have been embedded in the basic social institutions and social roles of our modern corporate social order, and to the extent these institutions and roles perpetuate the formulae and prescriptions of this science of man and influence the patterns of existence experienced by so many millions of Americans.[67]

All economic institutions presuppose validity of this conception of the world. This has led to vast technological advances around which economic organizations have been formed, in this way continually presenting nature as a mechanical order of matter to be controlled. These organizations are based on the ideal of achieving the most efficient means of achieving each end, with social structures modelled on the image of the machine in which each functional part is seen as a replaceable cog.[68] As a consequence, 'instrumental efficiency' is embodied in the products of human agency so that cities, buildings, machines and means of transport have come to incorporate the same one dimensional functionalism. In the life-worlds of people everything has a predefined function which in turn defines people as functionaries. The coordination of these functions with all other lives, with institutions, organizations, industry, the State, and with the rest of the world by means of maps and clocks constitutes lived space and time as flat, uniform order which dominates every aspect of people's lives.[69] This functional uniformity is reiterated in the architecture deriving from the Bauhaus, in city planning under the influence of Le Corbusier (despite recent reactions against this), in hyperplanned suburbs, in the incorporation of machines into the domestic economy, in the development of factory farms, and in the mass production of everything from battery hens to public opinion.[70] Mechanistic materialism has become all encompassing. It has become the mode of cognition of capitalism, and like the Azande culture in which every strand depends upon every other strand, people cannot get out of its meshes because it is the only world they know.

Since the mechanistic and Social Darwinian orientation to the world is embodied as a habitus, in the relationship between the sexes and in the institutions of society, the lack of explicit affirmation of it by members of the ruling élites is of little significance in judging the extent of its domination.[71] The lack of defence reflects how it is so completely taken for granted that no

---

67. Hamilton Cravens, *The Triumph of Evolution*, Philadelphia: University of Pennsylvania Press, 1978, p.273.

68. See Jacques Ellul, *The Technological Order*, ed. Carl F. Strover, Proceedings of the Encyclopaedia Britannica Conference, Detroit: Wayne State University Press, 1963.

69. The development of standardized space and time from the individual to the global level has been described by Robert K. Schaeffer, 'The Standardization of Time and Space' in Edward Freidman ed. *Ascent and Decline in the World-System*, Beverly Hills: Sage, 1982, pp.69-90.

70. See Siegfried Giedion, *Mechanization Takes Command*, Oxford: O.U.P., 1948; Charles A. Jencks, *The Language of Post-Modern Architecture*, N.Y.: Rizzoli, 1977, Part One: 'The Death of Modern Architecture,' and Edward Relph, *Rational Landscapes and Humanistic Geography*, Totowa: Barnes & Noble, 1981, esp. Ch.3 for an analysis of this process and its effects.

71. For this reason one should not take seriously Robert Bannister's claim in *Social Darwinism: Science and Myth in Anglo-American Social Thought*, Philadelphia: Temple Univ. Press, 1979, supported by Donald C. Bellomy in '"Social Darwinism" Revisited', *Perspectives in American History*, 1984, pp.1-129, that Social Darwinism has had negligible influence on conservative thought. Against Bannister and his supporters see Jim Moore, 'Socializing Darwinism', in Les Levidow ed. *Science as Politics*, London: Free Association Books, 1986, pp.38-80. Moore points out the spurious nature

alternative is seriously conceivable. What is important is that the assumption on which people base their important decisions is that to be hard-headed one must acknowledge that what life is really about is the struggle for survival and power, and everything that gets in the way of this struggle is an expendable luxury. A relaxation of this struggle can be tolerated for a time but it will eventually lead to biological degeneration, to economic inefficiency and to a loss of the ability of society to effectively meet challenges either from rival societies or from challenges within.

So while people espouse all sorts of different ideas, these have become irrelevant to people's world-orientation - which is revealed in the way they live and the decisions they make. While pollution, the greenhouse effect and the hole in the ozone layer over the South Pole led to a new upsurge in interest in the environment, people continued to take for granted the primary and over-riding importance of 'economic progress', in the need to increase productivity, to increase investment, to develop new technology and new consumer products, to maintain full employment, to get people to work harder and to work longer hours. And while some people might even admit that high levels of consumption are inimical to environmental conservation, they will still exalt the West for this very reason and disparage non-capitalist societies because they do not provide people with the same opportunities for conspicuous consumption.

There are some individuals and groups who do genuinely oppose the dominant world-orientation. This reflects the legacy of individual conscience. But it is these people who experience just how entrenched the dominant forms of thinking are, and how difficult it is to go against them. Unable to find an intellectual niche for their views, they tend to be easily demoralized. The wider the breach between the dominant forms of thinking and the opposing views, the more difficult it is for opposing views to find expression by which they could acquire formulation, clarity and vigour. Without finding such expression, the motives founded on these viewpoints tend to wither. Whole ideological movements have been destroyed in this way. This was the case with the New Left radicals of the late 1960's and early 1970's. Shocked by, among other things, the brutality and injustice of the Vietnam War, large numbers of people began questioning and seeing through the whole facade of late capitalist society, the extent and oppressive nature of its imperialism, the corrupt nature of its democratic and judicial institutions, the complicity of universities in its dynamics, the mind warping nature of its mass culture and the mass media in particular, the emptiness of life as a functionary in this system, and so on. But while intellectuals such as Marcuse provided students with insights into the nature and extent of oppression, there was no viable foundation on which an opposing direction could be established. The New Left was reduced to a media event, and its protest spluttered out into a mindless hedonism, furthering entrenching the forms of life on which the existing order is based.

## *The Postmodern Condition as Nihilism*

Not even the embodiment by people of mechanistic materialism and Social Darwinism can explain the destructive aggressiveness of Europeans. Underlying this is the recognition that the images and ideas which formerly inspired people have lost their meaning. In 1921 G.B. Shaw wrote that:

> ...the Darwinian process may be described as a chapter of accidents... There is a hideous fatalism about it ... If it be ... a truth of science, then the stars of heaven, the showers of dew, and winter and summer, the fire and heat, the mountains and hills, may no longer be called to

---

of the efforts to separate Darwinian evolutionary theory and Social Darwinism, and has revealed the extent to which people's thinking is dominated by the language of evolutionary theory. See esp. pp.68-75.

exalt the Lord with us by praise: their work is to modify all things by blindly starving and murdering everything that is not lucky enough to survive the universal struggle for hogwash.[72]

The Great Depression and the Second World War further shook people's faith in the ideal of mechanical efficiency and evolutionary progress. Twenty-five years of economic prosperity in at least some nations seemed to provide some grounds for optimism about the future, despite the constant threat of nuclear war; but with the subsequent economic decline, with increasing evidence that affluence in the centres of the world-economy is intimately tied to the impoverishment of vast numbers of people in the Third World, and then the growing evidence of global environmental destruction, this optimism is evaporating. The dominant thematic motif of European civilization has lost its power to orient people in life. And as Archibald MacLeish pointed out in his poem *The Metaphor:*

A world ends when its metaphor has died.
An age becomes an age, all else beside,
When sensuous poets in their pride invent
Emblems for the soul's consent
That speak the meanings men will never know
But man-imagined images can show:
It perishes when those images, though seen,
No longer mean.

Nietzsche argued at the end of the nineteenth century, 'Nihilism, this weirdest of all guests, ... stands before the door.'[73] This nihilism has now taken over the house. The final outcome of Western culture is a society of lonely people, perpetually insecure, denied a sense of their own significance, and too disoriented to begin to understand the cause of their situation or the possibility of its being different.[74] Correspondingly, the free floating resentment of the nineteenth century has evolved into the free floating malice of the twentieth century.

Nihilism has revealed itself in the loss of commitment to truth and, where intellectual life has not been totally reduced to the development of technology, by the predominance of intellectual game playing and the dilettantish pursuit of the latest intellectual fashions.[75] Academic writing has been reduced to a frenetic churning out of papers to publish before the fashions change and some new intellectual guru is proclaimed.[76] The effect of this is evident in the state of ethical and legal discourse and practice. Alasdair MacIntyre argued that:

... in the actual world in which we inhabit the language of morality is in ... grave disorder... What we possess, ... are the fragments of a conceptual scheme, parts which now lack those contexts from which their significance derived. We possess indeed simulacra of morality, we

---

72. George Bernard Shaw, *Back to Methuselah: A Metabiological Pentateuch*, N.Y.: Brentanao, 1921, pp.xlv-xlvi.

73. Friedrich Nietzsche, *The Will to Power*, §1. The first Russian to use the term nihilism (in the 1820's), Nadhezin, used it to designate scientific materialists - particularly those influenced by Locke.

74. One of the best descriptions of this state is still David Riesman's description of 'other-directed' people in *The Lonely Crowd*, N.Y.: Anchor Books, 1954.

75. This situation has been described in different ways by both radicals and conservatives. Despite the description of Herbert Marcuse's *One Dimensional Man* by Allan Bloom in *The Closing of the American Mind*, p.226 as 'trashy culture criticism', this work is essentially a conservative version of Marcuse's argument, the basic argument of the Frankfurt Institute philosophers, that reason has been eclipsed, and the subsequent nihilism has permeated everyday life. The state of the humanities has worsened considerably since the publication of *Crisis in the Humanities* ed. J.H. Plumb, Harmondsworth, Penguin, 1964.

continue to use many of the key expressions. But we have - very largely, if not entirely - lost our comprehension, both theoretical and practical, of morality.[77]

And Harold Berman in his magnificent study of the development of European legal thought described a similar situation:

> ... 'public policy' has come dangerously close to meaning the will of those who are currently in control: 'social justice' and 'substantive rationality' have become identified with pragmatism; 'fairness' has lost its historical and philosophical roots and is blown about by every wind of fashionable doctrine. The language of law is viewed not only as necessarily complex, ambiguous, and rhetorical ... but also as wholly contingent, contemporary, and arbitrary ... These are harbingers not only of a 'post-liberal' age but also of a 'post-Western' age.[78]

In everyday life, nihilism is experienced as disorientation, as loss of direction, as emptiness. As progress is seen as increasing control over the world, as making everything in it, including other people, into predictable instruments for human purposes, individuals are increasingly experiencing life as fragmentary, disorganized and uncontrollable, as a flux within which all boundaries are dissolving. Christopher Lasch attempted to describe the response to this situation in his book *The Minimal Self*. He argued that in the modern world:

> People take one day at a time. They seldom look back, lest they succumb to a debilitating "nostalgia"; and if they look ahead, it is to see how they can insure themselves against the disasters almost everybody now expects. Under these conditions, selfhood becomes a kind of luxury, out of place in an age of impending austerity. Selfhood implies a personal history, friends, family, a sense of place. Under siege, the self contracts to a defensive core, armed against adversity.[79]

However Lasch's account does not go far enough. It fails to capture the fragmentation of the self and the associated emotional flatness, the drifting quality of everyday life. Being 'armed against adversity' implies an heroic quality which not only lacking, but which is being denied as a possibility. Life in the late twentieth century is better described as the 'condition of postmodernity', a condition wherein people can only respond to the speed of change by accepting disorientation as a normal condition - even celebrating it as 'fun'.[80]

The postmodern condition involves a retreat from differentiation - whether between the aesthetic and the social, between high and low culture, or between superior and inferior forms of life. It is a regression from discursive signification which gives priority to words and narratives over images and which operates through critical reflection, to figural signification which is visual rather than literary, which juxtaposes signifiers taken from everyday life and operates through the spectator's unmediated immersion in the spectacle. As Fredric Jameson has noted, this is akin to

---

76. Hilary and Steven Rose, *Science and Society*, Harmondsworth, Penguin, 1970, p.xv estimated that if scientific publications continued to increase at the rate of the time, in 100 years they would weigh more than the earth.

77. Alasdair MacIntyre, *After Virtue: A Study in Moral Theory*, 2nd ed., Notre Dame, Indiana: University of Notre Dame Press, p.2.

78. Harold J. Berman, *Law and Revolution: The Formation of the Western Legal Tradition*, Cambridge, Massachusetts: Harvard University Press, 1983, p.41.

79. Christopher Lasch, *The Minimal Self: Psychic Survival in Troubled Times*, London: Norton, 1985, p.15.

80. This is best described by the Marxist David Harvey in *The Condition of Postmodernity*, Oxford: Blackwell, 1989, or in a postmodern style by the proponents of postmodernism, Arthur Kroker and David Cook in *The Postmodern Scene*, 2nd ed., N.Y.: St. Martin's Press, 1988.

schizophrenia, a condition characterized by a loss by the subject of 'its capacity actively to extend its pro-tensions and re-tensions across the temporal manifold and to organize its past and future into coherent experience' so that its cultural productions can be nothing but 'heaps of fragments'.[81] People now live through an ever recurring present, and the fading memories of pasts which seemed to have a future are now regarded as nothing but aspects of the present.

Many people have taken refuge from this chaos by returning to old gods, in being born again Christians or Moslems or Shintos, while others pack into the psychiatric clinics, working their way through, and successively placing their loyalty in, a series of psychiatric therapies. However most have surrendered entirely to the flux of the present, distracting themselves into mental oblivion with television and home videos, seeking ever more intense stimulii or living vicariously through the mass media constructions of the sleazy, pseudo-glamorous lives of celebrities. Increasing numbers of teenagers and young adults are voting against the postmodern condition with their lives, relieving the world of their unwanted creative potential by suiciding.

The free floating malice which characterizes twentieth century nihilism is dangerously manifest in the fascination with violence and war. The devastating wars fought by Europeans this century were not anomalies but clear expressions of a malignant civilization. Hitler was only unusual in that he treated some Europeans as Europeans had been treating people of other races in the Americas, Africa, Asia and Australia; he brought the nihilism of Western culture to fruition. In the postmodern scene, images of violence are mass produced in films such as *Bladerunner*, *The Cook, the Thief, his Wife and Her Lover* and *The Wild Ones*, and in films of past wars and projected future wars. However it is the small to medium scale wars have displayed the greatest power to capture audiences. Such wars, the Falklands War, the invasion of Grenada, Panama and Haiti, the bombing of Libya and the war against Iraq have become the most popular form of entertainment, and bombing raids are timed to coincide with prime television time.

The exhilaration experienced from violence associated with military might is not confined to the general public, but is clearly evident in the sadistic glee of those developing the technology of destruction. Thus Feigenbaum and McCorduck exuberantly proclaimed the possibilities for computer technology:

> The so-called smart weapons of 1982, for all their sophisticated modern electronics, are really just extremely complex wind-up toys compared to the weapon systems that will be possible in a decade if intelligent information processing systems are applied to the defense problems of the 1990's.[82]

Similar sentiments were evident in the comment of the head of ARPA's information processing research office on smart robot weapons: 'This is a very sexy area to the military, because you can imagine all kinds of neat, interesting things you could send off on their own little missions around the world.'[83]

While nuclear war strategy is now concerned with the survival of missiles rather than people and projected victory is measured by a hypothetical body count, the power of nuclear weapons elicits a morbid fascination, a fascination clearly evident in the unpopularity of any group in Britain, France or the United States which proposes nuclear disarmament. With a contracting of people's temporal horizons this fascination is particularly gruesome. Even the Nazis were concerned with the long-term future of humanity and saw their own struggle for power in terms of

81. Fredric Jameson, *Postmodernism*, Durham: Duke University Press, 1991, p.25.

82. Cited from Edward Feigenbaum and Pamela McCorduck, *The Fifth Generation: Artificial Intelligence and Japan's Computer Challenge to the World*, Reading: Addison-Wesley, 1983 by Joe Weizenbaum in 'The Myths of Artificial Intelligence' in Tom Forester ed. *The Information Technology Revolution*, Cambridge: MIT Press, p.90.

83. Cited ibid. p.91.

this. Today's megalomaniacs are only interested in the exercise of power for its own sake, or as a means to augment this power.

Those rebelling against this nihilism are losing the means to do so. Language itself is now a hindrance. Not only does it lead people to think in terms of things with properties, privileging being over what is becoming, but it has been debased by modern social life. Despite the efforts of such figures as Karl Kraus and George Orwell to alert people to the importance of language, the potential of language to facilitate expression has been steadily eroded by the mass media, by politicians, by bureaucrats, by academics, and by the mechanistic world-orientation which allows no significance to expression. It has been reduced to a means for recording information and for manipulating people and it is now difficult to use language for any other purpose. Conversation not serving such purposes has been reduced to idle chatter, or dried up completely. People live lives of quiet desperation because they no longer have the means to express their desperation. What Nietzsche noted in the nineteenth century has become even more true today:

> Man can no longer make his misery known to others by means of language; thus he cannot really express himself anymore... The results of this inability to communicate is that the creations of common action... all bear the stamp of mutual non-comprehension.[84]

We are fast heading towards the form of society projected by Samir Amin, where:

> There are no more individuals, neither men nor women. These beings - one does not know what to call them - are neither human nor animal, neither liberated nor alienated, neither conscious nor animated by false consciousness. They are perfectly plastic. Their nature is no longer determined by other men but by the perfect machine... These beings no longer speak - they have nothing to say, since they have nothing to think or feel. They no longer produce anything, neither objects or emotions. No more art. No more anything. The electronic machine produces - the word itself has lost all meaning - everything, these beings included.[85]

## *Western Culture Against the Environment*

By revealing the Platonistic, mechanistic, Social Darwinian, and ultimately, nihilistic ideological underpinnings of Western civilization, and in particular, of the Anglophone nations, by showing how they are incorporated by society and individuals, it is now possible to understand why, except in the case of a few cosmetic issues, the problems of the environment are so inadequately addressed. And through this, it is possible to see what the environmentalists are up against.

To begin with, notions of morality when they do play a part in discourse are merely a surface decoration. As Erving Goffman pointed out:

> In their capacity as performers, individuals will be concerned with maintaining the impression that they are living up to the many standards by which they and their products are judged... But, qua performers, individuals are concerned not with the moral issue of realizing these standards, but with impressions that these standards are being realized. Our activity, then, is

---

84. Friedrich Nietzsche, *Richard Wagner in Beyreuth*, 5 tr. Tracy B. Strong in 'Language and Nihilism: Nietzsche's Critique of Epistemology', *Theory and Society*, Vol.3, 1976, pp.239-263, p.248.

85. Ibid. p.12.

> largely concerned with moral matters, but as performers we do not have a moral concern in these moral matters. As performers we are merchants of morality.[86]

And underlying this superficial morality, the concepts and modes of thinking which actually affect the way people choose to live are inimical to environmental preservation.

Money plays a major role in this respect. Since for most people in Western capitalist societies the significance of anything is perceived in terms of money, anything outside the realm of the monetary economy cannot be taken into account without a great deal of effort. People tend to be as blind to the significance of unpriceable phenomena as were the feudal aristocrats to the significance of commoners who did not participate in the forms of aristocratic virtue. This does not mean that anything which cannot be priced will not be valued at all, but when it comes to decision making, the significance of what cannot be priced will appear shadowy, unreal, nothing but the product of emotion. This includes not only unique species of life and wilderness areas, but also people outside the monetary economy, people who are not seen as actual or potential resources. In Anglo-Saxon cultures the plight of the unemployed can barely be recognized as of any significance by their more fortunate members. Where the people suffering are the poor of the Third World, it is generally beyond the capacity of most members of the wealthy nations to acknowledge that they are real. If such people are considered, their plight tends to be defined in terms of the monetary economies of these nations. For instance Dr Stephen Enke and Mr Richard A. Brown, two members of the U.S. Cosmos Club argued:

> Why should publicly financed resources be devoted to preventing infant mortality when the economic worth of such marginal infants is negative? The economy would be better off without them. The burden of proof is surely on those who recommend diversion of health resources from caring for producing adults to caring for consuming children.[87]

However the significance of money goes deeper than this. The multi-millionaire mining magnate, Lang Hancock, characterized environmentalists as 'those unwashed ... dole bludging drop-outs' and described the environmental movement as the 'number one enemy of civilization'.[88] To understand such outbursts it is necessary to recognize that there is more involved than simply failing to see the significance of environmental problems. The explanation for such attitudes lies in the importance of money to people's identity. Money is the sign of election to a superior order of being, and is the basis of the moral order through which people are defined as significant and respected. As Marx wrote, 'That which exists for me through the medium of money, that which I can pay for ... that am I...'[89] Or as Barbara Kruger put it: 'I buy, therefore I am.' Money is sex appeal. Consequently to attempt to interfere with people's money making is not simply to deprive them of a certain amount of purchasing power (although increasing spending on consumption is the appropriate display of their money); it is to attack their very being, the only way in which their significance can be recognized. This applies not only to the wealthy but to small businessmen and to labourers. To be a labourer threatened with unemployment is to be threatened with being defined as a parasite. Just how money is important to working people has been shown by Robert Frank in his study of the role of status seeking as a motivating factor in economic life. To make his point he looked at people working in the nuclear power industry cleaning up radiation spills. While workers were fully aware of the risks of radiation exposure, there was no shortage of workers willing to

86. Erving Goffman, *The Presentation of Self In Everyday Life*, Edinburgh: Edinburgh University Press, 1956, p.156.

87. Stephen Enke and Richard Brown, 'Economic Worth of Preventing Death at Different Ages in Developing Countries', *Journal of Biosocial Science*, Vol.4, No.3, July, 1972; cited by David Dickson, *Alternative Technology and the Politics of Technical Change*, Glasgow: Fontana/Collins, 1974, p.161.

88. *The Australian*, April 10, 1978. This was brought to my attention by Dr Patsy Hallen, Murdoch University.

89. Marx *The Economic and Philosophical Manuscripts of 1844*, p.167.

accept as much exposure as their employers were willing to pay them for, stating as their reason: 'We need the money.' Frank argued from this that:

> The acceptance of such terms of employment may appear to signal a careless, if not totally irrational, disregard for the future. Yet if concerns about relative standing are an important motivating force for individuals, such behaviour need not be *individually* irrational at all. The forward move in the income hierarchy it enables may be more than enough to compensate for the future damage it will cause.[90]

Where the concern to be honourable has lost its meaning and been replaced by the quest for status, where having money is the only way to gain status, to enter the magical world portrayed in advertisements, to be seen and to feel oneself to be 'somebody', and where having more money than others is a zero-sum game in which the people to whom one wishes to be seen as significant are those who have more status than oneself, people will sacrifice their futures to increase their incomes in the present. And where people are sacrificing their own lives, they are not going to be overly concerned about the plight of others or the future of the environment. Consequently it is generally those people who have been able to gain some sense of their significance outside the monetary economy who have contributed to the environmental movement, and they are a small minority.

Underlying the fetishism of money is the mechanistic world-view. Mechanistic materialism undermines any alternative ideas in terms of which people could define their significance, while at the same time rendering people as blind to the degradation of life as were the Australian Aboriginals to Joseph Banks' ship. Achievements are seen as the degree of success with which everything and everyone are reduced to predictable instruments, until the world is made to run like a well-oiled machine (the 'ideal Platonic form' which individuals and society should conform to). People, especially when they live in the Third World, are of no significance unless they serve the economic machine. It is this which Leon Rosselson describes in *Who Reaps the Profit, Who Pays the Price?:*

> You take the earth from out of the earth
> You throw the corpses in
> One crop is as good as another
> As long as the cash comes pouring in
>
> The wheels must never stop turning
> The machine must be obeyed
> The future has got to be fuelled
> And there's a price to be paid

The fact that making the world totally predictable means obliterating all spontaneity, ultimately destroying life itself, is not acknowledged since spontaneity and life are incomprehensible from the perspective of a mechanistic world-orientation. So long as the greatest efficiency is achieved in the means to each defined end, nothing better can be conceived.

This outlook is buttressed by the mechanistic vision of evolutionary theory. According to this, progress, seen as essentially improvements in organizational efficiency for survival and expansion, has been achieved through the struggle for survival. Consequently it is inevitable that the behaviour of people who are the product of this evolutionary struggle will be based on self-interest and that people will be evaluated in terms of how efficiently they pursue their interests. A few negative

90. Robert H. Frank, *Choosing the Right Pond: Human Behaviour and the Quest for Status*, N.Y.: Oxford University Press, 1985, p.143.

side-effects are of no great significance in relation to the general trend of progress engendered by this self-interested struggle. And in a world in which evolutionary progress is the consequence of such a struggle it must be accepted that there will be continual transformations of the world, with older species making way for the new, and changes destructive to one form of life providing the conditions for the development of new forms. The development of humans and their institutions is a continuation of this evolution and it is inevitable that humanity, as the most highly evolved species, will radically transform its environment. The treatment of animals by humans is the inevitable process of exploitation of one species of organism by another characteristic of all nature. The subjugation of wilderness areas and the extinction of species must be seen as a continuation of evolutionary progress as the less fit make way for the more fit. The destruction of non-European cultures and the rise of capitalism is the continuation of evolution at the level of societies. European civilization and capitalism in particular have revealed themselves to be superior by their technological advances which have enabled them to dominate the rest of humanity.[91] Based on the recognition of the role of competition in generating progress, with its scientific and technological advances, the capitalist economy is the acme of civilization and the ultimate product of evolution. Even the destruction by affluent governments of Third World democracies in the name of Freedom, and the imposition of grotesquely oppressive dictatorships to facilitate exploitation of their countries' resources, is fully justified if all life is essentially the struggle for the means to survival. The imminent starvation of large numbers of people is the inevitable consequence of over-population, the natural means by which the inferior members of the species are eliminated. And this in itself is useful in that it can provide a weapon to superior societies in the struggle for power. As a report from the Central Intelligence Agency noted in August, 1974, the shortage of grain 'could give the United States a measure of power it had never had before... Washington would acquire virtual life and death power over the multitude of the needy ...'[92] Pollution of the human environment is an inevitable by-product of development to which the human organism must adapt. Where some people die as a consequence of pollution this must be seen as indicative of their inferior genetic endowment. As a freelance British consultant, F.J.C. Roe wrote in February, 1978: 'Cancer in its many forms is undoubtedly a natural disease. It is probably one of nature's ways of eliminating sexually effete individuals who would otherwise, in nature's view, compete for available food resources without advantage to the species as a whole.'[93] The exhaustion of the resources necessary for sustaining existing industries is a challenge which will, as it has in the past, stimulate the development of new forms of technology based on the exploitation of different resources.

The definition of the significance of people in terms of their money is tacitly recognized as simultaneously a measure of their significance for the economic life of society; that is, for economic progress, and thereby for evolutionary progress. This assumption has been brought out in the film *Wall Street* in which the takeover wizard Gorden Gekko, a character based in part on Ivan F. Boesky whose insider trading on the stock exchange landed him in gaol, stated in an address: 'The point is, ladies and gentlemen, greed is good. Greed works, greed is right. Greed clarifies, cuts through and captures the essence of the evolutionary spirit.' Environmentalists who would interfere in money making are seen as attempting to go against the course of nature. They are interfering with progress, the highest development of which is manifest in the technological advances of modern capitalism. They are threatening to weaken the capacity of these capitalist countries to survive in their struggle with other capitalist nations, particularly those peopled by

---

91. This is essentially the argument of one of the main theorists of the New Right, Friedrich Hayek. See *The Fatal Contract*, London: Routledge, 1988.

92. Cited by Susan George in *How the Other Half Dies*, Penguin, Harmondsworth, 1977, p.211. This power was used in the same year to punish Bangladesh for trading with Cuba. After the withdrawal of food aid 100,000 people died of starvation.

93. Cited by Samuel Epstein in *The Politics of Cancer*, San Francisco: Sierre Club Books, 1978, p.299.

different races. Environmentalism can be understood as the expression of resentment by those who have not been able to succeed in the economic struggle, who have fallen out of the monetary economy to become parasites on those who have been successful, the 'winners' in life.

What can be said about this conception of things? Environmentalists if they confronted these ideas might dismiss them as simply a rationalization of self-interest. But this is to accept the prevailing view that people are essentially egoists, and ideas are simply means to further individual interests. However there appears to be more involved. The notions on which the ruling élite of capitalism are justified tend not to be thought out. They are the ideas which the members of the most successful culture in the world are socialized to accept. However they have a solid foundation in a well worked out world-orientation backed up by the mainstream of science. The efficacy of this science is demonstrated by its technological achievements and by the success of Western societies in dominating civilizations based on different ways of conceiving the world. The Chinese may have had a more benign civilization with a more attractive conception of the world, but then look what happened to it.

Because of the nature of its inculcation and the way the dominant world-orientation is disguised, it has been difficult to question it. But if it is questioned, there are answers forthcoming. For instance, if the world is conceived of as nothing but configurations of matter whose motion is governed by immutable laws there is no reason to regard any one configuration as superior just because it happens to have survived. However Social Darwinists do not have to refer to any other criteria than capacity for survival, and on this basis it appears clear that humans are a higher form of life than those forms from which they evolved, and that modern civilization is a higher form of humanity than the primitive societies from which it has developed. With these developments explained in terms of the struggle for survival, there are compelling reasons to evaluate everything in terms of survival value. To maximise these advances in survival power it is necessary to accept the demise of the less fit. The way to continued progress in this direction it is to allow the struggle for survival to continue unimpeded.

The notion that people are always moved by self-interest can also be questioned. Firstly, there are examples of people whose behaviour cannot be accounted for in terms of self-interest, and secondly, if people were deterministically moved by their appetites and aversions, there would be no need to justify self-interested behaviour since such behaviour, and the evolutionary progress it leads to, would be automatic. However socio-biologists have explained apparently altruistic behaviour by arguing that it is not the individual as such which is the unit of evolutionary struggle but the genes.[94] Altruistic behaviour is selected for because it increases the chances of survival of the gene types which produce it. On the other hand where altruistic behaviour is excessive, as when it is extended beyond the members sharing the same genes or even beyond the members of the same species, this can be explained as a genetic defect in the individual, an excess of altruistic characteristics which, like physical deformities, will be eliminated by evolutionary necessity.[95] And adherents to Social Darwinism need not regard themselves as choosing to be self-interested, or as defending this choice. They can regard their use of everything and everyone as instruments for their own selfish ends as a fact about the world which is explained by Social Darwinism, and if they attack opponents of such selfish behaviour, this can be seen as clearing away the debris standing in the way of their achieving these ends.

But supposing all talk of evolutionary progress is dismissed as simply the residue of the Christian notion of providence which is really incompatible with a fully consistent mechanistic materialism, and the nihilism implicit within mechanistic materialism is accepted. In this case individuals must see their lives as a brief moment between two infinities of nothingness with only one chance to experience what life has to offer. Even if people who accepted such a view of life felt

94. See for example Richard Dawkins, *The Selfish Gene*, Oxford: Oxford University Press, 1976.

95. See Garrett Hardin, 'Discriminating Altruisms' in Michael Tobias, *Deep Ecology* San Diego: Avant Books, pp.182-205 for such an argument.

any concern for the fate of life in the world, and there is no reason why they should, they would be disinclined to sacrifice or even risk sacrificing any significant part of their lives to meet the challenge of its problems. People who have come to believe that subjective experience is the only real value in life, who have come to live by the principle 'if it feels good, do it,' are not going to waste much of their lives pondering the fate of the earth. As a distinguished Professor of Political Economy at the University of London wrote in *Business and Society Review*: 'Suppose that, as a result of using up all the world's resources, human life did come to an end. So what?'[96] More commonly the acceptance of this nihilism is associated with an emotional shallowness, an indifference to or vindictive enjoyment in the suffering of those who are being subjugated, and an obsession with money, power games, and conspicuous consumption. The attitude of the affluent to the environmental crisis is perhaps best conveyed by a full page advertisement in the Australian *Financial Review by Commodity Technical Trading Ltd.*[97] Titled 'How to trade the Greenhouse Effect', and quoting the predictions of scientists on how disastrous will be the effects of environmental destruction on food supplies, the advertisement describes how huge speculative profits could be made on the food commodities futures markets. So, we are left with the question asked by Robert Heilbroner in his book *An Inquiry Into the Human Prospect*:

> When men can generally acquiesce in, even relish, the destruction of their living contemporaries, when they can regard with indifference or irritation the fate of those who live in slums, rot in prison, or starve in lands that have meaning only insofar as they are vacation resorts, why should they be expected to take the painful actions needed to prevent the destruction of future generations whose faces they will never live to see?[98]

In short, so long as people experience the world and organize their lives through the categories of mechanistic materialism, while these are not replaced as the basis of the objective, scientific picture of the world, environmentalists have no grounds to justify their concerns or to expect others to take them seriously. The present behaviour of individuals and nations is all that can be expected. This does not mean that environmentalists are wrong in much of what they have attempted to bring to the attention of the public. The world is becoming over-populated. People are starving and dying from the effects of pollution. Ecosystems are being destroyed. We may be heading for the destruction of civilization. But so what? This is an inevitable consequence of human nature and a natural part of evolution.

## *The Irrelevance of Moral and Political Philosophy*

In Chapter II I tried to reveal the extent of the failure of the attempts by environmentalists to defend their position. I argued that the main problem was their tendency to base their arguments on assumptions which have been responsible for the problems in the first place, or else they have failed to fully transcend these assumptions. With the perspective provided by the analysis of ideology in Western civilization it should now be possible to bring these intellectual failures into sharper focus.

To begin with, the nature of the dominant ideology throws further light on why it is hopeless to attempt a defence of environmentalism in terms of the prevailing ethics and social philosophy. What has underlain the mainstream of this philosophy is the Hobbesian view of humanity and its place in the world. This is most clearly manifest in the assumption that while moral obligations are problematic, there is no problem with self-interest. Rights theory, utilitarianism and Kantian ethical

96. Cited by Robert L. Heilbroner, *An Inquiry into The Human Prospect*, N.Y.: W.W. Norton, 1975, p.170.

97. *Financial Review (Australia)*, Oct.5, 1988, p.11.

98. Ibid. p.143.

philosophy are all predicated on this assumption. The formulation of Social Darwinism on the basis of a fully developed mechanistic materialism has produced a disjunction between the 'official' ethics, based on extensions of rights theory, utilitarianism and Kantian categorical imperatives demanding constraints on egoism, and the effective ethics extolling egoism as the mainspring of evolutionary progress.

This situation made the very meaning of ethics problematic, and rather than attacking the foundations of Social Darwinist ethics, most philosophers in the English speaking world withdrew into technical issues. Thus A.J. Ayer wrote of his own work that it:

> ...is entirely on the level of analysis: it is an attempt to show what people are doing when they make moral judgements; it is not a set of suggestions as to what moral judgements they are to make. And this is true of all moral philosophy as I understand it. All moral theories... in so far as they are philosophical theories, are neutral as regards actual content.[99]

Ayer argued that ethical assertions are simply expressions of emotions aimed at influencing other people. As pointed out by MacIntyre, this emotivist theory of ethics expressed the pedantic recognition by Oxbridge philosophers what had been already recognized and brilliantly conveyed by Kierkegaard and Nietzsche: that the project of establishing morality within the framework of a world-orientation which denies the reality of a human potential to be realized, has failed; and that therefore all moral conflicts are nothing but the struggle by individuals to make their own subjective viewpoints prevail.[100]

Moral philosophers thus abrogated their traditional role, and ultimate duty, of orienting people in the world and showing them how to live. Philosophers conceived rationality in such a way as to make it inapplicable to ethical decision-making. To prefer justice to injustice or dignity to degradation is presented as having no more rational foundation than preferring red to blue or chocolate to ice-cream. And with a few notable exceptions the state of philosophy and attitude of philosophers today is much as R.G. Collingwood described them in 1939:

> The pupils, whether or not they expected a philosophy that would give them ... ideals to live for and principles to live by, did not get it; and were told that no philosopher (except of course a bogus philosopher) would even try to give it. The inference which any pupil could draw for himself was that for guidance in the problems of life, since one must not seek it from thinkers or thinking ... one must look to people who are not thinkers (but fools), to processes that were not thinking (but passion), to ideals that were not ideals (but caprice).[101]

As analytic philosophy came to dominate philosophy and philosophers turned their backs on questions about the nature of the world and of humanity and with how to live, philosophy lost its cognitive status to the sciences. It is to scientific experts that people now turn for the concepts to orient themselves for action in the world. In particular it is the economists and psychologists who provide the most influential and important of these concepts, becoming in effect the equivalent of the medieval casuists spelling out the implications of the prevailing world-view for how people should live. Of these, economists have the greater influence - the concepts 'economic' and 'uneconomic', 'profitable' and 'unprofitable' and 'economically efficient' are the most important concepts of evaluation in the modern world. How people should live is represented in the image of 'economic man' - the image of humans as efficiently functioning cogs in the economic machine, and the most important index for judging society as a whole is the rate of growth of GNP. Even when economic concepts are not evaluative their use immediately relates things and people to an

---

99. A.J. Ayer, 'On the Analysis of Moral Judgement' in *Philosophical Essays*, London: Macmillan, 1954, p.245f.

100. MacIntyre, *After Virtue*, Chs 2-4.

101. R.G. Collingwood, *An Autobiography*, London: Oxford University Press, 1939, p.48.

evaluative context. For instance to see anything as a resource is to see it as something economically useful which should be exploited as efficiently as possible. Economists provide society with the equivalent of Plato's *Republic* as the ideal form in which society must then strive to participate. In this ethical role economists are complemented by psychologists with their notions of 'abnormal', 'subnormal', 'neurotic', 'deviant', 'inadequate personality' etc. defining negatively the positive ideal of the 'psychologically healthy' or 'normal' person: the contented, emotionless, mindless, efficiently functioning cog in the economic system. Biologists provide the Neo-Darwinian framework (and the ultimate evaluative notions - 'fitness for survival' and 'survival value') for these human sciences, and this in turn is supported by the physical scientists who represent nature as totally devoid of meaning.

The extent to which ethical doctrines are acceptable is the extent to which they accord with and can support the ethical doctrines implicit within these sciences. Emotivism has come to be taken as common-sense since it is consistent with the mechanistic view of the world and with empiricism to see all value as merely subjective, and to see the relations between people as a struggle for supremacy without rational foundation. This also accords with a market economy in which people are seen to be free when they can express their subjective impulses in the market place; and it is instrumentally useful to see people's supposed convictions as emotions, since this implies that they are of no more significance than any other emotion, and that such convictions can be manipulated and controlled. But other ethical doctrines have not been entirely excluded. Rights theory and utilitarianism to some extent have retained their influence in the wealthy nations of the world, though apart from rhetorical purposes only in their original crude forms. The notion of rights has been subsumed under Social Darwinism where it is used by individuals, companies and nations to justify pursuit of their own interests at the expense of others without hindrance from superordinate institutions, and utilitarianism survives in the behavioural sciences designed for the efficient manipulation and control of people, and in the narcissistic, compulsive consumerism of the privileged members of the wealthy nations and the comprador classes in the Third World

It is the extensions of rights theory and utilitarianism which are excluded by Social Darwinism. If philosophers develop rights theory to justify the freedom of individuals, business organizations or nations to pursue their interests independently of any constraints from superordinate authorities, to accumulate as much wealth as they can without taking responsibility for others, they are likely to get a hearing outside philosophy. This is what elevated Robert Nozick to fame. To oppose freedom on the basis of rights theory in order to facilitate the struggle for survival as did Garrett Hardin is also a way of gaining attention. But if rights theory is developed in a way which does not accord with Social Darwinism, as with John Rawls' *Theory of Justice*, it will appear as nothing but an intellectual exercise, of value in the intellectual world only as a subject for another intellectual parlour game. In a world in which all living things are struggling for survival, where reason can only be the product of evolution and therefore an instrument in the struggle for survival, how can there be any rational justification for notions of rights which would interfere with this struggle? Similarly in the case of utilitarianism. Cost-benefit analysis provides the basis for the efficient scientific organization of society. It complements the prevailing economic doctrines, and is therefore acceptable. But the idea of extending the notion of the greatest happiness for the greatest number to people who are not functional members of the economy and have no capacity to threaten it, to the unemployed and people in Third World nations or even to animals, can hardly be taken seriously.

What then can be said about environmental philosophy? Much of the work in this field represents the efforts of fairly conventional analytic philosophers to be relevant. Such philosophers attempt to develop rights theory and utilitarianism to deal with environmental problems. But by invoking these largely discredited doctrines they are implicitly invoking and reinforcing the Hobbesian view of humans, the view formulated to accord with the new mechanical philosophy which was the ultimate expression of the medieval orientation towards aggressive domination of nature and other people, and the view which is most fully developed in Social Darwinism. This is equivalent to John of Salisbury's invocation in the twelfth century of the analogy of the body to describe society in his effort to defend its lower orders. His development of this analogy provided one of the most important constituents of aristocratic ideology in their defence of privilege.

As for traditionalists such as Passmore, it should now be even clearer that they simply have not identified the sorts of ideas which move most people to action. They have attempted to find a niche for their concerns in the ideas which disguise the dominant ideology rather than within this ideology itself. Looking back to what is said in the Bible or by various Christian thinkers to find seeds for a new attitude to nature misses the point stressed by Lynn White that what is important with a world-orientation is not what is explicitly confirmed, but people's sub-verbal assumptions about who they are, about their relation to the rest of the world, and about their destiny. In elaborating and developing Lynn White's point I have tried to support White's contention that Christianity is the source of the destructive attitude to the environment. However the source is no longer the cause, and it is now mechanistic materialism vouchsafed by science and institutionalized within capitalist society which is the ideological root of environmental problems. Christianity survives, but except for a fairly small minority this merely disguises the more basic commitment to a mechanistic and Social Darwinist world-orientation. Ideals deriving from it gain plausibility from the general population only insofar as they do not effectively contradict this underlying conception of the world.[102]

This means that the only hope for the future of humanity - and most other species of life on earth - lies in the development of a radically new way of thinking about the world, and associated with this, a radical reformulation of ethics and political thinking which overcomes the mechanistic world-orientation. But while the deep ecologists proposed such changes, the previous two chapters should reinforce what was argued in Chapter II. Deep ecology has tended to be a further expression of the 'impractical, irrational feminine side' of Western culture.[103] As such it complements rather than challenges the dominant metaphysics, providing an outlet for the sentimentality of the affluent, serving as the foundation for movements to convert some wilderness areas into parks, but nothing more serious than this.

## *Conclusion*

Given the nature of the culture promoting environmental destruction, addressing the environmental crisis cannot be achieved by piecemeal and fragmentary efforts. What is required is a fundamental critique of society which challenges its most basic assumptions. The main contender in this regard is the tradition of Marxism.

---

102. The way in which Christianity has been adapted to the modern world-orientation is clearest in Southern USA. A leading southern evangelist, Gerry Falwell, expressed the sentiments of this movement when he argued: 'Jesus was not a pacifist. He was not a sissy.' (Cited by A. Crawford, *Thunder on the Right*, Pantheon: New York, 1980, p.159f.)

103. This masculine/feminine opposition between the dominant culture and the deep ecologists corresponds to the two modes of defence against separation anxiety identified by Freudians: omnipotent domination of and regressive fusion with nature.

# 8

# MARXISM AND THE ENVIRONMENT

Marxism is the main tradition of radical opposition to the dominant culture within European civilization. It is committed to a total transformation of society, replacing capitalism with communism, as the only way to solve its problems.[1] Until the collapse of the Soviet Union, the retreat from Marxism by China and the rise of Islamic fundamentalism, mainstream Western culture and Marxism were rivals for world domination. Despite the recent defeats of Marxism, it remains the main focus of opposition to the dominant Western culture. Before any other challenges to the hegemony of mainstream Western culture can be considered it is necessary to evaluate Marxism and its potential for resolving the environmental crisis.

With the worsening of the environmental crisis Marxists claimed that this finally demonstrated the necessity for replacing capitalism by socialism. The editor of *Philosophy and the Ecological Problems of Civilization* argued:

> As many Marxists in all countries have observed, the crisis of the environment, which is reaching extreme development almost everywhere, coincides with the last stage of the general crisis of capitalism. This is evidence that it is inseparable from capitalism and is an integral element of it. A conviction is growing throughout the world that only collapse of the capitalist system and victory of socialism throughout the world will create a general, fundamental, social opportunity for rational use of natural resources and the highest degree of optimum interaction with nature... Convincing evidence that socialism is a necessary condition for optimising relations between society and nature is socialism as it actually exists, and the policy of socialist countries in respect of the environment.[2]

Was this claim justified?

The central idea behind Marx's work is that market relations, imposed and supported by its main beneficiaries, the bourgeoisie, have come to develop a life of their own which forces people to constantly revolutionize their mode of production and their way of life. First coming to dominate human relationships within Europe (where people first came to be treated as labour power to be bought and sold as a commodity), it then expanded to dominate the rest of world. Existing antagonistic social relations are not a reflection of human nature but are the product of an historically unique socio-economic formation. This is deforming people, reproducing not only these antagonistic social relations but also the conceptions people have of themselves. While in terms of the prevailing view of the world humans are such that relationships between them cannot be based on anything but a struggle for individual gain, Marx argued that capitalism is generating the conditions for the realization of a social order transcending such egoism in which the freedom of each will be the condition for the freedom of all.

These ideas were most fully developed in *Capital* as an immanent critique of capitalism. While this involved both revealing the defectiveness of the assumptions of the prevailing economic theory and implicitly thereby the framework of ideas supporting it, showing how these assumptions and

---

1. Despite the carping nature of the criticisms in the second and third volumes, Leszek Kolakowski's three volumed *Main Currents of Marxism* (Oxford: O.U.P., 1978) provides the best overview of this tradition. Of particular value is the first volume, describing the tradition of radicalism from which Marxism emerged.

2. A.D. Ursal ed. *Philosophy and the Ecological Problems of Civilisation*, tr. H. Cambell Creighton, Moscow: Progress Publishers, 1983, pp.10f.

ideas were generated and how they have been sustained, it did not explicitly set out to replace these assumptions. Marxism can be understood as the world-view which has developed to sustain Marx's critique, to explicate and defend his assumptions, and to generalize his analysis to new situations. This has led to the elaboration of both a general theory of history (historical materialism) and a general philosophy (dialectical materialism) to challenge the world-view on which capitalism is based. Though some Marxists, Karl Korsch for example, have criticised these efforts, the development of this world-view has been absolutely essential for the extension of Marx's ideas to changing historical circumstances. The questions which must be considered by environmentalists are whether Marx's critique of the prevailing socio-economic formation is justified, whether the Marxist critique extends to or can be extended to environmental issues, whether Marxism is capable of superseding the nihilistic world-view of mechanistic materialism, whether Marx has in fact revealed the way to a new social order, whether the new social order projected by Marxists would be such as to ameliorate environmental problems, and whether Marxism is adequate to sustain Marx's insights.

Answering these questions is a complex task. Marxism has been constructed out of various minor works, asides and polemical statements of both Marx and Engels, many of which are inconsistent with each other. Consequently there have emerged almost as many versions of Marxism and interpretations of Marx as there are avowed Marxists, and the Marxism of Communist countries was radically different from Western Marxism. Furthermore Marxian analyses of environmental problems have frequently been undertaken by people who are not avowedly Marxists, while most avowed Western Marxists have been hostile to environmentalists. I will proceed by first outlining Marx's analysis of capitalism, focussing on the place of the environment in Marx's thought and showing the relevance of this analysis for environmental problems, and then describe the efforts of Marxists to extend Marx's insights. Whether Marxism points the way to a social order which is not environmentally destructive will be answered in a preliminary way by examining the state of the environment in the Soviet Union before its collapse.

## *Marx, Capitalism and the Environment*

There can be no doubt that Marx considered nature as of no significance except from the point of view of human development. He was utterly contemptuous of the nature enthusiasm of the True Socialists, and the emancipation of humanity was seen by him in terms of the mastery of the whole of society over the mastery of nature. As Alfred Schmidt wrote of Marx's notion of communism:

> The new society is to benefit man alone, and there can be no doubt that this is to be at the expense of external nature. Nature is to be mastered with gigantic technological aids, and the smallest possible expenditure of time and labour. It is to serve all men as the material substratum for all conceivable consumption goods.[3]

Despite this, Marx's framework of analysis reveals the most important cause of humanity's recent destructive relationship to its environment.

The starting point for Marx was the conception of humans as a conscious part of nature in the process of forming themselves through their transformations of nature. In *Capital* he proclaimed:

3. Alfred Schmidt, *The Concept of Nature in Marx*, [1962] London: New Left Books, 1971, p.155.

> [Man] opposes himself to Nature as one of her own forces ... in order to appropriate Nature's productions in a form adapted to his own wants. By thus acting on the external world and changing it, he at the same time changes his own nature.[4]

However Marx pointed out that humans are only capable of reorganizing matter, and that labour is assisted by the forces of nature:

> The use-values, coat, linen, & c., i.e., the bodies of commodities, are combinations of two elements - matter and labour. If we take away the useful labour expended upon them, a material substratum is always left, which is furnished by Nature without the help of man. The latter can work only as Nature does, that is by changing the form of matter. Nay more, in this work of changing the form he is constantly helped by natural forces. We see, then, that labour is not the only source of material wealth, of use-values produced by labour. As William Petty puts it, labour is its father and the earth its mother.[5]

Seeing labour in such terms points to the limitations of human exploitation. However it was Engels rather than Marx who emphasised these limitations. Engels declared:

> Let us not ... flatter ourselves overmuch on account of our human victories over nature. For each such victory nature takes its revenge on us. Each victory, it is true, in the first place brings about the results we expected, but in the second and third places it has quite different, unforeseen effects which only too often cancel the first. The people who, in Mesopotamia, Greece, Asia Minor and elsewhere, destroyed the forests to obtain cultivable land, never dreamed that by removing along with the forests the collecting centres and reservoirs of moisture they were laying the basis for the present forlorn state of these countries. When the Italians of the Alps used up the pine forests on the southern slopes, so carefully cherished on the northern slopes, they had no inkling that by doing so they were cutting the roots of the dairy industry in their region; they had still less inkling that they were thereby depriving their mountain springs of water for the greater part of the year, and making it possible for them to pour still more furious torrents on the plains during the rainy seasons. Those who spread the potato in Europe were not aware that with these farinaceous tubers they were at the same time spreading scrofula. Thus at every step we are reminded that we by no means rule over nature like a conqueror over a foreign people, like someone standing outside nature - but that we, with flesh, blood and brain, belong to nature, and exist in its midst... [6]

While assuming that humans are part of nature, the specific problem Marx was concerned with was the emergence and development of capitalism. He described how capitalism originated, how it was developing according to its own laws independently of people's intentions, and why it was far more dynamic than any previous social organization, why it is breaking through all boundaries, both physical and social, to dominate the world. It is in relation to this dynamism that he considered the effect of capitalism on the environment.

The starting point for the emergence of capitalism from feudalism was the development of market relations to a stage in which people themselves were forced to sell their creative potential as labour-power, and exchange value came to take precedence over use value in defining people's relations to their products. This commodity fetishism engendered the process whereby capitalism attained a dynamics independent of people's intentions. As Marx argued:

---

4. Karl Marx, *Capital* tr. Samuel Moore and Edward Aveling, ed. Frederick Engels, Moscow: Progress Publishers, 1974, Vol.1, p.173.

5. Ibid. p.50.

6. Frederick Engels, *Dialectics of Nature*, Moscow: Progress Publishers, 1976, p.180.

> [T]he exchange of commodities breaks through all local and personal bonds inseparable from direct barter, and develops the circulation of the products of social labour, [developing] a whole network of social relations spontaneous in their growth and entirely beyond the control of the actors.[7]

This system produces 'not only commodities, not only surplus-value, but it also produces the capitalist relation; on the one side the capitalist, on the other the wage labourer.'[8] And as he emphasised: 'the capitalist is just as enslaved by the relationships of capitalism as is his opposite pole, the worker, albeit in a quite different manner.'[9]:

> Only as personified capital is the capitalist respectable. As such, he shares with the miser the passion for wealth as wealth. But that which in the miser is mere idiosyncrasy, is, in the capitalist, the effect of the social mechanism, of which he is but one of the wheels. Moreover, the development of capitalist production makes it constantly necessary to keep increasing the amount of the capital laid out in a given industrial undertaking, and competition makes the immanent laws of capitalist production to be felt by each individual capitalist, as external coercive laws. It compels him to keep constantly extending his capital, in order to preserve it, but extend it he cannot except by means of progressive accumulation... To accumulate is to conquer the world of social wealth, to increase the mass of human beings exploited by him, and thus to extend both the direct and the indirect sway of capitalism.[10]

It is this self-perpetuating expansion of the market which has also produced and reproduces ways of thinking and conditions conducive to environmental destruction. It has produced the conception of people as labour-power to be bought and sold and reduced nature to a mere resource to be exploited. It has produced general insecurity by creating a continuing reserve of unemployed, impelling the short term economic orientation which is one of the most important causes of environmental destruction. And it has generated population growth. In relation to this degradation of humans and nature to nothing but means of production Marx wrote:

> Thus, just as production founded on capital creates universal industriousness on one side - i.e. surplus labour, value-creating labour - so does it create on the other side a system of general exploitation of the natural and human qualities, while there appears nothing *higher in itself*, nothing legitimate for itself, outside the circle of social production and exchange. Thus capital creates the bourgeois society, and the universal appropriation of nature as well as of the social bond itself by the members of society. ... For the first time, nature becomes purely an object for humankind, purely a matter of utility; ceases to be recognized as a power for itself; and the theoretical discovery of its autonomous laws appears merely as a ruse so as to subjugate it under human needs, whether as an object of consumption or as a means of production. In accord with this tendency, capital drives beyond national barriers and prejudices as much as beyond nature worship, as well as all traditional, confined, complacent, encrusted satisfactions of present needs, and reproductions of old ways of life. It is destructive towards all this, and constantly revolutionizes it, tearing down all the barriers which hem in the development of the forces of production, and the exploitation and exchange of natural and mental forces.[11]

---

7. Marx, *Capital*, Vol.1, p.114.

8. Ibid., p.542.

9. Karl Marx, 'Results of the Immediate Process of Production' in appendix to the Penguin edition of *Capital* Vol.1, tr. Ben Fowkes, Harmondsworth, 1976, p.990.

10. Marx, *Capital* Vol.1, Progress Publishers, p.555.

11. Marx, *Grundrisse*, p.409f.

And he argued that this is associated with an inherent tendency to upset the balance of nature:

> Capitalist production, by collecting the population in great centres, and causing an ever increasing preponderance of town population, on the one hand concentrates the historical motive power of society; on the other hand, it disturbs the circulation of matter between man and the soil, i.e., prevents the return of the soil of its elements consumed by man in the form of food and clothing; it therefore violates the conditions necessary to the lasting fertility of the soil ... [A]ll progress in capitalist agriculture is a progress in the art, not only of robbing the labourer, but of robbing the soil; all progress in increasing the fertility of the soil of a given time, is a progress towards ruining the lasting sources of that fertility. The more a country starts its development on the foundation of modern industry, like the United States, for example, the more rapid is this process of destruction. Capitalist production, therefore, develops technology, and the combining together of various processes into a social whole, only by sapping the original sources of all wealth - the soil and the labourer.[12]

In relation to population growth he noted:

> In fact, not only the number of births and deaths, but the absolute size of the families stand in inverse proportion to the height of wages, and therefore to the amount of means of subsistence of which the different categories of labourers dispose. This law of capitalist society would sound absurd to savages, or even civilised colonists. It calls to mind the boundless reproduction of animals individually weak and constantly hunted down.[13]

However Marx offered no real explanation for this phenomenon.

Finally Marx revealed how this system, with all its destructive characteristics, has immanent within it the tendency to continue expansion until the entire world has been dominated: 'The tendency to create the *world market* is directly given in the concept of capital itself.'[14] And so:

> In history up to the present it is ... an empirical fact that separate individuals have, with the broadening of their activity into world-historical activity, become more and more enslaved under a power alien to them ... a power which has become more and more enormous and, in the last instance, turns out to be the *world market*.[15]

Marx's attitude to this expansion of the market was ambiguous. While he saw the overthrow of pre-capitalist modes of production such as those in India as progressive, in the case of Ireland he recognized a tendency for this expansion to lead to exploitation of one region by another, and he saw that the effects of this could be to divide and weaken the opponents of capitalism. And in a lecture on the free trade issue, he pointed out that:

> All the destructive phenomena which unlimited competition gives rise to within one country are reproduced in more gigantic proportions on the world market... If the free-traders cannot understand how one nation can grow rich at the expense of another, we need not wonder, since

---

12. Marx, *Capital*, Vol. 1, p.474f.

13. Marx *Capital* Vol. 1, p.602.

14. Karl Marx, *Grundrisse* tr. Martin Nicolaus, Harmondsworth: Penguin Books, 1973, p.408.

15. Karl Marx and Frederick Engels, 'The German Ideology', in *Karl Marx, Frederick Engels: Collected Works*, N.Y.: International Publishers, Vol.5, 1976, p.51.

these same gentlemen also refuse to understand how within one country one class can enrich itself at the expense of another.[16]

Apart from revealing the exploitative and destructive dynamics of capitalism, Marx was concerned to expose the debasement of humanity by capitalism, and through this, of the possibility of life in which people will realize higher potentialities than they are able to recognize within a capitalist society. He argued in the *1844 Manuscripts*: 'Production does not produce man only as a *commodity*, the *human commodity*, man in the form of a *commodity*; it also produces him as a *mentally* and physically *dehumanized* being.'[17] And his revulsion against this dehumanisation informed all his later work. *Capital* began with an analysis of 'the categories which make up the inner structure of bourgeois society', the categories of 'commodity, 'capital', 'wage labour' and so on which under capitalism mediate people's relationships, are the 'forms of being'.[18] In doing so he revealed how people are more than they are conceiving themselves to be - they are beings who to some extent form themselves through the way they conceive their social relationships. Rather than being nothing but commodities, commodity producers and commodity consumers, or labour power moved to work by their appetites, this analysis reveals people to be creative, social beings, and it implies that people have the potential to form relationships based on different categories in which their true nature is recognized and valued.

Marx, as opposed to his followers, clearly recognized that these categories are also blinding people to nature. In his *Critique of the Gotha Program*, Marx asserted: 'Labour is not the source of all wealth. Nature is just as much the source of use values... as labour, which itself is only the manifestation of a force of nature, human labour power.'[19] He then pointed out that by ascribing a supernatural creative powers to labour the bourgeoisie are able to avoid acknowledging that by reducing nature to something belonging to individuals they are depriving people of access to the primary conditions of labour, and are thereby forcing people to sell their labour power as a commodity in order to live.

Marx went on to show how the conceptions people are constrained to adopt about themselves and their relationships to each other and to nature mystify the real nature of these relationships, a mystification which is essential for the reproduction and development of capitalism independently of people's intentions. At the same time he showed how this development was creating the conditions for a revolution: the development of the means of production from which a new social order based on different principles of organization could be based - increasing instability in the economy and the growth of a disciplined social class, the proletariat, which could take advantage of this instability to create the new social order. So, revealing the cause of the nihilistic conceptions people hold about the world and themselves, he pointed a way beyond this nihilism.

In this way Marx provided a framework for analysing and explaining environmental degradation and pointed a way to overcoming it. However environmental problems were not nearly as severe when Marx was writing as they are now, and neither he nor Engels offered more than brief comments on these problems. Research in this area was left to his followers.

---

16. Karl Marx, *The Poverty of Philosophy*, Moscow: Progress Publishers, 1973,p.193f.

17. Karl Marx, 'Economic and Philosophical Manuscripts' in *Early Writings*, Harmondsworth: Penguin, 1975, pp.279-400, p.336.

18. Marx, *Grundrisse*, p.108, 106.

19. Karl Marx, 'Critique of the Gotha Program', in Robert C. Tucker ed., *The Marx-Engels Reader*, 2nd ed., N.Y.: W.W. Norton & Co., 1978, pp.525-541, p.525.

## *Marxist Social Theory Since Marx*

Marx's identification of capitalism as historically unique and his characterisation of its dynamics have been accepted by virtually all major social theorists (although few economists). Even a minor apologist for capitalism like Herman Kahn quoted Marx's ideas on this with approval.[20] Where anti-Marxist social thinkers disagree with Marx is about capitalism's oppressiveness, over the role of culture in the formation of capitalism, about the possibilities of its transformation, and more recently, about the relevance of Marx's analyses to modern societies. However Marxist social theorists have extended Marx's ideas to meet these challenges. The most important developments of Marxism since Marx have been studies of the expansionist, imperialistic tendencies of capitalism and its effects, studies of ideologies and culture, and more recently, studies of the State.

The most important of the early Marxist theorists of imperialism were Hilferding, Luxemburg and Bukharin.[21] Hilferding's main contribution to the theory of imperialism was his elaboration of the concept of finance capital - the product of the fusion of industrial and financial capital into huge interlocking groups which then competed with each other not by price cutting, but by enlisting State support to gain control of whole industries, this leading to inter-imperialist rivalries.[22] Luxemburg argued on doubtful grounds that capitalism can only overcome its contradictions by expanding into the non-capitalist world, but her real importance was to have revived Marx's concern with the way capitalism expands and breaks up non-capitalist social formations. Bukharin transformed previous studies of imperialism by setting them in the context of a world economy within which two tendencies were seen to be at work: the tendency towards monopoly and the integration of finance capital, and the tendency towards the acceleration of the geographical spread of capitalism and its integration into a single world capitalist economy.[23] Competition then becomes competition between State capitalist trusts within a world economy, with annexation and war being its instruments. None of these theorists of imperialism ever doubted that capitalism, despite its oppressiveness, was anything but a force for progress in economic development. The one person who did question this, the Indian Marxist Nath Roy who argued at congresses of the International in Moscow in the 1920's that the most important form of exploitation by capitalism was of its colonial territories rather than of its proletariat, disappeared into oblivion.

When imperialism became a major topic of Marxism again after the Second World War, the progressive nature of capitalism came under question. In 1957 Paul Baran published his *Political Economy of Growth*, an analysis of the dynamics of monopoly capitalism which included an argument that Western Europe was responsible for the poverty of Third World nations, having organized them into suppliers of cash crops. Baran's ideas concurred with the South American dependency theorists, the most notable of whom, Raul Prebisch, had argued that the poor countries of the world were being held in a state of dependent poverty by the affluent centres of the world-economy. Later Marxist theorists of imperialism attempted to assimilate the ideas of the Latin American dependency theorists to develop a Marxist version of dependency theory. The most influential of these were Paul Sweezy, Andre Gunder Frank, Arghiri Emmanuel and Samir Amin.[24] This Marxist notion of dependency was then reformulated in terms of a more general theory of global economics by Immanuel Wallerstein who argued that the economy of the world must be seen as a system, dominated by one mode of production: capitalism, and that the differentiation of the

---

20. Herman Kahn et.al. *The Next 2000 Years*, London: Associated Business Programs, p.47n.

21. Lenin is also regarded as a major theorist of imperialism, but his ideas derive almost entirely from Hilferding, Bukharin and the non-Marxist underconsumption theorist Hobson.

22. Rudolf Hilferding, *Finance Capital*, [1910] tr. Tom Bottomore, London: Routledge & Kegan Paul, 1985.

23. N. Bukharin, *Imperialism and World-Economy,* [1917] London: Allen Lane, 1972.

24. Most of these have been examined by Anthony Brewer in *Marxist Theories of Imperialism: A Critical Survey*, London: Routledge & Kegan Paul, 1980.

world system into affluent, semi-affluent and impoverished regions organized into a network of nation states, must be understood as a product of the dynamics of this system.[25] The different regions of the world were characterized as economic zones: the core, semiperipheral and peripheral zones. The core, the industrialized centres, contains everything which is most advanced and diversified and exploits the rest of the world, the semiperiphery possesses only some of these features and is both exploited and is exploiting, while the huge periphery represents backwardness, archaism, coerced cash-crop labour and raw materials, and exploitation. In opposition to Marx's claim that 'The country that is more developed industrially only shows, to the less developed, the image of its own future',[26] Wallerstein argued that nations will develop differently according to their position within the world-system.

While there has been considerable empirical evidence brought forward to support the claim that the economic centres have exploited and impoverished the Third World,[27] and a number of efforts to account for this evidence theoretically, there has been a revival by neo-orthodox Marxists of the view that capitalism is generally progressive. Bill Warren rejected the arguments of the dependency theorists as inconsistent with Marxism and offered counter-evidence to suggest that capitalism is progressive, that it undermines pre-capitalist modes of production and having done so, leads to rapid economic growth. He argued that to the extent that there is any backwardness in the world, this is due to the failure of capitalism to penetrate these regions and thereby to undermine the pre-capitalist modes of production.[28] Wallerstein in particular has been criticised for simply assuming the existence of a world-system without defining this theoretically, for over-emphasizing the role of the market and ignoring the particular modes of production, social relations, class struggles, power structures and cultures of the nations involved in this system.[29] His work, like that of most other Marxist dependency theorists, is characterized as empirical generalization rather than a theory accounting for the differentiations in this world economy, and he has been criticised for taking the world-market as the dynamic force of history rather than the capitalist mode of production.

While there is some substance to the criticisms of Wallerstein's methodology, notably his over-emphasis on the market, failure to take into account the importance of class struggle, failure to consider local conditions, failure to appreciate the degree of autonomy of States, and a tendency towards functionalism, a number of theorists aligned with Wallerstein have attempted to overcome these theoretical deficiencies and to describe the actual relations which constitute the global system of differentiation and exploitation described by the dependency theorists. Wallerstein has defended the primacy of the world-system over local modes of production as an object of analysis, arguing

25. I. Wallerstein, *The Modern World System*, 3 Volumes, N.Y.: Academic Press, 1974 - 1989; *The Capitalist World-Economy*, Cambridge: Cambridge University Press, 1979; and *The Politics of the World-Economy*, Cambridge: Cambridge University Press, 1984. The influence of Wallerstein's concepts is immense, not least through the two series *Explorations in the World-Economy* and *Political Economy of the World-System Annuals* (both published by Sage) both of which are edited by Wallerstein.

26. Karl Marx, *Capital*, Vol.1, p.19.

27. See for instance L.S. Stavrianos, *Global Rift: The Third World Comes of Age*, N.Y.: Morrow, 1981.

28. Bill Warren's major works are: 'Imperialism and Capitalist Industrialization,' in *New Left Review* Vol.81, Sept/Oct 1973, pp.3-44; and *Imperialism, Pioneer of Capitalism*, London: New Left Books, 1980. For a review of the arguments and literature surrounding this debate see James H. Weaver and Marguerite Berger, 'The Marxist Critique of Dependency Theory: An Introduction,' in Wilber ed. *The Political Economy of Development and Underdevelopment*, 3rd ed. N.Y: Random House, 1984, pp.45-64.

29. Robert Brenner, 'The origins of capitalist development: a critique of neo-Smithian Marxism', *New Left Review*, 104, July/August, 1977; Theda Skocpol, 'Wallerstein's world capitalist system; a theoretical and historical critique', *American Journal of Sociology*, Vol.32, No.5, 1977, pp.1075-1089; Ernesto Laclau, *Politics and Ideology in Marxist Theory*, London: Verso, 1979, pp.42-50; S. Aronowitz, 'On Wallerstein's thesis', in *Theory and Society*, July, 1981, pp.503-520; and Peter Worsley, 'One World or Three? A Critique of the World-system Theory of Emmanuel Wallerstein', in David Held et. al. eds, *States and Societies*, Oxford: Basil Blackwell, 1985, pp.504-525.

that non-capitalist modes of production are maintained by the world-system driven by the capitalist mode of production. Capitalism provides the conditions for the continued existence of non-capitalist modes of production by providing markets for goods produced, and by providing military support to oppress groups who attempt to undermine these modes of production. This has clearly been the case in Latin America where cash crops have been produced under a feudalistic mode of production, and efforts to redistribute land to the peasants have been violently opposed with strong backing from the United States. Wallerstein's argument in this regard has been supported by P.P. Rey and G. Arrighi who have argued that such a situation also pertains in Africa. As Rey argued: 'Throughout the world, capitalism to-day plays a fundamentally counter-revolutionary role: it keeps the most archaic forms in existence; it restores them when they are threatened (see for example the sultanates of Chad).'[30] With further developments in the world-systems approach the issue has become not whether what happens in each part of the world is determined by the world economic system, but how each country and region is constrained by the dynamics of the world-system and how have they responded to these constraints.

The nature and significance of ideology was made a focus of interest by those Marxists of the 1920's influenced directly or indirectly by Hegel, notably Georg Lukács, Karl Korsch and Antonio Gramsci, although many of their ideas were anticipated by Stanislav Brzozowski in Poland and Aleksandr Bogdanov in Russia more than a decade earlier.[31] Since then the study of ideology has made rapid advances in a number of directions, engendering the sociology of knowledge, the sociology of science and the whole field of Marxist aesthetics.[32] Ideology was studied to clarify the effects of capitalism on consciousness, to reveal how radical action is stifled or prevented by the dominant ideology, and to show what role consciousness must play if society is to be transformed. Lukács developed and extended Marx's notion of reification, arguing that under capitalism not only do we fetishise commodities, but that also 'time sheds its qualitative, variable, flowing nature; it freezes into an exactly delimited, quantifiable continuum filled with quantifiable "things"', and that '[i]n this environment where time is transformed into abstract, exactly measurable, physical space, an environment at once the cause and effect of the scientifically and mechanically fragmented production of the object of labour, the subjects of labour must likewise be rationally fragmented.'[33] Also going beyond Marx, Antonio Gramsci developed a more complex, multidimensional and concrete analysis of the role of ideology based on his notion of cultural and ideological hegemony - the organization of consent by a dominant class. Ideological hegemony was seen to encompass the whole range of values, attitudes, beliefs, cultural norms and legal precepts which are transmitted through the State, the legal system, the schools, the churches, bureaucracies, the media, the family - as well as the workplace, solidifying the class structure and the multiple forms of domination associated with it.

Since the end of the Second World War the study of ideology has expanded as the 'industrialization of the mind',[34] the systematic control of what people think, has become ever more ubiquitous. The most significant studies have been the Frankfurt Institute philosophers' work on the domination of instrumental reason and the development and mind warping nature of mass culture, Marxist studies of science, in particular, of the origins of mechanistic science, Darwinism

30. P.P Rey, *Colonialisme, Néo-colonialsme et Transition au Capitalisme*, Paris: Maspero, 1971, p.463, translated and cited by Brewer in *Marxist Theories of Imperialism,* p.198.

31. On Brzozowski see Andrzej Walicki, *Stanislav Brzozowski and the Polish Beginnings of 'Western Marxism'*, Oxford: O.U.P., 1989. Bogdanov's ideas will be discussed in Chapter 10 of this work.

32. For some idea of the extent of this burgeoning field of study see Jorge Larrain, *The Concept of Ideology*, London: Hutchinson, 1979; and John B. Thompson, *Studies in the Theory of Ideology*, Cambridge: Cambridge University Press, 1984.

33. Georg Lukács, *History and Class Consciousness*, tr. Rodney Livingstone, London: Merlin Press, 1971, p.90.

34. From Hans Magnus Enzensberger, 'The Industrialization of the Mind' in Stanley Hoffmann and Paschalis Kitromilades eds. *Culture and Society in Contemporary Europe*, London: George Allen & Unwin, 1981, pp.83-97.

and Social Darwinism, and Marxist theories of education focussing on how modes of thinking on which capitalism is based are reproduced from generation to generation. More recently attention has focussed on the effect of advertising, of public relations, and of new media, most importantly television, in forming the way people think. It is argued on the basis of such studies that it is because of the reproduction of the dominant modes of experiencing and thinking that socialism has been unable to establish itself, and that what is required is the emancipation of humanity from misconceptions reproduced by capitalist society. Extending this to the Third World, it has been argued that it is through the cultural imperialism of the economic centres and the destruction of local cultural traditions that the exploitation of the peripheries of the world-economy has been possible.[35]

The final area in which Marxist thought has been developed is in the study of the State.[36] Marx himself never developed an adequate theory of the State, and Marxists have striven to fill this gap. To begin with, the State was represented as an instrument of the ruling class of the capitalist economy. As such it was seen to have become increasingly important in capitalism's final monopolistic phase in which there appeared to be a fusion of monopoly forces with the State, forming a single mechanism of economic exploitation. It was this form of the State which is supposed to have climaxed in two world wars and with the rise of the military-industrial complex. However this orthodox view of the State (an alternative to which had already been developed by Gramsci) has recently been severely criticised.[37] It has been argued that the relationship between economic and State institutions is far more complex; that the State is a battleground for opposing classes, that it has interests of its own independently of any class, that it consists of a diversity of conflicting institutions, that it is part of a world-system of nation-States, and that it is now caught in an increasingly untenable position as social relations, international relations and the international economy become more complex. It is argued by Marxist theorists of the State that what we now have within the affluent nations of the world is a crisis of the State accentuated by the internationalization of capital.[38]

## *Marxist Environmentalism*

Marxist environmentalism has a long history. It began in Marx's own lifetime as efforts were made to take into account the second law of thermodynamics and give an account of Marx's labour theory of value and surplus value in terms of the accumulation of useful energy. The first to argue along these lines, a Ukrainian socialist named Serhii Podolinskii, proposed this to Marx personally, and Marx and Engels corresponded on Podolinskii's proposals.[39] A succession of socialist or

35. A good example of this is Renato Constantino, *Neocolonial Identity and Counter Consciousness*, London: Merlin Press, 1978. See also Armand Mattelart ed. *Communication and Class Struggle: 1. Capitalism, Imperialism,* N.Y.: International Genergal, 1979 and Armamd Mattelart, *Transnationals and the Third World: The Struggle for Culture*, South Headley, Mass.: Bergin and Garvey, 1983.

36. An excellent short introduction to these debates is Martin Carnoy's *The State and Political Theory*, Princeton: Princeton University Press, 1984. See also David Held et. al. eds, *States and Societies,* Oxford: Basil Blackwell, 1985; and James Anderson ed. *The Rise of the Modern State*, Brighton: Wheatsheaf Books, 1986.

37. See Bob Jessop, *The Capitalist State,* Oxford: Basil Blackwell, 1984; and *State Theory: Putting Capitalist States in their Place*, Cambridge: Polity Press, 1990.

38. For an analysis of the crisis of the State in modern capitalist societies see J. O'Connor, *Fiscal Crisis of the State*, New York: St Martin's Press, 1973; Claus Offe, *Contradictions of the Welfare State*, J. Keane ed., London, 1984; and Scott Lash and John Urry, *The End of Organized Capitalism*, Cambridge: Polity Press, 1987.

39. See J. Martinez-Alier and J.M. Naredo, 'A Marxist Precursor of Energy Economics: Podolinski', *Journal of Peasant Studies*, Vol.8, Jan., pp.207-224. Engels wrote disparagingly of the proposal, and argued against Podolinsky in a number of articles. See J. Martinez-Alier, *Ecological Economics*, Oxford: Blackwell, 1987, p.222.

otherwise radical thinkers put forward similar ideas at regular intervals up until the 1920's, but then because 'energism' was associated with the empirio-criticism of Bogdanov which Lenin attacked so vehemently, and probably because it implied limits to the economic growth which most Marxists believed would be the salvation of humanity, these thinkers were forgotten about until recently.[40] When the environment first began to become a major issue in the West in the early 1970's, most Marxists dismissed environmentalists,[41] and the late 1980's, Marxist environmentalists were marginal to both the Western tradition of Marxism and to the environmentalist movement. However with the publication of a number of significant works and the establishment in 1990 by James O'Connor of the journal *Capitalism, Nature, Socialism,* Marxist environmentalism is now one of the most dynamic fields of Marxist research.[42] The environmental crisis is portrayed as the 'second contradiction' of capitalism.[43]

In their concern to reveal behind the glittering facade of capitalism the environmental irrationalism it engenders, Marxist and Marxist influenced environmentalists have not only extended Marx's own ideas on the environment, but have drawn on and developed the more recent extensions of Marxism. In essence, environmental problems are seen to be produced because in a capitalist society economic activity is production of commodities for the market, with production for profits having replaced production for consumption as the primary goal of activity.[44] This system cannot take into account anything which cannot be expressed as a demand on present markets (such as the needs of future generations) which at its best can anticipate demand ten years into the future,[45] it reduces nature and people to mere factors of production and it opens the possibility of increasing production and exploitation indefinitely until the environment is destroyed. Where economic decisions are made on the basis of what will produce the greatest profit by business enterprises struggling for survival in a competitive environment, it is in the interests of, and in fact imperative for decision-makers to create scarcities to drive up prices, to produce in a way which deprives people of control of their lives and forces them to attain their needs and their livelihoods through markets over which monopoly control can be established, and to produce commodities which do not satisfy demand but which generate new demands, either by wearing out, by becoming obsolete, or by imposing new requirements on people. Thus Susan George noted in her study of the causes of Third World hunger:

> This is where the question of the individual sincerity of industry leaders is answered: they themselves - even if they are corporation presidents with the best will in the world - are not free agents. They *must*, under the logic of their system, market produce in countries that can best pay for it; they must get the best possible return on investment, which means either cheap labour or less labour and more amortizable machinery; they *must* control all the facets of food production and distribution for maximum profitability from field to supermarket to shelf.[46]

---

40. See ibid. for a study of these thinkers.

41. See for instance Hans Magnus Enzensburger, 'A Critique of Political Ecology' in *The Political Economy of Science*, ed. Hilary Rose and Stephen Rose, London: Macmillan, pp.161-195.

42. See for instance Reiner Grundmann, *Marxism and Ecology*, Oxford: O.U.P., 1991, and Ted Benton, *Natural Relations: Ecology, Animal Rights and Socialist Thought*, London: Verso, 1993.

43. This phrase was coined by James O'Connor. See 'The second contradiction of capitalism: causes and consequences', *Conference Papers*, Santa Cruz, CES/CNS Pamphlet 1. It is the theme of the recently published anthology, *Is Capitalism Sustainable?* ed. Martin O'Connor, N.Y.: Guilford University Press, 1994.

44. On some of the imperfections of the market mechanism in relation to resources, see Richard Lecomber, *The Economics of Natural Resources*, London: The Macmillan Press, 1979.

45. Theoretically anticipated future profits should be discounted at the rate of interest, but because of uncertainties, they are discounted at a higher rate.

46. Susan George, *How the Other Half Dies*, Harmondsworth, 1977, p.234.

In such a system most business enterprises are compelled to use up reserves as quickly as possible. Since investments amount to interest foregone and immediate income is necessary to return interest, they are compelled to exploit renewable resources in such a way that they are destroyed if this generates only slightly greater profits,[47] and to pollute their environments. Beyond this it actually pays firms to degrade the environment, to waste reserves and to destroy resources, to pollute the air and the water, since it is through the production of scarcities and the generation of needs that profits can be made. If timber companies can destroy most of the forests of the world, their profits will increase rather than decrease since scarcity will lead to escalating prices. If air is unpolluted, there is no room for capitalist enterprise; but if it becomes so polluted that people have to use respirators to breath, a whole new profitable industry will come into being. Where people are made ill by pollution G.N.P. can grow as drug companies and medical practicioners find new markets for their products and services.

Capitalism itself can be guaranteed to inspire solutions to some environmental problems, and in fact this is likely to be big business in the future. But by the very nature of the capitalist mode of production, the production of solutions will always fall behind the generation of the problems. To begin with, it is only where problems are recognized and there are people able to pay for solutions that capitalism will ever generate industries to solve these problems. Given the time required for problems to be recognized and along with vast numbers of impoverished people with no market power, this will always be only a small fraction of the problems. We now have cures for some of the cancers caused by pollution but it is only the affluent of the world who can afford them, and the cures hardly match the increased incidence of cancer generated by pollution.[48] And beyond this there are theoretical limits to how many problems can be solved. All activity generates at least as much disorder as it creates. It is clear that it requires far more usable energy to purify the world of pollutants than to pollute the world in the first place, and using up such energy must create even more pollutants.

The destructive nature of capitalism is particularly evident in its effects on agriculture. This has been more clearly manifest in the United States than in Europe because capitalism has reigned with less dilution from older traditions, and it is here that its dynamics are revealed. An exemplary work revealing these dynamics is Donald Worster's study of the creation of the Dust Bowl in USA through farming for profit.[49] The southern plains of the United States have been and continue to be used in a way which destroys their fertility, and the resultant dust bowl, along with the deafforestation of China's uplands about 3000 B.C. and the destruction of the Mediterranean vegetation by livestock, is frequently cited as one of the three worst ecological blunders in history. However the Dust Bowl took only 50 years to create and was not the work of illiterates or the product of over-population, but was the 'inevitable outcome of a culture that deliberately, self-consciously, set itself that task of dominating and exploiting the land for all it was worth.'[50] Since the publication of Worster's book, the dynamics of capitalism have further advanced the industrialization and concentration in control of agriculture.[51] While before the war farmers in the United States spent half their income on capital investments, they now must spend over 80%. The effect of this has been that by 1987 a third of the farmers in the Bread-Basket states of the Mid-West were facing bankruptcy. The family farm is being driven to extinction and farming is being completely taken over by transnational agribusiness companies, destroying whole farming

47. It would be more profitable to farm in a way which totally destroyed the land in seven years if this generated a profit of 20% per year on the cost of the land as opposed to farming sustainably, preserving the land for a billion years, for a profit of only 10% per year.

48. See Samuel S. Epstein, 'Losing the War Against Cancer', *The Ecologist*, Vol.17, No.2/3, 1987, pp.91-101.

49. Donald Worster, *Dust Bowl: The Southern Plains in the 1930's*, N.Y., Oxford: Oxford University Press, 1979.

50. Ibid. p.4.

51. See Bennett, *The Hunger Machine*, Ch.6, and Michael W. Fox, *Agricide*, New York: Schocken Books, 1986 for accounts of what is happening.

communities. This means a complete separation between workers and the ownership of land, which is treated by big business solely in terms of its capacity to make short-term profits. The effect on the land is worse than ever. Farms now lose two bushels of topsoil for every bushel of corn produced. By 1985, the USA had lost one third of its topsoil. Continual cropping has also reduced soil fertility, and agrochemicals have caused an actual fall in productivity since the mid 1970's. Crop mono-culture has played havoc with natural ecosystems: bird, fungi and insect species disappear while others multiply, increasing the need for pesticides. And the narrowing of the genetic base of crops makes them increasingly susceptible to disease. The aquifers which are used to supply water for irrigation are being depleted, and will be practically exhausted by 2030. Finally, such agriculture uses huge amounts of energy. The energy required to feed one person amounts to more than 310 gallons of petroleum a year. And the situation in USA is if anything better than in other countries: Canada, Australia, New Zealand and Argentina for instance.[52]

This industrialization of agriculture has been associated with the rapid growth in secondary industry to supply its new needs for fertilizers, pesticides, seeds and machinery, and this has produced some of the greatest irrationalities of capitalism. For instance the pesticide industry is now an extremely profitable industry, as pesticides destroy predators of pests, weaken the defences of crops, and thereby create an ever greater need for pesticides.[53] While the use of pesticides increased twelve fold between 1950 and 1980, losses to pests doubled,[54] while alternative, far more promising approaches to pest control based on ecological principles, have been ignored and their proponents have been hounded out of their jobs.[55] Not only has the use of pesticides been counter-productive, but it is poisoning wildlife, farm animals and half a million people throughout the world each year. Since all this involves continual production of new opportunities for profit making, the pesticide industry must be regarded as the archetypical successful enterprise within the capitalist economy.

Capitalism also leads to irrational behaviour when it comes to pollution. Pollution collectively affects the whole society adversely, but the pollution produced by each business enterprise scarcely affects its profits at all, which means that it pays individual firms to pollute their environment to the detriment of all. One of the best illustrations of this effect is the production of the chlorofluorocarbons which are destroying the ozone layer. It has been calculated that it would cost the United States $4 billion to reduce CFCs by 20% which, it is estimated, would save the lives of 993,000 people in USA over the next 90 years who would otherwise die from skin cancer and other diseases related to the loss of ozone (the lives which would be saved outside USA have not been calculated, but there would presumable be more than 20 million given that USA now has 5% of the world's population, and that this proportion is falling).[56] And as it has turned out, the amount of ozone depletion is greater than expected. But while these deaths would, among other things, cost the country $1.3 trillion, it is not profitable for business companies to spend this money, and it is against the philosophy of the New Right which now dominates politics, and which has been promoted mainly by the business community, to attempt to interfere with the functioning of the market. In Western Europe there has been even greater resistance to any controls over CFCs than in the United States. This resistance is spearheaded by chemical companies such as ICI. Similarly,

---

52. For the situation in Australia see Geoffrey Lawrence, *Capitalism and the Countryside: The Rural Crisis in Australia*, Sydney and London: Pluto, 1987.

53. See for example John H. Perkins, *Insects, Experts and the Insecticide Crisis*, N.Y.: Plenum, 1982.

54. See Edward Goldsmith, 'Pesticides Generate Pests', *The Ecologist*, Vol.10, No.3, 1980, pp.94-97.

55. The most notable case of this were the efforts to discredit Robert Van Den Bosch, author of *The Pesticide Conspiracy*. On the way environmentalist scientists generally have been discriminated against, see Brian Martin, 'The Scientific Straightjacket', *The Ecologist*, Vol.11, No.1, 1986, pp.33-43.

56. See Jeremy Cherfas, 'Ecology Invades a New Environment' in *New Scientist*, 22nd October, 1987, pp. 42-46.

efforts to control carcinogenic pollution have failed, and as a result, some 20% of the US population will die of cancer.[57]

While the development of agribusiness has depopulated the countryside and concentrated populations in the cities and megalopolises, the nature of these have been largely determined by the demands of industrialists, the interests of property developers in profits and the interests of States in maintaining a docile population.[58] To begin with, industry has been concentrated in a small number of centres since, due to availability of other products and trained personnel, ease of communication and so on, it is more profitable for firms to locate industries close to each other; and it is not firms which bear the costs of the infrastructure required for this, the consequent pollution and housing shortages. It is the employees who are subjected to higher taxes to pay for the infrastructure of cities and industries, who are forced to move to big cities to find work, who must then pay excessively for accommodation and then spend major proportions of their lives travelling to and from work. With the subsequent expansion of cities, a characteristic structure emerges. Old buildings, particularly old housing and apartments in the centres of cities conducive to the life of culture are knocked down and replaced with high-rise office blocks. There is a further movement inward and upward, as smaller office blocks, losing their customers as they are dwarfed by new buildings, are demolished. Industries take over the areas with the best access to transport, irrespective of the pollution they cause, and exclude housing. People, excluded from the city centres are concentrated in dormitory suburbs, with transport being organized to get their labour-power efficiently to the city or industrial areas then back to the suburbs for regeneration. Thus cities come to embody a one-dimensional functionalism, destroying the conditions for cultural life, fragmenting communities and isolating individuals.[59] With capitalist enterprises holding people and governments to ransom the efforts of people to prevent such developments, to control these cities in the interests of their populations, have been stymied, especially in the New World where pre-capitalist traditions are less strong or non-existent. City planning itself reflects the power of both industrialists and property developers to over-ride people's interests, and the determination of governments to assuage business interests.[60] These developments reinforce cultural changes which are obliterating any critical understanding by people of the world and its problems.

Responsibility for side-effects of profit making or for the future could only be taken by a superordinate authority able to force individuals to take them into account. But State institutions in capitalist societies, subject to subversion by sectional economic interests or lurching from one crisis to the next in a struggle to keep the economy going, are seldom able to enforce such accountability.[61] The criterion of success of modern capitalist economies, that societies maintain growth rates of 4% in order to maintain full employment, implies that there must be an exponential increase in the use of non-renewable resources and in the production of pollution.[62] In recent years national governments have lost power as transnational companies and international financial institutions have freed themselves from national controls and held governments to ransom to hold down taxes and provide incentives to invest. Continually grappling with immediate crises,

57. See Samuel S. Epstein, *The Politics of Cancer*, San Francisco: Sierre Club Books, 1978, p.8.

58. The classic Marxist work on urbanization is David Harvey's *Social Justice and the City*, London: Edward Arnold, 1973. For a major collection of Marxist analyses of urbanization examining these issues see Michael Dear and Allen J. Scott eds, *Urbanization & Urban Planning in Capitalist Society*, London: Methuen, 1981.

59. On this see Samir Amin, 'In Praise of Socialism', *Monthly Review*, Sept. 1974, No.24, Vol.26, pp.1-16.

60. Leonie Sandercock and Michael Berry have examined the way property developers and industrialists operate in Australia, virtually completely controlling the way cities develop. See *Urban Political Economy*, Sydney: George Allen & Unwin, 1983.

61. On this see, R.J. Johnston, *Environmental Problems: Nature, Economy and the State*, London: Belhaven Press, 1989.

62. As R. England and B. Bluestone pointed out in, 'Ecology and Class Conflict,' in H. Daly ed., *Towards a Steady State Economy*, San Francisco: Freeman, 1973, pp.190-214. To the extent that the dynamics of the market are successfully regulated by State institutions in the interests of its members, the society ceases to be capitalist.

particularly with the threat of disinvestment and unemployment, governments now have a stronger incentive than ever to block efforts to confront environmental problems, and even where governments have departments devoted to environmental issues (for example, the EPA in USA), their staff have been stymied and the information they have brought to light suppressed. In these circumstances, political power has become virtually unattainable by those who are concerned about fundamental, long-term problems of society.

## *The Environment and the Third World*

The regional differentiation of world-capitalism ensures that it is not in the affluent North that the worst environmental degradation occurs; the worst environmental degradation is taking place in the peripheries of the world-economy.[63] Since the sixteenth century peripheral regions have lost control over their best agricultural land and have been bled of their most valuable mineral reserves, resulting in general environmental degradation. Regional exploitation accelerated in the nineteenth century with the expansion first of British, then of European capitalism, and accelerated even further in twentieth century with the development of U.S. dominated neo-imperialism. By promoting comprador classes and addicting them to luxuries and military hardware, peripheral regions have become indebted to the economic centres and have been impelled to sell off their raw materials on a competitive market and to devote their agricultural land to cash crops for export to raise foreign currency. With transnationals based in USA and Europe controlling most of the markets, and with large numbers of Third World countries in similar positions, prices for these raw materials and cash crops have been kept low. The terms of trade for exporters of raw materials have declined almost continuously over the last hundred years, and with the exception of oil, even more rapidly over the last forty years.[64] Environmental exploitation has intensified in recent years with the international debt crisis. By focussing on the environment, environmentalist Marxists have revealed the full extent of the exploitation between regions in the world-economy, and in doing so have provided strong support for the world-systems approach and put paid to the arguments of those neo-orthodox Marxist defenders of capitalism and imperialism.[65]

While the global economy emerged in the sixteenth century, environmental destruction began to be produced on a global scale through the expansion of capitalism in the nineteenth century when the demands of the metropolitan societies for foodstuffs, fibres and raw materials led to land clearance for cash crop production and accelerating exploitation of forests in the colonies of capitalism.[66] Then, with the backing of colonial governments, indigenous constraints were swept away by market principles. Such destruction of indigenous constraints has accelerated rapidly in the twentieth century. Studies of the present situation have revealed both how transnational companies are able to destroy rainforests, and frequently the livelihoods of those dependent upon them, and how the preservation of forests in the wealthy nations along with the acquisition of cheap agricultural products are achieved at the expense of the land and forests of the Third World

---

63. The global nature of environmental problems and the consequent difficulty in confronting them has been well argued by Michael Redclift in *Sustainable Development: Exploring the Contradictions*, London: Methuen, 1987.

64. See L. Rangarajan, 'The Politics of International Trade', in Susan Strange ed., *Paths to International Political Economy*, London: George Allen & Unwin, 1984, pp.126-164.

65. See for instance Jacobo Schatan, *World Debt: Who is to Pay?* London: Zed Books, 1987. An excellent anthology in this area is R.J. Johnston and P.J. Taylor (eds), *A World in Crisis*, Oxford: Basil Blackwell, 1986.

66. P. Tucker and J.F. Richards (eds) *Global Deforestation and the Nineteenth Century World Economy*, Durham, NC.: Duke Press Policy Studies, 1983. The findings of these papers have been summarized by Tucker and Richards in 'The Global Economy and Forest Clearance in the Nineteenth Century' in Kendall E. Bailes (ed.) *Environmental History*, Lanham: University Press of America, 1985, pp.577-85.

countries.[67] As a consequence of this, 40% of the world's tropical rainforests were cleared between 1968 and 1988.

The best theoretical analysis of the process by which the peripheries of the world economy have been degraded into suppliers of raw materials to the economic centres of the world - and of the consequences of this, is Stephen Bunker's study of the Amazon. Bunker argued that:

> ... production models cannot explain the internal dynamics of extractive economies because the exploitation of natural resources uses and destroys values in energy and material which cannot be calculated in terms of labour or capital. When natural resources are extracted from one regional ecosystem to be transformed and consumed in another, the resource exporting region loses values that occur in its physical environment. These losses eventually decelerate the extractive region's economy, while the resource-consuming communities gain value and their economies accelerate.[68]

Bunker argued that orthodox Marxist analyses of the reproduction of modes of production and of the relationship between global and regional economies must be revised to take account of the ecological interdependencies between extractive and resource consuming economies, and to take account of the impacts of these relationships on natural ecosystems. He revealed how the increased energy and material flows to productive societies have facilitated the substitution of human for non-human energies to increase their complexity and power, while the consequent reduced energy flows in peripheral societies have simplified them and reduced their power. Increased energy flows in the productive centres has made possible increases in scale, complexity and coordination of human activities, greater division of labour, and the expansion of specialized fields of information. This has facilitated the development of increasingly complex systems of transport and communication and engendered the means for technological and administrative innovation, enabling these centres to change their technologies and thereby find substitutes for essential resources as these have been depleted. Conversely extractive economies have lost energy and so become economically and socially simpler, less diversified, and subject to the changes in market demand associated with new technologies produced by the centres. Under these circumstances, strategies tend to be adopted which maximize the short-term return to labour and capital, and which are little concerned with long-term social reproduction. Once the profit maximising logic of extraction for trade takes over, exploitation is concentrated on a limited number of resources at rates which disrupt the regeneration of these resources, the biotic community and associated geological and hydrological regimes. The development of modern State organizations in peripheral regions, being subject to manipulation by the productive centres of the world-economy, merely increases the rapidity of destructive exploitation of these regions. By exploiting such extractive economies, the industrial modes of production inevitably undermine the resource bases on which they depend; but they have evolved the social organizational and infrastructural capacity to change their own technologies and thereby to find substitutes for resources as they are depleted. However this process is finite as each new technology requires other resources from what is ultimately a limited stock.

The most striking environmental destruction in the Third World is caused by the transformation of agriculture wrought by capitalism, a transformation which has resulted in the massive impoverishment of local populations. In all, to feed and clothe themselves, Europeans and North

67. See for example *The Vanishing Forest*, A Report for the Independent Commission on International Humanitarian Issues, London: Zed Books, 1986; Norman Myers *The Sinking Ark* Oxford, N.Y.: Pergamon Press, 1979; V. Plumwood and R. Routley, 'World Rainforest Destruction - the Social Factors' in *The Ecologist*, Vol.12, 1982, pp.4-22; and Emilio F. Moran, 'Current Development Efforts in the Amazon Basin' in Michael Tobias (ed.) *Deep Ecology*, San Diego: Avant Books, 1984, pp.58-73.

68. See Stephen G. Bunker, *Underdeveloping the Amazon: Extraction, Unequal Exchange, and the Failure of the Modern State*, Urbana and Chicago, University of Illinois Press, 1985, p.22.

Americans have been using around 20% more of the world's agricultural land than their own.[69] This exploitation of Third World land has been at its worst where the penetration by the world market is associated with coercion by those in a position to benefit through its extension. This has been shown most clearly in the work of Susan George.[70] George described how where the capitalist market operates, land ownership is rapidly concentrated, and how the local élites then redirect the use of land which had been devoted to producing food for local consumption to the production of more commercially profitable crops, that is, crops for export to the wealthy nations. In South America 17% of the landowners control 90% of the land, and one third of the rural population must make do with 1% of the land. In Africa three quarters of the agricultural population have less than 4% of the land. The largest holdings produce the least food. In Brazil and Argentina the smallest properties produce eight times as much per hectare as the largest estates, while in Columbia they produce fourteen times as much. Cash crops take up the best land and most of the scarce inputs into farming. Fifty-five percent of the agricultural land of the Philippines and 80% of Mauritius are devoted to cash crops, while 50% of Senegal is devoted to peanuts alone.

In these countries the transport systems are all directed to transporting cash-crops to USA and Europe. In their study of agribusiness in Africa, Dinam and Hines noted:

> At present, industrial countries import about 90 per cent of all traded horticultural products, of which Third World countries ship 30-40 per cent. Trade is dominated by citrus fruits, potatoes and tomatoes, but in Africa an increasing amount of land is being cultivated to supply European markets with a variety of fresh flowers and out-of-season vegetables and fruits - either dried or flown fresh.[71]

The effect of these developments, associated with the control of markets by transnational agribusiness, has been to force down prices of such cash crops. As a consequence such countries have attempted to further increase their production of cash crops at the expense of subsistence crops, glutting their markets even further. As P.N. Bradley summed up the situation:

> The different processes: monetization, commoditisation, the manipulation of trade, control of the means of production through state apparatuses, penetration of foreign companies in allegiance with a comprador bourgeoisie, a global financial structure refereed by the IMF; all point to the same conclusion. We observe the transformation of rural societies, whose economies were based on some form of reciprocity in their exchange relationship, to a capitalist model of which the central characteristic is one of surplus value extraction and profit. The net result is that, by being more or less forcibly wedded to this capitalist suitor, peasant societies of the Third World have lost the freedom to determine their own futures... The power to grow food and ensure adequate nutrition has been wrested from them, while the meagre

69. Andre Gorz, *Ecology as Politics*, p.65. See also Andre Gorz, 'Their Famine, Our Food' in *Paths to Paradise*, London: Pluto Press, 1985, pp.92-100.

70. Susan George, *How the Other Half Dies: The Real Reasons for World Hunger*, Harmondsworth: Penguin, 1977; *Ill Fares the Land*, Institute for Policy Studies, Washington DC, 1984; Jon Bennett with Susan George, *The Hunger Machine*, Cambridge: Polity Press, 1987; and Susan George, *The Debt Boomerang*, London: Pluto Press, 1992. George's work has been corroborated by C. Tudge, *The Famine Business*, Harmondsworth: Penguin, 1979; F.M Lappe and J. Collins, *Food First*, London: Abacus, 1980; R. Burbach and P. Flynn, *Agribusiness in the Americas*, N.Y.: Monthly Review Press, 1980. For a more recent study, see P. N. Bradley, 'Food Production and Distribution - and Hunger', R.J. Johnston and P.J. Taylor eds, *A World in Crisis?*, Oxford: Basil Blackwell, 1986, pp.89-106.

71. B. Dinham and C. Hines, *Agribusiness in Africa*, London: Earth Resources Research, 1983, p.30.

rewards they earn for accommodating to a profit-based exchange system leave them too poor to purchase the very commodities they have been obliged to produce.[72]

With these developments not only are hundreds of millions of people being driven to the verge of starvation or beyond, but the form of agriculture being developed is more unreliable and more resource inefficient. In their effort to dominate world agriculture and expand business, agribusinesses have promoted crops which tie farmers into the mainstream of economic life.[73] This is the so called Green Revolution in which hybrid varieties of crops are being used which, having less adaptive ability, require far greater amounts of fertilizer, pesticides, weed control chemicals and irrigation, all of which have to be precisely controlled to avoid poorer outputs than with the old varieties. While producing crops with far lower protein content and making farmers dependent upon the transnational producers of seeds, agricultural chemicals and machinery (the prices of which have increased dramatically as a result of the oligo- or monopolistic control by transnational companies of these industries), the Green Revolution has also committed farmers to using far more resources for a given amount of output, produced a form of agriculture which is highly prone to failure (after disease attacked the new strains of rice in the Philippines in 1971 the crops were so devastated that rice had to be imported) and is more destructive of the soil. While at present crop yields are much higher than they were, yields are falling, and the acidification of the soil through the use of fertilizers will make the cultivation of rice increasingly difficult. And in the meantime, the genetic resources of crops are being impoverished as old, replaced strains die out, destroying the potential for adaptation to changing conditions.

All these problems have intensified in recent years as countries have struggled to increase exports to pay off massive foreign debts, which in 1992 stood at $1.2 trillion. As Susan George has pointed out, quoting a former IMF economist, '"environmental issues become totally marginal" when governments face huge debts...'.[74] Whatever conservation had been practiced in the past has been obliterated. As George continued:

> Brazil, contrary to appearances, does have the equivalent of an environmental protection agency, but its budget has been cut to the point that it can barely pay its employees. Outnumbered fire-fighters of the Brazilian national park system can no longer cope with the blazes. Costa Rica is asking for private donations to maintain its national parks. Mexico is draining irreplaceable groundwater to produce export vegetables for the US market. It will be depleted in a few years. Peru has fished its anchovy banks nearly to the point of extinction. Bolivia (aside from the drug trade) is actively engaged in massive exports of endangered wildlife. Mexico recently eliminated fifteen governmental under-secretaries, four of them environment-linked.[75]

Destruction of forests in particular has been accelerating as the Third World countries with the principle tropical forests: Brazil, Indonesia, Zaïre, Peru and Columbia, accounting for 60% of what is left of tropical forests, have been caught up in international debt.[76] Under these circumstances, the pressures to decimate these forests have become almost irresistible. And it is not only in the Third World that the pressure of debt has undermined efforts to conserve and preserve the environment in semi-peripheral regions of the world-economy. Australia also is wrecking its environment in the hope of reducing its international debt.

---

72. P.N. Bradley, 'Food Production and Distribution - and Hunger' in Johnston and Taylor, *A World in Crisis?*, p.104.

73. George, *How the Other Half Dies*, Ch.5.

74. Susan George, 'Debt and the Environment: Financing Ecocide,' in *A Fate Worse Than Debt*, Harmondsworth: Penguin, 1988, pp.155-168, p.167.

75. Loc.cit.

76. Ibid. p.166.

Along with the exploitation of agricultural land, the Third World has also been a source of minerals. The subjugation of the Third World has meant that minerals have been chronically undervalued. The exploitation of Third World minerals has increased dramatically since the Second World War, and for the most part, this has been associated with a decline in mineral prices. The greatest pressure on Third World countries to export their non-renewable resources has also been their international debts. After having been persuaded to borrow money to finance production for export (up to half of which was siphoned straight back to USA and Europe by corrupt politicians and officials with the help of the same banks who had lent the money), interest rates and prices of raw materials have been manipulated to the advantage of the First World, particularly the United States. As Jacobo Schatan concluded his analysis of the role of US government policy in the present plight of the Third World:

> The strangling pincer effect of the opposite trends in the cost of money and the prices of raw materials is forcing a steady increase in the physical resource outflow from South to North. A truly infernal circle is created: defence expenses and fiscal deficits in the US go up, rates of interest increase (or do not decline sufficiently) and debtor countries are forced to augment their remittances of raw materials; at the same time, such export volume increases press commodity prices down, pushing debtor nations to further increase their exports and request additional loans, in order to meet their service obligations... Latin American are donating the metals that serve to manufacture the chains that keep them tied to the yoke of the dominant Northern power.[77]

Schatan pointed out that the Third World was being forced to remit twice as much of its resources to the economic centres to pay off its debts as would have been required if interest rates had been held constant at the level at which they were originally contracted, and prices had remained constant at 1980 levels. As a consequence, the Third World is exporting the cheaper fractions of its resources while keeping for its own population the poorest and costliest mineral strata.

It is in these Third World cities that the most frightening built-up environments are being created. In recent years the peripheral regions of the world economy have become increasingly important for industrial production as suppliers of cheap labour willing to take on jobs in dangerous and polluting industries. Virtually all Marxist analyses of environmental degradation in built-up environments: of pollution, impoverishment of people's life-worlds and so on are applicable with greater force to the Third World, since here people have even less power to resist the forces of capitalism. The wealthy nations are exporting their polluting and health destroying industries to the Third World because people are so desperate for a livelihood (usually after having been forced off their land by the capitalist transformation of agriculture) that they will accept any risks. The cities expanding as a consequence: Mexico City, Sao Paulo, Rio de Janeiro for example, are developing into high-rise nightmares for their poorer inhabitants. For instance during Brazil's 'miracle decade' of export-oriented economic expansion, infant mortality in Sao Paulo increased by 45%.[78] As inflation soared, people were forced to work longer hours to compensate for the reduced purchasing power of their wages. Many were pushed to the periphery of the city by the rise in property values, and deprived of a minimum standard of nutrition, sanitation, and health care for their families.

It is in the Third World that the hegemony of the ruling culture is sustained with the most destructive effects. Dudley Seers argued, after noting the increasing numbers of oppressive dictatorships in the world:

> The explanation seems to be, in brief, that the bureaucrats, traders, and white-collar (as well as some blue-collar) employees in the modern sector, public and private, have become

77. Schatan, *World Debt: Who is to Pay*, p.42.

> increasingly determined that they and their children shall continue to enjoy the modern lifestyle, largely imported, whatever the brutality and whatever the inflows of aid and private capital needed to ensure this.[79]

This is the effect of extension of the culture of the economic core to the periphery, and has been one of the major achievements of the consciousness industry. As Richard Peet wrote in 'The Destruction of Regional Culture':

> [I]n the interaction between the centre culture and local culture, there can be little doubt which is more dynamic, and what direction cultural synthesis is taking. The tendency is towards the production of one world mind, one world culture, and the consequent disappearance of regional consciousness flowing from the specificities of the human past.[80]

This not only leads to the loss to the world of consciousness of these specificities, but it accounts for the blindness to regional problems, the loss of any ability to work out alternative strategies of economic action other than those imposed by the economic centres, and also the domination of these regions by comprador classes. Control over the consciousness of regional societies is effected at a number of levels. Hollywood factories, presenting dreams invested with all the technical and economic power of the centre, keep half the cinemas in the non-socialist world supplied with films. The mass media thus creates the images of what is the good life, generally creating a contempt for local lifestyles, and the aspiration to participate in the forms of life of the economic centres. Then the intellectual world is dominated by economic centres as it is necessary for scholars to legitimate their work by studying at the major universities of the core zones, or by publishing in their journals. Only when people think in ways acceptable to the culture of the centres are they able to attain respect, credibility and academic positions. Consequently there can be little sustained research on the problems and issues relevant to local regions, or development of quality regional culture which could challenge the prestige of the economic centres. Finally, imported technology and the models of professional training and organization required to use it, especially those associated with mass communications, reproduce in Third World countries the world-orientation of the economic centres.[81] The effect of such cultural control is that the affluent of these regions identify with the interests and aspirations of the economic centres and are completely indifferent to the original aspirations of their compatriots, and it is this more than anything else which makes them willing to use all the technology of repression available against their compatriots in order to continue living in the manner of their models in the core zones.

States developed in the economic peripheries have generally proved incapable of preventing their destructive exploitation by the economic core zones, and in fact usually become instruments for the expansion of international capital. This has been demonstrated by Stephen Bunker in his study of the Amazon which revealed how the development of the Brazilian State, itself subject to unequal exchange and resource exploitation from the economic core zones, intensified destructive exploitation of its own peripheries in order to overcome its balance of payments problems. He showed how: 'By irrationally extending energy-expensive structures and operating procedures into the energy-poor social formations of Amazonia, the state undermined existing but fragile human communities, devastated the ecosystem in which they subsisted, and severely distorted its own developmental projects. ... The state's self-legitimating claims that it can transform the

---

78. Bennett, *The Hunger Machine*, p.141.

79. Dudley Seers, *The Political Economy of Nationalism,* Oxford: O.U.P., 1983, p.7.

80. Richard Peet, 'The Destruction of Regional Cultures', in Johnston and Taylor, *A World in Crisis*, p.169.

81. As Rita Cruise O'Brien has argued in 'Mass Communications: Social Mechanisms of Incorporation and Dependence', in *Transnational Capitalism and National Development*, ed. José J. Villamil, Atlantic Highlands: Humanities Press, 1979, pp.129-144.

underdeveloped society are revealed as an illusion which can be maintained only when there is fundamental consonance of state policy with the evolutionary directions of the central social formation.'[82] In this process the State apparatus comes to be used to oppress the general population to maintain order, and it is supported in this by the economic core zones. So, backing up and complementing the consciousness industry there has emerged the burgeoning industry of oppression.

## *Capitalism, Socialism and the Environment*

The analyses by Marxist environmentalists exploit different facets of the Marxist research programme. Together they are sufficient to demonstrate that it is the dynamics of capitalism which are immediately responsible for the greater part of the world's environmental problems. However this leaves open the question of what caused capitalism, and whether it is possible to replace it. Capitalism could be explained in Social Darwinian terms as the social system which, by facilitating a faster though-flow of energy and materials, has been the most successful system in the struggle for survival, in which case it is likely that the only way in which capitalism will ever be challenged is by developing a system which will be even more environmentally destructive. Alternatively, capitalism could be explained as a cultural innovation, making it just possible that with a further cultural transformation a less oppressive and destructive social order could be created. If this is the case, the most important question is whether Marxism is able of providing a solution to the environmental crisis. Marx was not trying to provide an alternative economic theory but was criticising the very existence of an autonomous economy and of the science of this economy. The real issue posed by Marxism is the existence of capitalism and the possibility of creating from the conditions produced by it a new socio-economic formation which will not have its irrational qualities, which will liberate the potential of humanity, overcoming the nihilism of capitalism and the alienation of people from their creative powers and their social relations, and which at the same time will be environmentally sustainable. To evaluate Marxism in this respect it is necessary to examine its success or otherwise in practice, and to see whether societies based on Marxist thought have fared any better in relation to environmental problems than capitalist societies.

The society which most fully realized Marx's prognostications for capitalism is Sweden. Marx stated on several occasions (which will be discussed later) that his work in *Capital* was an analysis of Western Europe and only fully applicable to it. He wished to show how Western European capitalism was becoming increasingly unstable and at the same time was producing the conditions whereby it could be transformed into a socialist society. As a result of the Great Depression predicted by Marx, the Swedish Social Democratic Party backed by the Swedish Confederation of Trade Unions, gained power in 1932 and pushed by the trade unions, proceeded to re-organize the economy to gain control over its dynamics and to ensure that the controls over nature were used in the interests of humanity. As distinct from other democratic socialist parties such as those of Denmark and the Netherlands, the Swedish socialists did not confine themselves to constructing a welfare state on the basis of a capitalist economy, but began slowly to change the relations of production. Thus, as Winton Higgins has argued, 'The transitional process in Sweden is our first concrete illustration of Marx's general observations about the dissolution of mature capitalism.'[83] This society in which until recently people were not threatened with unemployment, with an

---

82. Bunker, *Underdeveloping the Amazon*, p's.56 & 242.

83. Winton Higgins, 'Working class Mobilization and Socialism in Sweden' in *Work and Inequality*, Vol.1, ed. Paul Boreham and Geoff Dow eds, Melbourne: Macmillan, 1980, pp.143-161. Higgins is highly critical of those conservatives and Marxists who have failed to see that what is happening in Sweden is 'an exceptionally powerful and sustained working-class mobilization which has put such pressure on the economy that the dominant mode of production has been undermined.'(p.143)

egalitarian distribution of income, and with the dynamics of society under democratic control had the best record of any industrialized nation on environmental problems. It had no population growth, strict controls on pollution, had introduced a wide range of resource conservation measures, and was phasing out nuclear power generation.[84] The success of these can be judged from its consumption of energy which, despite its higher standard of living and colder climate is about half that of the United States per head of population. Most of its pollution comes from other countries. Also Sweden is not an imperialist power and has assisted liberation movements in Third World countries.

But most capitalist nations did not embrace socialism during the Great Depression, Sweden has retreated from socialism (although the Social Democratic Party is now back in power), and those countries which have identified themselves with Marxism have been economically backward. Marxism must be judged at least partially in the light of the failure to develop socialism in Western Europe and its affluent colonies, and in relation to what has happened in those countries which have embraced Marxism as their official philosophy.

To begin with, the failure of Marxists in Western Europe during the Great Depression and after has revealed two things. Firstly, that while Marx's analysis of the destructive imperatives built into capitalism may have been correct, these will not necessarily pave the way for socialism. Under present circumstances it is more likely that the collapse of capitalism will pave the way for fascism or military-industrial complexes associated with far more powerful organizations of social control and oppression than exist at present. Secondly, it revealed the defects in Marx's analysis of society. It showed that there is more sustaining the capitalist socio-economic formation than a self-reproducing production process, that Marx underestimated the importance and complexity of States and the relationships between nations, and the importance of culture and ideology in maintaining the existing order and in determining which way people will respond to economic crises, and consequently, which political forces they will back.

Secondly, more recent failures have shown that the world politico-economic system has become more complex since the Great Depression. Apart from the development of neo-imperialism, capitalist societies have engendered institutions and forms of thought which are now moved by principles beyond the imperatives of increasing profits. The military-industrial complex has emerged from industrial capitalism as a new self-sustaining process as capitalism emerged from feudal society. Similarly the enormous size of organizations for such developments as dams associated with hydro-electric power and irrigation has enabled these to some extent to impose their own imperatives on the economies of countries.[85] While the dynamics of the military-industrial complex and other large scale organizations have not transformed the underlying capitalist organization of society, they have added new dimensions to it. They have given capitalism a new lease of life, and have created conditions which make it even more problematic how the failures of such a system could pave the way for the development of socialism. More recently transnational corporations have grown to such an extent that they have undermined almost all the potential the State previously had to ameliorate the oppressive effects of capitalism. With the power to move capital at astonishing speed, to utilize labour and exploit resources anywhere in the world, they have the ability to hold local organizations to ransom. There is now no longer a unified class opposed to the existing socio-economic system which is in a position to be a real challenge to either the military industrial complex, or to transnational corporations. If Marxism is to be taken seriously as the foundation for a challenge to the prevailing order it must be reformulated to deal with its past failures and with these new conditions. It is not at all clear that the developments in

---

84. On the Swedish effort to address environmental problems see Peter Bunyard, 'Sweden: Choosing the Right Energy Path?' *The Ecologist*, Vol. 16, No.1, 1986, pp.24-28.

85. See for instance Don Worster's study of hydro-electricity in *Rivers of Empire: Water, Aridity, and the Growth of the American West*, N.Y.: Pantheon, 1985.

the theory of imperialism and in the theory of ideology, nor even in the more recent work on the nature of the State, have yet been successful in revealing new paths to a better society.

## *The Soviet Union*

Even more problematic for Marxism is the poor record of those countries which embraced it. The first and pre-eminent society to embrace Marxism was the Soviet Union, and the Soviet experiment provides the best measure of the success or otherwise in relation to the environment of a society which has explicitly adopted Marxism as its creed.[86] Damning evidence against the Soviet Union came to light in 1978 in a pseudonymous work, *The Destruction of Nature in the Soviet Union*, written under the name of Boris Komarov as a contribution to the underground *samazadat* literature of the Soviet Union.[87] Komarov described high levels of air pollution, water pollution, destruction of soil fertility and wild-life, wastage of resources, and government inertia in relation to environmental problems. He argued that in relation to its population the Soviet Union produced twice as many air pollutants of all sorts as in the West, and that each Soviet car produced four times as much pollution.[88] A consequence of this was that between 1967 and 1977 the number of people with lung cancer doubled, 5 to 6 percent more children were born each year with genetic defects, and birth traumas and abortions increased at a rate of 6 to 7 percent a year.[89] There was severe chemical pollution of the Baltic sea, more mineral fertilizers were leached from the soil and ended up as pollutants than in any other country, and lakes, in particular Lake Baikal, the Caspian and Black Seas and the Sea of Azov, were being heavily polluted by oil. 100,000 tons of oil each year were dumped into these seas. The Sea of Azov, the worst polluted, now yielded a fish catch only one ninetieth of what it was just after the war.[90] Irrigation was reducing the levels of most of these bodies of water. Land reclamation schemes had succeeded in turning swamps into deserts. The Hydrological Planning Agency acquired enormous power and undertook projects causing far more destruction than benefit, largely by flooding fertile lands and depriving lands down-stream of water. The Kapchagai Power Plant only irrigated one hundredth of the 700,000 hectares it was supposed to while causing the lower Ili and half Lake Balkhash to dry up.[91] Open-cut mining also took huge areas of agricultural land out of production, and the fragile ecologies of the far north were being disrupted by attempts to economically exploit them. As in the West there was a massive switch to aluminium from steel despite the increased costs in energy usage and pollution. And so on.

Komarov's claims have been supported by other sources, particularly after *glasnost* and then the collapse of the Soviet Union. Philip Pryde in his books *Conservation in the Soviet Union* and *Environmental Management in the Soviet Union* pointed to a few successes in conservation, but

---

86. This is not to say that the situation of all Marxist societies is identical, although Eastern European countries were very similar. While China diverged from the Soviet Union after the cultural revolution of 1968 in a way which improved the environment, since the overthrow of the Maoist ethic environmental destruction has been catastrophic. On Eastern Europe see I. Volges ed. *Environmental Deterioration in the Soviet Union and Eastern Europe*, N.Y.: Praeger Press, 1966 and Ingmar Oldberg, 'Planned Economy and Environmental Problems: Eastern Europe from a Comparative Perspective' in Kendall E. Bailes ed. *Environmental History*, Lanham: University Press of America, 1985, pp.293-333. On China see Qu Geping & Woyen Lee eds, *Managing the Environment in China*, Dublin: Tycooly, 1984, and Vaclav Smil, *The Bad Earth: Environmental Degradation in China*, London: Zed Books, 1984.

87. Boris Komarov, *The Destruction of Nature in the Soviet Union*, London: Pluto Press, 1978. The author now lives in Israel.

88. Ibid. p.30.

89. Ibid. p.25.

90. Ibid. p.37f.

91. Ibid. p.58.

also to great destructiveness and enormous waste.[92] Between 1963 and 1968 reserves of coniferous timbers decreased by 3,300,000,000 cubic metres, or 5% of the total stand. This was taken from those lightly forested areas where deafforestation matters most.[93] Much of this was transported in streams, which not only was poisoning the streams but resulted in much of the timber being lost. Between 1958 and 1961, 825,000 cubic metres sank in the Kama alone.[94] Zhores Medvedev claims that before its collapse the Soviet Union was losing its forests at the same rate as was Brazil.[95] Thane Gustafson reported research revealing how much damage had been done to agriculture by reservoir flooding. The amount of land affected had doubled during the 1960's and was continuing to grow as the surveys were made. By the mid-1970's nearly 2.3 million hectares had been flooded, one fifth consisting of highly productive land, mostly in European USSR.[96] By 1983 little had changed despite the commotion produced by environmentalists.[97] The USSR was committed to the development of nuclear power, despite having had one major disaster in 1957 in which several hundred square miles were contaminated by radioactive material.[98] The Chernobyl disaster did not affect this resolve. Perhaps the grimmest indication of the seriousness of environmental destruction in the Soviet Union was the widespread pesticide poisoning in Uzbekistan and Moldavia which led to such high rates of mental retardation that secondary and tertiary educational institutions had to simplify their curricula.[99]

The most celebrated instance of environmental action in the Soviet Union, the fight to save Lake Baikal from industrial destruction which really began the post-Stalinist environmental movement, had only temporary success. As Gustafson wrote of this:

> [T]he lake's defenders can boast of no mean achievement, for they raised a nationwide scandal, gained top-level attention for the lake for a span of more than fifteen years, and turned the lake's preservation into a Soviet showpiece that the government now eagerly displays to foreigners. Yet the paradox of Baikal is that its defenders, for all their victories, are gradually losing the war. Large-scale economic development has now come to the entire region. ... These threats to the lake are far more serious than the Baikalsk plant, and yet there has not been the same storm of protest that there was fifteen years ago.[100]

So despite the existence of widespread public concern and support for environmental causes, both in relation to preservation and conservation, a government at least superficially influenced by this concern and forced out of economic necessity to face the problems of environmental destruction, there was an almost complete lack of effective action. The destructive dynamics of this economy were beyond the control of both environmentalists and the government.[101] It appears the environmentally destructive imperatives of the Soviet economy were at least as powerful as those of

---

92. Philip R. Pryde, *Conservation in the Soviet Union*, Cambridge: Cambridge University Press, 1972; and Philip R. Pryde, *Environmental Management in the Soviet Union*, Cambridge: Cambridge University Press, 1991.

93. Pryde, *Conservation in the Soviet Union*, p.97.

94. Ibid. p.98.

95 Zhores A. Medvedev, 'Environmental Destruction of the Soviet Union,' *The Ecologist*, Vol.20, No.1, Jan./Feb., 1990, pp.24-29, p.24.

96. Thane Gustafson, *Reform in Soviet Politics: Lessons and Recent Policies on Land and Water*, Cambridge: Cambridge University Press, 1981,p.47.

97. Philip R. Pryde, 'The "Decade of the Environment" in the USSR', *Science*, Vol. 220, 15th April, 1983, pp.274-279.

98. This is described by Medvedev in 'The Environmental Destruction of the Soviet Union,' p.24f.

99. This is claimed by Medvedev, 'The Environmental Destruction of the Soviet Union,' p.24.

100. Ibid. p.45.

101. The literature and state of the debate was reviewed in 1985 by Kristian Gerner and Lars J. Lundgren in 'Nature's Revenge: the Soviet Debate over Nature and Society, 1960-1979', Kendall E. Bailes (ed.) *Environmental History*, Latham: University Press of America, 1985, pp.412-436.

capitalism. The Soviet Union not only had higher energy consumption per unit of output, but while in Western Europe, USA and Japan this has been decreasing since 1970, it continued to rise in the Soviet Union.[102]

## *Conclusion*

While Marxism provides a powerful framework for analysing the environmentally destructive imperatives of capitalism, it appears that it has not yet provided a solution to environmental problems. If Marxist ideas are to be utilized by environmentalists, the failure of Marxism in practice must be explained.

---

102. These claims are made by William U. Chandler, 'Designing Sustainable Economies', in Lester R. Brown ed. *State of the World 1987*, N.Y. and London: W.W. Norton & Co., 1987, p.193f.

# 9

# MARXISM AND METAPHYSICS

The questions which now must be answered are: To what extent was Marxism responsible for the environmental problems of the Soviet Union? Were these problems a manifestation of basic deficiencies within Marxism? Or can Marxism be augmented to effectively confront not only the environmental problems in the West but also environmental problems of Russia? And these questions raise the more fundamental question of just what is Marxism.

Marx, like most radical thinkers writing in the latter half of the nineteenth century opposed science to metaphysics and rejected metaphysics in favour of science, despite his high regard for the philosophy of Aristotle and his continuing respect for Hegel. For this reason he never fully clarified his ontological commitments. This task was left to Engels who bequeathed to posterity a theory of being which was essentially Heraclitean, but which also combined elements of Hegelianism and mechanistic materialism. Consequently Marx has been interpreted in terms of different metaphysical frameworks. He was first interpreted mechanistically, since especially among radicals, this was the prevailing conception of being. On this basis he was seen as having discovered the laws of the development of humanity. This version of Marxism was itself developed in a number of different directions. Plekhanov in Russia interpreted Marxism in terms of the philosophy of Spinoza, Kautsky interpreted Marxism in terms of Darwinian evolutionary theory, Vorländer and Bernstein attempted to supplement a mechanistic interpretation of Marx with Kantian ethics, the Austro-Marxists developed an original synthesis of their own and Bukharin interpreted Marxism in terms of an early version of systems theory. Lenin developed another version of Marxism which was both voluntarist and materialist, and this became the basis of Soviet Marxism. Marx was also reinterpreted in terms of Hegelian philosophy, first by Gramsci, Lukács and Korsch, and then by the social philosophers associated with the Frankfurt Institute for Social Research. This conception of Marxism came to predominate in Western Europe after the publication of Marx's early works, in particular the *1844 Manuscripts* which was published in 1932. Soviet Marxism underwent a number of transformations with the evolution of Soviet society while new versions of Marxism have been produced in the West at more or less regular intervals, usually to accord with changing intellectual fashions.

To get any perspective on this, to see whether there is any substance to Marxism or whether it is a confused family of ideas whose only coherence derives from their serving to oppose capitalism, it is necessary to understand the historical background against which Marx developed his ideas. And since the background of metaphysical assumptions are part of long term history, this means going back to the origins of Marxian themes in the early Middle Ages.

## *The Neoplatonic Background*

Leszek Kolakowski traced the roots of Marxist eschatology to Neoplatonism, and characterized this in terms of the branching of Western European Neoplatonism which began with John Scotus Eriugena (c.810-c.877) in the ninth century.[1] However what Kolakowski did not point out (which is significant for understanding how Marxism was assimilated into Russia) was that John Scotus

1. Leszek Kolakowski, *Main Currents of Marxism* (3 vols), Vol. 1 *The Founders*, tr. P.S. Falla, Oxford: Oxford University Press, 1981. The basic tenets of Neoplatonism are described in Volume I, Ch.3 of Kolakowski's work. For a general account of and history of Neoplatonism see R.J. Wallis, *Neo-Platonism*, N.Y.: Charles Scribner's & Sons, 1970.

was not familiar with the works of the founder of Neoplatonism, Plotinus, nor with the other early Neoplatonist philosophers.[2] His starting point was the works of the thinkers whose ideas formed the foundation of the Orthodox Church: the Greek Christian Fathers - Origin, Basil and his brother Gregory of Nyssa, John Chrisostom, the sixth century Syrian monk known as the Pseudo-Dionysius whose work was inspired by the last great Alexandrian Neoplatonist, Proclus, Maximus the Confessor - and the work of the most influential father of Latin Christianity, St Augustine, whom John quoted more than anyone else. The only work of Plato with which he was familiar was the *Timaeus*. John translated into Latin the works of the pseudo-Dionysius, Maximus the Confessor and Gregory of Nyssa. The basic aim of his major work *Periphyseon* (or *De Divisione Naturae*) was to synthesize the Eastern and the Western traditions of Christianity.[3]

Neoplatonism is the synthesis of Plato's ideas into a system along with aspects of Aristotelianism and Stoicism, which is generally held to have culminated in the work of Plotinus (204-70 A.D.). According to Plotinus, the world consists of a hierarchy of hypostases. The first hypostasis, the indefinable One or the Good, was seen as the source of all defined and limited realities. The second hypostasis, the Intellect or Spirit, seen as the first creation of the One, is the realm of forms or Ideas, together with the unchanging thought of these forms. Soul was seen as the intermediary between the realm of the Intellect and the realm of the senses. It is produced by Intellect, as Intellect is produced by the One, by a double movement of outgoing and return in contemplation. Soul is movement, and the cause of movement. It has different levels, a higher level in touch with the Intellect, which forms and rules the universe from above, and a lower level, nature, which acts as an immanent principle in living and growth. The lowest hypostasis is the realm of matter where the One diminishes into nothingness. The material universe was seen as a living, organic whole in which each part of the universe is in harmony with every other part. The levels of the universe are not spatially separated, and the One, while being nowhere in particular, is present everywhere so that each part of the universe contains the whole. We, as embodied souls, can live at any level of the soul's experience and activity. We can turn away from the desires of the body to return, through intellectual discipline, to the wholeness of the Intellect, or even to a mystical union with the One itself.

As opposed to the Western tradition of Neoplatonic Christianity inspired by St Augustine for whom God is transcendent, the sensible world is the fallen world doomed to destruction, and the final goal of salvation of the soul involves returning to God in heaven, the theology of the Greek Christian fathers represented the world as an emanation of the One which they identified as God. They described two paths to knowledge of God: 'the way of negation' and 'the way of union'. According to the way of negation, the only way God can be defined is by stating what he is not, since it is impossible to establish a ratio between God and anything in the world. 'God is infinite and incomprehensible,' wrote John of Damascus, typifying this point of view, 'and all that is comprehensible about Him is His infinity and incomprehensibility.... God does not belong to the class of existing things: not that He has no existence, but that He is above all existing things, nay, even above existence itself.'[4] Nevertheless the goal of life is union with God, and it is believed possible to have an immediate experience of Him through the Prayer of the Heart, a form of prayer involving the whole person, body and soul together, in which people enter into direct relation with His energies. Rather than seeing the material world as in a state of perpetual decay as did Western Christianity, the Orthodox believed that the whole of God's creation, material as well as spiritual, is to be redeemed and glorified. Where the Latins talked of salvation the Greeks spoke of

---

2. For studies of John Scotus Eriugena and his background see S. Gersh, *From Iamblichus to Eriugena*, Leiden: Brill, 1978; and Dermot Moran, *The Philosophy of John Scotus Eriugena*, Cambridge: Cambridge University Press, 1989.

3. This has been argued by Francis Yates in 'Ramon Lull and John Scotus Eriugena' in Francis A. Yates, *Lull & Bruno*: Collected Essays, Vol. I, London: Routledge & Kegan Paul, 1982.

4. Cited by Timothy Ware, *The Orthodox Church*, Baltimore: Penguin, 1967, p.73 from *On the Orthodox Faith*, 1,4, (P.G. xciv, 800B).

redemption and deification. If humans are to share in God's glory, if they are to be 'perfectly one' with God, this means in effect that humans must be deified, to become by grace what God is by nature. Accordingly St Athenasius summed up the purpose of the Incarnation by saying: 'God became man that we might be made god.'[5]

The radical innovation of John Scotus Eriugena, designed to effect the reconciliation between the East and the West, was to account for God's creation of the world by postulating in Him an original deficiency. God must create the world characterized by transience, contingency and evil so that the fullness and immensity of His goodness could be manifested and adored. In developing this theme John conceived creation as a natural unfolding of the divine unity, and declared in consequence that in creating other things God is equally creating Himself. As he put it:

> [T]he divine nature both creates and is created. For it is created by itself in the primordial causes and thereby creates itself, that is to say it begins to manifest itself in its own theophanies, desiring to pass beyond the most secret boundaries of its nature, in which it is as yet unknown to itself and recognizes itself in nothing, inasmuch as it is unlimited, supernatural and supereternal and is above all things that can and cannot be understood.[6]

In this creative process there are four types of being: the creating and not created (God as the source of all); the created and creating (what the Greeks call forms, the primal causes of everything in the whole universe, the Divine Word or Logos which is the instrument of God's creative power); the created and not creating (the created universe, all that is known in the sensible world); and the not creating and not created (God as the end to which the universe is progressing). Human-kind has a special place in this manifestation of the Deity as the microcosm of the creation with attributes of both the sensible and the invisible world, and they must lead the cosmos, participating in the depths of creation from which they will return to unite with the divine source of all Being. In this return what has been created will not be destroyed but will be absorbed by the higher. Thus the corporeal will be ennobled by becoming spiritual, and the individuality of the soul will be preserved while being united with God. In other words, redemption will be attained by a transformation of life on earth.

The themes developed by John Scotus Eriugena have reappeared again and again throughout the history of European thought, with Hegel arguing in the nineteenth century that with Eriugena, 'true philosophy first begins...'[7] As a structural transformation of the basic Christian Neoplatonist tradition which dominated Europe, they have represented one of the two metaphysical foundations for opposition to the dominant culture, the other being Gnosticism. The most significant group to be influenced by John were the Heretics of the Free Spirit, who also drew on the mystical Neoplatonism of Meister Eckhart.[8] It was because of these Heretics that John's works were condemned as heretical in 1210 and 1225. The Heretics of the Free Spirit argued that history has three stages, the first being the original unity, the second, the present, being the age of the Fall, and the third being the new age about to be realized in which a paradise will be created on earth. This millenialism was reinforced through the appropriation of the ideas of Joachim de Fiore (1145-1202). Joachim argued that history is an ascent through three successive ages, each presided over by one the Persons of the Trinity, with the first age, the age of the Father or the Law being one of fear and servitude, the second and present age, the age of the Son being one of faith and filial submission, and the third age, the age to come, being one of love, joy and freedom - where the

5. Cited ibid. p.29. from *On the Incarnation*, 54.

6. John Scotus Eriugena, *De Divisione Naturae* III, 23, tr. Kolakowski *Main Currents of Marxism*, Vol.1, p.25.

7. G. W. F. Hegel, *Hegel's Lectures on the History of Philosophy*, (3 vols) ed. and trans. E.S. Haldane and Frances H. Simon, London: Kegan Paul, Trench, Trubner & Co., 1896, Vol.III, p.59. For a review of the literature on the influence of John Scotus Eriugena in the Middle Ages, see Moran, *The Philosophy of John Scotus Eriugena*, Ch.13.

8. See Norman Cohn, *The Pursuit of the Millenium*, London: Paladin, 1970, Chs. 8 & 9.

knowledge of God will be revealed directly to the hearts of all men.[9] Conceiving of God as immanent in the world, the Heretics of the Free Spirit saw themselves and their actions as expressions of God, as the agents through which God's goals will be realized.

This radical form of Neoplatonism received some legitimation in the fifteenth century with the work of Nicholas of Cusa (1401-1464) who was also influenced by Eastern Christianity. Nicholas had received his first education at the school at Deventer conducted by the Brothers of the Common Life and had been introduced to the works of the Pseudo-Dionysius, John Scotus Eriugena and Meister Eckhart, but went to university at Heidelberg, and then Padua, receiving the standard scholastic philosophy of the fifteenth century. However later in life he was sent on a papal mission to escort the Patriarch of Constantinople and a large number of bishops and theologians to Venice to negotiate the reunion of the Churches. Strongly influenced by this contact he wrote his influential *De Docta Ignoratia*, developing the tradition of ideas he had been introduced to in his school days. Following the Pseudo-Dionysius he argued that it is impossible to define the nature of God except by defining what He is not. Following John Scotus Eriugena, he describes the universe as an outflow from and return to the Deity. Developing ideas of Eckhart he argued that God is the infinite in which all opposites are reconciled; and that the universe and all bodies in it are the result of a contraction of the infinite so that what was enfolded within the Deity, is made finite, and thereby explicit. Nature was seen as the finite spirit, the movement diffused throughout the universe and all its parts through which form and matter is connected.[10] All that can be known positively is known through establishing ratios with other beings, mathematics being the ultimate means for this task, making God unknowable except by negation. On this basis Nicholas concluded that the universe as a whole cannot be conceived as a determinate object, but must be conceived as a sphere whose centre is everywhere and whose circumference is nowhere.[11]

Such Neoplatonic ideas set the stage for the development of Copernicus's astronomy in which the sun was seen and exalted as the centre of the solar system,[12] and for the rise of the Hermetic Neoplatonism of Pico della Mirandola and Marsilio Ficino based on the translations of Hermes Trismegistus and the Jewish Kabbala.[13] The most original thinker of the Hermetic movement was Giordano Bruno who, largely inspired by Nicholas of Cusa, went beyond him to identify the indeterminate One or God of Neoplatonism with matter, and to characterize the universe as not merely the indeterminate manifestation of the Deity but as the Deity itself.[14] For Bruno the universe was a composite of universal matter and universal form, but the matter in this scheme was conceived of as containing the forms and as the source of activity or motion, and thus of being. Matter was then identified with nature, effecting a complete reversal of the relative status given to matter and form in Plotinus and St. Augustine. The universe was seen to consist of a plurality of inhabited worlds. The most important features of this conception of the world was that the hierarchical conception of the world was rejected, nature was elevated and conceived of as divine, living and creative, and humanity was seen as a participant in this creative world.

---

9. See Marjorie Reeves, *Joachim of Fiore and the Prophetic Future*, N.Y.: Harper & Row, 1977.

10. See Nicolas Cusanus, *Of Learned Ignorance*, tr. Germain Heron, New Haven: Yale University Press, 1954, II, x, p.103ff.

11. Ibid. II, xii, p.111.

12. On the relationship between Neoplatonism and astronomy, see Thomas S. Kuhn, *The Copernican Revolution*, Cambridge: Harvard University Press, 1957, esp. p.131. Copernicus was introduced to Neoplatonism by his associate and teacher at Bologna, Domenico Maria de Novara who knew the Florentine Neoplatonists and who translated Proclus and Hermes Trismegistus.

13. The Hermetics are described by Francis Yates in *Giordano Bruno and the Hermetic Tradition*, [1964] London: Routledge & Kegan Paul, 1964.

14. See Wallis, *Neo-Platonism*, p.70. For an analysis of the relationship between Nicholas Cusanus and Giordano Bruno and an analysis of their metaphysical ideas see Ivor Leclerc, *The Nature of Physical Existence*, London: George Allen & Unwin, 1972, Part III.

Another important thinker for the future of this tradition of thought was Jacob Boehme (1575-1624), the Silesian mystic who was designated by Hegel as the forerunner of German idealism.[15] Influenced by both Eckhart and Bruno as well as a number of politically radical German thinkers, Boehme developed his system from within the tradition of the Lutheran Church.[16] As a Lutheran his concern was to find a place for evil in the world. Boehme held all the universe, including evil, to be expressions of God, an infinitely transcendent, yet omnipresent force who had created the universe out of His own essence. All human longings: sexual, intellectual and social were seen as expressions of 'homesickness' for the lost unity produced by the diremption of God in His effort to know Himself. This thirst for reunification is present in God's own longing for Sophia which is not merely the Holy Wisdom, but the principle of 'eternal femininity'. This can only be achieved through human thought which extracts structures from the beings of the world and expresses them in language.

While Hermetic philosophy was widespread, it was largely driven underground by political developments in the seventeenth century associated with the triumph of the rising bourgeoisie, and survived only among the Rosicrucians and some elements of the Masons.[17] The conception of nature as active rather than dead and inert remained a feature of radical thinkers in the eighteenth century, both Diderot and Priestly being examples, though the relationship between this opposition and political commitments became increasingly confused.[18] However a different situation prevailed in Germany. Germany which had been broken up after the Thirty Years War of 1618-1648 only began to develop capitalist social relations in the nineteenth century. German philosophers, who for the most part worked within universities, played little part in the development of mechanistic materialism and preserved the intellectual environment within which the themes of the radical Neoplatonists could be revived and developed to challenge the atomistic ideas of the Enlightenment.[19] The German revival of radical Neoplatonism as a major movement occurred at the end of the eighteenth century and the beginning of the nineteenth century as Germans struggled to gain a national identity, to unite Germany and to come to terms with the advance of commerce and the dissolution of old forms of relationships. While these thinkers looked back to Radical Neoplatonists such as Boehme and Bruno for inspiration, three philosophers who cannot be identified with such Neoplatonism but who were nevertheless influenced by it, who developed original ideas on this basis, made major contributions to this intellectual movement. They were Spinoza, Leibniz and Herder.

Spinoza's philosophy was an attempt to overcome the dualism of Descartes' philosophy and to provide an answer to the mechanistic political philosophy of Hobbes, while providing a new foundation for a Platonic - Stoic ethics within mechanistic science. It was firmly rooted in ancient and the medieval tradition of thought.[20] Most importantly from the point of view of the German thinkers influenced by him, Spinoza accepted Bruno's identification of the One and matter to conceive the world pantheistically, with extension and thought being seen as two of an infinite number of attributes by which it could be known. As later understood and appropriated by Herder, Schelling and Hegel, the world for Spinoza was a self-causing, unified totality in which the subject could be seen to fit into a universal current of life.

---

15. G. W. F. Hegel, *Hegel's Lectures on the History of Philosophy*, (3 vols) ed. and trans. E.S. Haldane and Frances H. Simon, London: Kegan Paul, Trench, Trubner & Co., 1896, Vol.III, p.188.

16. See Robert F. Brown, *The Later Philosophy of Schelling: The Influence of Boehme on the Works of 1809-1815*, Lewisburg: Bucknell University Press, 1977, p.35f.

17. On this, see Margaret Jacob, *The Radical Englightenment,* London: George Allen & Unwin, 1980.

18. See S. Shapin, 'The Social Uses of Science' in G. Rousseau and R. Porter eds. *The Ferment of Knowledge,* Cambridge: C.U.P., 1980, pp.93-139.

19. See Lewis White Beck, *Early German Philosophy*, Cambridge, Mass.: Belknap Press, 1969.

20. As clearly shown by Harry Austryn Wolfson in *The Philosophy of Spinoza* [1934], Cleveland and N.Y.: Meridian Books, 1958.

After reading the work of Spinoza, Leibniz, who had originally accepted the new mechanical philosophy and rejected the reality of substantial forms, came to the conclusion that the new physics was not adequately grounded and attempted to develop a new metaphysics to fulfil this task. Echoing Plotinus, Leibniz rejected the characterization of being as 'extension' and conceived the universe as a pre-established harmony in which everything reflects everything else. He argued that if change in the world is to be explained then being must be conceived of as active rather than inert and argued that the ultimate constituents of the universe must be conceived of on the analogy of minds, referring to them as monads. Rather than substance being taken as the unchanging aspect of the world with changeability pertaining only to accidents, the enduring and the mutable were seen to be mutually dependent. Mutability was no longer subordinated to the changeless, and the stability of the monad was seen as the constant rule of its progression. Thus substance was seen as dynamic, being directly active and revealing its nature in the sequence of its activities. Its stability lies in this capacity to emanate what is preformed within it without cessation. Though composed of an infinity of monads, the pre-established harmony of the world ensures that they will develop as a unity so that: 'In the universe all things are closely knit together, they are in one piece, like an ocean: the slightest movement transmits its influence far and wide all around.'[21] Leibniz temporalized the Great Chain of Being, conceiving development of the universe as an evolution towards greater and greater perfection.

Herder drew on the ideas of a wide range of philosophers in the development of his philosophy. While the most immediate influences were Kant, Hamann, Spinoza and Leibniz, he was familiar with the work of the radical Neoplatonists, and was influenced by Bruno whose opinions he had encountered through the writings of the pantheist John Toland.[22] Following these philosophers, Herder argued that nature is a great stream of life of which we are part. It consists of a great creative force composed of dynamic, purpose seeking forces which clash, combine and coalesce to constitute all movement and growth. But while the dynamism of beings is provided by these vital forces, they require the right environment to flourish. This dynamic world is evolving (not through the evolution of species, but through the successive emergence of new species) to produce higher and higher levels of being with humanity a special creation at the top of the scale.

In developing his notion of humanity, Herder elaborated Leibniz's notion of the monad and conceived of civilizations, societies and individuals as defining themselves by unfolding and expressing their inner essences, with humanity as a whole evolving to higher levels of being.[23] While human life was always understood with reference to its physical and geographical environment, all human activity was seen as the expression of individuals or groups striving to actualize their own unique natures. The challenge for each society and civilization is to discover its own centre of gravity and then to actualize its potential. All aspects of a particular people, the way they speak, move, eat, drink; their laws, architecture, theology and social outlook, their music and dance forms, and in particular their language, were seen to be pervaded by and to express patterns and qualities unique to their cultures, each aspect of a culture therefore reflecting the whole culture. In opposition to the individualism of the Enlightenment philosophers, Herder argued that individuality is only achieved by participation in and expressing the particular culture through which one's humanity has been attained. All human activity was seen as expressing the total personality of individuals or groups, with self-realization being the richest and most harmonious form of self-expression, which is what all people, whether they are aware of it or not, live for.

---

21. Quoted by Hegel in *Hegel's Lectures on the History of Philosophy*, tr. E.S. Haldane and Frances H. Simson, [1896], (3 vols) Atlantic Heights, N.J.: The Humanities Press, 1974, Vol.III, p.344, from *Essais de Theodicee*.

22. See H. B. Nisbet, *Herder and the Philosophy and History of Science*, Cambridge: The Modern Humanities Research Association, 1970 for an excellent account of all the influences on Herder.

23. For a succinct statement of Herder's' ideas on humanity see Isaiah Berlin, *Vico and Herder*, London, Chatto & Windus, 1976.

With this notion of humanity it was the creativity of people which was emphasised, and people were seen to be living in worlds which they themselves had largely created.

There was another reaction against the mechanical world-view which was not directly connected with Neoplatonism, but which was important for its later development. This was the effort by Rousseau and Kant to resurrect the status of reason as a capacity transcending the mechanical world and the calculation of self-interest. Rousseau was trying to show how people could subordinate themselves to society without losing their freedom, and argued that in fact it is through willing subordination to impersonal rules which are rational in the sense of being in accordance with the General Will, that real freedom is attained. The General Will is the Will of the whole society, transcending the sum of particular, egoistic wills, and true freedom is doing what one ought to do from the perspective of the whole society. Kant accepted this idea, referring to Rousseau as the Newton of moral science. He argued that people are truly free only when they act according to the categorical imperative, that is, when they act according to principles which can be willed to be universal laws. To further legitimate this notion of freedom, Kant represented the mechanically determined sensible world as only the world of appearances, ordered first by the imagination and then according to forms of intuition and the categories of the understanding deriving from the transcendental ego. Only by acting according to moral principles are people direct manifestations of the real world.

The ideas of Herder and Kant, along with the advances of the Romantic thinkers inspired by Herder, were assimilated into an expanded Neoplatonism by Fichte, Schelling and particularly of Hegel whose main concern was to integrate the vision of the world as an expressive totality with the notion of the rationally autonomous will.[24] Fichte used Kant's philosophy to reformulate Neoplatonism, reducing Nature to something posited by the subject, while representing the real world as the moral order founded in God, so that each of us exists only in God and through God.[25] He then identified God with the one, true and unchangeable Being or Absolute which manifests the appearance of a world with all its diversity to become visible to itself through the moral life of individuals. Fichte's philosophy was the point of departure for Schelling who, in his early work was mainly concerned with the nature of the physical world, and was both influenced by and influenced the natural philosophy of Herder. He was the main inspiration for the *Naturphilosophen*, the philosophers who attempted to replace the mechanistic conception of the physical world by one which stressed the inter-dependence within and the dynamic activity of nature.[26] He also began the search among previous Neoplatonists, writing a book on Bruno and coming under the influence of Boehme.[27] However the greatest Neoplatonic systematiser of the German reaction against the mechanistic, utilitarian philosophy of France and Britain was Schelling's colleague, Hegel.[28] Hegel's philosophical system continues the whole tradition beginning with John Scotus Eriugena of conceiving the world as a creation of the Deity in His effort to attain full self-consciousness, a process in which humanity plays the leading part; but in accordance with German Neoplatonism this development was represented as the unfolding of the inner essence or self-actualization of what Hegel variously called the Absolute, the Idea or the World-Spirit.

Hegel's system begins with the Logic, conceived to be the ground-plan of the whole of reality which 'shows forth God as he is in his eternal essence before the creation of Nature and finite

---

24. On this see Charles Taylor, *Hegel,* Cambridge: Cambridge University Press, 1975, Ch.1.

25. For the most accessible account of Fichte's philosophy see Johann Gottlieb Fichte, *The Vocation of Man*, [1800] ed. Roderick M. Chisholm, Indianapolis: Bobbs-Merrill, 1956.

26. See F.W.J. Schelling, *Ideas for a Philosophy of Nature*, 2nd ed. [1803] tr. Errol E. Harris and Peter Heath, Cambridge: C.U.P., 1988.

27. For a study of this, see Brown, *The Later Philosophy of Schelling*.

28. As Ludwig Feuerbach noted: 'Hegel is not the German or Christian Aristotle - he is the German Proclus. The "absolute philosophy" is the resurrection of Alexandrianism.' in *Samliche Werke*, ed. Bolin and Jodl, Stuttgart: Frommann Verlag, Vol. II, 1905, p.291.

Spirit.'[29] Beginning with 'Being', which was taken on its own evidence only, but then revealed to be a mere abstraction, Hegel deduced categories to cover matter, life and mind through a process of immanent critique whereby each finite, limited category was seen to suppress itself and pass over into its own negation, engendering a negation of the negation which transcends and preserves the original category in a less limited category. The full development of the categories enable Being to be grasped in a concrete way in the 'Concept', the universal, self-conscious and self-identical inner principle of the diverse totality of Being. These categories correspond to Plato's hierarchy of forms, with Hegel's 'Concept' corresponding to Plato's 'Form of the Good', and though the relationship of the categories to the rational order within the world was conceived differently, change in the world was still in effect regarded as 'the moving image of eternity'.

According to Hegel, Logic requires the positing of Nature, and presupposes Nature as its being, and the Philosophy of Nature followed the Logic as the science of the Absolute self-externalized. Since Nature is the Idea estranged from itself and thus unmindful of itself, the study of Nature is required to liberate Spirit in Nature. Spirit emerges as the truth of Nature, the negation of Nature's negativity. However this development is not a development in time but in space. Hegel rejected the idea of evolution in Nature. Though Spirit emerges from Nature, Nature is posited by Spirit and Spirit is logically prior to Nature. Spirit which is presupposed by and develops out of Nature, cognizes the Logical Idea in Nature and thus raises it to its essence.

The Philosophy of Mind or Spirit describes the moral as opposed to the physical aspect of reality. It displays humanity in its development from Subjective Spirit in which it struggles to overcome the vestiges of its natural heritage with its bonds of individualism, to Objective Spirit in which humanity battles to construct objective institutions: the family, civil society and the State. There has been a sequence of forms of Objective Spirit each inspired by the basic principle of a national spirit, a principle which 'defines the common features of [a nation's] religion, its political constitution, its morality, its system of law, its mores, even its science, art, and technical skill.'[30] These forms of Objective Spirit have flourished then decayed as the contradictions of the principles inspiring them became manifest. Since new forms are built on the failures of preceding forms, there has been a constant tendency towards progress, leading from a form of the State in which only one person was free, through forms in which only some were free, to the form of the Prussian State in which all are free. This is the social order in which the State, transcending the realm of civil society governed by egoism, provides individuals with a basis for attaining freedom by doing their duty in accordance with the laws and usages of the State. The sequence of social forms have provided increasingly better vantage points for Spirit to attain a view of itself as Absolute Spirit through Art, Revealed Religion and Philosophy. The whole of world history is thus seen as a rational progression by which the World-Spirit, through individuals participating in the socio-cultural development of humanity, struggles to attain full consciousness of itself. Through this historical scheme Hegel attempted to synthesize Herder's conception of humans as expressions of an integral totality with the enlightenment notion of freedom through living according to reason as this had been developed by Rousseau and Kant.

In this account of history, Hegel was only concerned to show the rationality of events after they had occurred. As he put it: 'The owl of Minerva spreads its wings only with the falling of the dusk.'[31] So while Hegel's philosophy incorporated a vision of humanity which was implicitly evaluative of social formations and implied the possibility of creating a new social formation in which the goal of humanity would be fulfilled, Hegel's philosophy was profoundly conservative. When it came to decisions on how to live, Hegel's philosophy left individuals in the lurch. This was the basis of the charge Kierkegaard made against Hegel in *The Present Age* that Hegel had

29. *Hegel's Science of Logic*, tr. W.H. Johnston and L.G. Struthers, (2 vols) London: Allen & Unwin, 1929, Vol.1, p.60.

30. G.W.F. Hegel, *Reason in History: A General Introduction of the Philosophy of History*, [1837] tr. Robert A. Hartman, Indianapolis: Bobbs-Merrill, 1953, p.79.

31. *Hegel's Philosophy of Right*, tr. T.M. Knox, Oxford: Clarendon, 1952, p.12f.

constructed a mighty palace and left the individual living in a hovel on the outside. The task Hegel set his radical followers (the 'Young Hegelians') was to show how philosophical theory could be related to practice,[32] and the most eminent of these radicals was the young Marx.

## *Neoplatonism and Mechanistic Materialism in Marx*

Marx, along with the other Young Hegelians, set about this task by rejecting Hegel's idealism while retaining the basic Neoplatonic eschatology and the conception of humanity as the creative agent through which the ultimate end of history will be realized. In this he was strongly influenced by Feuerbach who, in his *Essence of Christianity*, argued that God is nothing but the highest qualities of Man projected onto an extra-mental realm and treated as a real power, leaving Man with an impoverished conception of himself. The task Feuerbach set humanity was to reappropriate these alienated highest qualities.

Following Feuerbach, Marx saw humans as part of nature rather than nature as being posited by Spirit:

> Nature is man's inorganic body, that is to say nature in so far as it is not the human body. Man lives from nature, i.e. nature is his body, and he must maintain a continuing dialogue with it if he is not to die. To say that man's physical and mental life is linked to nature simply means that nature is linked to itself, for man is a part of nature.[33]

Then in place of Spirit struggling to attain full consciousness of itself through the development of the State and philosophy, Marx took Man as the subject-object which forms itself through the transformation of nature, with civil society and economic life being placed at the centre of the stage. As he put it: '...the whole of what is called world history is nothing more than the creation of man through human labour, and the development of nature for man...'[34]

Marx conceived this as taking place in stages. Man's original state was understood to one of immediate involvement in the world with non-antagonistic social relations: primitive communism. This was followed by Man's progressive domination of nature. However this was seen to involve the alienation of Man from his creative activity, his Species Being, with the separation of labour from its means of production and the emergence of class societies. Instead of experiencing an increasing control over nature, Man's creative essence becomes something alien, an external power to which individuals must submit, while labour and its products become something external to the individual so that he denies himself rather than affirms himself in his work and creations. This situation also involves increasingly antagonistic social relations and class conflict. The evolution of society is seen to take place through a series of revolutions in which new classes emerge representing their own interests as the interests of society as a whole. These take power from the old classes whose interests have become manifestly at odds with the interests of the whole society. Each new class represents more universal interests than those which preceded it, but each class eventually reveals its own interests to be particular and limited, thereby generating new class antagonisms. However with capitalism there has emerged a new class, the proletariat, which by virtue of its total alienation from its creative essence, represents the universal interests of humanity. It is the universal class, and its coming to power will be the full appropriation of Man's now perfected creative essence and the final supersession of antagonistic social relations. The social

---

32. On Hegel, his opponents and different followers, see John Edward Toews, *Hegelianism*, Cambridge: Cambridge University Press, 1980.

33. Karl Marx, 'Economic and Philosophical Manuscripts', in Karl Marx, *Early Writings*, tr. R. Livingstone and G. Benton, Harmondsworth: Penguin Books, 1975, p.328.

34. Ibid. p.357.

order in which this will be achieved, the final stage of history, is communism. In describing this as the culmination of history, Marx clearly revealed the domination of his thinking by Neoplatonic eschatology:

> Communism is the positive supersession of private property as human self-estrangement, and hence the true appropriation of the human essence through and for man; it is the complete restoration of man to himself as a social, i.e. human being, a restoration which has become conscious and which takes place within the entire wealth of previous periods of development. This communism, as fully developed humanism equals naturalism, and as fully developed naturalism equals humanism; it is the genuine resolution of the conflict between man and nature, and between man and man, the true resolution of the conflict between existence and being, between objectification and self-affirmation, between freedom and necessity, between individual and species. It is the solution of the riddle of history and knows itself to be the solution.[35]

This system makes the ultimate goal of history Man's self-actualization in communism, and conceives this actualization as yet to be achieved.

The oddity of this conception of the world is that it retains the radical Neoplatonist Christian eschatology while having rejected the framework which justifies it. There is no Supreme Being whose end pre-exists the beginning of history and whose rationality guarantees the rationality of history. The young Marx avoided this problem in two ways. Firstly Man is hypostatised and treated as the subject-object of history in place of people. In this way the problem of how a multiplicity of separate individuals can generate rational progress is avoided. Secondly nature is reduced to a mere abstraction - that on which Man works in the process of his self-formation. Only as humanized is this nature seen to be really knowable or to be of any significance. Only by such evasions of the real complexity of human history could Marx's optimism about the future be sustained on the basis of his early works.

But Marx did not retain this basis. Stung by Max Stirner's critique of Feuerbach's essentialist conception of Man which had largely formed the basis of his own work, Marx moved rapidly beyond his Young Hegelian origins. He did this by immersing himself in the works of the political economists and attempting to find in them a basis for sustaining his hope for the future, hope that the divisions within existing society could be overcome through the establishment of a communist society.

To this end he first appropriated the general theory of history of the Scottish philosophical historians and political economists, namely, the theory that history passes through stages driven by the quest for fuller control over nature and greater surplus product, each stage having its unique configuration of institutions appropriate to that mode of subsistence.[36] Marx gave the most forceful expression to this theory of society and its history, which after his death came to be known as 'historical materialism', in his preface to *A Contribution to the Critique of Political Economy*:

> In the social production of their existence, men inevitably enter into definite relations, which are independent of their will, namely relations of production appropriate to a given stage in the development of their material forces of production. The totality of these relations of production constitutes the economic structure of society, the real foundation on which arises a legal and political superstructure and to which corresponds definite forms of social consciousness. The

35. Ibid. p.348.

36. See Ronald L. Meek, 'The Scottish Contribution to Marxist Sociology' in *Economics and Ideology and Other Essays*, London: Chapman and Hall, 1967; and Andrew Skinner, 'A Scottish Contribution to Marxist Sociology?' in Ian Bradley and Michael Howard eds, *Classical and Marxian Political Economy* London: Macmillan, 1982, pp.79-114. The main

> mode of production of material life conditions the general process of social, political and intellectual life. It is not the consciousness of men that determines their existence, but their social existence that determines their consciousness. At a certain stage of development, the material productive forces of society come into conflict with the existing relations of production or - this merely expresses the same thing in legal terms - with the property relations within the framework of which they have operated hitherto. From forms of development of the productive forces these relations turn into their fetters. Then begins an era of social revolution. The changes in the economic foundations lead sooner or later to the transformation of the whole immense superstructure.[37]

However in opposition to the Scottish thinkers on this subject, Marx argued that commercial society is not the last form of society but are paving the way, through the development of industrial capitalism, for a social order in which class oppression will be eliminated. The material forces of production are driving humanity from one form of society to another, ultimately progressing to communism:

> In broad outline, the Asiatic, ancient, feudal and modern bourgeois mode of production may be designated as epochs making progress in the economic development of society. The bourgeois mode of production is the last antagonistic form in the social process of production and the productive forces developing within bourgeois society create also the material conditions for the solution of this antagonism. The prehistory of human society accordingly closes with this social formation.[38]

The defence of this view of history also drew heavily on British thought. It involved the appropriation and development of the labour theory of value as the basis for the analysis of capitalist society. Ricardo's ideas were most important in this respect. Marx's achievement in the domain of political economy was to develop a more dynamic model of the economy than hitherto to deal with the complexities of an increasingly industrialized society, to reveal the basis of capitalism's instability, and to show why in the long term it is likely to become more so. He also analysed the tendencies of capitalism leading to imperialism, and the by-products of the development of capitalism such as the growth of population and destruction of the soil described in the previous chapter.

Marx's concepts of nature and of humanity changed with his development of political economy. He no longer saw nature purely in relation to humanity. This was especially so after the publication of Darwin's *Origin of Species*. Nature pre-existed humanity and had to be seen as having an independent existence. But in accepting the labour theory of value as a means to understand capitalism, Marx gave the impression that he viewed nature not as a co-participant in human creative activity, but as merely the material for humans to work upon. Everything it offers humanity was thus represented as a spontaneous and free gift. Only when it is formed by labour does it have any value.

With the conception of the struggle for more efficient control over nature as the force pushing society from one socio-economic formation to another and with the labour theory of value, the conception of humans as creative, social beings struggling to overcome their alienation and their antagonistic social relations tended to be displaced by an almost Hobbesian view of people. Only the ascription of motivation and agency to classes rather than to individuals or nations really distinguished the more mechanistic formulations of Marx's ideas from Social Darwinism. The

---

figures involved in the development of this theory were Smith, Millar, Ferguson, Robertson, Dalrymple and Lord Kames. The school was originally inspired by Montesquieu.

37. Karl Marx, *A Contribution to the Critique of Political Economy*, Moscow: Progress Publishers, 1977, pp.20-21.

38. Ibid. pp.21-22.

motive force for the creation of communism is then the struggle to improve the productive power of humanity, presumably to facilitate higher levels of consumption, and the only deficiency of capitalism is that it will no longer be the best means for developing the forces of production. All ideas apart from science were viewed as disguises or instruments for the self-interested struggles of classes. Understood in this way Marx was seen by his followers to have discovered the laws of development of society, demonstrating the inevitability of progress. As Engels put it in his funeral oration for Marx: 'Just as Darwin discovered the law of development of organic nature, so Marx discovered the law of development of human history.'[39] With such a Hobbesian view of humanity there was a tendency to think that only by producing such a super-abundance of goods could the conflicts between people be overcome. Communism is then a system where 'the process of material production ... is treated as production by freely associated men, and is consciously regulated by them in accordance with a settled plan.'[40]

This general theory of history and the theory of capitalist society together with a slightly modified mechanistic view of nature and Hobbesian view of humans formed the elements from which orthodox Marxism was constructed (originally by Kautsky and Plekhanov).[41] According to orthodox Marxism the aim of science is the discovery of causal laws. These are the general laws of nature including those associated with the Darwinian theory of evolution, special laws associated with the development of humanity, and even more particular laws associated with the dynamics of capitalism. The theory designating the most general laws of nature came to be known as 'dialectical materialism', while the theory designating the laws of the development of humanity came to be known as 'historical materialism.' (Each of these terms was coined after the death of Marx.) The whole universe was seen to operate deterministically, making the progress of humanity to communism inevitable, and it was held that there were no scientific value judgments involved in the understanding of these laws. This basic position was succinctly stated by Hilferding in the preface to one of the most important contributions to Marxism, *Finance Capital* published in 1910:

> In logical terms Marxism, considered only as a scientific system, and disregarding its historical effects, is only a theory of the laws of motion of society. The Marxist conception of history formulates these laws in general terms, and Marxist economics then applies them to the period of commodity production. The socialist outcome is a result of tendencies which operate in the commodity producing society. But acceptance of the validity of Marxism, including a recognition of the necessity of socialism, is no more a matter of value judgment than it is a guide to practical action.[42]

## *Marx and Process Philosophy*

However despite appearances, Marx and Engels never really deserted the tradition of German thought which emphasised the creativity and sociality of humans, the value of art and literature, and the importance of liberating creative potentiality of people. While struggling for a firmer foundation for their vision of the future which would liberate human creativity, Marx and Engels

---

39. Frederick Engels, 'Speech at the Graveside of Karl Marx' in *Karl Marx and Frederick Engels: Selected Works in Two Volumes*, Vol.2, Moscow: Foreign Languages Publishing House, 1962, p.167. This interpretation of Marx's work is supported by Marx himself in the 'Afterword to the Second German Edition', *Capital*, 3 volumes, tr. Edward Moore and Edward Aveling, Moscow: Progress Publishers, 1974, Vol.1, pp.22-29.

40. Marx, *Capital*, Vol.1, p.84.

41. A slightly polished version of such orthodox Marxism has been argued for by G.A. Cohen in *Karl Marx's Theory of History - A Defence*, Oxford, Oxford University Press, 1979, and has been widely acclaimed by Marxists.

42. Rudolf Hilferding, *Finance Capital: A Study of the Latest Phase of Capitalist Development*, tr. Morris Watnick and Sam Gordon, London: Routledge & Kegan Paul, p.23.

developed ideas which went beyond both radical Neoplatonism and mechanistic materialism, ideas which only cohere when understood in terms of the process view of the world.

The first advance made by Marx in this direction was to develop a new notion of humanity by rejecting the hypostatization of 'man' and 'history' of his unpublished, essentially Neoplatonic works. In relation to history he wrote:

> History does nothing, it 'owns no tremendous wealth', it 'fights no battles'. Instead it is man, real, living man that does all this, owns and struggles; there is no such thing as 'history' that uses man as its means in order to attain its ends - as if it were a separate person - for history is nothing but the activity of man pursuing his ends.[43]

But then in relation to the meta-subject-object 'man' he wrote:

> The individuals, who are no longer subject to the division of labour, have been conceived by the philosophers as an ideal, under the name 'man', and the whole process which we have outlined has been regarded by them as the evolutionary process of 'man', so that at every historical stage 'man' was substituted for the individuals existing hitherto and shown as the motive force of history. The whole process was thus conceived as a process of the self-estrangement of 'man'...[44]

Marx strove to see history not as the development of an hypostatized subject 'man' but as the creation of social individuals struggling within particular historical conditions.

But this did not involve the acceptance of Hobbes' atomic individualism. Marx retained his relational conception of humans as essentially social and creative beings, attacking political economists for seeing commodities, labour, capital, exchange and value as naturally occurring 'things' rather than as historically specific 'relations'. Marx's commitment to a conception of people as creative and social is not only immediately evident from the *Grundrisse*, but is also clear from Marx's analysis of commodity fetishism in the first chapter of *Capital*. It is clear from this that Marx believed that people are forced by the capitalist socio-economic formation to conceive themselves as atomic individuals only interested in the world and other people insofar as they are useful to their own egoistic interests, that it is not 'natural' to be this way. In particular the labour theory of value itself is seen as a category only fully applicable to capitalist society, and as with all economic categories of capitalist society, as a 'form of being' which not only constitutes social relations, but also partly hides the true nature of these relations.[45] Marx was concerned throughout his work to go behind appearances in which people and their products appear as quantifiable things to the underlying durational processes, the labouring process, the processes of production and exchange and the actual consumption of commodities, from which these appearances are generated.[46] Explaining how people have come to conceive their relations in mechanistic, utilitarian terms, how they define all value through money, and how their conceptions have been sustained and reproduced, is a major part of Marx's research programme and a major part of his explanation of capitalism's existence.

---

43. Karl Marx and Frederick Engels, *Werke*, Dietz-Ausgabe, 1958ff. Vol.2, p.98, tr. in Helmuth Fleischer, *Marxism and History*, [1969], N.Y.: Harper & Row, 1973, p.17

44. Karl Marx, 'The German Ideology', in *Karl Marx, Frederick Engels: Collected Works,* Vol. 5, 1845-1847, tr. Clemens Dutt et.al., N.Y.: International Publishers, 1976, p.88.

45. Marx, *Grundrisse*, p.106.

46. See for example ibid. p.'s 91 & 255. Marx's critical analysis of such reification corresponds to Alfred North Whitehead's analysis of the 'Fallacy of Misplaced Concreteness' - the tendency to abstract, to ignore the level of abstraction and then to treat the abstraction as a concrete reality.

Correspondingly Marx vehemently rejected conceptions of humans which reduced them to mechanisms satisfying their appetites. Thus in *Capital* he referred to Jeremy Bentham as 'that insipid, pedantic, leather-tongued oracle of the ordinary bourgeois intelligence of the 19th century', and, condemning the implications drawn from the utilitarian principle that art criticism is of no value because it interferes in the enjoyment of works of art, as 'a genius in the way of bourgeois stupidity.'[47] Marx suggested that Bentham's utilitarianism amounted to assuming that what is useful for that queer normal man, the modern English shopkeeper, is absolutely useful. In opposition to this he argued that to discover what is good for humanity it is necessary to begin with 'human nature in general, and then with human nature as modified in each historical epoch.'[48]

One of the most important features of Marx's work was to stress the uniqueness of capitalism as an emergent phenomenon which must be understood in its own terms. He stressed that capitalism is more than the conditions of its emergence. As he wrote in the *Grundrisse*: 'These conditions and presuppositions of the *becoming*, of the *arising*, of capital presuppose precisely that it is not yet in being but merely in becoming; they therefore disappear as real capital arises, capital which itself, on the basis of its own reality, posits the conditions of its realization.'[49] To comprehend the dynamics of capitalism once it had been established, Marx developed a new scheme for understanding the relationship between purposive activity and social dynamics, which involved the development of a new notion of contradiction. Marx conceived of the relationships between people as conceptual relations, yet held that the dynamics of capitalism are not reducible to these conceptual relations. The self-reproduction of capitalism involves constraining people to define their relations in terms of the categories of 'commodity', 'capital' and so on, and to define their goals accordingly. But by so doing it can constrain people to act in pursuit of goals, the conditions for achieving which they undermine in the process. Thus there is a contradiction between the capitalists' efforts to maximise their profits by keeping wage levels down, since this deprives consumers of the means to buy what is produced. There is a further contradiction between the effort to expand profits by further investing in fixed capital, since when all capitalists do this there must eventually be a further reduction in profits. If machinery produces everything there will be no wage-earners to buy what is produced; the whole basis of the system will be destroyed. Finally there is a contradiction in the drive to succeed since while concentrating ownership of the means of production, this reduces greater and greater proportions of the population to wage labourers, creating and rending more powerful a class which has the potential to wrest power from the capitalists and to expropriate the means of production. In each case the goals of the capitalists cannot be freely chosen but must be pursued in order to avoid losing the game, which would mean becoming a wage-labourer. According to Marx while such contradictions can be contained for some time, they will eventually render capitalism unviable and provide the conditions which make it possible for the working class to replace capitalism with a socialist organization of production.

Marx pointed out that the failure to appreciate the uniqueness of capitalism is a consequence of the tendency to project the categories which dominate the present on the past. Thus, in *A Contribution to the Critique of Political Economy*, in the same work in which the famous passage occurs in which all history is seen as progress in the development of the forces of production, Marx wrote:

> What is called historical evolution depends in general on the fact that the latest form regards earlier ones as stages in the development of itself, and conceives them always in a one-sided

---

47. Marx *Capital*, Vol. 1, p.570 & 571n.

48. Ibid. p.571n.

49. Marx, *Grundrisse*, p.459.

> manner, since only rarely and under quite special conditions is a society able to adopt a critical attitude towards itself...[50]

He elaborated on this in the *Grundrisse*. Quoting the adage 'Human anatomy contains a key to the anatomy of the ape' he argued in relation to the categories of capitalist economy which appear in past societies that

> this must be taken with a grain of salt. They can contain them in a developed, or stunted, or caricatured form etc., but always with an essential difference. The so-called historical presentation of development is founded, as a rule, on the fact that the latest form regards the previous ones as steps leading up to itself ... since it is only rarely and only under quite specific conditions able to criticise itself...[51]

Such an affirmation of genuine emergence and rejection of the idea of historical evolution is completely at odds with the Neoplatonic eschatology of the *1844 Manuscripts* and the scientific materialist formulation of this as a technological determinism. This break with historical determinism is not simply a late development in Marx's thought and appeared throughout his writings. In the *Communist Manifesto* Marx and Engels had pointed out the historical possibility of the common ruin of contending classes in a society rather than the emergence of a new order. In his characterization of the types of socio-economic formations in the *Grundrisse* Marx included the Asiatic mode of production without giving it any place in an evolutionary scheme. But most importantly in considering societies other than Western Europe, he at times resolutely rejected the notion that there is one unique course of development for all societies.

In this regard his comments on Russia are most illuminating and reveal the extent to which Marx transcended both his early works and the historical scheme of the 'Preface' to *A Contribution to the Critique of Political Economy*. Engels had engaged in a polemic with the Russian Populist Petr Tkachev in 1874-75 concerning the possibility of establishing a communist society within Russia. Tkachev had argued on the basis of Marx's analysis of socio-economic formations that it is only when capitalism is established that one must wait until it has fulfilled its potentialities. However Russia was in an epoch of transition, and this made it possible to skip capitalism and move straight on to communism. Engels rejected this: 'The bourgeoisie,' he wrote, 'is just as necessary a precondition of the socialist revolution as the proletariat itself. Hence a man who will say that this revolution can be more easily carried out in a country because, although it has no proletariat, it has no bourgeoisie either, only proves that he has still to learn the ABC of socialism.'[52]

Marx took exactly the opposite view. In a letter written in November, 1877 and addressed but not sent to the editor of a Russian literary-political journal in reply to a charge made against Marxism by Nicolai Mikhailovski that it condemned Russians to the oppression of capitalism, Marx rejected the notion that his theory implied anything of the sort. Criticising Mikhailovski's interpretation he wrote:

> He must by all means transform my historical sketch of the development of capitalism in Western Europe into a historical-philosophical theory of universal development predetermined by fate for all nations, whatever their historical circumstances in which they find themselves may be, in order finally to achieve that economic formation which with the highest upswing of the productive forces of social work assures mankind its most universal development. But I beg his pardon. (That [view] does me too much honour and too much insult.)

---

50. Marx, *A Contribution to the Critique of Political Economy*, p.211. See also *The German Ideology*, p.89.

51. Marx, Grundrisse, p.105f.

52. Karl Marx and Frederick Engels *Selected Works*, London: 1950, Vol.2, pp.46-47.

Marx then went on to describe how ancient Rome had produced a situation very similar to that of late feudal Europe: peasants were expropriated from the means of production and subsistence in an economic formation consisting of large landownership and large-scale capitalism. But instead of the dispossessed selling their labour-power they became an idle mob, and instead of a capitalist production system, a system developed based on slave labour. Marx concluded from this:

> Thus events of a striking analogy, because they took place in a different historical milieu, led to entirely different results. If one studies each of these developments by itself and then compares them with each other, one will easily find the key to each phenomenon, but one would never thereby attain a universal key to a general historical-philosophical theory, whose greatest advantage lies in its being beyond history.[53]

This conclusion was reiterated in a letter to Vera Zasulich in March, 1881 concerning the possibility of establishing communism on the basis of the Russian village commune. Marx concluded against the avowed Marxist Zasulich and in favour of the populists 'that this village commune is the fulcrum for the social regeneration of Russia...', emphasising that his analysis in *Capital* referred only to Western Europe.[54]

Marx also developed an epistemology appropriate to his relational conception of humans as beings in the process of becoming in which the course of history is crucially dependent upon human agency. In the *Theses on Feuerbach* he argued against the contemplative materialism of Feuerbach, holding that knowers and knowledge must be considered as part of the material world being understood, that the educators themselves must be educated. Correspondingly he argued that it is impossible to judge the validity of theories until they have been acted upon and made part of social reality: 'Man must prove the truth, i.e., the reality and power, the this-worldliness of his thinking in practice. The dispute over the reality or non-reality of thinking which is isolated from practice is a purely scholastic question.'[55] There can then be no question of discovering the laws of history which will determine the future. So much for historical materialism. And this flexible interpretation of history, assuming a relational view of people as essentially social and creative agents, in which capitalism is seen to have developed as an emergent process with its own unique dynamics, and in which people have no guarantee that there will be a happy end to their struggles, fully accords with the framework of the process view of the world.

## *Engels' Effort to Supplement Marx*

Albeit in a very paradoxical way, Engels elaborated a process view of the world in order to provide a general philosophy for Marx's conception of history. In his efforts to conceive the world so that Marx's ideas would be fully intelligible Engels presented a picture of the world:

> ... in which nothing remains what, where and as it was, but everything moves, changes, comes into being and passes out of existence. This primitive, naive, yet intrinsically correct conception of the world was that of the ancient Greek philosophy, and was first clearly

---

53. *The Letters of Karl Marx*, ed. and tr. Saul K. Padover, Englewood Cliffs, N.J.: Prentice-Hall, 1979, p.321f.

54. Ibid. p.335f.

55. Karl Marx 'Theses of Feuerbach' (Original Version) in *Karl Marx, Frederick Engels: Collected Works*, Vol.5, 1845-47, tr. Clemens Dutt et.al. N.Y.: International Publishers, 1976, p.3.

> formulated by Heraclitus: everything is and also is not, for everything is in flux, is constantly changing, constantly coming into being and passing away.[56]

Engels argued that despite the achievements of the analytical approach to understanding the world which has displaced this primitive vision of the ancient Greeks:

> ... this method of work has left us as a legacy the habit of observing natural objects and processes in isolation, apart from their connection with the vast whole; of observing them in repose, not in motion; as constants, not as essentially variables; in their death, not in their life.[57]

But this static conception of the world is now being transcended, and the view of the world as a dynamic totality is being reinstated, but at a higher level. It is not understood in its primitive naivete, but as enriched by the achievements of science and as the conclusion of strictly scientific research. In opposition to mechanistic materialism of the old science, Engels argued that 'the world is not to be comprehended as a complex of ready made things, but as a complex of processes.'[58]

But Engels did not develop this conception of being consistently. Far more than Marx, he was committed to a deterministic view of the universe in which the coming of communism would be inevitable. To combine a process view of the world with determinism he developed this conception of being in terms of 'dialectics' which he claimed he and Marx had rescued from Hegel's idealist philosophy. He defined dialectics as 'the science of the general laws of motion and development of nature, human society, and thought'[59] and argued that there are three such laws:

> The law of transformation of quantity into quality and vice versa;
> The law of interpenetration of opposites;
> The law of the negation of the negation.[60]

It is this conception of the world which after the death of Engels, Plekhanov designated dialectical materialism.

This list of 'laws of motion and development' describes features one would expect in a world consisting of processes. There would be qualitatively novel processes emerging, processes which are opposed to each other while being dependent upon each other, and developments involving sequences of transformations. But to present such characteristics of being in terms of 'laws of dialects' represents a gross confusion between the logic of the relations between categories or concepts by which the world is conceived and the causal relationships within the world as it is conceived.

The last two of these laws express the relations existing between categories or concepts. The second law expresses the insight common to Neoplatonists that opposites such as A and not A imply a category or concept in terms of which they are related as contradictories. The third law expresses Hegel's conception of the development of thought as the production of new categories which transcend the limitations of old categories. The concepts of the first law: quality and

---

56. Frederick Engels, *Herr Eugen Dühring's Revolution in Science (Anti-Dühring)*, N.Y.: International Publishers, 1939, p.26f.

57. Frederick Engels, 'Socialism: Utopian and Scientific' in *Karl Marx and Frederick Engels: Selected Works*, Vol.II, p.130.

58. Frederick Engels, 'Feuerbach and the End of Classical German Philosophy' in *Karl Marx and Frederick Engels: Selected Works*, Vol.II, p.387.

59. Engels, *Anti-Dühring*, p.155.

60. Frederick Engels, *Dialectics of Nature*, tr. Clemens Dutt, Moscow: Progress Publishers, 1954, p.62. The laws are essentially the same in *Anti-Dühring*.

quantity, are categories which Hegel deduced by means of his dialectical method as being essential for understanding the world. But simply presenting these without any framework of support provides no basis for accepting their validity, and most modern physical scientists would disagree with Engels and agree with Galileo that the world can be understood entirely in quantitative terms.

In actual fact the notion of dialectics pertains to discussion and argument. It derives from the Greek word for discourse: *dialektike*, and was originally applied to the question and answer approach to the exploration and development of ideas. It could be applied to nature by Idealists such as Hegel because they saw nature as posited by Spirit, and therefore as having a logical structure reflecting the structure of the development of thought. But a materialist has no grounds for assuming that logical relations are in the world itself - except insofar as logic is a creation of humans who are part of the material world.

However while it might be invalid for a professed materialist such as Engels to refer to these laws of dialectics as laws of nature, he might be justified in ascribing such laws to the development of society and to the development of thought. To think of society as evolving dialectically he would merely have to accept Hegel's arguments that history moves from the embodiment of one set of ideas, categories or concepts to another. But this flies in the face of Marx's struggle to transcend Hegel by showing to what extent the dynamics of history is not the dynamics of a subject-object moving from one conceptual structure to another in its struggle to develop itself, but engenders social forms, notably capitalism, with an autonomy which transcends the intentions of all subjects and which confronts people as a second nature to which they must conform. Marx pointed out that contradictions in society in fact frequently do not impel any developments. As he argued in relation to the contradictory conditions associated with the exchange of commodities, these develop 'a *modus vivendi*, a form in which they can exist side by side.' And he went on: 'This is generally the way in which real contradictions are reconciled.'[61] Engels' characterization of the development of society as a dialectical development cannot be justified as an account of Marx's work.

Lastly, Engels might be justified in claiming that at least his last two dialectical laws apply to thought. It is in this domain that Engels would seem to be on strongest grounds given the inadequacy of empiricist and conventional rationalist theories of knowledge to take into account what is involved in the development of concepts. But this brings to the fore the question of what Engels means by 'laws'. It was seen in Chapter 5 that the notion of law was originally applied to nature on the assumption that it was ordered by God. Engels can hardly claim this conception of law. Secondly the existence of laws implying intelligibility was based on the assumption that these laws were logical laws. The world could therefore be understood as ordered by logical necessity. Even without a God such a notion of law might be justified if reinterpreted subjectively to imply a convenient way of ordering experience to make predictions. But the real superiority of dialectical notions of thought over logicist accounts is that they give a place to the originality involved in developing and creating new concepts. For this reason it is highly misleading to refer to a dialectical account of the development of thought in terms of laws. So even here Engels' ideas are questionable.

But this brings to light the fundamental ambiguity of Engels' ontology. It is an ontology which implies the possibility of new processes forming which cannot be understood in terms of the conditions of their formation. Since these are not determined by such conditions, the future cannot be entirely determined by the past. The idea of a general set of laws of being governing the development of the whole of reality in any conventional sense of the notion of law is therefore simply out of place. Similarly in relation to knowledge. Marx's achievements were genuinely original and cannot be thought of as simply the product of the operation of specifiable laws of thought. In his effort to capture Marx's achievements, Engels was forced to develop an ontology

61. Marx *Capital*, Vol.I, p.106. Marx castigated the 'dialectical' account of economic categories and development in his critique of Proudon in 'The Metaphysics of Political Economy', *The Poverty of Philosophy*, Moscow: Progress Publishers, 1973, Ch.2.

which would allow for the emergence of novelty, for an open future. But by presenting this ontology first in terms of the categories appropriate to an Idealist conception of being, and then formulating this in terms of a notion of law deriving from mechanistic materialism, Engels disguised the anti-deterministic implications of a process view of the world. When this disguise is shed, then the applicability of the Heraclitean or process conception of being for understanding Marx's most important insights can be appreciated.

## *Conclusion*

To evaluate Marxism it is therefore necessary to consider whether it is Marxism as understood in terms of Neoplatonism, in terms of mechanistic materialism or in terms of process philosophy. In the following chapters the orthodox Marxism of the Soviet Union will be analysed, and it will be argued that its defects derive from its Neoplatonic and mechanistic elements. But at the same time a rival tradition of Russian Marxism committed to reformulating Marx's ideas in accordance with a process view of the world will be revealed.[62] This tradition was inspired by Aleksandr Bogdanov and became a major force after the October revolution in the form of the *Proletkul't* movement. According to Bogdanov, the creation of a new form of society will require a new culture based on a dynamic view of nature and of humans as organized and organizing forms of energy. The dynamics of Russian culture will be shown to have affected the way in which Marxism was appropriated and developed in Russia and then in the Soviet Union, suggesting the importance of culture in constraining what can be achieved in any society. Also, that there are other processes in society than those associated with the market that can take on a life of their own inimical to the ends of the people who are its constituents. Together these arguments will be used to suggest that if Marxism is to be developed to deal with environmental problems and to adequately comprehend history, it must give a far greater role than orthodox Marxists have allowed to the so-called superstructure of society. This will provide a defence for Bogdanov's Marxism and the *Proletkul't* movement and their project of reformulating Marxism in accordance with process philosophy.

& myth

62. While the first thinker to attempt to reformulate Marx along these lines was Bogdanov, in the 1930's Joseph Needham also attempt to synthesize Marxism with a process view of nature. More recently Cornelius Castoriadis, building on the work of Maurice Merleau-Ponty, has contributed to this task. See his *The Imaginary Institution of Society*, [1975] Cambridge: Polity Press, 1987. Taking a very different approach, Russell L. Kleinbach has shown the relevance of Whitehead for the development of Marxian social theory in *Marx via Process*, Washington: University Press of America, 1982.

# 10

# MARXISM AND THE DYNAMICS OF RUSSIAN CULTURE

Throughout most of its history, the vast majority of Russia's population have been illiterate, with a relatively small number of noblemen and monks living in a sea of peasants. As a consequence it is even more difficult to understand the nature of the cultural dynamics of Russia than to understand the cultural dynamics of Western Europe. For the most part, it is the forms of thinking of a small minority of the population which have been given expression to, and the forms of thinking of the majority are manifest only insofar as they have positively or negatively influenced this minority. Apart from folktales and the like, the orientation of the vast majority has only directly manifested itself in their revolts against the established order.

Russia is distinguished from the rest of Europe first of all by the harshness of its environment. It has featured vast, thinly populated spaces full of forests and swamps, low rainfall, freezing winters, and in the north, poor soil. The impression made by this country on its visitors was described by Braudel: 'A traveller to Persia entering Russian territory at Smolensk in 1602, found Muscovy a "great and vast" country, "wild, deserted, marshy and covered in scrub" and forests, "interspersed with swamps which one crosses by paths made of fallen tree-trunks"...; a country like nowhere else on earth, empty ..., with appalling roads, difficult even in summer, a country in short "so resistant to access that it is impossible to enter or leave it discreetly, without permission or a safe-conduct from the Grand Duke".'[1] Apart from this, Russia lay beyond the Roman Empire. Its Christianization in the tenth century involved the assimilation of forms of thinking which were far more different from its original culture than was the case with those areas of Europe which had been part of the Roman Empire. And Russia was Christianized to the Orthodoxy of Constantinople which separated it further from Western Europe. And yet Russia has been Christian, and consequently its culture has many surface characteristics in common with those of the rest of Europe, though these generally form part of basically different cultural configurations.

Partly because of the nature of the physical environment, people's relationship to it has been fundamentally different than in Western Europe. Traditionally, nature was not conceived of as something to be subjugated, but as the source of both life and hardship. As in the West nature has been conceived of as feminine; but it has not been conceived of as a female to be conquered. Rather it has been seen as the great 'damp mother earth.' As G. P. Fedotov put it, 'Earth is the Russian "Eternal Womanhood," not the celestial image of it; mother, not virgin; fertile, not pure; and black, for the best Russian soil is black.'[2] Nature has symbolized the Russian feminine virtues of endurance, non-resistance to evil and voluntary suffering.

In a land of forests, the axe was the essential tool of men. They cleared the forests with it, cut up, carved and planed wood with it, defended themselves with it against both animals and people, fought with it against Teutonic knights and Mongols, and beheaded prisoners of war with it. It was used by tsars to suppress urban rebellions and by peasant rebels to terrorize the provincial nobility. There was no impetus to develop new technology. But while the Russians' axes cleared the land, provided shelter from the elements and provided defence against large animals and people, the omnipresent insects and rodents constantly gnawed away at their crops, their buildings and at the people themselves. Mosquitoes swarmed over people and lice got into their clothing, infecting them with disease. Cockroaches invaded their dwellings while mice and rats devoured crops and spread

---

1. Fernand Braudel, *Civilization & Capitalism 15th-18th Century, Volume III, The Perspective of the World* [1979] tr. Sian Reynolds, London, Fontana, 1985, p.26.

2. G.P. Fedotov, *The Russian Religious Mind* (2 vols) Vol. 1. Cambridge Mass.: Harvard University Press, 1946, p.13.

disease. The first official English ambassador in the mid-seventeenth century was advised by Russian officials to sleep with his servants 'lest the Rats run away with them being single.'[3]

The harsh conditions fostered not the individualism characteristic of Western Europe, but strong communities in which the virtues emphasised were the capacity to endure hardship and to subordinate oneself to the group. Life was not a struggle to conquer nature, but a struggle to survive within it. The family was of central importance in this struggle, and 'small' or nuclear families frequently recruited new members without blood ties into their households to form extended or 'great' families.

While the eternal womanhood of the earth, the axe and the family have been pervasive symbols, the ultimate symbol and metaphor, the thematic motif of Russian culture which has dominated it as the machine has dominated the West since the seventeenth century, is *fire*. In Russia's bleak, icy winters, fire provided warmth and light. It was revered, requiring cleanliness in its presence and reverent silence when being lit or extinguished. But fires also swept through forests and towns, burning the wooden houses and buildings, and Moscow had seventeen major fires between 1330 and 1453. It was also feared. Fire played a major part in Russian religious symbolism. Perun, the god of thunder and creator of fire held the pre-eminent place in the pre-Christian galaxy of deities, and the firebird a special place in mythology. Russians accentuated all references to or analogies with fire in scripture and in religious philosophy, and their onion domed churches symbolise the purifying power of flames. The basic metaphor for explaining the combination of God and man in Christ has been that of fire infusing itself into iron, and a popular definition of Christian commitment portrayed the committed person as 'having become all fire in the soul, he transmits the inner radiance gained by him also to the body, just as physical fire transmits its effect to iron.'[4] Russians tended to see the heavenly orders in terms of the writings of the Pseudo-Dionysius for whom angels are 'living creatures of fire, men flashing with lightning, streams of flame ... thrones are fire and the seraphims ... blazing with fire.' And they accepted his conception of the world according to which: 'Fire is in all things ... manifesting its presence only when it can find material on which to work ... renewing all things with its life-giving heat ... changeless always as it lifts that which it gathers to the skies, never held back by servile baseness.'[5] Christ's statement that 'I have come to send fire on the earth' was frequently cited, as was the fact that the Holy Spirit first came down to man through 'tongues of fire'. In the seventeenth century fire was the weapon of the fundamentalists who burned musical instruments, foreign style paintings and the buildings of the foreign community itself. After the Old Believers had been anathematized in 1667 they burned themselves in oil soaked wooden churches. Rebels and revolutionaries retained this fascination with fire. Bakunin prophesied during the revolutionary crisis of 1848-49 that 'tongues of flame' would shortly appear all over Europe to bring down the old gods, the symbol of fire featured centrally in the early twentieth century revolution in music (in such works as Scriabin's 'Poem of Fire'), and Lenin titled his revolutionary journal the *Spark (Iskra)*.

## *Christianity in Medieval Russia*

The conversion to Christianity began with the conversion of Grand Prince Vladimir of Kiev in Byzantium in 988. There followed a mass baptism of Russians in the Dnieper. Though the actual permeation of Christianity through the whole of Russia took centuries, the embracing of Christianity by the ruling elite was characterized by its unreservedness. If a society voluntarily adopts a vast new culture then there must be reasons for doing so, and the pre-existing culture must have been such

---

3. Cited by James H. Billington, *The Icon and the Axe: An Interpretive History of Russian Culture*, N.Y.: Vintage Books, 1970, p.23.

4. Cited ibid. p.24.

5. Cited, loc. cit.

that it was capable of absorbing a whole set of new ideas. Understanding how such a mass conversion was possible illuminates one of the most enduring features of Russian culture.

Russian culture has been characterized by a duality.[6] As distinct from the West, Russians have divided everything into the elevated (sacred) and low (profane) without leaving any room for an intermediate, neutral realm. Reflecting the total subordination of the individual to the group, actions have been seen as either good or bad, holy or sinful, pro-state or anti-state. In the religious sphere, Russians allowed no purgatory; only heaven and hell. Politically, they have been either revolutionary or reactionary. There has been no middle ground for conservatism nor for a realm within which a succeeding system could be developed. But this duality has been such that whatever has been rejected remains an active part of the culture. This has facilitated complete cultural inversions with what was previously conceived to be sacred becoming profane and vice-versa. As Lotman and Uspenskii wrote of Prince Vladimir, 'He did not simply accept a new system of values, replacing the old with the new, but rather wrote the old into the new - with a minus sign.'[7] The old culture was preserved in the system of proscriptions and by renaming the pagan gods as saints or devils.

In the Orthodox Christianity to which they were converted, the 'radical' Neoplatonism of Clement, Origen, Gregory of Nyssa, the Pseudo-Dionysius and Maximus the Confessor dominated the religion rather than being a minor tradition, as it had been in Western Europe. Consequently, unlike Western Christianity there was no rejection of the here and now or contempt for the body but a belief in the potential holiness of matter. As Fedotov put it: 'The distance between the two worlds is not the gulf between the flesh and the spirit ... but between the fallen and the transfigured ... flesh.'[8] The end of history was described in the twelfth century by a Russian monk as a conflagration purifying the just, burning the sinners, and transforming the world:

> Afterwards, the earth will be new and flat (the ideal antique landscape) as it was in the beginning, and whiter than snow; it will be changed by the order of God, and will be like gold; there will grow upon it various grasses and flowers, never fading, because spiritual; and trees will come forth, not similar to those visible now; their height, beauty, and splendour the lips of men are unable to express, because spiritual.[9]

In this transfigured world, humans would be deified, becoming by grace what God is by nature.

The most important function served by this form of Christianity was to provide an ideology to unify Russia's multinational empire. Its focus was not on morality, on the need to transform one's humanity to attain personal salvation, but on cosmic redemption in which the Christian Empire of the East would be transformed into the final heavenly kingdom. All that was needed by Christians was 'right praising' through the forms of worship handed down from the Apostolic Council and defined for all time by the ecumenical councils. Muscovites spoke of following or serving Christ rather than imitating Christ, and put greater stress on the suffering which such service entailed. What was important was Christ's mission rather than his teachings - which were little known anyway in the absence of a complete Slavonic New Testament. The function of the Christian was to serve God by enlisting in that mission, by beating off his enemies and following Christ in his personal compassion and willingness to suffer.

This world-orientation enabled Russians to see themselves in historical perspective, just as did their Christian and Islamic opponents. Constantinople had thought of itself as the New Rome:

---

6. Iurii M. Lotman and Boris A. Uspenskii, 'Binary Models in the Dynamics of Russian Culture' in *The Semiotics of Russian Cultural History* Alexander D. Nakhimovsky and Alice Stone Nakhimovsky eds., Ithaca and London: Cornell University Press, 1985, pp.30-66.

7. Ibid. p.34.

8. Fedotov, *The Russian Religious Mind*, p.129.

9. Cited ibid. p.175.

capital not of 'a' but 'the' Christian Empire, specially chosen to guide men along the path, marked out by the chronicles, from Christ's incarnation to His Second Coming. The Russians represented themselves as superior to Byzantium by virtue of the 'newness' of Russia in comparison with 'old' Byzantium.[10] With the fall of Constantinople and the liberation of Russia from the Mongols, Moscow came to conceive of itself as the 'Third Rome', the only bulwark of Orthodoxy. Since Byzantium had fallen to Islam while Russia had liberated itself from Islam, the two had changed places, and Russia had become the centre of the Orthodox, and therefore of the Christian World. This vision of Russia's place in the world was vividly conveyed in a fifteenth century letter by Filofei of Pskov to Ivan the Great:

> The church of the first Rome fell because of the godless heresy of Apollinaris. The gates of the second Rome at Constantinople were smashed by the Ishmaelites. Today the holy apostolic church of the third Rome in thy Empire shines in the glory of Christian faith throughout the world. Know you, O pious Tsar, that all empires of the orthodox Christian have converged into thine own. You are the sole autocrat of the universe, the only tsar of all Christians ... According to the prophetic books all Christian empires have an end and will converge on one empire, that of our gossadar, that is, into the Empire of Russia. Two Romes have fallen, but the third will last, and there will not be a fourth one.[11]

Individuals were related to this Christianity through the lives of the saints and through icons. Each of these was far more important to people in Russia than to people in the West. These presented the Platonic forms for individuals to participate in and to identify themselves by. As such they reflected the way Russians adapted Christianity to their own culture, with its emphasis on community and endurance. The nature of this adaptation was particularly evident in the evolution of the stories of the martyred princes in Russia.[12]

Originally the stories of the Kievan princes, Boris and Gleb, killed by their brother Svjatopolk, were modelled on and strongly resembled the stories of Saint Wenceslas (Vaclav) of Bohemia. In each case innocents, who refused to defend themselves against their brother, were murdered. But in contrast to the Czech stories which, in accordance with Western thought emphasised individual self mastery, the Russian stories strongly emphasised the themes of quiet, humble submission to one's fate, and of brotherly love and the proper relation of younger to elder brothers. Through their murder the brothers are transfigured and are ecstatically reunited in heaven. As Norman Ingham wrote of this:

> Rusian (sic) writers ... did not just 'borrow' ready made formulations; they absorbed ideas and freely reshaped them. They continued some themes and developed others whose seeds they found in the Bohemian texts. Their original contributions were major ones: the kenotic brand of nonresistance; the combined religious and political principle of brother-love....Wencelas's image evolved along different lines. In the religious aspect he came to be portrayed as an ascetic, and in the political, as the rex perpetuus of the Czech kingdom.[13]

Icons or 'forms' (*obraz*) provided an external expression of the transfigured state of humanity and of the material embodiment of an inconceivable Deity. Virtually every peasant possessed and

10. Lotman and Uspenskii, 'Binary Models in the Dynamics of Russian Culture', p.35.

11. Cited by Eric Voeglin in *The New Science of Politics: An Introduction*, Chicago & London, The University of Chicago Press, 1952, p.114f.

12. Norman W. Ingham, 'The Martyred Prince and the Question of Slavic Cultural Continuity in the Early Middle Ages' in Henrik Birnbaum and Michael S. Flier, eds, *Medieval Russian Culture*, Berkeley: University of California Press, 1984, pp.31-53.

13. Ibid. p.53.

venerated an icon, and screens of icons adorned the churches. Through the icons, people recognized and acknowledged themselves as part of Christian society. Such perceptible identifying signs were far more important in Russia with its poor development of individualism and abstract philosophy than in Western Europe, and paintings rather than philosophical tracts were the medium through which religious confrontations were expressed and fought.[14] The icon screens modelled the hierarchically ordered Russian society, with each figure occupying a prescribed position in a prescribed way, unified by their common distance from the God of the sanctuary and their dependent relationship to the central panel of Christ enthroned. The vertical hierarchy of saints connected the heavenly with the earthly Church, and a hierarchical pyramid of patron saints from the *'saints militaires'* of the ruling elite down to the patrons of trade, agriculture, and cattle raising venerated by merchants and peasants, enabled people to identify themselves within the order of things and provided them with models of ethical-social behaviour. The highest development of this art, the iconostasis produced in the fourteenth century provided 'no less than a pictorial "*Summa Theologiae*" of the Eastern Church, an iconic representation of the conceptual-imperceptible cosmos...'[15] Correspondingly, it was believed that the Christian society ordered on this basis was itself an icon, that 'the Tsar is, as it were, the living icon of God, just as the whole Orthodox Empire is the icon of the heavenly world.'[16] When during the seventeenth century the Patriarch Nikon attempted to increase his power in relation to the Tsar in accordance with Western Christianity, his accuser, Ligardes 'summoned up the distinctively Russian symbol of the icon screen as the model for an ordered hierarchical society to challenge Nikon's concept of a symphony of powers between civil and ecclesiastical authority.'[17]

Russian culture was not such as to generate innovation among its members. Despite the Platonism of Russian Christianity through which the ideal prince was conceived to be the living icon of God, this ideal was not a philosopher but a guardian of tradition. The highest good in Muscovy society was not knowledge, but memory, '*pamiat*'. Rather than saying 'I know', the Russian would say 'I remember'. There was no higher appeal in a dispute than the important, good and firm memory of the oldest available authority. Thus Muscovy was 'bound together not primarily by formal codes and definitions or rational procedures, but by an uncritical and unreflective collective memory.'[18] This general attitude was confirmed and supported by Orthodox Christianity since what was most important for the Orthodox was, precisely, being orthodox. This religion was mystical rather than rationalistic and was strongly influenced by the anti-scholastic Hesychasts. Hesychast mysticism encouraged the belief among the Orthodox that the transformation of the Christian Empire of the East into the final heavenly kingdom was possible through a spiritual intensification of their own lives. Generally God's incomprehensibility to the limited human intellect was emphasised. While in Western Europe people believed they could achieve a deeper knowledge of God by investigating the nature of his creation, there was no impulse to investigate nature in medieval Russia.[19] In fact in 1350 the Patriarch of the Church banned the study of mathematics and astronomy. Correspondingly, despite the common Neoplatonic emphasis on hierarchy, the Eastern conception of the endurance of humanity in a natural world beyond their control reflected in the image of God as beyond human comprehension was radically different from the Western image of human domination of nature reflected in the image of God as having created the universe by an act of will. And there was no conception that people were participating with God in His creation of the

---

14. These are described by Konrad Onasch, in 'Identity Models of Old Rusian Sacred Art' in Birnbaum and Flier eds, *Medieval Russian Culture*, pp.175-205.

15. Ibid. p.186.

16. Billington, *The Icon and the Axe*, p.35.

17. Ibid. p.155f.

18. Ibid. p.62.

19. On the slow development of science is Russia, see A. Vucinich, *Science in Russian Culture. A History to 1860*, Stanford: Stanford University Press, 1963.

world. While there was a burst of artistic activity in the production of holy pictures in the fourteenth and fifteenth centuries, 'Russia was moving not toward a renaissance, a new release of emancipated creativity and individual self awareness, but toward a synthetic reaffirmation of tradition.'[20]

The government of Russia has correspondingly developed into a far more autocratic structure than existed anywhere in Western Europe. There were historical reasons for this.[21] The Tartars had contributed to this autocratic tendency by destroying all vestiges of democracy, promoting the rule of oppressive princes, and providing a model of total subordination of subjects to a ruler and by their insisting that the subjugated Russians pray for only one ruler, the Tartar khan. Kiev had been a far more democratic society than Moscow. So also had Novgorod in the north with its close relations with the West, commercial cosmopolitanism, representative government and philosophic rationalism. But it was Moscow with its xenophobic autocracy which emerged as the dominant city of Russia in the fight against the Tartars. The rise of autocracy was also facilitated by the very lack of dynamism of the general population.

Centralization reached its peak with Ivan IV (the Terrible) who ruled from 1533-84 the first ruler to be crowned tsar (caesar) in Russia. Ivan conceived of himself as head of a monolithic religious civilization, never simply as a military or political leader, and brutally suppressed the Russian hereditary aristocracy, the boyars. The leading apologist for Ivan's rule, Ivan Peresvetsov, argued that 'A realm without dread is like a horse beneath a Tsar without a bridle.'[22] While succeeding tsars were not as brutal as Ivan, with the exception of Boris Godunov (1598-1605) they followed Ivan's precedent of absolutism, even after Peter the Great had discarded the religious garb which had legitimated it.

This autocracy was exercised to appropriate the military and technological innovations of Western Europe so as to be able to effectively confront Russia's western neighbours. In the 1550's Ivan the Terrible began to employ foreign mercenaries and adopted Swedish and Dutch military innovations in the struggle against the Poles, a struggle which only ended with Poland's defeat in the war of 1654-67. Western measurement began to impose itself with the development of military maps, the erection of a gigantic English built clock on the Moscow Kremlin in 1625 (popularly opposed as contamination of eternity with time) and the appearance of weather vanes atop the crosses of churches. In 1632 the Dutch built the first modern Russian arms plant and arsenal, and in 1647 they printed the first military manual and drill book for Russian foot soldiers. The modernization of the army was associated with the growth of bureaucracy and the formalization of peasant serfdom as a means of guaranteeing the state a supply of food and service manpower. The struggle with Poland was followed a half century later by war with Sweden. For this Russia was aided by the Danes to develop a navy.

The tsars' concern with Western culture was almost entirely practical. They were interested in military and administrative techniques. Symptomatic of this, the word '*nauka*' later used for 'science' and 'learning' in Russia was introduced in a military manual in 1647 as a synonym for 'military skill.' However dealing diplomatically with the West, using its technology and conquering the Westernized populations formerly governed by Poland and Sweden, undermined the unity of Russian culture. As Billington noted of Ivan the Terrible:

> The mounting fury of Ivan IV's last years seems less a product of his paranoia than of a kind of schizophrenia. Ivan was, in effect, two people: a true believer in an exclusivist, traditional ideology and a successful practitioner of experimental modern statecraft. Because the two roles were frequently in conflict, his reign became a tissue of contradictions. His personality was

---

20. Billington, *The Icon and the Axe,* p.35.

21. These have been analysed by Richard Pipes in *Russia Under the Old Regime*, [1974], Penguin: Harmondsworth, 1990, Part I. See also Dinko Tomasic, *The Impact of Russian Culture on Soviet Communism*, Glencoe: The Free Press, 1953.

22. Billington, *The Icon and the Axe*, p.76.

increasingly ravaged by those alternations of violent outburst and total withdrawal that occur in those who are divided against themselves.[23]

## *Peter the Great and the Well Ordered Police State*

This contradiction was eventually overcome by discarding the exclusivist, religious character of Russian society. Peter the Great set about reforming Russia into a well ordered, secular state, able to efficiently make use of its natural and human resources, in accordance with the precepts which had been elaborated by cameralist theorists and German rulers since the first half of the seventeenth century.[24] Peter inverted Russian culture, just as Vladimir had done in the tenth century. Whereas previously, the old, identified with Nature and with the Church, were extolled as the sacred, and the new which was identified with Culture was denigrated as profane, Peter rejected the old and Nature as profane, and embraced the new, including the culture of Western Europe, as sacred. The new capital built by Peter, St Petersburg with its Dutch name and geometrical layout became the icon of a new world. As it was described in the early years of Peter's reign:

> geometry has appeared,
> land surveying encompasses everything.
> Nothing on earth lies beyond measurement.[25]

But Russia did not become a Western culture. As Lotman and Uspenskii pointed out, 'A close examination reveals convincingly... that the new (post-Petrine) culture was significantly more traditional than is usually thought. The new culture was constructed not so much on models from "Western" culture (although it was subjectively experienced as "Western") as on an "inverted" structural plan of the old culture.'[26] But this destroyed the unity of the culture. While it involved the assimilation of much of the achievements of Western European culture, the foundations of this culture which had produced these achievements were not assimilated, and what was assimilated did not fit in easily with other modes of thought and institutions of Russian society. Forms of behaviour and institutions were left floating in a vacuum. As Marc Raeff wrote, 'The effect of Peter the Great's reign was to tear Russian society apart, leaving behind a legacy of uncertainty and insecurity that ultimately led to an identity crisis among the Russian elite.'[27]

In particular, the detached, activist individualism of Western Europe did not replace the Russian tradition of passivity and subordination of the individual to the community. This passivity was described by the Danish envoy, Just Juel in his description of the fire fighting efforts of Muscovites:

> Being endowed with an exceptionally quick intelligence the Tsar [Peter the Great] sees at once what needs to be done to contain the fire. He climbs up on the roof, moves to the most dangerous spots, encourages people and nobles alike to lend a hand, and does not rest until the fire is out. If, however, the ruler is not present, things are totally different. Then the people just watch, often with total indifference, and no one helps. It is entirely useless to berate them

23. Ibid. p.99.

24. For a full study of this see Marc Raeff, *The Well Ordered Police State: Social and Institutional Change Through Law in the Germanies and Russia, 1600-1800*, New Haven: Yale University Press, 1983.

25. Cited from the corrector of books by Billington, *The Icon and the Axe*, p.184.

26. Lotman and Uspenskii, 'Binary Models in the Dynamics of Russian Culture', p.54.

27. Marc Raeff, *Understanding Imperial Russia: State and Society in the Old Regime* tr. Arthur Goldhammer, N.Y.: Columbia University Press, 1984, p.35.

> or to offer them money to help; they merely wait for the moment when they can steal something.[28]

Only gradually was the conception of the individual as an earthly being with personal attributes, private interests and responsibilities developed. The Russian language only appropriated a future tense in the sixteenth century, and it was only in the late seventeenth century was the word *persona* applied to individuals - and then only to important or strong individuals. The word 'personal' and precise terms for 'private' and 'particular' did not enter the Russian language until the eighteenth century, and only then did the words used for 'law' and 'crime' enter into Russian jurisprudence with their modern meaning.[29] This lack of individualism was characteristic of all classes. Among the peasants it was manifest in the persistence of the communal organization of agriculture until the late nineteenth century. Among the aristocracy it was manifest in the late eighteenth century in the vehemence with which Russian deputies to a legislative commission established by Catherine the Great opposed a proposal by the Baltic nobility to draw up and submit to Catherine a code of laws spelling out the rights and privileges of each individual. The ruling elite preferred relations based on a personalized form of ultimate authority to a system based on a legal code and impersonal regulations. Debates revealed: 'a conception of society as an "organic" structure based on a heredity division of functions, a vision of a stable, harmonious society in which, by its very nature, conflict had no place.'[30] And Russians remained unable to compete against Westerners in commerce; as Braudel noted:

> In competition with foreign merchants, in Moscow and later in St Petersburg, Muscovite merchants rarely proved much of a challenge. It is surely curious that the richest merchants in Siberia in the 1730s - a man who had travelled to Peking as agent for Lange - was probably a Dane. Similarly, when after 1748 Russia began direct trading with the Black Sea, once again this was handled by foreign intermediaries.[31]

The opposition to the Western form of individualism continued even after capitalism emerged in the late nineteenth century, and many wealthy heirs of business fortunes turned against their fathers' values.[32]

Symptomatic of the lack of cultural integration and the identity crisis produced by partially adopting Western culture was the way Russians imitated Western forms of behaviour. Russians continued to identify themselves in terms of the role they were playing, rather than conceiving themselves as autonomous individuals, just as previous Russians had identified themselves through the roles represented by icons. But this identification with Western roles lacked integration into a perceived order of things. The image of European life was reduplicated in a ritualized play-acting of European life so that Russian gentry felt as though they were forever on a stage, which in many cases led to a bizarre confusion of life and fiction.[33]

The period from the reign of Peter the Great to the revolution of 1917 was characterized by a struggle to reunify Russian society both by its tsarist rulers and by various ideological factions of the intellectual elite. The tsars struggled to develop the institutions and to educate its population to

---

28. Cited by Christian Schmidt-Häuer, *Gorbachev: The Path to Power*, tr. Ewald Osers and Chris Romberg, London, Pan, 1986, p.23.
29. Billington, *The Icon and the Axe*, p.189.
30. Raeff, *Understanding Imperial Russia*, p.93.
31. Braudel, *The Perspective of the World*, p.462.
32. See Raeff, *Understanding Imperial Russia*, p.217.
33. This has been brilliantly analysed by Iurii M. Lotman in 'The Poetics of Everyday Behaviour in Eighteenth-Century Russian Culture', 'The Decembrist in Daily Life' and 'Concerning Khlestakov', in Nakhimovsky and Nakhimovsky eds, *The Semiotics of Russian Cultural History*.

consolidate Peter's reforms and to keep abreast of Western scientific, technological and military developments. But all the tsars were determined to maintain unlimited autocracy without the benefit of the traditional ideology which had legitimated it. They varied according to how much they also wanted to free people to think and organize within the framework of this autocracy. Peter III (1762), Paul I (1796-1801) and Nicholas I (1825-55) attempted to impose Prussian discipline on Russians, while Catherine the Great (1762-1796), Alexander I (1801-25) and Alexander II (1855-81) were relatively liberal. Catherine was a Francophile, and was the first to confront the dilemma of wanting rational rule based on natural laws while being unwilling to give up any power. After the Pugachev rebellion of 1773-74 and the French revolution she clamped down on free speech and banished one of the foremost Enlightenment thinkers within Russia, Aleksandr Radishchev, to Siberia. The conflict between the implications of Western thought and Russian autocracy came to a head in 1825 with the Decembrist revolt in the reign of Alexander I. The tsar who abolished serfdom, Alexander II, was assassinated. Tsardom culminated and ultimately failed with the oppressive, reactionary, nationalist rule of Alexander III (1881-94) and Nicholas II (1894-1917). The failure of the tsars in their struggle to develop science, technology and a professional administration and to industrialize was manifest by their defeat in the Crimean War in 1856, the Russo-Japanese War of 1905 and the First World War. The last two wars were to some extent the outcome of the final ideology, Social Darwinist Pan-Slavism, by which the supporters of the tsars attempted to legitimate their rule. By their own criteria, the fate of the last tsar: Nicholas II was justified.

## *Opposition to the Police State*

The first opposition to Peter the Great's reforms came from Old Believer communalism, the Cossack-led peasant insurrectionists, and the monastic revival within the official Church. These were entirely reactionary. The Old Believers appealed to instinct rather than intellect, and communal honour rather than individual reason. Their ideal order was an organic religious civilization of Great Russian Christians united by traditional forms of ritual worship and communal activity. The peasant uprisings which were a response to their increasing subordination to facilitate the advance of Russian military strength also wished to return to the old organic religious civilization ruled by the true tsar. Pugachev 'claimed to be the surviving tsar, Peter III, and promised the peasants "land, meadows, and woods," as well as "beards" - in other words, a return to the old traditions of pre-Petrine Russia.'[34] Such rebels offered no political program and simply attacked violently anyone symbolizing the new order. Less dramatic than the other forms of reaction, the monastic revival involved the rediscovery of the traditions of patristic theology and inner spirituality.

However the most important opponents of the Tsarist Police State were the intelligentsia who emerged as a distinct group and who became the main bearers of radical ideology in the late eighteenth century. These were the educated, generally French speaking intellectual elite of society who were trained to fill positions within the government. The intelligentsia experienced within their own lives the incoherence of the prevailing order, they were alienated from and felt guilty towards the even more oppressed peasantry and especially after the 1850's they suffered increasing oppression at the hands of the tsars. The members of this group embarked on an intellectual saga which eventually proved successful in forging an ideology able to overthrow the old order. This saga began in the Masonic lodges, fraternal societies and philosophic 'circles', but it was the circles among university students who came to play the greatest role in radicalizing each generation of students, advancing the ideological opposition to tsardom, and assimilating and disseminating new ideas. In this way the intelligentsia came to constitute itself as a self-conscious class.

---

34. Andrzej Walicki, *A History of Russian Thought: From the Enlightenment to Marxism*, [1973] tr. Hilda Andrews-Rusiecka, Stanford: Stanford University Press, 1979, p.7.

Philosophy was central in these ideological struggles, and supporters and opponents of the tsars drew on virtually every major philosophy developed in Western Europe. There was a constant battle between rationalists and romantics, French and German influences, universalists and nationalists, St Petersburg and Moscow. It was out of the dialectical conflict between these different positions that there slowly emerged a new vision of the world and the place of Russians within it. However while these ideas were developed through engagement with Western philosophy, they were also rooted in Russian culture.

The philosophical doctrines which took root in Russian society were those which resonated with the Neoplatonic world-vision of Orthodox Christianity, and associated with this, with the strong sense Russians had of being part of an historically significant community. The Russian tradition of philosophy really began with Maxim the Greek who moved to Russia in 1518. Having studied in Renaissance Italy he had absorbed the doctrines of the Neoplatonic revival associated with he rise of Hermeticism, which he effectively espoused in Russia. Among his students were Kurbsky, Karpov, and Ermolai-Erazm, the intellectual leaders of mid-sixteenth century Russia. Another injection of Western radical Neoplatonism occurred when the ideas of Jacob Boehme were brought to Russia in 1689 by Quirinus Kuhlmann in his attempt to prepare Russia for transformation into the apocalyptic fifth monarchy.[35] Though Kuhlmann was burnt for heresy in the same year he arrived, Boehme's basic ideas took root and influenced the Old Believers. Boehme's ideas also influenced Russia through the highly influential higher order masonry, and in particular through the works of such religious philosophers as Eckhartshousen, Schwartz and Saint-Martin. Inspired by the masons, young Russians flocked to Germany to study the works of the Rosicrucians. The most significant figure in Russian masonry was Novikov who until his arrest was the most influential intellectual figure in the Russian Enlightenment under Catherine the Great. He managed to combine within himself the practical philanthropy, normally associated with Enlightenment philosophy, and Neoplatonist, theoretical mysticism. But his orientation was more towards mysticism and to the development of a new religion based on the theosophy of Boehme and the older religious traditions of Russia, and it was these ideas which he disseminated most widely through the Moscow University Press and two private presses which he set up.

When Schelling's philosophy with its conception of the organic unity of all nature and the presence therein of a 'world soul' was introduced into this intellectual environment it was embraced with enthusiasm and immediately displaced the atomistic thinking of philosophers such as Locke. Schelling in turn paved the way for an even more enthusiastic reception of Hegel. Hegel's works, Herzen wrote,

> were discussed ... incessantly; there was not a paragraph in the three parts of the *Logic*, in the two of the *Aesthetics*, in the *Encyclopaedia*, etc. that had not been the subject of desperate disputes for several nights running. People who loved each other avoided each other for weeks at a time because they disagreed about the definition of 'all-embracing spirit,' or had taken as a personal insult an opinion on the 'absolute personality and its existence in itself.' Every insignificant pamphlet of German philosophy published in Berlin or even a provincial district town was ordered and read to tatters and smudges; the leaves fell out in a few days if only there was a mention of Hegel in it.[36]

Hegel's thought in turn provided the basis for the reception of the works of the Young Hegelians such as Feuerbach, whose materialistic humanism had far more radical implications. With God portrayed as 'merely the projected essence of Man',[37] Man was presented with the task of

35. Ibid. p.171ff. and 310ff.

36. Alexander Herzen, *My Past and Thoughts*, tr. Constance Garnett, London: 1927, Vol.2, p.115.

37. Ludwig Feuerbach, *The Essence of Christianity*, ed. E. Graham Waring and F.W. Strothmann, N.Y.: Frederick Ungar, p.65.

appropriating from religion his alienated essence. Or as Bakunin formulated this in the tradition of Russian culture: 'Jesus Christ began as a man-animal and finished as a man-god, such as we all must be.'[38]

The notion that Russia was in a privileged position, and by virtue of this was capable of serving as the saviour of European civilization had been a recurring theme of Russian thought ever since Moscow had been conceived of as the 'third Rome'. Thus Russians were susceptible to new explanations of their unique status, as addressed by Leibniz to Peter the Great, the Encyclopaedists to Catherine the Great, and the Pietists to Alexander I. The substance of these arguments were that it was an advantage for Russia to have been absent from the stage of history since it was uncommitted to the follies of Europe. This would enable Russia to play a unique role in the next stage of history. Such notions were reinforced by the philosophy of Schelling with his emphasis on the becoming of the world, and on this basis received their most forceful expression in the philosophical letters of Chaadaev, published in 1836 but widely discussed before then. But with the Hegelianization of Russian thought from 1838-48 the radicals introduced a new dimension to this notion. They began talking of the total destruction of the existing state and its replacement by a socialist society, 'the idea of ideas' which according to Belinsky 'has absorbed history, religion, and philosophy.'[39] The essence of this radical Neoplatonic revolutionary vision was described by Billington:

> Truth was to be found within rather than beyond history. Russia had some special destiny to realize in the coming redemption of humanity. A new, prophetic art was to announce and guide men to this destiny. The golden age 'lay not behind us but ahead': in a time when man's Promethean labours will end and he will come to rest both physically and spiritually in eternal and ecstatic union with the elusive feminine principles of truth and beauty.[40]

Where philosophers who were not Neoplatonists were widely embraced it was generally because their thought resonated in some way with its assumptions and supplemented it where it was inadequate. For instance the ideas of Saint Simon and Comte had a similar teleological view of history and presented the ideal of a new religion of humanity. The nihilists who embraced the materialism of Moleschott and Darwin were struggling for a more concrete grasp of the world around them to facilitate effective action, but remained committed to radical Neoplatonist eschatology. As Billington wrote, they were convinced 'that a direct reconstitution of society was morally necessary, logically implied by the progress of science, and uniquely among the Russian people.'[41]

Along with this general eschatology, Russian philosophical thought was distinguished by three other features. Firstly it was appropriated and developed in accordance with the traditional forms of Russian culture. To a considerable extent Russian philosophical ideas were developed through literature and literary criticism. As the lives of the saints and icons had provided models for people to live by in traditional Russian society, the intelligentsia produced literature and art, the main feature of which was the provision of such models for the people of the day.[42] As Nadhezhin, the literary critic under whom Vissarion Belinsky (1811-48) served his apprenticeship, wrote in 1818: 'To teach people the good is the duty of the poet.'[43] This notion was taken up and developed by

---

38. Cited from a letter, Ginsburg 'The "Human Document" and the Formation of Character,' Nakhimovsky and Nakhimovsky eds, *The Semiotics of Russian Cultural History*, p.205.

39. Billington, *The Icon and the Axe*, p.377.

40. Ibid. p.351.

41. Ibid. p.392.

42. See Katerina Clark, *The Soviet Novel: History as Ritual*, The University of Chicago Press, 1981, Ch.2. See also Lidiia Ia. Ginsburg 'The "Human Document" and the Formation of Character'.

43. Cited Billington, *The Icon and the Axe* p.330.

almost all other theorists of aesthetics and in almost all literature: that of Gogol, Turgenev, Dostoyevsky and Tolstoy; and of the minor tracts produced by revolutionaries in their efforts to influence the peasants and the proletariat. In this way various forms of being in the world were explored, tried out and evaluated.

Secondly, all Russian thought was coloured by its emphasis on the community. Even Radishchev (1749-1802), the foremost exponent in the eighteenth century of the philosophy of the Enlightenment and its doctrine of individual rights, emphasised the social nature of humanity. He criticised Rousseau's notion that humans are by nature reclusive and defended civic rights as a means for becoming a genuine 'son of the fatherland.'[44] Herzen (1812-70), oriented towards French rather than German thought and concerned to defend the autonomy of the personality and the rationalization of social relations, rejected the atomic individualism of the West just as vehemently as anti-Western Slavophiles. He subsequently became the founder of 'Russian Socialism'. And the nihilists of the 1860's who were determined to recognize nothing that could not be rationally justified - bonds imposed by family, society and religion, saw themselves as 'fighting for the happiness of mankind.'[45] However it was in the radical tracts echoing the lives of the saints and prefiguring Socialist Realist literature that the anti-individualist, communalist orientation of Russians was most clearly manifest.[46] These were characterized by three basic features. First, the political movement being championed was identified with a 'family.' This family was frequently to supplant members' natural families. Second, some naive individual was brought to see the light by an emissary of the movement. Third, this individual became a martyr, leading an ascetic life of extraordinary dedication, and frequently dying for the cause, whereby the hero was resurrected in the ongoing movement, often symbolized by one of his comrades picking up the fallen banner.

While these two features of Russian thought were reflections of the traditional Russian culture, the other distinguishing feature of Russian thought was a struggle against this tradition. With a culture characterized by a lack of orientation to individual initiative, the Russians opposed to the prevailing order found themselves engaged in an extraordinary struggle to overcome this deficiency. Early radicals were essentially divided in themselves between their lives of dissolute carousing and their romantic aspirations to transform the world. Awareness of this state of being was manifest in the fascination with the Hamlet theme in the late eighteenth century. Hamlet, the privileged court figure torn between the mission he was called upon to perform and his own private world of indecision and poetic brooding, symbolized life for the Russian intellectual elite.[47]

It became an over-riding preoccupation of these intelligentsia to achieve their romantic ideals. But while their dissolute lives were the traditional, socially acceptable mode of being which they understood unreflectively as a habitus, what they aspired to become was an alien form of behaviour which they had to struggle to realize. The members of the Decembrist movement attempted to overcome this dualism by acting as though every action and gesture had significance, like the descriptions of characters in a novel, and in this way totally excluded their traditional habitus. They succeeded in creating a new type of character capable of self-respect, but despite this they were peculiarly inept. On the morning of December 14, 1825 when the Decembrists came out onto Senate square, before the uprising had begun and while there was every chance of success, Aleksandr Odoevskii cried out: 'We are going to die, brothers, oh, how gloriously we are going to die.'[48] After their arrest and during their investigation they were utterly bewildered. There were no literary role models for their situation, since death without monologues in the vacuum of a military bureaucracy had not yet become the subject of art.

---

44. Walicki, *A History of Russian Thought*, p.40 & 45.

45. V. Zaitsev, quoted ibid. p.210.

46. Clark *The Soviet Novel*, p.49.

47. See Billington, *The Icon and the Axe*, p.351ff.

48. Lotman, 'The Poetics of Everyday Behaviour' in Nakhimovsky and Nakhimovsky eds *The Semiotics of Russian Cultural History*, p.87.

The first group of radicals inspired by the Decembrists were the generation of disaffected aristocrats of the 1830's and 1840's. These included Herzen, Belinsky (though not an aristocrat) and Bakunin. These were followed by less aristocratic generation in the 1850's who paved the way for the much broader group of the 1860's. In the 1860's the intelligentsia, based in the universities, emerged as a self-conscious class and developed the most original of the radical social movements within Russia, the populist movement. The new movement of radicals, manifesting again the Russian tradition of inverting everything while retaining the same basic orientation, totally rejected everything valued by the previous generation of radicals: poetry, literature, etc. and embraced the title 'nihilists'. They outfitted themselves in bizarre forms of dress designed to distinguish their members from the past, practiced free love, and attempted to live and work communally. But their major preoccupation was to succeed where the radicals of the past had failed.

The struggle for efficacy took place on many fronts, including literature, both major and minor. In the major literature it took the form analyses of character deficiencies and attempts to develop models of the efficacious personality. Turgenev in particular participated in these efforts. He developed the concept of the 'superfluous individual' in *Rudin* based on Bakhunin. In *On the Eve* he presented the ideal of the 'strong nature' capable of acting effectively in Insarov, a resident Bulgarian fighting for his country's freedom from the Turks. In *Fathers and Sons* Turgenev explored the nature of the new generation of radicals, with their rejection of art and their extolling of science, in the character of Bazarov. The analysis of these characters became a major concern of radical intellectuals, with Dobroliubov championing Insarov and the leading nihilist of the 60's, Pisarev, championing Bazarov. However the character which had greatest influence on the radicals was Rakhmetov, Chernyshevsky's main character in the novel *What is to be Done*. A scion of the wealthy gentry turned revolutionary, Rakhmetov is familiar with the people's lot. He has measured the whole of Russia on foot, and has worked at cutting timber, quarrying stone and hauling riverboats. In order to train his will-power and resistance to pain he even sleeps on a bed of nails. Of Rakhmetov and his kind, Chernyshevsky writes: 'They are few in number, but through them flourishes the life of all; without them it would die out and go sour. ... They are the flower of the best people, the movers of the movers, the salt of the salt of the earth.'[49] Lenin, who according to his wife Krupskaya recalled this work in every slight detail, stated: 'Under his influence hundreds of young people became revolutionaries ... he cast his spell over my brother, for instance, and over me too. He cut a very deep furrow in me.'[50] Lenin went on to explain that Chernyshevsky showed 'what sort of person a revolutionary should be, what rules of conduct he should follow, how he should proceed to his goal, and by what means he should attain it.'[51]

Along with this image of the strong-natured person, intellectuals struggled to place the individual in social context. The intelligentsia of the 30's and 40's moved from an idealization of the personality to an intensive investigation of the personality in terms of philosophical categories, then to realistic determinism - the analysis of humans in relation to their social conditioning. The shift in the 50's away from idealistic philosophy, leading to the nihilists' attacks on their predecessors, on any works of art not serving a political function, to the exaltation of science as the means to liberate humanity, the embracing of a simplistic materialism and a radical utilitarianism, was not merely a manifestation of the changing class background and the changing social position of the intelligentsia (although it was partly this), but was part of the struggle to attain a more realistic understanding of the world. These radicals wished to differentiate themselves from what they regarded as the 'superfluous generation' of the 1840's. They were struggling to be 'practical rather than "superfluous" people: students of science and servants of history.'[52] Their nihilism was not a rejection of all meaning in the world. As I have already pointed out, they remained radical

---

49. Cited in Robert C. Tucker ed. *The Lenin Anthology* N.Y.: Norton, 1975, 'Introduction: Lenin and Revolution' p.xxx.

50. N. Valentinov, *Reminiscences of Lenin*, tr. Paul Rosta and Brian Pearce, London, 1968.

51. Ibid.

52. Billington, *The Icon and the Axe*, p.390.

Neoplatonists in their conception of history. What they rejected was everything which did not serve their ambitions to transform the world.

Finally, the intelligentsia struggled to find a formula for effective organization. While the early populists thought in terms of spontaneous, decentralized activities held together by the justice of their aims, later populists attempted to develop a more unified movement. Pyotr Lavrov argued in his *Philosophical Letters* published in 1868-69 that the prime movers of history were critically thinking, justice seeking individuals who became a force through effective organization. Sergei Nechaev in his *Revolutionary Catechism* advocated the formation of a closely organized professional revolutionary cadre ready to employ ruthless and unscrupulous methods. Such methods were justified by Nechaev on the grounds that the revolutionary must despise and hate the existing ethic: 'for him, everything that allows the triumph of the revolution is moral, and everything that stands in its way is immoral.'[53] Peter Tkachev, a former associate of Nechaev became the foremost exponent of this position after the triumph of reactionary Pan-Slavism. He argued in his journal between 1875 and 1881 for the formation out of the rootless intelligentsia of a disciplined, revolutionary military organization capable of destroying the existing regime, attaining power, and effecting a revolution from above.

But effectuality was not achieved. The culmination of populism occurred in the 'mad summer' of 1874 and with the assassination of the tsar in 1881 by the 'People's Will'. In 1874 more than 2000 students dressed as peasants and set out from the cities to live among them, to join in their daily lives and to bring them the good news that a new age was dawning. They were totally rejected by the peasantry who turned many of them over to the police, and 770 were arrested. The assassination of the tsar achieved nothing but an even more repressive reaction, and all the members of the People's Will were arrested and executed. The significant feature of the populist movement was that despite their supposedly Western orientation, they were profoundly reactionary. They were simply combining elements of the three original forms of protest against Peter the Great's reforms. As Billington wrote:

> [P]opulism was a loose tradition rather than an organized movement. Like most of the Old Believers, the populists believed in preserving the old communal forms of economic life and in the imminent possibility of sudden historical change. Like the peasant insurrectionaries, the populists believed in violent action against police and bureaucrats and in the ultimate benevolence of the 'true tsar.' Even after killing Alexander II in 1881, the populists could conceive of no other program than to address utopian appeals to his successor. Like the monastic revivalists, the populists believed in ascetic self-denial and in humbling oneself before the innocently suffering Russian people.[54]

It was in this social and intellectual environment that Marxism was introduced.

## *The Reception of Marx's Ideas*

Marx's writings were received enthusiastically, but critically in Russia by the populists. Marx was seen as an economist who had revealed the exploitative, oppressive nature of Western capitalism. As such, his ideas were embraced as justification for the rejection of capitalism and the attempt to base Russian socialism not on the development of the means of production but on the peasant commune. This led to the polemical debate between Tkachev and Engels in 1874-75, in which Tkachev argued that Marx had only shown the inexorable nature of the development of capitalism once it was established, and that Russia had the opportunity to avoid capitalism before it

---

53. Cited by Walicki, *A History of Russian Thought*, p.177n.

54. Billington, *The Icon and the Axe*, p.204.

got underway and to establish communism immediately. However with the failure of the populist program and the growth of capitalism in Russia the idea that society was determined by economic development and must go through a capitalist stage was embraced and systematically argued for by the former populist Georgy Plekhanov (1856-1918). The significance of this was that a radical was arguing not for a reactionary opposition to the tsars but for the need to destroy old forms of relationships and to develop technology. The great mission of the working class, he argued, is to complete the Westernization of Russia begun by Peter the Great.[55]

Plekhanov's viewpoint was generally supported by the Legal Marxists whose defence of capitalism and opposition to populism had enabled them to legally disseminate Marx's ideas throughout Russia. The significance of Lenin was to have used Marxism to give a new direction to the activist orientation developed by the populists and symbolized by the martyrdom not only of the members of the People's Will who had assassinated Alexander II, but also by Lenin's older brother who had attempted to assassinate Alexander III. Lenin developed a voluntarist form of Marxism by rejecting the distinction between subjective and objective factors in history. He did not see Marxism as a theory of the stages of economic development, but as a theory of class struggle intimately related to praxis. For Lenin a materialist discloses class contradictions and in so doing defines his or her own stand-point. In opposition to the Legal Marxists, Lenin argued that capitalism was already definitely and irrevocably established in Russia since, despite Russia's backwardness, it was an economy based on commodity production through hired labour. The class antagonisms were those of a capitalist society and it would therefore be possible to effect a socialist revolution. The achievement of a revolution would then facilitate the development of the means of production. As Lenin put it: 'Communism is Soviet power plus the electrification of the whole country.'[56]

The communist revolution was similar to the cultural inversions which occurred in Russia with the adoption of Christianity in the tenth century and Peter the Great's Westernization of Russia in the late seventeenth century. What had been previously rejected was embraced, while what had been embraced was rejected. The unique feature of the inversion achieved by the Bolsheviks was that it was the opponents of the ruling class who had effected this inversion, and there was a double inversion - against the old ruling class and its political relations to the West, and against the old opponents of the regime. Until Lenin's inversion, revolutionaries had been, despite appearances, essentially reactionary in orientation, looking backward to traditional communalism rather than forward to the development of technology. But as in previous inversions the culture retained a fundamental continuity with its past. Assimilating Marxism to a culture pervaded by Neoplatonism involved the accentuation of the Neoplatonism of Marxism and the transforming of basic concepts of the existing Neoplatonism. In this transformation many of the forms of thinking of traditional Russian culture were also assimilated into Russian Marxism. These imposed themselves on communism like a force field constraining the possibilities open to the revolutionaries.

Marxism realigned the radical opponents of the prevailing order within Russia both in relation to other Russians and to the rest of the world. It aligned them with the new Russian proletariat who were proving to be more radical than the peasants, and the main opponents of the dynamics of the West European societies which were threatening Russia. However the most significant feature of this cultural inversion was that it provided the basis for assimilating the orientation towards action, science and technological development of Western Europe to Russian culture. Marxism represented itself as the culmination of Western European culture and extolled its scientific and technological achievements as the means to emancipate humanity, while being profoundly antithetical to the socio-political order of Western Europe, thus enabling Russians to retain their traditional hostility to Western Europe while appropriating its achievements. But more importantly Marxism fused this scientific-technological orientation with the form of Neoplatonism which underlay Russian culture. Marx was the thinker who had assimilated to the forms of thinking of Eastern Christian

55. See Walicki *A History of Russian Thought*, p.414.

56. Lenin, 'Communism and Electrification' in Tucker ed. *The Lenin Anthology*, p.492-95, p.494.

Neoplatonism, which had been taken up in the West in the ninth century by John Scotus Eriugena and developed there for a thousand years, the highly activist and technological orientation of Western Europe. The general scheme of history offered by Marx thus accorded with the basic Orthodox Neoplatonic Christian eschatology. Communism was to be the final transfiguration of the material world and the development of technology was now seen as part of the realization of heaven on earth, the process by which, according to the 'God-builders' among the Marxists, humans would become gods. This spirit of Russian Marxism was perhaps best expressed by another founder of Russian Marxism, Pavel Axelrod in a letter to Plekhanov in 1898:

> ...we shall pave the way for a race of gods on earth, of beings endowed with an all-powerful reason and will, consciousness and self-consciousness, and capable of grasping the world with their thoughts and ruling it. This is the psychological foundation of my spiritual and social strivings, of my ideas and my deeds.[57]

Furthermore since Lenin argued that Russia was in a position to begin the revolution which would sweep the world, the movement towards communism in Russia resonated with the traditional Russian conception of Russia's special historical destiny, its divine mission to consummate world history. Instead of becoming the Third Rome, Russia would become the host to the Third International. And Marx conceived humans to be both essentially social and essentially the creators of their world, thus enabling Russians to assimilate to their traditional tendency to subordinate the individual to the group the drive for technological development. As the Marxist Lunacharskii wrote in 1903:

> Man moves towards the radiant sun; he stumbles and falls into the grave. But ... in the ringing clatter of the grave-diggers' spades he hears creative labour, the great technology of man whose beginning and symbol is fire. *Mankind* will carry out his plans ... realise his desired ideal.[58]

Thus Russian Marxism integrated the traditional communalism of Russians and a conception of world-history, abandoned by the tsars since Peter the Great, with the drive to technological and scientific development.

In appropriating Marxism, the Russian Marxists also developed it. Lenin's most important conceptual innovation facilitating the assimilation of Marxism to Russian society, the innovation in terms of which all other aspects of Lenin's thought and the subsequent development of post-revolutionary Russian culture must be understood, was to conceive development in terms of the opposition between consciousness and spontaneity.[59] In his most influential work before the revolution, *What is to be Done*, Lenin described and justified the development of a revolutionary vanguard as the means of giving conscious direction to the spontaneous impulses of the oppressed workers of Russia. In this context spontaneity was equated with wildcat strikes, mass uprisings etc. without the guidance of politically aware bodies. However the whole of history was conceived in terms of the struggle between consciousness and spontaneity, between deliberate action and impersonal historical forces, progressing through a series of ever higher order syntheses towards the ultimate culmination in communism in which the opposition will be reconciled. The connotations of the Russian concept of spontaneity, *stixijnost*, which is formed from the root *stixija*, meaning 'the elements' enabled this consciousness/spontaneity dialectic to be extended to cover humanity's struggle with nature.

---

57. Cited by Ladis K.D. Kristof, 'Francis Bacon and the Marxists: Faith in the Glorious Future of Mankind' in *Society and History: Essays in Honour of Karl August Witfogel*, The Hague: Mouton, 1978, pp.233-257, p.246f.

58. Cited by Billington, *The Icon and the Axe*, p.488.

59. See Lenin 'What is to be Done' in Tucker, *The Lenin Anthology*, pp.12-115. The significance of this dialectic for the development of Russian culture is shown by Clark in *The Soviet Novel*.

This form of Marxism provided the intelligentsia of Russia with a framework for the activist asceticism on which the dynamics of Western Europe had been based and which the Russian intelligentsia had been struggling to achieve.[60] Activist asceticism was achieved and symbolized in the personality of Lenin who demanded of his followers an absolute dedication, also conceived of as a struggle of consciousness or disciplined rational awareness over spontaneity: impulse, passion and ego-centric wilfulness. This became an attractive orientation to the intelligentsia who, struggling for control of their destiny within the rapidly industrializing society of late nineteenth and early twentieth century Russia corresponded in social position to the rising bourgeoisie of early capitalist Europe who had converted to Protestantism; though unlike the Protestants who were oriented to self-advancement alone, the Russian intelligentsia were oriented towards the emancipation of the downtrodden of society. Through self-renunciation and ascetic self-discipline, Lenin's followers could experience themselves as transfigured into instruments of Providence through which the millenia would be achieved.

## *Marxism After the Revolution*

The way Marxism was understood after the revolution evolved with the problems confronted by Soviet society and with the ideological conflicts between the different factions of the Bolsheviks. In the early years, Lenin's Marxism was challenged. The concept of material existence was a particular point of contention in ideological struggles, although this was confused by the conflation of epistemological and ontological questions. Lenin's celebrated defence of materialism in *Materialism and Empirio-criticism* is in fact an epistemological argument: a defence of representational realism against the empiricists, specifically as this trend of thought was represented by the empirio-monism of Bogdanov. Defining matter, he wrote: 'Matter is a philosophical category which refers to the objective reality given to man in his sensations, - a reality which is copied, photographed, and reflected by our sensations, but which exists independently of them.'[61] He characterized idealism as a doctrine in which 'the mental is taken as the starting-point; from it external nature is inferred or constructed; and in short order the consciousness is deduced from nature.'[62] Lenin was indifferent to which theory of being is correct, being quite happy to accept that the old notion of matter defined by its impenetrability, inertia, mass and so on had been superseded and explained as relative to the behaviour of electricity.[63] The important point was that consciousness was conceived to be separate from material existence and oriented towards its control.

The dualism argued for by Lenin, which accorded with his basic conceptual dichotomy of consciousness and spontaneity, was similar to the Cartesian dualism which developed in Western Europe at the beginning of the emergence of capitalism. In both cases an activist orientation to the world led to the development of the conception of people as centres of action acting on an essentially passive world existing independently of them. This resonance reveals the extent to which Lenin's thought was an manifestation of the striving by Russians' to industrialize Russia, and the kinship between Russian Marxists and the ruling classes of capitalist societies.[64] However the rise of Marxism and the revolution was associated with the elaboration of more radical ideas and ideals. It was Bogdanov and his followers who thought out what it would mean to create a socialist society, and in doing so, they transcended the Neoplatonism of Russian culture.

---

60. An excellent study of this role of Marxism in the Soviet Union is Timothy W. Luke, 'The Proletarian Ethic and Soviet Industrialization' in *American Political Science Review*, September, 1983, Volume 77, no.3. pp.588-601.

61. V.I. Lenin, *Materialism and Empirio-Criticism*, N.Y.: International Publishers, 1927, p.101f.

62. Ibid. p.190.

63. Ibid. p.220.

64. This has been argued by Anton Pannekoek in *Lenin as Philosopher*, [1938] Merlin Press, London, 1975.

Like Western Marxists (and unlike Lenin), Bogdanov was primarily interested in people's alienation from the world and from each other and the cultural conditions for creating a socialist society, rather than in the struggle for political power.[65] To provide a philosophy appropriate for socialism, he developed the ideas of the energeticists who had been concerned to transcend the dualism between the material and the mental aspects of reality. In his work *Empiriomonism*, Bogdanov added a social dimension their epistemological ideas. He argued that the experience of the mental world was the product of individually organized experience, while the physical world was the product of socially organized experience. These two worlds reveal two different biological-organizational tendencies.[66] The conflicts of value associated with the sphere of individually organized experience are manifestations of the divisions within society based on class, race, sex, language, nationality, work specialization, and relations of domination and subordination of all kinds. It was necessary to overcome these conflicts for a new communal consciousness to emerge in which basic values could be agreed upon. But while Bogdanov accepted that it was important to transform class relations to achieve this, he argued that the importance of this had been over-emphasized by Marx. Other conflicts, including organization relations and unequal relations between the sexes, also had to be overcome. And to achieve this, it was necessary for the proletariat to transcend bourgeois culture, which he argued could only be done by creating a new culture to organize their experience.[67]

Bogdanov's critique of bourgeois culture extended to science. Anticipating later Marxist critiques, he saw the mechanical view of the world, the split between mind and matter, idealism and materialism, as expressions of the social practices of capitalist society, of the fetishism of commodities involved in market relationships and of the split between the organizational and the executive functions in the labour process. Bogdanov called for a cultural regeneration based on the modes of understanding appropriate for a society in which the divisions in society, including the division between manual and mental labour, had been overcome. The key to this was presented by him in his three volumed work, *Tektology: The Universal Oganizational Science*.[68] Tektology, for Bogdanov, was designed to provide a harmonious unity between the spiritual cultural and the physical experience of the 'working collective' in whose interest all science and activity were to be organized and all past culture, including bourgeois science, reworked. By uniting the most disparate phenomena under one conceptual scheme, tektology would allow human beings torn apart by strife to find a common language. Since the sources of strife were larger than the merely economic, the common language had to be larger than traditional Marxism, although Marxism was included as a special case.

Bogdanov's new proletarian science was a precursor to, and arguably a superior version of, the process oriented systems theory of von Bertalanffy.[69] The focus was not on what the world was made of, but on the nature of organization. Objects are distinguishable as different degrees of organization. Organized complexes or systems are composed of inter-related elements, conceived of

---

65. See Zenovia A. Sochor, *Revolution and Culture: The Bogdanov-Lenin Controversy*, Cornell University, 1988. See also Robert C. Williams, *The Other Bolsheviks: Lenin and His Critics, 1904 - 1914*, Bloomington and Indianapolis: Indiana University Press, 1986.

66. See Kenneth Jensen, *Beyond Marx and Mach: Aleksandr Bogdanov's Philosophy of Living Experience*. Dordrecht, Holland: D. Reidel, 1978.

67. It appears that Gramsci's ideas ultimately derived from Bogdanov. See Zenovia A. Sochor, 'Was Bogdanov Russia's Answer to Gramsci?' *Studies in Soviet Thought*, Vol. 22, (Feb. 1981), pp.59-81.

68. This has not been translated. However a good idea of his philosophy can be gained from *Essays in Tektology: The General Science of Organization*, tr. George Gorelik, Seaside, Calif.: Intersystems Publications, 1980.

69. See Ilmari Susiluoto, *The Origins and Development of Systems Thinking in the Soviet Union: Political and Philosophical Controversies from Bogdanov and Bukharin to Present-Day Re-evaluations*, Helsinki: Suomalainen Tiedeakatemia, 1982; George Gorelik, 'Bogdanov's Tektology: Its Nature, Development and Influence', *Studies in Soviet Thought*, Vol. 26, (1983).

as activities, such that the whole is greater than the sum of its parts. Living beings and automatic machines are dynamically structured complexes in which 'bi-regulators' provide for the maintenance of order. Bogdanov argued that no matter how different the various elements of the universe - electrons, atoms, things, people, ideas, planets, stars - and regardless of the considerable differences in their combinations, it is possible to establish a small number of general methods by which any of these elements joins with another.

In the early years of the revolution, Bogdanov inspired and largely organized the *Proletkul't* movement which gained 400,000 members, published twenty journals, and attracted the support of a wide section of the Russia's artists, musicians and writers.[70] Bogdanov also established a proletarian university in Moscow with 450 students. He defended such activity in a time of crisis on the grounds that only through a cultural transformation could socialism be achieved. The differences between Leninism and the ideals of Proletkul't were most clearly manifest in the efforts to develop a work ethic. All Russian Marxists were concerned to develop an activist orientation in everyday life, to overcome the slovenliness of Russian workers. Lenin called upon the cadres of the communist party to 'teach people how to work', to develop a new proletarian work ethic in which work would be undertaken as virtuous habit, a transformation which, according to Trotsky, was to seal 'the people's final break with the Asiatic, with the seventeenth century, with Holy Russia, with icons and cockroaches'.[71] The work ethic was propagated through mass educational offensives, with the *Central Labour Institute* organized by A.K. Gastev supported by Lenin promoting Western practices, and the *Time League*, striving to create a new orientation to work appropriate for a socialist society.[72]

The *Central Labour Institute* aimed at a total mechanization of human life on the foundations of Taylorism and Pavlovian psychology. Gastev wanted to reform human psychology, merge Marxism with American practicality, eliminate education in the humanities in favour of technical, practical knowledge, replace universalism with specialism, and adjust individuals to make them into suitable machine parts for the total organization by conditioning people's wills, minds, and bodies. As he described his ideal of scientific organization:

> Before us there is the prospect not only of an individual mechanized worker, but of a mechanized system of labour management. Not a person, not an authority, but a 'type' - a group - will manage other 'types' or groups. Or even a machine, in the literal sense of the word, will manage living people. Machines, from being managed, will become managers.[73]

Members of the *Time League* criticised such measures, arguing that they would facilitate a new kind of subordination, and promoted the application of a new kind of scientific organization to all spheres of human endeavour. Stressing the need for self-discipline rather than the reduction of people to objects to be controlled, they focussed on the organization of time, entreating workers to: 'Measure your time, control it! Do everything on time! exactly, on the minute! Save time, make time count, work fast! Divide your time correctly, time for work and time for leisure! Utilize your leisure so as to work better afterwards!'[74]

The *Proletkul't* movement was attacked by Lenin who republished his *Materialism and Empirio-criticism* to undermine Bogdanov's authority. With Lenin's support, the *Central Labour Institute* prevailed over the *Time League*. Late in 1920 Lenin forced the subordination of the hitherto free-

70. On this movement see Lynn Mally, *Culture of the Future: The Proletkult Movement in Revolutionary Russia*, Berkeley: University of California Press, 1990.

71. Cited Luke, 'The Proletarian Ethic...' p.596.

72. On the conflict between these two, see Zenovia A. Sochor, 'Soviet Taylorism Revisited', *Soviet Studies*, Vol. XXXIII, no.2, April 1981, pp.246-264.

73. Cited by Susiluoto, *The Origins and Development of Systems Thinking in the Soviet Union*, p.106.

74. Cited Luke, 'The Proletarian Ethic...' p.598.

wheeling *Proletkul't* to the People's Commissariat of Education (or Enlightenment) (*Narkompros*), and it was later abolished altogether. As it became evident that the rest of Europe was not going to follow Russia and that a socialist organization could not easily be imposed on the peasantry, Lenin shelved utopian ideas, and to consolidate the revolution promoted the New Economic Policy which was adopted in 1921. This was characterized by limited capitalism controlled by the State. To maintain control of the State in a capitalist society, all other political parties and all factions within the Communist Party were banned. While people associated with the 'Worker's Opposition' were inspired by *Proletkul't* to oppose the 'return to capitalism' of the N.E.P., and also Trotsky's call for a militarization of society based on the principles of war communism, and called for workers' control in the factories, they had little success. By the time *Tektology* was completed in 1922, Bogdanov's prestige had been almost destroyed, though he continued to have some influence, particularly through Lunacharsky, a supporter of Bogdanov's philosophy, who until 1929 was the Commissar of Education.[75]

Lenin's backtracking from socialism was justified by arguing that it is necessary to work in accordance with the dynamics of the world. In terms of his philosophy, spontaneity was given pre-eminence over consciousness. During this period, a mechanistic world-view was promoted within educational institutions, and Marxism was interpreted accordingly. The prevailing interpretation of Marxism was Bukharin's, essentially a mechanistic version of systems theory emphasising the equilibrium of systems. According to this, oppression and class conflict are caused by the economic base of society and are therefore eliminable through its transformation. In psychology, physiological and behaviourist approaches to humans were adopted almost exclusively. Pavlov's ideas on the reflex arc dominated, and those psychologists focussing on consciousness were condemned as idealists. People were seen as products of their environments and biological constitutions, and there was no acknowledgement of the possibility of individuals transcending the conditions of their existence. As Raymond Bauer wrote: 'In the psychologies of the twenties, man was an adaptive mechanism that responded to external forces in such a way as to maintain an equilibrium between himself and his environment.'[76] Those supporting the revolution upheld the primacy of the environment as the determinant of abilities. In relation to the dynamics of society, the future was thus seen to be determined by forces external to individuals. However in the new educational and research institutes established by the Commissariat of Education, provided bases for the proponents of an essentially proletarian culture based on dialectics.[77]

## *The Rise of Stalin*

While many radical breaks were being made with the past, traditional Russian culture continued to influence both Marxists and the general population. This was evident in the struggles of Lenin to oppose the religious terminology of the God-builders on the one hand, and the tendency for people to treat him as a new tsar on the other. Though Lenin imposed and upheld the dictatorship of a small revolutionary elite, suppressing both parties opposing the Communist Party and factions within it, while he was alive there was no office of supreme leader in the Soviet system. The highest party organs were the Central Committee and its subcommittee, the Politburo, and Lenin was officially an ordinary member of these. Decisions in each were taken by majority vote. Lenin had no more than one vote and did not expect people to agree with him. He advocated this system and took care to uphold it in practice, and he would resolve differences between himself and subordinate

75. On Lunacharsky's policies and influence, see Sheila Fitzpatrick, *The Commissariat for Enlightenment*, Cambridge: C.U.P., 1970. This also contains a chapter on *Proletkul't*.

76. Raymond A. Bauer, *The New Man in Soviet Psychology*, Cambridge, Harvard University Press, 1952, p.75.

77. This development, and the struggle between the mechanists and the dialecticians has been described by David Joravsky in *Soviet Marxism and Natural Science*, London: Routledge & Kegan Paul, 1961.

government leaders by referring the issue to the Politburo for a decision by majority vote. At the Tenth Party Congress in 1921, Lenin gave his party office as 'member of the Central Committee'. But to the people Lenin was the personification of political power, the source of divine light and the icon of the Deity. A reporter to the *New York Times*, Walter Duranty wrote: 'I have seen Lenin speak to his followers. ... I turned round and their faces were shining, like men who looked on God.'[78] Ignazio Silone who saw Lenin in 1921 recalls that 'whenever he came into the hall, the atmosphere changed, became electric. It was a physical, almost a palpable phenomenon. He generated contagious enthusiasm the way the faithful in St. Peter's, when they crowd round the Sedia, emanate a fervour that spreads like a wave throughout the basilica.'[79]

Lenin abhorred this. Recovering in 1918 after an attempted assassination, he was horrified by what had been printed in the press. His attitude is evident in his exclamation to his aide, V.D. Bonch-Bruevich:

> What is this? How could you permit it? Look what they are saying in the papers. Makes one ashamed to read it. They write that I'm such-and-such, exaggerate everything, call me a genius, a special kind of man. And look at this piece of mysticism: they collectively wish, demand, and desire that I get well. Next they'll be holding public prayers for my health. Why, this is horrible![80]

But the fact was that Lenin himself, with all his protestations against old forms of thinking, was being assimilated into the basic forms of traditional Russian culture.

This, along with Lenin's destruction of the *Proletkul't* movement, made it very easy for Stalin to re-invert Russian culture, to embrace the traditional Russian culture as sacred and to condemn Western forms of thinking as profane. In contrast to Lenin, Stalin was always prepared to exploit the legitimating power of traditional Russian culture to the full. A former seminarian educated in the catechistic theology of Orthodoxy, he was much more in tune with this traditional Russian culture than the other Bolshevik leaders - many of whom were of Jewish, Polish or Baltic origin, and his rise to, and maintenance of power was at least partly due to his willingness and ability to accord with and use traditional cultural forms to legitimate himself and his actions. To begin with, Stalin's way of arguing accorded with the way of thinking of ordinary people. As Martin McCauley wrote about him in his struggle with Trotsky, 'He had a knack of communicating easily with the run-of-the-mill party member, whereas Trotsky appeared to be addressing the angels most of the time as no one on earth could follow him.'[81] The symbolic universes were contrived to accord with Russian tradition. Lenin was embalmed and laid out for public veneration with hands folded in the manner of the saints in the monastery of the caves of Kiev, something which embarrassed all the leading Bolsheviks except Stalin. As Stalin rose to power, his pictures took the place of holy icons, and art and literature were cultivated to take the place of icons and the lives of the saints, with literature being placed under Party control to ensure that it served the revolution.[82] Socialist realist works were expected to provide the ideal forms for people to strive to imitate, and artists and writers were directed what to produce. The evolution of the Socialist Realist novel thus came to reflect the evolution of Soviet ideology.[83]

Having used traditional Russian culture to gain power, Stalin was then able to use it to redirect the revolution. The communists had stressed the conception of society as a 'great family' in

78. Cited by Tucker in his Introduction to *The Lenin Anthology*, p.lix.

79. Cited loc.cit.

80. Cited ibid. p.lx.

81. Martin McCauley, *The Soviet Union Since 1917*, London: Longman, 1981, p.60.

82. For a description of literature in the 1920's leading up to its co-optation by the Party see Robert A. Macguire, *Red Virgin Soil: Soviet Literature in the 1920's*, Princeton: Princeton University Press, 1968.

83. As shown by Clark in *The Soviet Novel*.

accordance with Russia's traditional communal orientation. But while originally the horizontal axis of brotherhood was emphasised, Stalin twisted this axis to emphasise the hierarchical aspects of the family. Socialist realist novels abounded in heroes whose lives have been changed by contact with the fatherly figure of a political leader within the Communist Party, while Stalin was presented as the great father. The aim presented to the general male population was to become 'good sons' to the almighty father of the 'Great Family'. This hierarchical conception of the family resonated with the hierarchical Neoplatonic framework of the culture as a whole inherited from the tsarist past.[84] Stalin as a supra-terrestrial being was held to have access to a higher order truth, a truth which had been passed on to him by the original father of the revolution, Lenin. Access to this truth could be attained by model sons, but they could only grasp intuitively and inchoately and with the father's guidance the forms of higher level knowledge to which the father had complete access. As Katarina Clark wrote of this:

> Lenin passed his 'light' and 'mystery' on to Stalin. Now Stalin was passing it on to his chosen few. The myth of the 'great family' provides not only for a succession of generations but a chain of kairotic moments akin to the laying-on of hands in a church adhering to the doctrine of the apostolic succession. ... For the time being, however, the chain is not infinite. Not all are able to receive the 'mystery' and 'light' that the leaders have to give. ... In Stalinist culture of the thirties there were, then, two orders of reality, ordinary and extraordinary, and, correspondingly, two orders of human being, of time, of place, and so on. Ordinary reality was considered valuable only as it could be seen to reflect some form, or ideal essence, found in higher-order reality.[85]

## *The Cultural Revolution*

The world-orientation of Soviet Marxism finally crystallized with the Cultural Revolution of 1928-1931 associated with the First Five-Year Plan, and what emerged from this was essentially a refurbished form of the nihilism of the radicals of the 1860's.

While the mechanistic conception of being dominated until 1928, the threat of war with England, a breakdown in food acquisition from the peasantry and pressure from the working class and students who saw the N.E.P. as a betrayal of the revolution, led Stalin to the conclusion that the collectivization of agriculture was necessary. He then attempted to have the economy organized on the basis of five year plans.[86] From this point onwards, the superiority of communism was seen to rest not on its having overcome a repressive society, but on its superiority for developing the means of production. This was associated with the Cultural Revolution in which the members of the Party struggled to attain control of the positions of power in the sciences and arts and to proletarianize and socialize culture. What they struggled to effect was another cultural inversion. While under the N.E.P. spontaneity was extolled and consciousness was denigrated, under the new order consciousness was extolled and spontaneity denigrated.

---

84. Leszek Kolakowski in *Main Currents of Marxism* represents Marxism and socialism as essentially Neoplatonist and blames them as such for the Stalinist developments in Russia.

85. Clark, *The Soviet Novel*, p.145f.

86. See Alec Nove, *An Economic History of the U.S.S.R.* Harmondsworth, 1982, Ch.'s 7 & 8; Maurice Dobb, *Soviet Economic Development Since 1917*, 6th ed. London: Routledge & Kegan Paul, 1966, Ch.10; Sheila Fitzpatrick, *The Russian Revolution: 1917-1932* Oxford: Oxford University Press, 1982, Ch.5. For the cultural revolution associated with this see Sheila Fitzpatrick ed. *Cultural Revolution in Russia: 1928-1931*, [1978] Bloomington: Indiana University Press, 1984.

To begin with Stalin remained wedded to the metaphor of the machine. He cultivated the form of the machine as the ideal to be realized. The machine was taken to stand for harmony, progress and control, while that which was not integrated into the machine was condemned as chaos, hard labour, primordial and lacking in rhythm. This ideal was used to justify the collectivization of agriculture into large scale, highly mechanized operations subject to central planning. Society was a 'train' rushing to catch up a hundred years of Western development in ten years, and a 'planned city' in which everything was scientifically coordinated and the latest technology used. But the machine did not fit in with traditional Russian culture. It aroused suspicion, it was too impersonal and it gave no place for the centralized, guiding role of the Party. It was also inconsistent with the dialectic of spontaneity and consciousness espoused by Lenin. For these reasons it was soon replaced by the image of the 'Struggle with Nature,' associated with which people were exhorted to overcome all obstacles, to storm and break traditional limits in order to achieve society's ends; that is, to make consciousness dominate over spontaneity. On the basis of this image the view was promulgated that anything can be accomplished; the laws of science are only blinkers imposed upon people to prevent them reaching their full potential.

With the Cultural Revolution, the dialectical materialist philosophers were able to gain positions of power and to make their views prevail over those of the mechanists. The term 'dialectical materialism' which has become the official theory of being of Soviet Marxism, was coined by Plekhanov, but apart from interpreting Marxism in terms of Spinoza, conceiving matter and thought as two aspects of the one reality, Plekhanov did not speculate on the nature of matter. However the term was taken up by other Marxists, led by Deborin, who set about elaborating on Engels' philosophy of nature. Attacking the mechanistic conception of being for its reductionist implications, they followed Engels in arguing that matter is essentially active and that it generates qualitatively new levels of being which must be understood according to their own specific laws. Their intellectual credentials were reinforced by the publication of Engels *Dialectics of Nature* in 1925 and then Lenin's *Philosophical Notebooks* in 1929, but the real reason why they were able to gain positions of power was that they gave a far greater role to consciousness than the mechanists. Consciousness was seen by them to be irreducible to biology or behaviour, and capable of acting according to its own principles. This legitimated the rejection of the principle of equilibrium, associated with Bukharin's defence of the N.E.P., as a projection of biology onto a higher level of being. The dialecticians justified the primacy of consciousness over spontaneity, the demands being made by the Party for a radical break with the past, and the struggle to consciously transform society and nature.

The significance of the Deborinites went beyond this. They had successfully promoted the view, which had originally been put forward by Bogdanov, that there is a socialist science different in character from bourgeois science, and that the Communist Party was entitled to ensure that scientists developed their ideas along Marxist lines. This paved the way for the attempt by the Party to effect far-reaching control over the sciences to create a specifically Soviet science in opposition to Western science.

The reign of the Deborinites was short-lived. With the failures of the first five-year plan and changes in the West increasingly threatening Russia, Stalin intensified the struggle for rapid economic development. This was associated with an increasingly anti-Western attitude and with a growing emphasis on Russian nationalism. His attitude was expressed in his famous 1931 speech calling for the full mobilization of society:

> One feature of old Russia was the continual beatings she suffered because of her backwardness. She was beaten by the Mongol khans. She was beaten by the Turkish beys. She was beaten by the Swedish feudal rulers. She was beaten by the Polish and Lithuanian gentry. She was beaten by British and French capitalists. She was beaten by Japanese barons. All beat her - because of her backwardness, because of her military backwardness, cultural backwardness, industrial backwardness, agricultural backwardness... We are fifty to one hundred years behind the

> advanced countries. We must make good this distance in ten years. Either we do it or we shall go under.[87]

Responding to this new climate, Deborin and his colleagues were attacked for showing insufficient party spirit by a band of younger party activists led by M.B. Mitin at the second philosophical conference in April 1930. They were charged in particular with 'separating theory from practice'. Stalin labelled their position 'menshevising idealism' and their fate was sealed. They were purged from the party in January, 1931.

While the mechanists had been knowledgeable about science but relatively ignorant about philosophy and the Deborinites had been knowledgeable about philosophy but relatively ignorant of science, Mitin and his colleagues constructed a version of dialectical materialism which synthesized the ignorance of each.[88] While they did formulate a version of dialectical materialism, it was not the conception of being which was taken as defining the socialist science as Deborin had believed. The real defining feature of socialist science and of the philosophy of Mitin and his colleagues was a revival of the views of the Russian nihilists of the 1860's who had argued for the total subordination of science to technology, and the elimination of everything which did not serve a strictly utilitarian function for the Revolution.

According to this version of the unity of theory and practice, practice is 'the basis of knowledge and the touchstone of truth'. Knowledge only has significance and is only to be pursued for practical and technological reasons, and practical efficacy is the ultimate test of the hypotheses on which action is based. This was formulated to accord with Lenin's reflection theory of knowledge in opposition to the 'hieroglyphic' theory of knowledge of Plekhanov which had been upheld by Deborin. This pragmatic theory of knowledge, which was very similar to that developed in America by William James under the influence of Darwinian evolutionary theory, ultimately led to the view that what is true is what is good for the development of communism. Everything, including truth, came to be measured in terms of its contribution to the goals of the Communist Party. As Ernst Kol'man, a leading agent of the great break in philosophy and science declaimed:

> Now it is clear to everyone that the basic lesson of the philosophical discussion is this: philosophy, and every other science as well, cannot exist in the conditions of the proletarian dictatorship separate from the Party leadership. Now it is clear to everyone that all efforts to think of any theory, of any scholarly discipline, as autonomous, as an independent discipline, objectively signify opposition to the Party's general line, opposition to the dictatorship of the proletariat.[89]

It is important to note in relation to evaluating the contribution of Marxism to the development of Stalinism that this theory of knowledge has little to do with either Marx's notion of the unity of theory and practice outlined in his *Eleven Theses on Feuerbach*, nor with Lenin's philosophy. Marx did not reduce the status of theory but pointed out that since people and their theories are part of the world and theories change the world by affecting people's behaviour, this must be taken into account in theory.[90] Social theory must struggle to articulate and express the problems and aspirations of people and reveal how, through this new consciousness, they can change the world. This is inconsistent with a reflection theory of knowledge with its implicit dualism between mind and world and its reduction of truth to a means for realizing the millenium. And the narrowly

---

87. J.V. Stalin, *Works*, Vol.13, Moscow: Progress Publishers, 1955, p.40-41.

88. See Leszek Kolakowski, *Main Currents of Marxism*, Volume 3, 'The Breakdown', p.65ff.

89. Cited by David Joravsky in 'The Construction of the Stalinist Psyche' in Sheila Fitzpatrick ed. *Cultural Revolution...* p.109.

90. See Leszek Kolakowski, 'Karl Marx and the Classical Definition of Truth' in *Marxism and Beyond* tr. Jane Zielonko Peel, London: Paladin 1971, pp.59-87 on this.

utilitarian view of science was at odds with the ideas of Lenin, who in opposition to Bogdanov also rejected the whole idea of a specifically socialist science. The reduction of knowledge to an instrument of power was condemned by Lenin's wife, Krupskaia, as 'a naive, idiotic conception of the matter.'[91]

These developments in philosophy inspired attacks on mechanistic psychology for its failure to deal with consciousness. Such psychology had presented a view of people which was far too passive for a society in which they were supposed to be transforming the world. This inverted the previous state of affairs where those psychologists who had focussed on consciousness had lost or were in danger of losing their positions. As one Soviet psychologist wrote: 'That which I had considered my virtue - regarding objective reality as the direct source of the laws of psychological development - became its opposite, or nearly so.'[92] However even psychologists who had argued for the reality and causal significance of consciousness such as Vygotskii were criticised for relying too heavily on concepts of adaptation and equilibrium. Vygotskii had argued that the child grows and develops in the process of accommodating to disequilibrating forces, while his critics argued that the initiative for action lies with the individual alone independently of his or her environment. The psychologists were to produce a theoretical model of a conscious, purposeful builder of socialism. The two factor theory of development according to which behaviour is determined by heredity and environment gave way to a three factor theory according to which behaviour is determined by heredity, environment and training, and then by a four factor theory which also included self-training, the shaping by people of their own character.[93]

Similarly, from 1930, onwards the natural sciences were also reduced to instruments of the Party. The attack against the scientific establishment was led by I. I. Prezent and his main follower, Lysenko, on grounds that they were promoting ideas which implied that there are limitations to the dominion of humans over nature. Most well known of the theories attacked was the Mendel and Morgan theory of heredity - which the Deborinites had supported. This theory implied that there are limits to what species can be acclimatized to Russian conditions. However the attack on these theories had been preceded by attacks on community ecology, for virtually the same reasons.

Associated with these developments there were new demands placed on education and training in industry. By 1931 the Central Committee of the Party was clamouring for 'completely educated men possessing a good foundation in the sciences'[94] in place men trained by rote in a restricted range of mechanical skills. In 1935 Stalin instructed that the old slogan 'Technique decides everything' be replaced by a new slogan, 'Cadres decide everything.'[95] In Socialist Realist literature individuals were extolled for showing initiative to battle against the elements and red tape to achieve outstanding developments in industry. Writers were instructed to create a literature of 'revolutionary romanticism' in place of bourgeois literature which depicts the small deeds of small people.

The general pattern for these novels has been described by Katarina Clark.[96] In brief they begin with the hero arriving at a microcosm, seeing that all is not right (the state plan is not being fulfilled) and concocting a scheme for righting the wrong which is then rejected by the local bureaucrats. The hero defies the bureaucrats and mobilizes the people, and work on the project begins. With snags in this and problems in the hero's love life the hero seeks help from a more authoritative figure. A dramatic/heroic obstacle associated with an actual, symbolic or near death leads to grave self-doubt on the part of the hero. The hero talks with his local mentor and this gives him the strength to carry on to the completion of the task. The completion is associated with the

---

91. Cited by Douglas Robert Weiner, *The History of the Conservation Movement in Russia and the U.S.S.R. from its Origin to the Stalin Period*, Ph.D. Thesis, Columbia University, 1983, p.362.

92. M. Ia. Basov, *Pedologiia*, 1931, nos. 5-6, p.16; cited by Bauer *The New Man in Soviet Psychology*, p.102.

93. Ibid. p.149. See also David Joravsky, *Russian Psychology: A Critical History*, Oxford: Blackwell, 1989.

94. Ibid. p.100.

95. Ibid. p.46.

96. Clark, *The Soviet Novel*, Appendix A, pp.255-260.

resolution of the emotional problems, the hero transcending his selfish impulses and acquiring an extrapersonal identity. A funeral is held for the victim killed during the climax, or alternatively the protagonists visit their fallen comrade's grave, and they make speeches. There is then a reshuffling of personnel in the microcosm with the hero frequently being promoted to the post formerly held by his mentor. The theme of regeneration and the glorious time that awaits future generations is introduced at the completion of the task as a thematic counterpoint to sacrifice and death.

These developments further accentuated the Neoplatonism of Marxism. History was hypostatised and treated as a subject using people as willing instruments in its struggle to attain the millenium. This was dramatically illustrated even by the opponents of Stalin, for instance in the 'confession' of Bukharin in 1937 before his execution. Bukharin had originally developed a version of Marxism in terms of a version of systems theory. However while on trial he defined his position from the point of view of the world-historical process. As he stated: 'World history is a world court of judgement ...'[97] Treating history as a judge, Bukharin was left to conceive himself as nothing but a rejected instrument of history. This means in effect that to be right is to be successful, precisely the same ethic as Social Darwinism.

The reformulation of Marxism into an anti-Western ideology was translated into work, educational and social practices, a process which led to the reassertion of many traditional Russian cultural practices against the efforts of those who had striven to transform Russians according to Western principles. In fact the main bearers of these Western principles, the radical intelligentsia, were a major component of the several hundred thousand people who were executed in the purges of the 1930's, and most of the remainder found themselves among the 4½ to 5 million prisoners who became a virtual class of slaves in the forced labour camps of the Gulag. And the new class of intellectuals and administrators who rose from the ranks of the working class and peasantry to take control of the society embodied a fusion of traditional Russian orientations and the technicist rationality of the nihilists both towards nature and towards people. This reversion manifested itself in the Stakhanovite movement, in which workers were made heroes for vastly overfilling production quotas, which began in 1935. It was associated with a rejection of the effort to develop a generalized work-discipline and was a return to a more traditional approach of exalting the exceptional. Similarly in 1936 pedologists with their batteries of tests were dropped from the education system. While failure had previously been blamed on heredity or environment, it was henceforth demanded of students that they succeed whatever the external limitations, although success had to be defined from above. While initiative was encouraged, this was supposed always to serve the party.

In accordance with this orientation, efforts were made to develop and to inculcate an ethics of service to society, to the country and ultimately, to the realization of socialism, an ethic which has persisted up to the rise to power of Gorbachev.[98] The account of communist morality by V. Afanasyev in his popular exposition of Marxist philosophy is a typical expression of this:

> Communist morality ... is subordinated to the interests of the proletariat's class struggle. *Its content and aim is to build and consolidate communism*. It is this idea which underlies the *moral code of the builder of communism*, formulated in the Programme of the C.P.S.U. *Devotion to the cause of communism, love for the socialist Motherland* which blazes for mankind the trail into the communist morrow, *love for all socialist countries*, is the first, cardinal demand in the moral code of the Soviet citizen.[99]

---

97. Maurice Merleau-Ponty, *Humanism and Terror*, [1947] tr. John O'Neill, Boston: Beacon, 1969, p.62.

98. The place of ethics in Soviet Marxism has been analysed by Herbert Marcuse in *Soviet Marxism: A Critical Analysis* [1958], Harmondsworth: Penguin, 1971, Part II.

99. V. Afanasyev, *Marxist Philosophy: A Popular Outline*, [1963], Moscow: Progress Publishers, 1968, p.338.

The most important feature of this moral service was seen to be 'conscientious labour for the good of society.'[100] In other words in accordance with the traditional Russian tendency to subordinate the individual to the group and with the Neoplatonic emphasis of Soviet Marxism, people were required to become willing instruments of Providence, represented by the Communist Party, for the creation of the order on earth to be achieved through the transfiguration of nature by industry.

With these developments, the brilliant intellectual life of Russia which had developed in the nineteenth century and had flowered in the early years of the revolution, was virtually snuffed out. As David Joravsky described this change:

> From autonomous critics of the existing system, seeking an integral understanding of the universe and of human destiny as the first step to reform or revolution, the intelligentsia has been transformed into a class of obedient servants of the exiting system, performing specialized mental labour for specified rates of pay.[101]

With the Second World War the nationalization of Soviet Marxism and the mobilization of the population for service to society was completed, with Stalin successfully appealing to Russians to fight for the Soviet motherland.

This mobilization was successful to the extent that the Soviet Union was able to defeat the Nazis in the Second World War (although given Stalin's massive blunders, including purging the army of its best commanders just before the war, then refusing to prepare for the German attack, despite precise intelligence reports on when it would take place, orders not to retreat, leading to the encirclement of Soviet troops, and so on, there is no reason to believe Stalinism as such was required for this victory). Then, after having endured the massive mobilization of society before the war with its associated famines and starvation, after having had twenty million people killed and much of the pre-war achievements destroyed during the war, under constant threat of nuclear attack from the United States, they were able to again rebuild their country.

To achieve this, they had not only developed an economic base, but appear to have had some success in changing the mode of being of Russians. As Bauer wrote of emigres in the early 1950's in comparison to emigres before or immediately after the Revolution:

> They are more practical and less contemplative; more concerned with results and less with the means whereby they are gained. They are more manipulative and better extemporizers. Rationality is more prominent and emotion less so. They are more militantly self-confident. They exhibit, in short, the 'reflex of purpose' which Pavlov found lacking in the Russian.[102]

However such changes must not be over-emphasised. In the conclusion to his study of the Russians, the journalist Hedrick Smith noted the continuity of the traditional Russian character: 'the centralized concentration of power, the fetish of rank, the xenophobia of simple people, the futile carping of alienated intelligentsia, the passionate attachment of the Russians to Mother Russia, the habitual submission of the masses to the Supreme Leader and their unquestioned acceptance of the yawning gulf between the Ruler and the Ruled.'[103] The contrast between Estonians and Russians provides a good measure of the limited success in the efforts of the Communists to transform the Russian habitus.

---

100. Loc.cit.

101. David Joravsky, 'The Construction of the Stalinist Psyche' in Fitzpatrick, *Cultural Revolution...*, p.106.

102. Bauer, *The New Man in Soviet Psychology*, p.182.

103. Hedrick Smith, *The Russians*, London: Sphere Books, 1976, p.614.

## *From Khrushchev to Gorbachev*

After the death of Stalin in 1953 ideological conflicts took the form of a struggle between Stalinists, the anti-semitic, xenophobic, essentially anti-Marxist nationalists oriented towards achieving central control over society and expanding the international power of the Soviet Union; and Leninists, the people who took the ideals of Marxism seriously and tried to liberalize society and decentralize power, and who were outward looking and tried to reduce the tensions of the Cold War.[104] The most extreme of these Stalinists have been the Russophiles or 'Russites', the heirs of Social Darwinist Slavophiles of the late 19th century, described by Christian Schmidt-Häuer:

> To them, Lenin is suspect. For him, the extension of the Russian empire was a means to achieve world revolution, not an end in itself. It was Stalin who re-asserted Russian hegemony, transforming the Comintern (the Communist International) into an instrument for the expansion of Russia. For this reason, the Russites see Stalin as a true expression of Russian history, and his purges as the cleansing of the homeland from Western, Marxist and Jewish subversion.[105]

However behind the more extreme form of Stalinism characteristic of the Russites, an increasingly hereditary class of bureaucrats tended towards Stalinism not as an expression of nationalism, but as a means of suppressing critics of their privileges and incompetence. Marxism allowed this class to present itself not as a privileged class, but as the representatives of the proletariat of the world, so that any criticism of them or of Russia's exploitation of national minorities in the Soviet Union or of Russia's Eastern European allies, could be condemned as treason against the international working class and against socialism.

The major events in the conflict between the Stalinists and the Leninists were the rise to power of Khrushchev and his denunciation of Stalin in 1956, the deposing of Khrushchev, which was engineered by the Russites, and the domination of political life by Suslov, Kosygin and Brezhnev, and then the deaths of Suslov, Kosygin and Brezhnev and the rise to power of Andropov and later, Gorbachev.

During the period when Khrushchev was premier between 1958 and 1964, the degree of control over literature, philosophy and the sciences was relaxed. Individuals were no longer expected to defy the laws of nature, and a more consumerist orientation developed in society. The Socialist Realist novels portrayed and celebrated better educated, better dressed and more senior members of the Communist Party. After the death of Stalin the main characters tended to be less heroic, and there was some exploration of more complex issues such as the relationship between individual initiative and discipline. Systems theory and cybernetics were slowly revived as the basis for management, just as they have been developed in the West.[106] A more polished version of the vulgarized dialectical materialism of Mitin and his colleagues became the orthodox position in Soviet philosophy and science, with a basic opposition emerging between the more orthodox 'Aristotelians' and the more radical 'Hegelians'.[107] The Hegelians promoted Lenin's *Philosophical*

---

104. See Christian Schmidt-Häuer, *Gorbachev: The Path to Power* tr. Ewald Osers and Chris Romberg, London: Pan Books, 1986, Ch.4. A good idea of the nature of the political struggles in the Soviet Union between 1964 and 1971 can be gained from the translations of the *samazadat* (self-publishing) *Political Diary* edited by Roy Medvedev. These are contained in Stephen F. Cohen ed. *An End to Silence*, tr. George Saunders, N.Y.: Norton, 1982.

105. Schmidt-Häuer, *Gorbachev: The Path to Power*, p.74f.

106. As noted by Susiluoto, *The Origins and Development of Systems Thinking in the Soviet Union*, Ch.11.

107. The basic position is espoused by Afanasyev, *Marxist Philosophy*. A more recent and polished version is presented in the works of E.V. Ilyenkov, in particular, *Dialectical Logic*, Progress Publishers, Moscow, 1977. The classic Western study of Soviet philosophy is Gustav Wetter, *Dialectical Materialism*, tr. Peter Heath, London, 1958. The developments and problems of dialectical materialism have been analysed by N. Lobkowicz in 'Materialism and Matter in Marxist-Leninism' in Ernan McMullin ed. *The Concept of Matter in Modern Philosophy*, Notre Dame: University of Notre Dame

*Notebooks* as against his more mechanistic *Materialism and Empirio-Criticism*. But science continued to be regarded as the means to control nature (though a role in this has been found for 'basic science' to replace 'pure science'), and dialectical materialism continued to be formulated to accord with the reflection theory of knowledge and the dualism between materialism and idealism promoted by Lenin. Since thought is seen to be reflecting the material world, and idealism is understood as denying that contents of consciousness are really representations of something existing independently of thought, this has led to the maintenance of the dualism between thought and matter.

In this philosophical environment, consciousness retained a more significant role in Soviet psychology than in Western psychology, although as in Western psychology cybernetics was incorporated into research.[108] Pavlov was rehabilitated as a hero of Soviet science, but as a biologist rather than as a psychologist, and a more positive attitude was taken to the work of Vygotskii. In economic thought there was a veritable revolution in thinking as the input/output models of the economy, originally developed by Leontief and Fel'dman in the 1920's, were revived as a basis for planning the economy, linear planning, which was also originally developed within the Soviet Union, came to be used as a general means for calculating the most efficient way of using resources, economists strove to work out how the criterion of utility could be incorporated into planning and how markets could be developed to improve efficiency, and the Stalinist dogma that steady growth requires that the investment sector of the economy grow faster than the consumption sector was laid to rest.[109] In social science, historical materialism came to be understood more in accordance with the systems theory of Bukharin, and the role of consciousness was downgraded. As Helmuth Fleisher described Soviet social science:

> We are not told that producers and managers are faced with professional politicians, legislators and administrators, but that 'the economy' determines 'politics', that the latter has repercussions on the economic base, and that base and superstructure influence each other with unequal determinative force. We are not told that men, who among other things work, consume and quarrel, and in the process develop theoretical ideas and make practical plans about the objects of their environment ... but that economic development produces ideas that in turn influence economic development and play an 'active role'.[110]

The state of social thought can be judged from the enormous popularity achieved by the sociological theories of Talcott Parsons.

Khrushchev attempted to liberalize Soviet society and achieve a rapprochement with the West. He successfully downgraded the status of Stalin and inaugurated a flowering of new ideas in virtually all intellectual fields. But he failed to break down the concentrations of power which had developed under Stalin, and the limitations of centralized planning became evident for the first time. The economy had become too complex. The expulsion of Khrushchev from power was engineered by Suslov, a patron of the Russites, who until his death in 1982 was a major driving force in insulating Soviet society from the West. The reign of Brezhnev saw the further entrenchment of the

---

Press, 1963, pp.154-188. The struggles within Soviet philosophy have been described by Eugene Kamenka in 'Soviet Philosophy 1917-67' in *Social Thought in the Soviet Union*, ed. Alex Simirenko, Chicago: Quadrangle Books, 1969. For the influence of dialectical materialism on Soviet science see Loren R. Graham *Science and Philosophy in the Soviet Union*, N.Y.: Knopf, 1972.

108. On psychology after 1953 see John A. Molino, 'Is There a New Soviet Psychology?' in *Social Thought in the Soviet Union*, pp.300-327.

109. See Howard J. Sherman, 'The "Revolution" in Soviet Economics' in Simirenko ed. *Social Thought in the Soviet Union*, pp.222-268.

110. Helmuth Fleischer, *Marxism and History* [1969], tr. Eric Mosbacher, Harper Torchbooks, N.Y.: Harper & Row, 1973, p.39f.

new privileged class of bureaucrats and officials of the Communist Party, with an extension of their special privileges and the power to pass on these privileges to their descendants by giving them preferential treatment in their careers. This was associated with the stagnation of the economy. But despite the high regard for Stalin and the suspicion of the ideals and Western orientation of Lenin, neither Suslov and those he supported, nor Brezhnev and his retinue were able to undo all that Khrushchev had achieved, while at the same time increasing numbers of people had come to recognize the impossibility of a totally planned economy. (As one sarcastic Soviet author remarked: 'Mathematicians have calculated that in order to draft an accurate and fully integrated plan for material supply just for the Ukraine for one year requires the labour of the entire world's population for 10 million years.'[111])

With the rise to power of Gorbachev in 1985 a new cultural revolution was inaugurated.[112] Gorbachev represented a new inversion of Russian culture, this time with spontaneity being exalted over conscious direction from above, and with traditional Russian culture being downgraded in favour of a Western outlook.[113] Under the slogans '*glasnost*' and '*perestroika*', Gorbachev attempted to democratize the political order and replace central planning of the economy by markets (although not to the same extent as in Hungary). However the liberalization of Soviet society led to economic breakdown, ethnic violence and the rise of disintegrating nationalisms. The radicals were divided among themselves, consisting of 'liberals' - essentially Moscow intellectuals who believed that the Soviet Union should emulate Western liberalism, various democratic socialist groups, and anarcho-syndicalists.[114] A number of independent populist movements also emerged, the most significant of which was that inspired and led by Boris Yeltsin. The growing chaos within the Soviet Union combined with the abandonment of communism and hostility towards Russians by Eastern European nations led to a resurgence of the conservatives, and a move towards the use of force to re-establish law and order. This culminated in the attempted coup in August, 1991. The defeat of this coup and the consolidation of Boris Yeltsin heralded the end of communism in Eastern Europe and the destruction of the Soviet Union.

What is the significance of this? The historical perspective offered here supports the diagnosis of the Soviet historian Yuri Afanasyev who argued that the current political and economic crisis must be seen as part of a greater problem. As he put it: 'The current crisis coincides with another, larger one, which began in the nineteenth century - the crisis, or perhaps the exhaustion, of this Eurasian civilization, with its egalitarian, statist ethic and its imperial forms and values. This civilization is no longer workable.'[115] The failure of Gorbachev must be seen as a failure of Soviet culture.

It is in the context of this history of Russian culture and of Soviet Marxism that the role of Marxism in the Soviet Union's relation to its environment must be understood and evaluated.

---

111. O. Antonov, cited by Alec Nove, *The Economics of Feasible Socialism,* London: George Allen & Unwin, 1983, p.33. For a full study of the Soviet economy and its problems, see Alec Nove, *The Soviet Economic System*, 3rd ed. Boston: Allen & Unwin, 1986. On East Europe in general, see Michael Ellman, *Socialist Planning*, Cambridge: Cambridge University Press, 1979.

112. Some idea of this can be gained from Gorbachev's manifesto, *Perestroika*, London: Collins, 1988. However the movement is best described by Alec Nove, *Glasnost in Action: Cultural Renaissance in Russia*, Boston: Unwin Hyman, 1989.

113. See Christian Schmidt-Häuer, *Gorbachev: The Path to Power* p.74f.

114. On the various movements within the Soviet Union, see Boris Kagarlitsky, *Fairwell, Perestroika: A Soviet Chronicle*, tr. Rick Simon, London: Verso, 1990.

115. Yuri Afansayev, 'The Coming Dictatorship', *The New York Review of Books*, Vol. XXXVIII, No.3, January 31st, 1991, pp.36-39, p.39.

# 11

# SOVIET ENVIRONMENTALISM AND THE FUTURE OF MARXISM

The effects of humans on the environment was not a significant issue in Russia until the reign of Peter the Great - when Russians set out systematically to appropriate Western technology to develop their economy. The effect of this was to reveal the reverence felt by Russians for nature. While Peter the Great was concerned with expanding the productivity of Russia's economy, he also acknowledged that there were limitations to this and that there was a need to conserve forests, and promulgated regulations accordingly. Though little action was taken on these regulations, developments within Russian thought indicate that this concern for conservation was widespread. The subsequent history of Russians' relationship to their environment can be seen as a conflict between a growing concern to develop science and technology to dominate nature to keep up with the West, and reactions to this based on a deep rooted reverence for nature. The concern to preserve the environment was manifest in the development of ecological thought which was frequently in advance of that in Western European and USA. This precocity was stimulated by battles over conservation, and the career of ecology is an index of the successes and failures of environmentalism in Russia.[1]

The orientation to nature of Russians involved in biological research is evident G.I. Dokhman's, *Istoriia geobotaniki v Rossii*.[2] Russians were far more inclined to see nature holistically and to recognize inter-relations within nature than Western European biologists. Ivan Komov was already treating the forest as a community in 1788, and in the early years of the nineteenth century Schelling's anti-reductionist philosophy of nature was received with great enthusiasm. This influence was reflected in studies of nature which presaged later developments in ecology. In 1835 Gilderman observed that nature prefers mixed forests to monocultures, and Semenov took forests as a unit in his study of forest self-renewal. In 1838 M.G. Pavlov argued that nature should be taken as a model for working out what crops should be rotated. In 1848 Teploukov noted that 'Virgin forests regenerate themselves continuously according to the laws of externally acting nature' and that, '[i]n the economy of nature, all trees are equally important, the willow, the aspen and birch ... are just as essential for natural forest renewal as the oak and ash.'[3]

An opposing tendency was inspired by French thought with its stress on empiricism, materialism and concern for practical applications. People influenced by this tradition, including biologists, evaluated science in terms of its technical benefits. Karl Rul'e (1814-1858), who in mid-nineteenth century had become the doyen of Russian zoology, under the influence of Geoffry Saint-Hilaire and Lamarck argued for the plasticity of species and conceived the notion that zoological knowledge could be applied to the acclimatization and domestication of new species into Russia.[4] Following the

---

1. This is the subject of Douglas, R. Weiner's superb study, *Models of Nature: Ecology, Conservation, and Cultural Revolution in Russia*, Bloomington: Indiana University Press, 1988.

2. G. I. Dokhman *Istoriia geobotaniki v Rossii*, Moscow: Nauka, 1973. The translations which follow have been made for me by Dr Douglas Weiner, at the time, a research fellow in the Center for Russian Research, Harvard University and at present (1996), professor of history at the University of Arizona. For this entire section I am deeply indebted to the work and advice of Professor Weiner.

3. Ibid. p.176 and 177.

4. For a description of this movement from its origins until the revolution see Douglas Weiner, 'The Roots of "Michurinism": Transformist Biology and Acclimatization as Currents in the Russian Life Sciences', *Annals of Science*, Vol.42 (1985), pp.243-260.

establishment of an acclimatization society in France in 1954 Rul'e and his supporters, most notably Bogdanov (1834-96) and Usov (1827-86), established an acclimatization society in Russia in 1857. This continued to grow in strength after the death of Rul'e and its members worked to acclimatize a number of new species into Russia. The golden age for acclimatization was the late 1850's and early 1860's. The movement's specific aims and general philosophy were supported by a wide cross-section of Russian society, from one of the Tsar's brothers and from ministers of the government to the nihilists of the 60's and 70's who were concerned to create a utilitarian paradise on earth through the scientific mastery of nature, and who accordingly promulgated the view that science must serve technology. While subsequent to this the movement suffered financial set-backs and slowly declined, efforts at acclimatization were still being made at the time of the communist revolution, most notably by Michurin.

However biologists influenced by *Naturphilosophie* developed a strong counter movement to the acclimatizers and their philosophy, and soon came to dominate the field. In 1863 A.M. Bazhanov, a Professor of Agronomy argued that people should look at natural meadows as a model for agriculture, providing humans had not interfered with the 'economy of nature' in these meadows. In the 1880's V.V. Dokuchaev developed soil science along lines inspired by the *Naturphilosophen*. He was extremely critical of Western geology which studied soil only for utilitarian reasons. In place of this he analysed the 'extremely close and everlasting inter-relationships between water, air, land, plant and animal organisms' as well as growth and changes in human society.[5] In 1883 Kravchinski argued that forests are communities, and in 1884 Ia. Medvedev used the adjective 'social' to describe forest structure for the first time. Kravchinski studied forests as communities, developing the distinction between pre-climax communities which pave the way for their own dissolution by changing the nature of the soil, and climax communities which sustain the conditions of their existence. This was more than two decades before Clements in USA developed similar ideas. While in the 1890's a group of scientists argued along mechanistic lines that the physical environment, and in particular the soil, determines vegetation; such thinkers were in a minority. In 1896 the term 'phyto-sociology' was coined, uniting the discipline studying the relationships between organisms under an explicitly anti-reductionist metaphor. Research proceeded rapidly and by 1898 P.N. Krylov had investigated the role of fauna in determining vegetation and examined the nature of the equilibrium which develops between different plants and local conditions. Between 1904 and 1910, G. F. Morozov used and fully elaborated the metaphor of 'organism' to describe plant communities, and developed a conception of abstract models as the means for analysing particular concrete communities. At the fiftieth jubilee congress of the acclimatization society in 1908, its new president, Kozhevnikov, elevated conservation to the status of paramount concern, and made not one reference to the need to acclimatize anything.[6]

Concern with nature was not confined to biologists, and it was a Russian (or rather, Ukrainian) Marxist who in the 1870's and 1880's first attempted to reformulate economic theory to accord with the second law of thermodynamics. Serhii Podolinskii (1850-91), a Ukrainian socialist and a friend of Lavrov, attempted to measure the input/output ratio in agriculture in terms of energy, beginning with the assumption that all physical and biological phenomena on earth are expressions of the transformation of usable energy from the sun.[7] Representing the task of labour as being to increase the accumulation of usable solar energy on earth, he attempted to combine an energy theory of value with the labour theory of value, and with only limited success, to gain Marx's endorsement for his project to give a foundation in natural science to the theory of surplus value. Podolinskii then used this framework to attack Social Darwinism, arguing that poverty is caused by social relations, and that 'in the countries where capitalism triumphs, a great part of work goes towards the production

5. Billington, *The Icon and the Axe*, p.444.

6. Weiner, 'The Roots of "Michurinism"', p.258.

7. On Podolinskii see J. Martinez-Alier and J.M. Naredo, 'A Marxist Precursor of Energy Economics: Podolinski', *Journal of Peasant Studies*, Vol.8, pp.207-224; and Juan Martinez-Alier, *Ecological Economics*, Oxford: Blackwell, 1987, Ch.3.

of luxury goods, that is to say, towards the gratuitous dissipation of energy instead of increasing the availability of energy.'[8]

At the beginning of the twentieth century with the growth of capitalism and with conservation being undertaken in the West, three orientations emerged in Russia towards environmental protection. The first group argued for conservation on utilitarian grounds, pointing out how non-renewable resources imposed limits to economic growth.[9] The second group, for whom Semenov-Tian-Shanskii was the most articulate spokesman, represented the romantic tradition which approximated the transcendentalists in USA. Semenov-Tian-Shanskii urged Russians to 'strive to realize ... not only a broad right for humans to live and develop in all of their spiritual variety, but also the right (upon which humanity now tramples) of all living things on Earth to their existence.'[10] However the most important group were the ecologists, led by Grigorii Aleksandrovich Kozhevnikov who was professor of invertebrate zoology at Moscow University. Kozhevnikov argued for the preservation of wilderness areas, *zapovedniki*, which could serve as standards of nature (*etalony*) against which human actions could be measured. This idea presupposed that existing ecosystems embodied a natural harmony and were to a certain extent self-regulating. By preserving such wilderness areas, the extent to which humans had disturbed the natural environment would be revealed - knowledge which would be invaluable for restoring areas damaged by humans to health.

## *Environmentalism After the Revolution*

The tsarist regime did very little for conservation, so environmentalists were generally happy to see it overthrown. The leading Bolsheviks, especially Lenin, enthusiastically embraced conservation, and in particular the ideas of Kozhevnikov,[11] and a large number of game reserves, monuments to nature and *zapovedniki* were set up. Weiner wrote of this period in Russian history: 'Only in the 1920's did the first truly popular conservation organization - the All-Russian Society for Conservation - emerge, and it is only in those years as well that the beginnings were laid for the creation of planned network of *zapovedniki* throughout the USSR.'[12] While game management and forest protection were administered by the People's Commissariat of Agriculture, Lenin put the Commissariat of Education in charge of protection and conservation of the environment, and in particular, of the *zapovedniki*. Having no interest in resource exploitation and headed by Lunacharskii who was strongly sympathetic to conservation (and also to the program of the *Proletkul't* movement to create a socialist culture, including a socialist science), this proved an effective defence for the environment against those concerned with purely economic goals, in particular the Commissariat of Agriculture and the Commissariat of Foreign Trade.

It is clear from this that Lenin interpreted Marxism in such a way as to acknowledge the limitations of the environment, of the existence of dynamics within nature with which humanity must accord. His 'consciousness/spontaneity dialectic' was not understood by him as implying the possibility of the total subordination of nature to human designs. In fact it suggests that Lenin recognized the limitations of such efforts and the impossibility of such total control. However in his

8. Cited ibid. p.48.

9. Bogdanov's concerns are evident in his novels *Red Star* and *Engineer Menni*. See Alexander Bogdanov, *Red Star: The First Bolshevik Utopia*, ed. Loren R. Graham and Richard Stites, tr. Charles Rougle, Bloomington: Indiana University Press, 1984 with an afterword on the subject 'Bogdanov's Inner Message', by Loren Graham.

10. Douglas R. Weiner, 'The Historical Origins of Soviet Environmentalism' in *Environmental History: Critical Issues in Comparative Perspective*, ed. Kendall E. Bailes, Lanham: University Press of America, 1985, p.383.

11. This is described in Douglas Robert Weiner, *The History of the Conservation Movement in Russia and the U.S.S.R. from its Origin to the Stalin Period*, Ph.D. Thesis, Columbia University, 1983. *Models of Nature* was based on this.

12. Ibid. p.161.

conflict with Bogdanov, Lenin not only rejected Mach's theory of knowledge, but also Ostwald's energetics. The domination by Lenin of Marxism virtually ruled out the possibility of assimilating Podolinskii's work to found the labour theory of value on an energy theory of value which might have related Marxist theory and environmental concerns in a systematic way, and led to a devaluation of all those socialist thinkers who had argued this position.[13] Furthermore, Lenin crippled the efforts of *Proletkul't* to create a new socialist culture, and promoted the acceptance of the instrumentalist rationality of capitalist societies.

Lenin's environmental policies had considerable success. By late 1927, 29 *zapovedniki* with a combined area of about three million hectares had been established, with twelve more, having been promoted by the State Committee on Conservation, at some stage of environmental review.[14] There were also hundreds of *zakazniki* or game reserves, and hundreds more 'monuments to nature'. Taken together these territories had a combined area of 7 million hectares; and beaver, saiga, moose and egrets were moving away from the brink of extinction. Associated with reforms in education inaugurated by the revolution and carried out by the Commissariat of Education, there was also a considerable amount of research undertaken. From 1924 to 1928 the budget of the Astrakahn *zapovednik* was increased from 950 to 27,200 roubles, that of the Caucasus, from 2,120 to 74,920 roubles.

Ecological research undertaken on the *zapovedniki* resulted in major theoretical advances. I. K. Pachoskii studied the division of labour within plant communities, V.N. Beklemishev articulated theories on the structures of ecological communities and S.A. Severtsov pioneered the study of population dynamics among wild mammals. However the most important work was done by V. V. Stanchinskii. Stanchinskii was strongly influenced by energetics and the work of Vernadskii on geochemistry and on the concept of the biosphere, which in turn had been partly inspired by the work of Podolinskii. Developing such ideas, Stanchinskii worked out mathematical models based on his research to show the nature of energy flows, and in particular, trophic levels in eco-systems a full decade before similar ideas were developed in USA by Hutchinson and Lindeman.

These ecologists, and Stanchinskii in particular, conceived their work to be important both for the advancement of science, and for the development of agriculture. S.A. Severtsov showed the importance of ecology in working out the best way to exploit nature through his studies of population dynamics, and N.A. Troitskii pointed out how overgrazing reduced yields. When the Five Year Plan was formulated, the ecologists attempted to make a contribution, and they spelt out the significance of the work they were undertaking. At the First All-Russian Congress for the Conservation of Nature held in September, 1929, V.V. Alekhin attempted to show how removal of land to the *zapovedniki* would increase agricultural production. Stanchinskii argued that a truly planned economy functioning within the sustainable limits of the productivity of nature could be achieved only with the active participation of conservationists. He pointed out how biocenotic research could aid in such areas as biotic protection, which would obviate 'the use of pesticides, which often contain toxic substances ... that not only kill the pests but cause injury to human and to useful organisms.'[15] His concern for the applicability of ecological research was manifest in his proposals for the siting of *zapovedniki*. He argued: 'We must select for *zapovedniki* the most typical territories which will have the greatest economic significance as natural *etalony* ... The network of *zapovedniki* must be linked with the Five Year Plan.'[16] It was also proposed at the congress that an inventory of all natural resources in the Russian Soviet Federal Socialist Republic be made, and it was argued that the conservation organizations must be able to review Plan targets and monitor Plan fulfilment.

Stanchinskii's arguments carried the day, and the Congress resolved:

---

13. As Juan Martinez-Alier has shown in *Ecological Economics*, Oxford: Blackwell, 1987.

14. See Weiner, 'The Historical Origins of Soviet Environmentalism', p.387f.

15. Cited by Weiner in *The History of the Conservation Movement...* p.334.

16. Cited ibid. p.338.

> The economic activity of man is always one form or another of the exploitation of natural resources ... The distinction and tempo of economic growth can be correctly determined only after the detailed study of the environment and the evaluation of its production capacities with the aim of its conservation, development and enrichment. This is what conservation is all about.[17]

The ecologists became trenchant critics of the implementation of collectivization. To the project of increasing harvests by 35% A.A. Teodorovich exclaimed: 'without conservation, without rational ... use of natural resources. there cannot be any talk about increasing the harvest.'[18] N.N. Pod"iapol'skiii, an agronomist warned in March, 1930 that the tractor and the combine would be environmentally destructive, imposing a uniformity hitherto unknown. And the ecologist Kashkarov slated the collectivization of traditional societies, arguing that:

> ... the entire life cycle of the Kirghiz is determined by ecological considerations ... The Kirghiz is the product of his habitat: His annual cycle of activity and his nomadic wanderings are dictated by ecological considerations, his psychology and practical philosophy of life as well.[19]

## *Environmentalism Under Attack*

However this was the period of Cultural Revolution associated with the struggle to raise economic output. The relative status of spontaneity and consciousness was inverted with the rejection of the N.E.P., and the image promoted was that of the 'Struggle Against Nature'. Typical of the new orientation of this period were the sentiments of a book written for students by a young Soviet engineer, M. Ilin. With titles of chapters such as 'Conquerors of Their Own Country', 'The Conquest of Water and Wind', 'On the March for Metal', and the 'The War with the Kilometres', Ilin pronounced:

> Within a few years all the maps of the U.S.S.R. will have to be revised. In one place there will be a new river... in another a new lake... A great new power has appeared in Nature - the power of human labour. Not only the blind forces of Nature, but also the conscious, organized, planned labour of man now fashions rivers and lakes, plants forests, and transforms deserts, moderates and accelerates the flow of waters, creates new substances and new species of plants and animals.[20]

The achievement of such ends was seen not as in the West as the subjugation of an essentially passive nature, but as a mighty struggle against an aggressive opponent. As Ilin wrote in relation to a section entitled 'The War with the River': 'Man must fight the river, as the animal-tamer fights wild beasts.'[21] This was not a propitious cultural environment for the promotion of environmental causes.

Criticism of environmentalism began at the Conservation Conference in 1929. Some delegates could not see why all land should not be used for economic production and conservationists were

---

17. Ibid. p.348.

18. Ibid. p.353.

19. Ibid. p.377.

20. M. Ilin, *New Russia's Primer: The Story of the Five-Year Plan*, tr. George S. Counts and Nucia P. Lodge; Boston and New York: Houghton Mifflin Co., 1931, p.141f.

21. Ibid. p.35.

labelled the 'old bourgeois professoriat'. One enthusiastic member of the Young Naturalist Organization declared that 'The naked idea of preservationism is organically alien to active youth and in particular to Soviet Youth, seized ... with the enthusiasm of socialist construction and reconstruction.'[22] and A. Kiselev argued that under the prevailing economic conditions, science for science's sake would not do, and that conservationists should not look on *zapovedniki* as sanctuaries for birds and animals. Conservation was also attacked in the press. On 30th June, 1930 a letter from V.V. Karpov was published in the journal of conservation *Okhrana prirody* attacking the organization for conservation. Karpov argued:

> It is clear ... that the old theory of conservation of nature for the sake of nature itself ... an idea which reeks of ancient cults of nature's deification ... stands in sharp opposition both to our economic as well as our scientific interests that there is no place for it in our land of socialism-in-the-making...[23]

Ecology first came under attack from the Deborinites. The Deborinites charged the mechanists with reducing everything to the conservation and transportation of matter and energy. Following this attack Kozhevnikov was deprived of his position at Moscow University and a number of other ecologists were upbraided for promoting ideas inconsistent with dialectical materialism. Kashkarov for instance was seen as being too mechanistic in his assessment of the situation of Kirghiz. V.N. Liubimenko was attacked as reductionist for arguing that 'Social problems are problems of a biological character, and therefore we must seek out biological laws which govern social phenomena and the life of all natural communities alike.'[24] It was argued that humans are cultural beings and cannot be reduced to the laws of biology. However Bugaev, the Deborinite who focused his attention on ecology, did not want to demolish it but to ensure that it was consistent with Marxism. Most ecologists were able to make the appropriate modifications to their theories. Stanchinskii took special pains to stress the historical, dynamic and dialectical nature of his concept of biocenosis, replacing the static notion of 'equilibrium' with the more acceptable 'proportionality' and emphasising the continuous self-creation of the biocenosis. He depicted this self-creation of the biocenosis as emerging from interactions between both its components the abiotic environment, with the result that new syntheses were continually arising in successional series.[25]

Furthermore the philosophy of the Deborinites made it possible to continue to justify conservation. Humans were seen as part of nature and the biological realm was seen as having laws irreducible to physical laws to which humanity must accord. Most importantly, Engels had already spelt out the implications of this in the *Dialectics of Nature*, the book the Deborinites revered.

But with the rise of I.I. Prezent and his associates, community ecology and the conservation cause came under sustained attack. Rejecting all science not immediately serving the development of technology, and committed to the wholesale importation and acclimatization of exotic species, they set out to demolish community ecology as a discipline standing in the way of their projects. Their general aim was expounded by Kashchenko:

> The final goal of acclimatization, understood in the broad sense, is a profound rearrangement of the entire living world - not only that portion which is now under the domination of man, but also that portion that has still remained wild. All living nature will live, thrive, and die at none

---

22. Cited by Weiner, *The History of the Conservation Movement...*, p.279.
23. Cited ibid. p.424.
24. Cited ibid. p.373.
25. Douglas R. Weiner, 'Community Ecology in Stalin's Russia', *Isis*, Vol.75, 1984, pp.684-696 p.692.

> other than the will of man and according to his designs. These are the grandiose perspectives that open up before us.[26]

To begin with they began attacking the holism of ecology. V.L. Komarov argued in his *The Vegetation of the USSR and Adjacent Countries* which appeared in 1931 that all reference to 'plant communities' should be expunged from biology. The conflict came to a head after the 1931 Anti-Drought Congress and the 1932 Faunistics Conference.[27] Prezent called upon the Soviet biologists to become 'engineers' and 'inventors' in a top-to-bottom transformation of nature. Among the first of the projects developed by his minions was the 'General Plan for the Reconstruction of Economically Important Fauna of European Russia and the Ukraine' drawn up by B.K. Fortunatov directed towards wholesale acclimatization of exotic species. When this plan was outlined at the All-Union Congress of 1933, it was attacked by Stanchinskii, Severtsov, Kozhevnikov and other distinguished figures. Their holistic views on ecology also led them to oppose projects that constituted the very centre of socialist reconstruction. For example projects to construct enormous hydroelectric installations were opposed by hydrobiologists, and projects to extend monocultural agriculture to the virgin steppes were opposed by zoologists and phytocenologists. Led by Prezent, ecologists were denounced as 'traitorous' opponents of the heroic projects of the five year plans.

By mid-1932, Prezent and his supporters had succeeded in closing down Stanchinskii's pathbreaking research at Askania-Nova, and converting the reserve to the All-Union Institute for Agricultural Hybridization and Acclimatization of Animals. By 1934 Stanchinskii and his supporters had been driven from Askania and vilified as 'mongrels of society' and 'saboteurs'. At the Academy of Sciences' Ecological Conference of January, 1934 Prezent explained that the holistic conception of the biocenosis implied natural limits to the ability of people to transform nature and was therefore in opposition to socialist construction. Following this, Prezent succeeded in putting an end to almost all the original theorizing on ecology in the Soviet Union: Alpatov's work on the role of density in regulating animal populations, Severtsov's statistically based attempts to correlate fertility with longevity in animals, and Gauze's experiments in population dynamics which led him to postulate the competitive-exclusion principle for which he is still known. And he aborted the publication of Stanchinskii's major work.

The demise of ecology did not coincide with the undoing of all that had been achieved by the environmentalists. Though the head of the Commissariat of Education, Lunacharskii, lost his position in 1929, a large number of Bolsheviks of the second order continued to support conservation issues. The deputy director of the Main Administration for *Zapovedniki*, Vasilii Nikitich Makarov actually managed to expand the network of *zapovedniki*. But there was a slow whittling away at the role of the *zapovedniki* associated with a general increase in environmental destruction, much of it due to the acclimatizers. Then in 1951 and 1952 there was a general attack on the *zapovedniki*, an attack in which the number of reserves was decreased from 128 with an area of 12.5 million hectares to 40 with an area of less than 1.5 million hectares.[28]

## *The Institutionalization of Anti-Environmentalism*

The fate of the ecologists symbolized the general state of environmentalism in the Soviet Union. Some associated movements, such as the movement of the architects to decentralize housing and industry, maintained their positions for longer, but they were all ultimately defeated.[29] Moscow was

---

26. Cited by Weiner, *The History of the Conservation Movement...* p.517.

27. For a description of this see Weiner, 'Community Ecology in Stalin's Russia', p.691f.

28. Weiner, 'The Historical Origins of Soviet Environmentalism' p.397f.

29. See S. Frederick Starr, 'Visionary Town Planning during the Cultural Revolution' in Fitzpatrick ed. *Cultural Revolution in Russia: 1928-1931*, pp.207-40.

rebuilt in stone, and people were packaged in fourteen storey apartment blocks. But what is most important from the point of view of ideological analysis is which ideas are incorporated into institutions. The defeat of the ecologists meant that the forms of thinking which came to underlie the Five Year Plans did not take into account the limits of the environment. This failure was consolidated with the liquidation of economists and by an adherence to the labour theory of value. Taken out of the context of the analysis of capitalist society and in a society in which economic ends were conceived of not in terms of reduction of necessary labour time but in terms of the development of industrial capacity, the labour theory of value led to the contributions of nature, capital and services being ignored.[30] Nature was seen as a free gift, and land, water and minerals were not counted as costs of production. The Five Year Plans incorporated the ideas promulgated by the ideologists such as Ilin and Prezent in support of Stalin's policies in the 1930's. And as institutionalized, they have taken pre-eminence over all other forms of thinking. They have presupposed a conception of humanity struggling to subdue nature, and have measured success in terms of material production and the rate of increase of this production.

These developments were associated with a reinterpretation of the meaning of socialism. While Lenin had thought of his policies as backtracking to develop the conditions for the achievement of socialism, Stalin defined his militarisation of society in the service of economic development as socialism achieved. Soviet Marxism was formulated as a technological reductionism underpinned by a Neoplatonic eschatology, with history being understood as a progression measured in terms of the development of the material base, that is, the means of production of societies. Stalin presented socialism as superior to capitalism by virtue of its greater capacity to develop the means of production. After the Second World War this came to be encapsulated in the slogan 'Catch up and pass the United States', and later, Khrushchev's claim that the Soviet Union with its faster growing economy would bury USA.

The destruction of nature wrought in this process was not measured or considered in the criteria by which the Five Year Plans and economic growth were judged. At the same time these Plans gave enormous power to those sectors of Soviet society involved in the development of the means of production. In particular, *Gidproekt*, the hydro-power engineering agency was almost unmatched for 'arrogance and seemingly unassailable political strength.'[31] The power of this agency, which through the use of convict labour was closely associated with the KGB was only curtailed in the 1950's when it was incorporated into the Ministry for Power and Electrification. However other such large scale organizations had entrenched powers which made them extremely difficult to regulate.

Although there were variations with the rise to power of different power groups, the direction in which Soviet socialism developed varied little. The central focus was on growth of material production. It was associated with the use of material incentives to get people to work harder, and the development of highly differentiated scales of income. As in the West, money became the measure of people's participation in historical progress, and the success of Soviet society was defined in terms of the commodities available to people on the market. Workers sold themselves as labour power to State enterprises, and the built-up environment was organized for the efficient movement of commodities, of labour power to and from work, and for the recuperation of labour power. The managerial elite of the West and the Soviet Union had essentially the same orientation to people: to control them efficiently, and this led to similar developments in the West and the East. As in the West, the ruling elite of the Soviet Union committed itself to the development of nuclear power rather than decentralized forms of energy production, and promoted distinctive consumption, rapidly expanding the number of privately owned cars despite the excellent public transport available. These developments were associated with increasing levels of corruption, what Brezhnev

30. See Gustafson, *Reform in Soviet Politics*, Ch.4.

31. Ibid. p.54.

in 1979 called 'negligence, lack of responsibility and stupid bungling'[32], and the loss of meaning in the lives of Soviet citizens which expressed itself, among other things, in the highest incidence of alcoholism in the world. It was also associated with the reintegration of the Soviet Union into the capitalist world economy.

## *The New Environmental Movement*

A new environmental movement began to develop as soon as Stalin died, and gained momentum in the sixties. Membership of the main environmentalist organization, the All-Russian Society for Conservation (VOOP) grew immensely, and has continued to grow, from 916,000 in 1959 to 32 million in 1981, along with other, more vigorous environmentalist groups.[33] With Stalin's death Lysenko came under attack, ecologist-activists inaugurated a war of liberation for the *zapovedniki*, and they demanded a return to fundamental ecological research in the reserves.[34] By 1961 this movement had increased the areas of the reserves to 6,360,000 hectares. In 1967, the Ministry of Agriculture finally got around to banning acclimatization in its reserves. The environmental movement manifest itself most dramatically in the early 1960's with a storm of protest over the building of paper and pulp industry on the shores of Lake Baikal, a protest which had some success in controlling economic activity. In the seventies there also emerged strong environmental sentiments associated with the 'village movement,' which depicted rural society as one of the great sources of virtue and strength in Russian society. This has found expression in literature. In general, the village, organic unity and nature are celebrated. For instance Boris Vasiliev's work *Don't Shoot at While Swans!* published in 1973 has the hero, Egor, exclaim: '...we are orphans. We are not at peace with our mother earth; we have quarrelled with our father the forest, and, with our sister the river, there has been a bitter separation.'[35] The theme of this work is the ending of innocence and wholeness when the city intrudes into the village, and throughout nature is exalted at the expense of technology. Each of these developments can be seen as part of a struggle to elevate the status of spontaneity in relation to the status of consciousness, and to free life from the rigid central control which had been the legacy of Stalin's reign.

But for the most part, this movement was only effective when its ends coincide with the economic and political aims of the government. The most effective pressure for environmental reform has come from the realization that environmental destruction is limiting present economic growth, and most of the pressure for taking account of environmental problems was utilitarian. For instance the debate over Lake Baikal were largely framed in terms its role as a filter of water.[36] The conservation program which was mounted was 'not a program designed primarily to preserve wilderness or protect natural beauty but to protect public health and facilitate further economic growth (particularly of irrigated agriculture) in the most highly developed regions of the country.'[37] The idea that nature is significant in itself independent of human goals was rejected by Soviet philosophers.[38]

---

32. Cited by Nigel Harris in *Of Bread and Guns*, Harmondsworth: Penguin, 1983, p.176.

33. See Douglas R. Weiner, 'The Changing Face of Soviet Conservation', *The Ends of the Earth: Perpsectives on Modern Environmental History*, Cambridge: Cambridge University Press, 1988, pp.252-273.

34. Douglas R. Weiner, *Prometheus Rechained: Ecology and Conservation in the Soviet Union* unpublished paper, 1987, p.8f.

35. Cited Clark, *The Soviet Novel*, p.247.

36. Pryde, 'The "Decade of the Environment" in the U.S.S.R.' p.275.

37. Gustafson, *Reform in Soviet Politics*, p.51.

38. See Yu. A. Sholenko in 'Human Ecology and Social Consciousness' in Ursal ed. *Philosophy and the Ecological Problems of Civilisation* p.398f.

In the mid-1970's the state of the world environment became a major focus of attention, and in the Soviet Union much was written on this topic. That which was published in translation was, for the most part, directed against capitalism, arguing the necessity for its replacement by socialism. The ecological crisis was portrayed as due to the way 'material production operates as production for the sake of profit' in capitalist societies and as 'a component of the general crisis of capitalism at today's stage, and one of its manifestations.'[39] Then when it came to the Soviet Union, environmental problems were presented as though they were merely minor and correctable malfunctions of socialism due to such factors as the attitudes of the managers of industries, or of particular workers. However within the Soviet Union the significance of environmentalism went far beyond this. It became one of the most active areas of intellectual debate, with some thinkers radically departing from orthodoxy in their efforts to confront the issues. The only real limitations were that the superiority of socialism over capitalism could not be questioned.

The most important feature of this debate was the change in focus in Soviet environmentalism from particular issues to global issues. Whereas the ecologists of the 1950's and the 1960's had echoed the ecologists of the 1920's with their focus on preserving ecological communities unsullied by human interference, this had proved indefensible. Their opponents were able to show the impossibility of identifying ecological communities as pristine, discrete, self-regulating ecological communities and of conceiving of nature in abstraction from human activity. To counter these arguments the proponents of environmentalism shifted their focus to the conditions of survival of populations of species living in particular areas. Focussing on these conditions led to concern with increasingly broader contexts, and ultimately, to the state of the world. The basic framework for the new analyses of environmental problems was the conception of humans as part of nature, but with their own unique laws of development. The two concepts which were most important in developing these analyses were the 'noosphere' developed by the geologist Vernadskii, and the 'biotechnosphere' developed by the biologist Khil'mi. According to Vernadskii, the noosphere is the 'final stage of evolution' in which 'man, taken as a whole, becomes a powerful geological force. And before him, before his thought and labour, stands the question of the rebuilding of the biosphere in the interests of freely thinking humanity as a unified whole.'[40] According to Khil'mi, the initial biosphere and urban surroundings created by man form a new system,' a symbiosis of nature and technology which includes 'the physical surroundings, living organisms, and technical equipment, in particular - the large-scale structures, transforming the atmosphere, the hydrosphere, and the lithosphere of the Earth.'[41] On this basis efforts were made to elaborate concepts to analyse humanity in the context of nature.[42]

The conclusions drawn on the basis of these ideas were often that more centralized technological control was needed. Ecology was seen to imply complex interdependence, and the only way to capture and manage this is through engineered, closed cycle cybernetic systems on a large scale. While most Soviet scholars rejected the complete replacement of the self-regulating features of the biosphere by a complex of technical mechanisms, there were those whose commitment to the underlying eschatology of Soviet Marxism led them to take the principle of subordinating spontaneity to consciousness even to this ultimate stage.[43] For example E.V. Girusov argued that

---

39. I.T. Frolov, 'The Marxist-Leninist Conception of the Ecological Problem', Ursal ed. *Philosophy and the Ecological Problems of Civilisation*, p.43.

40. Cited Joan DeBardeleben, *The Environment and Marxism-Leninism: The Soviet and East German Experience*, Boulder: Westview Press, 1985, p.93.

41. Loc. cit.

42. See for example Yu. P. Trusov, 'The Ecological Approach and Problems of Moulding the Noosphere' and I.P. Gerasimov 'Methodological Problems of the Ecologising of Modern Science' in Ursal ed. *Philosophy and the Ecological Problems of Civilisation*, pp. 58-78 and 189-206.

43. The dominant attitude is typified by E. Fedorov who concludes his book *Man and Nature*, Moscow: Progress Publishers, 1980 with a call for a transformation 'From Spontaneously Developing to Consciously Directed Civilization'.

human progress has so far passed through three stages: firstly the overcoming of natural limitations of men's use of the material of nature through the development of tools; secondly the overcoming of the natural limitations of the use of energy resources; and thirdly, the stage we are going through in which the constraints of natural information processing on production are being overcome by the development of artificial means of processing information. He then projects the fourth stage, a new revolution in human history as the 'ecological revolution':

> The ecological constraint is a very real one. It consists in the limiting values of natural resources, including the limits of environmental pollution, the limits of territory, and the limits of biospheral equilibrium. In the long run removal of all these constraints will constitute a transition to artificial means of ensuring all the natural conditions of men's existence up to and including artificial means of maintaining equilibrium of the a, which means that we will have to pass, in that case, to what may be called artificial reproduction of the environment.... [This] will be a matter of a radical change in the very mode of society's development. In place of the mode spontaneously built up there will be a mode of development consciously controlled in accordance with scientifically developed theories of progress.[44]

Other writers with similar commitments have called for the industrialization of outer space.

Other significant ideas on the environment were developed in geography. Geography had a major place within the Soviet academic world, and geographers were among the first to concern themselves with environmental problems. In doing so they went against fundamental tenets of their discipline. Under Stalin, economic and physical geography were held to be distinct, as the laws of society were held to be irreducible to the laws of nature. This division was attacked in 1960 by Anuchin on the basis of a perceived ecological threat. Though his ideas were attacked as 'bourgeois determinist', (while Anuchin countercharged that his opponents were 'voluntarists' i.e. Stalinists) it gained official favour, an indication that at least some elements of the Soviet leadership were unconvinced by the technological optimism of scholars such as Girusov and were concerned with the limits of conscious control revealed by environmental problems.

The scholars who were most important for their influence on immediate practices however were the economists. Most of these remained wedded to the labour theory of value and rejected the idea that nature has any value until labour is added. What was disputed was the character of labour expended to make resources usable for production. While the labour theory of value had some good effects in that the attention paid to the real productivity of labour brought to light the deleterious effects of environmental destruction on the economy, it also had some negative features. One negative consequence was that there was a bias towards material production in evaluating economic success. But more importantly the labour theory of value meant that until recently there were no direct charges for resources. It was affirmed in 1968 that the 'Use of land free of charge is one of the greatest achievements of the Great October Socialist Revolution' and it was not until 1982 that charges were introduced for water.[45] While DeBardeleben has pointed out that there are a number of indirect ways by which resources have been charged for, these have still been inadequate to force efficient use of them.[46]

However some economists attempted to strike out in new directions. The most radical position in the field was taken by P. G. Oldak. Oldak along with some other economists argued for the establishment of the field of bioeconomics to 'study the productive environment - to study the relationship between rates of growth, level of technology, and the quality of the environment', and he rejected the simple coordination of scientific disciplines as inadequate for this task.[47]

---

44. E.V. Girusov, 'The Normative Character of Ecological Knowledge' ibid. p.215.

45. Cited DeBardeleben, *The Environment and Marxism-Leninism*, p.244.

46. Ibid. Ch.8, 'Natural Resource Pricing and the Labour Theory of Value'

47. Ibid. p.116.

Bioeconomics was to be a completely new discipline. Using systems analysis, Oldak himself tried to demonstrate the applicability of this modelling to the analysis of the optimal use of Lake Baikal. He also proposed replacing the notion of 'gross social product' by 'gross social wealth' as the basis for evaluating the economy. Social wealth was defined as: (1) accumulated material wealth; (2) the flow of services; (3) accumulated knowledge; (4) the condition of reproduced natural resources; and (5) the condition of the health of the population. As Joan DeBardeleben wrote of this:

> Oldak clearly intends a nearly revolutionary critique of existing economic concepts, a tendency expressed not only in regard to environmental issues but also in his demand for inclusion of social activities like education and scientific research in gross social wealth. He explicitly rejects the notion underlying the productive-unproductive distinction - the idea that material production serves as the basis for expansion of non-material services. On the contrary, Oldak sees knowledge as the decisive element of natural wealth.[48]

But while Oldak was taken more seriously by his colleagues than comparable economists in the West (N. Georgescu-Roegen or H. Daly) he still had little influence on government policy.

Other thinkers made radical departures from the prevailing doctrine by basing their environmental critiques on the humanism of Marxism-Leninism. For instance I.T. Frolov condemned capitalism for being uncoordinated 'either with the needs and wants of a real individual or the possibilities and constraints of external nature.' The consequence of this is that it 'leads to man and society beginning to relate to material production (i.e. the process of the "exchange of matter" between man and nature) as to a field of "absolute freedom" passing into gross despotism, in which nature functions as a defenceless, passive material and man as its omnipotent "demiurge".'[49] In a later paper, written with Viktor Los, he came to the revolutionary conclusion that:

> Under the influence of the crisis nature of the developing socio-ecological situation man is gradually moving away from the illusion of anthropocentricism and rejecting the traditional hegemonistic relationship to nature. His thinking has ceased to limit itself to notions centring around needs and designs of him and him alone. His activity is acquiring an ever broader a orientation, and his thinking is drawn to 'biocentrism.'[50]

This clearly involved the rejection of the striving for total power which is characteristic not only of those committed to mechanistic materialism in capitalist societies, but also to Marxist-Leninism dominated by a Neoplatonic eschatology formulated in terms of the struggle to totally subordinate spontaneity to consciousness. In opposition to this it implies the ideal of achieving a situated freedom in place of the ideal of an absolute freedom which must inevitably turn against itself by reducing everything: nature and people, to instruments for this abstract ideal; and Frolov also argued for the need to reintegrate both aesthetics and values into our way of relating to the world. He argued against thinking of people in the present as mere means to the achievement of some superior future state. Frolov's position also involved some acknowledgement of a convergence between the problems of the East and the West which not only went against the orthodox affirmation of the qualitative superiority of socialism, but could be seen as an effort to divert the struggle between social systems to a confrontation with what he regarded as a world problem. He suggested that international cooperation may further 'a general rapproachment of peoples and the strengthening of peace in the world.'[51]

---

48. Ibid. p.214.
49. Frolov 'The Marxist-Leninist Conception of the Ecological Problem' p.43.
50. Cited by Weiner in *Prometheus Rechained*, p.25.
51. Cited by DeBardeleben, *The Environment and Marxism-Leninism*, p.125.

The thesis of a convergence between East and West was upheld even more radically by other scholars, notably Rychkov and Arutiunov. Arguing for a convergence in both problems and cultures, these scholars rejected the model which reduced culture to a superstructural aspect which must be explained in terms of the forces and relations of production. Arutiunov went so far as to argue for the need to undertake ethnic studies to find out which cultures are ecologically sound. He attributes much of environmental destruction to 'ritual-prestige consumption.'[52]

These published intellectual debates were far more significant within the Soviet Union than similar debates in the West would be. They were much more closely related to what the Soviet leadership was actually thinking. Furthermore, the way environmental problems were addressed made their arguments more effective. There was not the separation of values and facts, of romantic idealists and realists, characteristic of the debates on the environment in the West. This was particularly important as having an explicit ideology underpinning the legitimacy of the government made the legitimacy of the ruling élite far more vulnerable than Western ruling classes. Therefore the fact that such departures from orthodoxy were aired indicated the seriousness with which the Soviet leadership were taking environmental problems.

However a major part of this could have been an effort to retain legitimacy in the face of a decline in economic growth which had previously been the main basis of their legitimacy. And when all the works in relation to environmental problems are examined, the radical forms of environmentalism were exceptional. The positions argued for were, at least until the rise of Gorbachev, still predominantly in accordance with the Neoplatonic Prometheanism of orthodox Soviet Marxism. There remained the emphasis on the development of material production as the subjugation by consciousness of spontaneity as the principle of the history of humanity, and on the need to retain or increase central control of society. While there were some individuals who challenged the extent to which consciousness can be made to prevail over spontaneity, the majority of Soviet environmentalists appeared to believe that environmental problems could be solved through technology.

Furthermore this orientation was institutionalized, making it difficult for environmentalists with opposing points of view to have any impact (although it is noteworthy that environmentalists were more successful in Estonia than elsewhere, revealing the extent to which failure is a matter of culture). The Soviet leadership defined themselves as consciousness striving to dominate spontaneity, as a central decision-making apparatus controlling society and nature. The idea of decentralizing power to deal with environmental problems was consequently anathema. And with a theory of history as progressive technological mastery of society over nature combined with the obvious failure of past central directives to prevent environmental destruction, there was a constant tendency to look to massive technological projects for which such centralized control of nature was appropriate as a means to deal with environmental problems. This included the project of diverting northward flowing rivers to the south.

Also, the leadership was deeply committed to the maximum growth of material production, associated by them with both international competition and meeting consumer demands, as a means to maintain their legitimacy. The criterion for success in all areas of the economy was meeting the production schedules of the Five Year Plans. It was for such successes that people were rewarded. As a leading Soviet economist explained: 'Since the economic mechanism is above all oriented toward the fulfilment of traditional plan indicators, the enterprises are not interested in the realization of nature-protecting legislative acts and plan targets of the regulation of nature-protection.'[53]

What the environmentalists were up against was manifest in the struggle for Lake Baikal. As I pointed out in Chapter 8, the environmentalists were partly successful in limiting the development of the original polluting industries. But the struggle revealed far more fully the way economic

52. Ibid. p.84.

53. Cited ibid. p.151.

organizations were able to over-ride directives even from the Communist Party, and were frequently able to ignore legal regulations with impunity. No success achieved by the environmentalists was final, and the economic managers were ready to continue on their path as soon as political and public pressure eased. And other industries were being developed which were even more destructive to the lake without any correspondingly outcry from the public.

While Soviet environmentalists attributed great theoretical importance to closed-cycle and low waste technology, it was difficult to get even these adopted. Since the amount of resources used did not enter into the criteria by which managers were evaluated, and since it was impossible for managers to develop uses and customers for their by-products, there was little incentive for them to consider such options. The attempt to ameliorate these conditions through the introduction of fines for environmental damage had even less effect than such extrinsic environmental regulation in capitalist economies. This was illustrated in the case of timber cutting. As Pryde wrote:

> No manager objects to paying a few hundred or thousand rubles of the firms money in fines, if the extra timber gained enables his firm to overfill its quota, thereby bringing in ten times the amount in bonuses payable to the individual employees themselves.[54]

And while fines were later increased and laws enforced with greater vigour, the overall effect was negligible (again with the exception of Estonia, where people had traditionally been more law-abiding).

So, it must by concluded that Soviet environmentalists were no more successful than Western environmentalists. Ecologically based conceptions of how the economy should be organized which were suppressed by Stalinism were only just beginning to be redeveloped, and had only minor influence. The aspects of Marxism on which such thinking could find support, those which accorded with the process conception of being, remained subordinate to Marxism's Neoplatonist and mechanist aspects. Dialectical materialism was anti-mechanistic and closer to process philosophy, thus providing a better ontological foundation for anti-reductionist ecology. But its liberating potential was almost completely neutralized by the reduction of science to a means for developing technology in accordance with the technological determinism of historical materialism. And this debasement of science was embedded in an economic, social and political order which embodied the instrumentalist orientation to both nature and people as fully as the West.

Before the ascent to power of Gorbachev, it was clear that the structure of Soviet society was inimical to facing up to environmental problems. As Nigel Harris argued in 1983: 'The blunt instrument of the State and a monopoly of power, so effective, if so cruel, in bludgeoning crude output out of an obdurate nature, now becomes a powerful obstacle ... [T]he physical planning targets which pay little attention to the relative scarcity of materials militate against economy...'[55] With Gorbachev's *perestroika*, this situation changed. The environmental movement played a significant part and symbolized the opposition to the project of total control by consciousness over spontaneity, and there was a close association between intellectuals such as Frolov and the struggle within the Communist Party which culminated in the rise to power of Gorbachev. Gorbachev himself expressed strong concern for environmental issues, and Frolov was later appointed chief editor of the Communist Party's theoretical journal *Kommunist*. The exaltation of spontaneity over conscious direction involved a determined attack on bureaucratic inertia with its heavy handed approach to the environment. Perhaps the biggest success for environmentalists was the shelving of plans to divert rivers flowing North to the South. The 'progressives' of the Popular Front, the social democrats and independent communists made environmental problems a central issue.

---

54. Pryde, *Conservation in the Soviet Union*, p.99.

55. Harris, *Of Bread and Guns*, p.176.

However it was the push for a market economy which dominated *perestroika*, driving the Soviet economy towards a Western style, market driven consumerism. And not the consumerism of the core zones of capitalism, but of the Third World. As Kagarlitsky pointed out:

> The guardians of old ideas can talk about the restoration of capitalism, but the fact is that this social milieu is incapable either of creating from within itself a modern Western-style bourgeoisie or of 'building' developed capitalism. The most of which it is capable is forming a dependent, poorly developed society with a parasitic ruling class combining all the negative features of both the 'Eastern' and 'Western' models... We have yet to realize fully that we are needed by the centres of contemporary advanced capitalism only as a supplier of cheap resources and as a massive (one sixth of the world!) rubbish heap for filthy technologies...[56]

Despite expressions of concern for the environment, *perestroika* did very little for the environment.

## *Conclusion*

What, then, can be concluded from this examination of Marxism in general and Soviet Marxism in particular in relation to the environment? To begin with, it is necessary to accept that Marx's work remains the single most important critical analysis of the distinctiveness, dynamism and destructiveness of capitalism. As part of this analysis, Marx revealed the inevitable environmental destructiveness of capitalism to the environment, and he tried to point the way to overcoming this system and its nihilistic modes of thinking. Some of these insights have since been developed further, and those influenced by his ideas have revealed the extent to which the dynamics of capitalism are responsible for the massive destruction of the environment in the Third World. It is clear from their work that an unregulated market will lead to enormous suffering and destruction in the short term, and total disaster in the long term. It will only by overcoming the fetishism of commodities, by controlling of the dynamics of the market and creating a socio-economic formation in which people and nations are not forced into continual competition with each other for economic survival, that it might be possible to live without destroying the conditions for humanity's continued existence.

However Marx under-estimated the limitations of the natural environment, and those who have used Marxist notions to analyse environmental problems have been marginal to the mainstream of Marxism. The failure to take up and carry through the initiative of Podolinskii in reformulating the theory of value to ground it in physical reality has meant that most Marxist thought has not taken the environment into account in any systematic way - with disastrous consequences for the environment in communist countries. As Juan Martinez-Alier argued:

> The ecological view of the conditions of human existence could have been easily connected with Marxism through an adequate definition of productive forces or productive powers. This was not done by Marx. Despite the superficial similarity between an ecological approach and an approach in terms of 'reproduction' of social systems, there has been a long-standing divorce between Marxism and ecology.[57]

The best framework of analysis incorporating Marxist notions to study environmental destruction, that of Stephen Bunker, broke fundamentally with many traditional Marxist ideas. To begin with, Bunker defended an energy theory of value in opposition to the labour theory of value, arguing:

---

56. Boris Kagarlitsky, *Fairwell; Perestroika*, London: Verso, 1990, p.204.

57. Juan Martinez-Alier, *Ecological Economics*, Oxford: Blackwell, 1987, p.5.

> A labour theory of value excludes from consideration the usefulness to continued social reproduction of energy transformations in the natural environment. Nor can it take into account the value of ideas, beliefs, and information which underlie human social organization. These and all other human experiences are formed out of previous dissipation of energy. ... Measures of energy and matter and their conversion, however, touch everything which is humanly useful. Rather than separating human activity from other ecosystemic processes, these measures allow us to see the interdependencies between human energy use and energy transformation processes which proceed naturally, i.e., without human intervention.[58]

From this starting point, he then went on to attack the blindness of various Marxist theories of development and underdevelopment:

> All these theories have assumed variants of labour theories of value; all have extended economic models based on the false notion that production systems in some sense are self-enclosed and can reproduce themselves; none has taken into account that production systems require extraction systems; that extraction systems subservient to present forms of industrial production inevitably deplete their own resource bases; and that this process is finite as the limited stock of matter and energy which is or will become convertible to human uses.[59]

On the basis of his research on the Amazon he attacked economic reductionism, arguing that: 'Marxist notions of the primacy of the economic in explaining the activity of the state ... must ... be qualified to include the motives of ideological consistency and bureaucratic facility within the state's political imperative to maintain itself and to expand its control.'[60] This is clearly a long way from mainstream Marxism.

While in the Soviet Union Marxism did originally provide the basis for the development of an approach to the environment in which the constraints of ecosystems were recognized, this was swept away with the Cultural Revolution and the rise of Stalinism. It appears that Marxism was developed and incorporated into Soviet society in such a way that it became very similar, though not identical, to Social Darwinism in the West. With the development of Marxism into Stalinism, Soviet society incorporated the Western fixation on progress through technological transformation of the physical world and the development of a consumer society in which the significance of people is measured in terms of money, the mode of thinking foreshadowed by the Russian nihilists of the 1860's. As in the West the ultimate goal of society was seen as economic progress in order to win out in the struggle for world power, and everything was reduced to a means to this end, though in the case of the Soviet Union this was supposedly part of a long term struggle to realize the immutable ideal of communism. In both forms of society an extreme instrumentalist orientation was institutionalized so that both nature and people came to be defined by the dominant institutions as instruments of economic progress. Soviet Marxism was then scarcely less nihilistic than the modes of thought dominating the West.

In fact it appears that the roots of Soviet Marxism, as with the culture of the West, lay in ancient Greek philosophy, specifically in Neoplatonism. This Neoplatonism had been developed so that the cultures of Eastern and Western Europe had become mere structural transformations of each other.[61]

---

58. Stephen Bunker, *Underdeveloping the Amazon*, Urbana and Chicago: Illinois Uni. Press, 1985, p.35.

59. Ibid. p.244.

60. Ibid. p.223.

61. This is similar to cultural transformations noted by anthropologists and historians. Levi-Strauss identified a number of these among different Indian groups in Brazil, and similar transformations have been revealed between the Maori and Hawaiian cosmologies by Marshall Sahlins and between the Roman historical epic and Indian cosmic myth by Georges Dumazil. See Marshall Sahlins, *Historical Metaphors and Mythical Realities*, Ann Arbor: The University of Michigan Press, 1983, p.14.

Following the Christian Neoplatonic tradition there was a common rejection of the changing world of the present for an eternal world. But while in the West, dominated by St Augustine's philosophy, this eternal world was seen as beyond the material world, in the East the eternal world was seen as something to be realized through the transfiguration of the material world. While in the West the rejection of the world ultimately produced mechanistic materialist science, capitalism, neo-classical economics, Darwinian evolutionary theory and Social Darwinism, in the East it produced a culture oriented to realizing a perfect world on earth. So both East and West, which between them dominated the world, were both ultimately founded on the Neoplatonic rejection of life and becoming for what is eternal, and they both produced societies within which everything came to be reduced to instruments for some abstract notion of progress. As in the West, political movements which contravened the metaphysical assumptions dominating society were unable to achieve the unity required to become effective.

Thus the important question, in what sense can Marxism be said to be an alternative to the prevailing world-view of the West of neo-classical economics and Social Darwinism based on mechanistic materialism, cannot be answered in any simply way. Its interpretative successes prevent its being completely ruled out, but these have generally been undertaken by non-orthodox Marxists. As a starting point in any attempted evaluation it is necessary to rule out orthodox or neo-orthodox Marxism, that is, an economic and/or class reductionist theory of society and history underpinned by a radical Christian Neoplatonist eschatology - but assuming humans to be egoists, in which technology is seen as the engine of progress leading inexorably towards a proletarian revolution and a socialist mode of production from which a socialist society will unfold itself. For the superiority of Marxism to be demonstrated it must be shown to provide a basis for comprehending both the successes and the failures of the prevailing ideology, and for going beyond these failures. But orthodox Marxism explains virtually nothing that cannot be explained by Social Darwinism, and Social Darwinism can explain much else beside. Social Darwinism justifies the contention of orthodox Marxists that history has been the progression of socio-economic formations from those which produce small surplus value to those which produce greater surplus value. But the driving force of this has been the struggle between societies, with those formations able to devote the greatest efforts to developing their war machines having subjugated the rest. From the perspective of Social Darwinism, political ideals, including communism, are simply instruments for mobilizing people to unite in their struggle against others. When a society is not threatened by outsiders, people will not subordinate themselves to the interests of society as a whole but will struggle for supremacy against each other, using political ideals as means to exploit each other. The liberation of humanity from such egoistic struggle is impossible, and where people do succeed in reducing competition, the result is decay and stagnation. All this is clearly manifest in the history of Russia, from when its struggle against the West began to the present, from the rise of Marxism to the stagnation of the Soviet Union under Brezhnev and Kosygin, to Gorbachev's *perestroika* when under pressure from a more dynamic West, markets were introduced to promote efficiency. If this is the case, there can be little hope for the future of the environment. All that can be expected is an intensification of the struggle between nations and power blocks for diminishing resources, which will inevitably increase the rate of environmental destruction and the conflict between nations until overwhelming international tensions culminate in all-out nuclear war. Orthodox Marxism provides no solution to this.

However I have also tried to show that there are tendencies within Marx's thought associated with the different metaphysical assumptions which have been suppressed by Marxists. It was because Marx did not fully emancipate himself from the prevailing intellectual environment that his thought is vitiated by tendencies towards a conception of history as a unilinear movement towards a final state of perfection and by tendencies towards technological determinism. And it was for this reason that forms of Marxism which freed themselves from these conceptions, such as that developed by Bogdanov and elaborated by the *Proletkul't* movement, could be condemned as heretical, and which allowed Soviet Marxism to be transfigured into Stalinism. This transfiguration was further facilitated by the original condition of Russia - its economic backwardness in relation to the Western powers threatening it, and the underlying Neoplatonism of Russia's Orthodox Christianity. It is for this reason that despite the orientation towards the liberation of human potential implicit in all Marx's work, and despite the development of dialectical materialism with many features in common with process philosophy, it has been the Neoplatonist and mechanist aspects of Marx's thought which have dominated and become institutionalised within Soviet society. In fact Marxism has been essentially a means of appropriating the orientation to the world developed in Western Europe by Russia, and then following Russia, by other areas of the world.

In opposition to both Social Darwinism and orthodox Marxism I have shown how significant is culture and the forms of thinking embodied by it in history, that culture cannot be explained as nothing but instruments of egoistic struggles and that ideas play a major role in determining the direction of development of societies and civilizations. Furthermore I have shown that Marx's most original ideas have never really been given a chance, although the *Proletkul't* movement took the first steps necessary for recreating a socialist society in the years immediately following the revolution. I have suggested that the dimensions of Marx's thought which show promise of challenging the dominant ideology, those in terms of which the analysis of the dynamics of capitalism was based, those on which the *Proletkul't* movement was based and the original successes of environmentalism in the Soviet Union were founded, and those which accord with the analysis of the dynamics of society presented here, are those which accord with a process view of the world. This process view of the world undermines the most fundamental assumption of both Western and Eastern culture which has underlain the aggressiveness and oppressiveness of each, that the end of history is more significant than the process of moving towards it. And in doing so it provides a framework to incorporate the insights of Podolinskii, Bogdanov's quest for a new science, and the anti-reductionist ecology of Stanchinskii, and to support the ideas of Soviet environmentalists such as Oldak, Frolov and Arutiunov and Western environmentalists such as Stephen Bunker. So while orthodox Marxism is little more than a variant of Social Darwinism, a consistent reformulation of Marx's ideas in terms of process philosophy might offer an alternative vision of the future to that of both Western capitalism and Soviet bureaucratic centralism, able to provide the coherence for new political movements and for new forms of social life, other than those based on the market and those based on a centrally planned economy.

However merely proposing an alternative to capitalism and bureaucratic centralism does not say anything about its viability. As Alec Nove asked of socialism: 'What if the vision is unrealisable, contradictory? Does it make sense to "blame" Stalin and his successors for not having achieved what cannot be achieved in the real world?'[62] It has also been pointed out how much more difficult than Marx anticipated will be the task of those struggling to create a more just world, and that the difficulties are increasing. Communism in Eastern Europe has failed, the proletariat is fragmented and we can no longer rely on a growing proletariat, the capitalist system has generated an immense apparatus of social control which functions to ensure its continued reproduction, it has engendered an international struggle for power so that any country which fails to keep pace with economic and technological developments of the most powerful nations is liable to be subjugated, and it has generated transnational corporations of enormous power which cannot be controlled by any State. Only by deepening our understanding of the world social order, developing an alternative world-orientation to that of mechanistic materialism and by developing new strategies for action can there be any hope for the future. And what will now be argued is that this will only be possible through the development of a process world-orientation, the development of a new ethics, political philosophy and science of humanity on this foundation, and a struggle to act and to live in accordance with this philosophy.

62. Alec Nove, *The Economics of Feasible Socialism,* London: George Allen & Unwin, 1983, p.ix.

# 12

# EPISTEMOLOGY, DIALECTICS AND METAPHYSICS

The analyses of the relationship between environmental problems and the dynamics of Western civilization and Russian and Soviet culture have been undertaken from a perspective outside both the dominant world-orientation and orthodox Marxism. The ordained disciplinary boundaries have been ignored and questions posed which are of a scope generally not considered academically respectable. This was necessary because only in this way has it been possible to reveal the metaphysical assumptions, encoded in disciplinary boundaries and in what is considered 'academically respectable', underlying Western civilization and Eastern European culture. Furthermore, relationships such as those between the concepts developed in practices and those which are articulated theoretically, socio-economic formations and geographical conditions, have been examined from a pre-formulated interpretative scheme. This interpretative scheme and the associated disregard for academic boundaries is justified on the basis of metaphysical notions, those of process philosophy, which I contend have the capacity to completely replace the dominant world-orientation and orthodox Marxism. The attempt to redefine academic boundaries and to analyse the ideological history of European civilization has been designed as a test for this metaphysical scheme. If I have been at all successful in these analyses, this should provide some evidence in favour of process philosophy.

In the following chapters this metaphysical scheme will be articulated more fully and the framework of analysis which until now I have assumed, will be defended. At the same time the approach used: of analysing the problems and dynamics of civilizations in terms of an implicitly accepted set of assumptions about the nature of the world and of humanity, and then concluding by defending these assumptions, will be justified. The nature of metaphysics will be clarified and it will be shown what sort of reasons could justify the claim that process philosophy is superior to prevailing metaphysical assumptions. This requires an examination of the relationship between metaphysics and epistemology, which will be the main subject of the present chapter. Then a defence of process philosophy will be made on the basis of its capacity to generate the concepts required to make intelligible both the phenomena revealed by recent advances in the physical sciences and the existence of life and mind; how concepts deriving from process philosophy have been vindicated within various domains of science, and how these offer support both for each other and for many of the 'creative redescriptions' of life and humanity developed within philosophy.[1] This work itself attempts a creative redescription of these ideas in an attempt to formulate the process world-orientation with greater coherence, and to improve its prospects as a research programme for the sciences and humanities, as a basis for action, and ultimately as the foundation for a new world order.

## *What is Metaphysics*

As was suggested in the introduction to this book, in the academic community the enterprise of developing metaphysical systems, that is, speculative theories about the nature of being or existence, is barely acknowledged to have any meaning, and the forms of reasoning associated with

1. The term 'creative redescription' comes from Charles Taylor, 'Philosophy and its History' in Richard Rorty et. al. eds, *Philosophy in History*, Cambridge: Cambridge University Press, 1984, pp.17-30. Taylor defends it as an essential part of philosophy.

it have no acknowledged status.[2] There are a number of layers of obfuscation involved in the denigration and rejection of metaphysics, some of which have been implicitly or explicitly argued against. Firstly, metaphysical speculation is represented as having been displaced by science. That in the seventeenth century 'science' was part of philosophy, and that its advance was only possible because philosophers had developed a coherent metaphysical theory which could serve as the basis for experimental research, is not generally acknowledged. Secondly, that what is now taken as common sense is largely the world-view based on this metaphysical theory is denied. Thirdly, scientific theories, insofar as they accord with the metaphysical assumptions which now dominate everyday life - such as Newtonian physics and Daltonian and Mendeleevian chemistry - are presented in a dogmatically realist way, while those developments of science which bring these assumptions into question - such as relativity theory, quantum theory and non-linear thermodynamics - are presented in educational institutions in a forbiddingly formalistic manner interpreted in a vaguely positivist way; as though all that matters is getting the predictions right. This has been reinforced in recent years by the tendency to conceive science only as a means for developing technology. Finally the nature of metaphysics, insofar as it is considered at all, is grossly misrepresented, 'metaphysics' being used as a term of denigration for everything from dealing with questions about the existence and nature of a transcendent realm of being beyond what is knowable empirically (by Kant and then by the logical empiricists), and belief in a reality independent of all interpretations and beliefs to which true beliefs correspond (by Hilary Putnam), belief in immediately given absolute knowledge (by Jacques Derrida), to scholastic nitpicking (by almost every other philosopher). Metaphysics is seen to be in a space of its own, and at very best a decoration to life.[3]

What then is metaphysics? The notion of metaphysics derives from Aristotle, and simply designates the work that came after physics in Aristotle's collected writings. Consequently it is how Aristotle defined his subject matter and what he analysed in his *Metaphysics* which must be given the pre-eminent position in defining the subject. Aristotle defined the subject thus:

> There is a science which takes up the theory of being as being and of what 'to be' means, taken by itself. It is identical with none of the sciences whose subjects are defined as special aspects of being. For none of them looks upon being on the whole or generally; but each, isolating some part, gets a view of the whole only incidentally, as do the mathematical sciences. Since we are searching for the first principles and most general factors of being, these must clearly be distinctive traits of some nature.[4]

Aristotle did not see himself as creating the subject matter of metaphysics but as clarifying what had always been the central problem of philosophy. As he put it: 'In short, the question that has always been asked and is still being asked today, the ever-puzzling question "What is being?" amounts to this: "What is primary being?"'[5] ('Primary being' here translates '*ousia*', usually and

---

2. This theme together with a defence of metaphysics has been further developed in Arran E. Gare, '"After Philosophy" After Metaphysics', (forthcoming).

3. There are exceptions to all this. The most clear-headed modern defences of metaphysics have been made by Charles Sanders Peirce in 'The Approach to Metaphysics' republished in Justus Bucher ed. *Philosophical Writings of Peirce*, N.Y.: Dover, 1955, pp.310-314, and by Alfred North Whitehead in 'Speculative Philosophy', *Process and Reality*, [1929], N.Y.: Free Press, 1978, Chapter 1, pp.3-17. More recent purported defences of metaphysics such as Stephan Körner, *Metaphysics: Its Structure and Function*, Cambridge: Cambridge University Press, 1984; Brian Carr, *Metaphysics: An Introduction*, Houndmills, Hampshire: Macmillan, 1987; and José A. Benardete, *Metaphysics: The Logical Approach*, Oxford: O.U.P., 1989, only defend a very attenuated form of metaphysics.

4. Aristotle, *Metaphysics*, 1003a21-28. Unless otherwise indicated, I have quoted the Richard Hope translation, University of Michigan Press, 1960.

5. Ibid. 1028b3-5.

very misleadingly translated as 'substance'.)[6] And he saw metaphysics as basic to all enquiry, both theoretical and practical, by facilitating the investigation of the world and by enabling the world to be understood in all its complex diversity:

> ... since any science deals chiefly with what is primary to its subject, other considerations being derived from and dependent upon the primary, the philosopher must have within his province the first principles and primary factors of primary beings. Furthermore, as any class of things is united in sense perception and in a science (for example, grammar is one science and unites in theory all articulate sounds), so the theoretical science of being as being includes as its parts the sciences of the species of being within the general class of being as being.[7]

At the same Aristotle included a definition of metaphysics as a theory of entities which are both independent and immovable, that is, as theology, which was then distinguished from the science of entities which are independent but changing - natural philosophy, and the science of entities which are immovable but dependent - mathematics. However this definition can be regarded as a particular answer to the question of what is being, and it is not the only answer proffered in the *Metaphysics*. Even while claiming that metaphysics is the science of immovable primary beings, the unmoved movers, Aristotle acknowledged that if there were no such beings then natural philosophy would be first philosophy;[8] and in Books *Zeta, Eta* and *Theta*, which may have been written later than the other books, Aristotle appears to have accepted this identification and characterized primary beings as individuals, the intelligible constitutions of which are the outcome of processes.

Metaphysical questions are only indirectly related to questions such as whether knowledge is obtainable by reason alone, whether there can be synthetic *a priori* knowledge, or whether the nature of the world beyond our experience can be known. These are epistemological questions. And there is no reason why a theory of being should not be based on experience, and ultimately verified in experience. One of the most important requirement of a theory of being is to be able to serve as the foundation for the sciences. As Kant argued in his *Metaphysical Foundations of Natural Science*, metaphysics must demonstrate the possibility of the theoretical concepts required by the different sciences.[9] A metaphysical system should be able to provide the basis for understanding inanimate nature, what is life, and in particular, what is human life. Science has a privileged place not only because of its past achievements, but also, as Aleksandr Bogdanov pointed out, because it is the one area in which the division between intellectual and manual labour has been transcended and theory and practice united. Then, as Aristotle argued, by providing the basis for an understanding of the place of humanity in the cosmos, a metaphysical theory serves as the foundation for the practical sciences: ethics and politics, and for the productive sciences: art and technology. By doing so, metaphysical ideas can be incorporated into the social and physical

---

6. 'Substance', implying that which 'stands under' changing accidents, is a totally inadequate translation of '*ousia*'. To overcome this, Leibniz used the term 'monad' and Whitehead the term 'actual entity'. 'Primary being' is used by Richard Hope in his translation of Aristotle's *Metaphysics*, and also by Edward Pols. For a discussion of this concept, see Ivor Leclerc, *Whitehead's Metaphysics*, [1958], Lanham: University Press of America, 1986, Ch.2.

7. Aristotle, *Metaphysics*, 1003b15-23.

8. Ibid. 10026a25-32.

9. See Immanuel Kant, *The Philosophy of Material Nature*, Indianapolis: Hackett Publishing Co., 1985 which contains the James W. Ellington translation of 'Metaphysical Foundations.' That metaphysics so understood is essential to science has been argued this century by Alfred North Whitehead, Alexandre Koyré and Edwin Arthur Burtt; and more recently by Gerd Buchdahl in *Metaphysics and the Philosophy of Science*, Oxford: O.U.P., 1969, and also by Ivor Leclerc in *The Nature of Physical Existence*, London: Allen & Unwin, 1972, and in *The Philosophy of Nature*, Washington: C.U.A. Press, 1986, (although Leclerc does not characterize natural philosophy as metaphysics).

worlds of people as a major component of what is taken as common sense, and so can become a major determinant of the way people live their lives.

## *Epistemology, Dialectics and Process Philosophy*

How then is any particular metaphysical theory to be judged? This brings us to the question of what is knowledge. The problem with most theories of knowledge is that they presuppose and assume the validity of a particular metaphysical theory. They do this not only by the criteria they present for judging the validity of claims to knowledge, but also by the very conception of knowledge they take for granted. This tendency arises from the nature of metaphysical systems which must be total perspectives on the world and must therefore include theories of knowledge as part of their domain. Theories of knowledge therefore must be able to be explained by the metaphysical systems which are legitimated by them. The consequence of the acceptance of metaphysical assumptions is that any alternative to the theory of knowledge which has been formulated in accordance with the dominant metaphysics will tend to be evaluated in terms of criteria based on the assumptions of the dominant metaphysics. Any theory of knowledge which implies that these assumptions are questionable will thereby be ruled out. What will be defended here is a theory of knowledge which is compatible with alternative metaphysical systems and which at least allows the possibility of cognizing the intrinsic significance of the world. As stated in the introduction to this book, a dialectical theory of knowledge in which the ultimate goal of disciplined inquiry is understanding will be argued for. To avoid the dogmatism which would follow from a closed circle in which a theory of being and a theory of knowledge mutually imply one another, a theory of knowledge, explicable by a philosophy of process, which can justify a claim to validity being made by this philosophy without assuming that it is valid, will be proposed and defended.

For most of this century the academic establishment in Anglophone countries have accepted some variant of logical empiricism as the correct account of the nature of knowledge.[10] The status of this doctrine can be attributed to its concordance with the dominant metaphysical assumptions of society. These have their roots in Platonism but are more immediately grounded in the acceptance of a mechanistic conception of the world. The most important and most taken for granted assumption of logical empiricism is that the objects of knowledge must be found outside time and be free of all particular viewpoints. Truth must therefore be unprovisional. This assumption derives from the arguments of Plato concerning the need for an omni-temporal object of knowledge if knowledge itself is to escape the flux of change. Then with the development of mechanistic materialism (itself grounded in Platonistic forms of thinking) the central problem became: How can individual minds, spatially enclosed within mechanical bodies, attain knowledge of the outside world? This problem was brought into focus by Descartes' replacement of the medieval notion of intentionality by the notion of knowledge as representation, and was explicitly formulated in these terms by Hobbes. Locke's representational realism according to which knowledge is conceived of as 'ideas' of primary qualities in the mind which actually represent the external world, the subjective idealism proposed by Berkeley and worked out consistently by Hume according to which we can only talk about such ideas (or sense impressions), their copies and their relationships, and Kant's transcendentalism according to which sensations are ordered by imagination, the forms of intuition and categories of the understanding, are all proposed solutions to this problem. Logical empiricism can be understood as the effort to represent the objects of knowledge as eternal (true propositions or facts and the logical relations between them), while being consistent with the form of empiricism engendered by the mechanistic conception of the world (true propositions are those

10. See Richard Rorty, *Philosophy and the Mirror of Nature*, Oxford: Basil Blackwell, 1980 for a critical analysis of modern epistemology.

which have been confirmed directly or indirectly by sense impressions). And this amounts to a research programme to characterize human knowledge and rationality in a way which is consistent with the conception of humans as complex machines - a project continued by most Anglophone philosophers of language. This was clearly evident in Russell's philosophy, though most of his epigoni have lost sight of this as their goal. In fact the goal has been hidden by the anti-psychologism of these epistemologists, while at the same time the problems which revealed themselves within this research project have to some extent undermined the programme. Still, the project's underlying *telos* has been revealed by its success in providing the foundation for the development of computers. Knowledge is now seen as information which can be stored and processed by machines. It is for this reason that logical empiricism appears as a 'hard-headed' conception of knowledge, despite its manifest failures.

Despite their overwhelming success within the philosophy of science, the more radical opponents of logical empiricism have been suspect and misinterpreted because they have rejected the prevailing assumptions and their associated problematics.[11] Historically oriented post-logical empiricist philosophers of science abandoned the attempt to characterize knowledge atemporally or specify in atemporal terms the criteria of validity of all inferences, and the conception of knowledge as a relationship between the individual consciousness and the external world. In effect they were no longer constrained by the effort to understand human thought in mechanistic terms. They were more likely to be concerned with scientific creativity and with freeing science from the constraints of over-rigid methodologies rather than formulating prescriptions to delimit science. Consequently they were inclined to formulate their arguments in accordance with this freer notion of rationality. To those committed to the dominant modes of thought, therefore, their work appeared to lack the features which are required for intellectual respectability.

To overcome this situation the ideas of the historically oriented opponents of logical empiricism will be supported by presenting a theory of knowledge specifically in terms of anti-mechanistic metaphysical assumptions.[12] My concern in doing so will not be to specify the eternal criteria for what is to count as knowledge and valid inference, which I believe to be impossible, but to address the more fundamental issue (addressed by Plato in *Theaitetus*) of what knowledge is; though without devoting an entire work to the subject neither all the ramifications of this answer nor all the reasons for accepting it can be presented.

The theory proposed here is meant as a continuation of the tradition which began in Ancient Greece with the establishment of the democratically ruled polis on the assumption that truth could be arrived at through discussion. Though this truth was seen to be continually open to revision and development by individuals in dialogue, it was given a status and significance above and beyond individuals. This is the tradition which was formalized in philosophy as dialectics (from the Greek *dialektos*, 'art of debate' and *dialektike*, 'discourse').[13] According to Aristotle, it was Zeno of Elea who 'invented' dialectic as a philosophical method.[14] In his hands it was a way of refuting an opponent's opinion by accepting it hypothetically and forcing his opponent to admit that it led to conclusions contradicting this opinion, or some other of his beliefs. Plato explored the possibilities

---

11. See Richard J. Bernstein, *Beyond Objectivism and Relativism*, Philadelphia: University of Pennsylvania Press, 1983, p.52ff. for an account of the misrepresentations of anti-positivist philosophers of science.

12. Some justification for such an approach is provided by Hilary Putnam who summed up the argument of his book *Reason, Truth and History* (Cambridge: C.U.P., p.215) as 'that theory of truth presupposes theory of rationality which in turn presupposes our theory of the good. "Theory of good", however, is not only programmatic, but is itself dependent upon assumptions about human nature, about society, about the universe (including theological and metaphysical assumptions).'

13. For a history of dialectics see Ronald H. McKinney, 'The Origins of Modern Dialectics' in *Journal of the History of Ideas*, Vol.44, No.2, 1983 pp.179-190.

14. Reported from Aristotle's lost *Sophist* by Diogenes Laertius VIII; see G.S. Kirk, J.E. Raven and M. Schofield, *The Presocratic Philosophers*, 2nd ed., Cambridge: Cambridge University Press, 1983, p.278.

of dialectics and its relation to truth more fully. For him it became a form of dialogue in which through a series of questions and answers, and ultimately through the method of division and synthesis, one attempted to find true definitions and an understanding of their relations to achieve a comprehensive knowledge of the diverse totality of the forms of the world. He was particularly concerned with which forms could and which could not be combined with each other. Plato himself denied having arrived at eternally valid truths, although he presented this as an ideal. Aristotle also accorded a primary place to dialectic conceived as the critical examination of reputable opinions to establish the first principles of any enquiry.[15] And in practice he developed all his ideas dialectically. But he also presented the ideal of knowledge as the attainment of eternal verities presentable through syllogistic logic. With the rise of Christianity and its development in the Middle Ages, the philosophies of Plato and Aristotle were assimilated in such a way that to begin with, genuine dialectics was eclipsed by the view of knowledge as doctrine to be passed on through disciplines from generation to generation. However dialectics of a kind was revived in the high Middle Ages as a means of settling disputes in scriptural exegesis. To resolve conflicts of interpretation a procedure was adopted in which the defender of a thesis would state his case, and then his opponent would offer a *prima facie* proof of the opposing thesis. The defender would then make a concession, but by positing a crucial distinction, would nullify the objection, after which the opponent would either contest this distinction or accept it and then go on to challenge the new premise.

Such dialectical thinking was crushed in the later Middle Ages, and was totally eclipsed with the rise of Protestantism, the counter-reformation and the rise of the mechanistic world-view. Descartes opposed dialectics in his effort to conceive knowledge in relation to the isolated individual subject and to provide an unquestionable foundation for all knowledge. This subjectivist turn was brought to fulfilment in the philosophy of Kant who revived the concept of dialectical reason, but reduced it to a way of affirming apparent contradictions. He argued that by positing the necessary distinctions of meaning, both of the opposing sides could be accepted. Then Hegel rejected Kant's approach for failing to acknowledge the existence of genuine contradictions in reality, and reformulated dialectics into an account of the self-movement of thought, identified with reality. This self-movement takes place through a process of *Aufhebung* by which inadequate ideas generate their negation, which is then overcome by a further negation, which at the same time affirms the opposing principles. As a procedure for analysing the historical development of ideas and forms of life in the *Phenomenology of Spirit* and in the historical lectures, this conception of dialectics proved extremely fruitful. However when used to elaborate a categorial scheme in the *Logic*, this procedure congealed into a closed system. Despite his concern to exhibit a 'logic of process', Hegel's orientation was towards the eternal.[16] As Mikhail Bakhtin described this form of dialectics: 'Take a dialogue and remove the voices ..., remove the intonations (emotional and individualizing ones), carve out abstract concepts and judgements from living words and responses, cram everything into one abstract consciousness - and that's how you get dialectics.'[17] In returning to the tradition of dialectics it is necessary to jettison completely its association with the search for eternal verities and to return to the Ancient spirit of dialectical thinking as critical and creative dialogue - incorporating into this Hegel's socio-historical dialectics. And in doing so it is necessary to reject Platonic realism in favour of conceptualism.

To overcome prevailing assumptions and to justify a return to dialectics the starting point taken here is not in a conception of the subject as the bearer of knowledge, but in a theory of the nature of the world, in a conception of humans as processes of becoming within an active, dynamic nature always beyond their full comprehension; as embodied subjects who are essentially social, and

---

15. Aristotle, *Topics*, I 2, 101a25-101b4.

16. On this, see Michael Rosen, *Hegel's Dialectic and its Criticism*, Cambridge: C.U.P., 1982.

17. M.M. Bakhtin, *Speech Genres & Other Late Essays*, tr. Vern W. McGee, ed. Caryl Emerson and Michael Holquist, Austin: University of Texas Press, 1986, p.147.

through whom the world is, at least to some extent, being brought to consciousness of itself, its uniformity, its creativity, and that it is able to become conscious of itself.[18] In opposition to the idea of enquiry as the accumulation of knowledge conceived of as bits of information, the ultimate aim of all enquiry is taken to be 'understanding'.[19] The ability to confirm particular propositions, which has a place in the process of enquiry, must always be understood as secondary to and as an extension of this. Understanding, as the word implies, is a mode of being in the world by which the world becomes in some degree intelligible, a way of experiencing our world as at least a partially comprehensible reality. One struggles for understanding as a participant in a social tradition, even if one is rebelling against traditional assumptions, and advances in understanding always involve a struggle to overcome the limits of one's tradition, including its language. As Whitehead put it: 'Words and phrases must be stretched towards a generality foreign to their ordinary usage; and however such elements of language be stabilized as technicalities, they remain metaphors mutely appealing for an imaginative leap.'[20] To advance understanding is not to know the eternal but to produce theories from the perspective of which new aspects of the world can be understood and the achievements and limitations of all rival theories can be comprehended.

Participating in the struggle for understanding raises individuals above the flux of their own immediate becoming in the world, which is what Plato validly recognized to be important; but this is achieved not by discovering the eternal but by participating in the creation of a temporal order transcending the perspectives of individual subjects. This involves fusing experiential horizons, thereby expanding the intellectual community, not only between contemporaries, but also between those who have expressed themselves in the past and those who will struggle for understanding in the future.[21] This takes place within institutions, the nature of which is to some extent constrained by a larger social context. As Hegel argued, the achievements in philosophy or science in any era are only possible through the stand-point provided by the 'Objective Spirit' of that era, and they are only fully comprehensible in terms of that stand-point. As participants in the development of understanding it is necessary to acknowledge that our engagement in the development of understanding takes place within contexts which provide specific possibilities and problems for understanding, but precisely through this acknowledgement it is possible, at least to some degree, to overcome the constraints of, and to change our situations.

## *Twentieth Century Epistemology: From Logical Empiricism to Dialectics*

Most of the critics of logical empiricism in the twentieth century have been explicitly or implicitly in the tradition of dialectics, and have contributed to this tradition. It has been developed explicitly by a number of Hegelian Marxist thinkers, notably Lukács, Goldmann, Adorno, Sartre and Merleau-Ponty and implicitly by theorists of hermeneutics and theorists of cultural

---

18. In giving the privileged place to metaphysics over epistemology, to being over the subject, I am following Whitehead, Heidegger, Merleau-Ponty, Rom Harré and Roy Bhaskar, although the position defended here differs in some ways from the ideas of all these thinkers.

19. That the great achievements of science can only be understood as advances in understanding has been argued in two papers by Maurice A. Finocchiaro: 'Cause, Explanation and Understanding in Science: Galileo's Case', in *Review of Metaphysics*, Vol.29, 1975, pp.117-128; and 'Scientific Discoveries as Growth of Understanding: The Case of Newton's Gravitation' in T. Nickles (ed.) *Scientific Discovery, Logic and Rationality*, Dordrecht: Reidel, 1980, pp.235-255. However it is implicit in much of the anti-positivist tradition of epistemology, particularly in Toulmin, Polanyi and Kuhn, and was argued for by Alfred North Whitehead in *Modes of Thought*, [1938] N.Y.: Free Press, 1968, Lecture Three, 'Understanding', pp.42-63. It was also argued for by Émile Meyerson who was commended for this by Einstein.

20. Whitehead, *Process and Reality*, 1978, p.4.

21. The notion of 'fusion of horizons' is developed by Hans-Georg Gadamer in *Truth and Method* [1960] N.Y.: Crossroads, 1984, passim.

development. However the most telling critics of logical empiricism have been historians of science and historically oriented philosophers of science, and these critics have contributed most to reformulating knowledge in accordance with the tradition of dialectics.[22] Disputes within the philosophy of science led to an evolution of ideas similar to the dialectical scheme described by Hegel in the beginning of the *Phenomenology of Spirit*.[23] The effort to ground knowledge in sense experience or observed objects through ostensive definitions and logically proper names gave way to a focus on propositions, statements or descriptive sentences as the primary epistemological units. The subsequent failure to distinguish empirically meaningful propositions by some criterion of cognitive meaning shifted the focus of interest to conceptual frameworksr. The problems generated by the considerations of the relationship between different conceptual frameworks and any conceptual framework and the world led to the realization that the interesting questions about the rationality of scientific inquiry can only be understood in terms of the conflict of theories, paradigms, research programmes and research traditions in their historical development. The study of this historical development has revealed the close relationship between science and its socio-economic and cultural contexts, which in turn can only be comprehended from the perspective of a totalizing world-view, that is, a theory of history, a philosophical anthropology, a philosophy of nature, all founded on a general theory of being. This evolution transformed our understanding of science - although there is now a new generation of philosophers of science who are reworking the lower stages in this dialectic.[24]

The most influential logical empiricists were the logical positivists of the Vienna Circle. According to the original members of this group, the aim of science is to accumulate and order knowledge to enable predictions to be made.[25] They argued accordingly that valid scientific knowledge is analysable into singular existential statements reporting sense-experiences, but is organized as systems of mathematically expressible laws generalizing on observed regularities to enable people to calculate the probability of having any future sense-experience. Mathematics was held to be a deductive scheme reducible to logic (actually, to logic and set theory), and theoretical entities to be nothing more than heuristic devices to support the mathematics. By only accepting what has been observed in controlled experiments, science is able to accumulate certain knowledge about the world, expressible in laws of successively greater generality, and to thereby improve people's ability to make predictions. No other knowledge is accorded any significance, and metaphysics can be assigned to the scrap heap. As Rudolph Carnap claimed: 'metaphysics can make no claim to possessing a scientific character.... *Philosophy is to be replaced by the logic of science* - that is to say, by the logical analysis of the concepts and sentences of the sciences, for *the logic of science is nothing other than the logical syntax of the language of science*.'[26]

The rigourous efforts to justify this position revealed its inadequacy as a theory of knowledge, as an account of science and as an account of metaphysics. Efforts to develop logical positivism

---

22. I have argued in Gare, '"After Philosophy" After Virtue"' that Alfred North Whitehead was a major inspiration for this development.

23. This accordance has been noted by Richard Bernstein. See *Praxis and Action*, London: Duckworth, 1972 p.24n. and *Beyond Objectivism and Relativism*, p.75ff.

24. Recent philosophers of science have been preoccupied with the opposition between logical empiricism, constructivism and realism. For a review of these developments, see Richard Boyd, 'Introductory Essay', *The Philosophy of Science*, ed. Richard Boyd, Philip Gasper, and J.D. Trout, Cambridge, Mass.: M.I.T. Press, 1991.

25. For an illuminating history of twentieth century philosophy of science up to the mid-70's see Frederick Suppe 'The Search for Philosophic Understanding of Scientific Theories' in Frederick Suppe ed., *The Structure of Scientific Theories*, 2nd ed., Urbana: University of Illinois Press, 1977, pp.3-241. The picture painted of logical positivism oversimplifies the work of the Austrians, particularly when Otto Neurath's contribution is considered, but Suppe fully captures the spirit of Anglo-American logical empiricism.

26. Rudolf Carnap, *The Logical Syntax of Language*, [1934] tr. Amathe Smeaton, London: Routledge & Kegan Paul, 1937, p.xiii.

forced its major proponents to recognize the problematic nature of sense-experience and that there is more to theories than generalizations, eventually resulting in the acceptance by most of them of the reality of both objects of perception and theoretical entities.[27] However the anti-positivist philosophers have gone far beyond this, inverting the relationship between theories and the apparatus of prediction, arguing that it is theories, understood as the means to make the world intelligible, which are primary. This new notion of theory is spelt out by David Bohm:

> The word 'theory' derives from the Greek '*theoria*', which has the same root as 'theatre', in a word meaning 'to view' or 'to make a spectacle'. Thus, it might be said that a theory is primarily a form of *insight*, i.e. a way of looking at the world, and not a form of *knowledge* of how the world is.[28]

Sense-experience could no longer be regarded as the ground from which knowledge is built, or as simply the point of departure and point of return for a predictive apparatus. Rather, one of the most important aims of science is to enrich experience. As Bohm put it: 'science is *primarily* an activity of extending perception into new contexts and into new forms, and only secondarily a means of obtaining what may be called reliable knowledge.'[29] Such extension of perception involves the use of technologies of observation made possible by scientific theories. It is through science and technology that people are able to see the structure and dynamics of molecules and galaxies. Whether it is physicists in their laboratories or ecologists in the wilderness, the scientifically literate (unless they are under the influence of logical empiricism) should see more and have a richer experience of the world, than ignoramuses.

This development of experience cannot be understood only in relation to individuals; it is essentially social. Scientists are involved in a struggle to reveal the limitations of and go beyond what has been perceived in the past and to validate their own observations and theories in the eyes of others, communicating them (making them common) by defining them in propositional form to relate them to what has been commonly experienced in the past. As Bohm argued: 'scientific research does not consist of *first* looking at something and then communicating it. Rather the very act of perception is shaped and formed by the intention to communicate, as well as by a general awareness of what has been communicated in the past, by oneself and by others.'[30] The technologies of observation have been increasingly designed to facilitate communication, often inscribing a visible, quantified record of the observed situation. It is generally only in communication that the whole meaning of what has been observed is comprehended.

The study of theories, particularly at their inception, has revealed them to be based on analogies or metaphors.[31] These do not represent reality but are the means for making sense of the world,

---

27. This is not to say that all philosophers have abandoned logical empiricism. In philosophy departments dominated by analytical philosophy, philosophers still defend induction. For a modern logical empiricist account of science, see Ronald N. Giere, *Understanding Scientific Reasoning*, New York: Holt, Rinehart and Winston, 1979. Modern logical empiricism underpins the development of decision theory.

28. David Bohm, *Wholeness and Implicate Order*, London: Routledge & Kegan Paul, 1980, p.3f.

29. David Bohm, 'Science as Perception-Communication', in Suppe ed. *The Structure of Scientific Theories*, pp.374-391, p.374.

30. Loc. cit.

31. As Richard Boyd argued, 'metaphors are *constitutive* of the theories they express...' ('Metaphor and Theory Change' in Andrew Ortony ed., *Metaphor and Thought*, Cambridge: C.U.P., 1979, pp.356-408, p.360.) For a general study of analogies in science together with a review of other work in this area, see W.H. Leatherdale, *The Role of Analogy, Model and Metaphor in Science*, Amsterdam, 1974. The most important philosophers of science to analyse the role of analogies/metaphors have been Mary Hesse and Rom Harré.

serving as foundations or 'hard cores' of 'research programmes'.[32] The world is seen as something (as an organism, a mechanism or a complex of force fields, for instance), and is made sense of accordingly; it is not just represented as having certain characteristics. There is no such thing as access to the world as it is independent of any metaphors, and major advances in understanding involve successive reinterpretations through new metaphors of the world as it had been interpreted by old metaphors. Such metaphors define the objects of scientific enquiry,[33] radically restructuring the perception of situations, creating new questions and largely determining the nature of the answers. For instance, as Judith Schlanger pointed out, when the technological metaphor of regulation is adopted in the study of cells, this 'establishes the field for which it sets the boundaries and is the coordinator.'[34] Even that which cannot be comprehended in terms of these metaphors is defined and understood in terms of them. For instance once the analogy of cybernetics is assumed in the attempt to understand the brain, its inadequacies are described in terms of this metaphor: as the impenetrable 'black boxes' of the brain, or as 'the mysterious nature of human encoding and decoding'. This is because it is only through the metaphor that there is something to think about. As Schlanger commented: 'The cybernetic analogue provokes and instigates its own theoretical elaboration.'[35] The use of metaphors can no longer be seen as merely a heuristic device for formulating predictions; it is central to science. It is the finding of regularities in nature which must be seen as an heuristic for the deployment of metaphors.[36] It is necessary to regain the sense of science as the imaginative and creative use of metaphor, and to overcome the deadening effect of metaphors which are no longer recognized as such; it is necessary to recognize the profundity of Nietzsche's insight into what is taken to be truth:

> What then is truth? A movable host of metaphors, metonymies, and anthropomorphisms: in short, a sum of human relations which have been poetically and rhetorically intensified, transferred, and embellished, and which, after long usage, seem to a people to be fixed, canonical, and binding. Truths are illusions we have forgotten are illusions; they are metaphors that have become worn out and have been drained of sensuous force, coins which have lost their embossing and are now considered metal and no longer coins.[37]

Through elaboration, metaphors are articulated into frameworks of concepts. To 'conceive', deriving from the Latin *concipere*, means 'to take hold, to take to oneself, to take in', that is to perceive (from the Latin *percipere* - to grasp) some aspect of the world and to appreciate its relevance to other things perceived and conceived - and concepts, from *conceptus*, are the means for such 'grasping together' and 'taking in'.[38] Thinking of cognition in these terms frees us from

---

32. The terms 'hard core' and 'research programme' derives from Imre Lakatos, *The Methodology of Scientific Research Programmes. Philosophical Papers Volume 1*. ed. John Worrall and Gregory Currie, Cambridge: Cambridge University Press, 1978. However it is unlikely that Lakatos would see the hard cores of research programmes as being related in any way to analogies or metaphors.

33. Gaston Bachelard pointed out the importance of object construction in science, using the existence of theoretically constructed objects of inquiry as opposed to common-sense objects as the defining characteristic of a true science.

34. Judith Schlanger, 'Metaphor and Invention,' *Diogenes*, vol.69, 1970, p.21.

35. Ibid. p.26.

36. As argued by Romano Harré, *The Principles of Scientific Thinking*, London: Macmillan, 1970.

37. Friedrich Nietzsche, 'On Truth and Lies in a Non-Moral Sense' in *Philosophy and Truth: Selections from Nietzsche's Notebooks of the Early 1870's*, ed. and tr. Daniel Breazeale, New Jersey: Humanities, 1990, p.84.

38. This view of concepts as means which only become objects of thought when the mind reflects on its own operations goes back to Thomas Aquinas at least. Those philosophers who reject the notion of concepts often assume that they are and always have been understood as ideas or representations in the mind. See for example Hilary Putnam, 'The Meaning of "Meaning"', in *Mind, Language and Reality: Philosophical Papers, Volume II*, Cambridge: Cambridge University Press,

the conception of knowledge as a reflection or representation of reality - and the corresponding tendency to treat abstract concepts as concrete entities - what Whitehead called the 'fallacy of misplaced concreteness'. Efforts to conceive the world, including ourselves, reveals it as more - and more complex and interdependent - than the metaphors and conceptual frameworks which are used to grasp it; it is the 'unprethinkable Being' which is before all thought and presupposed by all thought and all enquiry.[39] The adequacy of such conceiving will be partly a function of the coherence with which concepts and their relations can be formulated and partly a function of what they reveal or fail to reveal of the world. For instance to define something as an acid is not just to identify and classify a kind of being which will always act in a certain way, which for instance will always dissolve metals. It is to relate it to an integrated framework of concepts, which means that the potentialities or powers of this kind of being, and the conditions required to realize them, can be recognized and explained in terms of 'atoms', 'valency', 'chemical bonding', 'electron', 'electrical charge', and so on, all of which allow the specific nature of acids to be distinguished from everything else.[40] The coherence of this framework ultimately derives from underlying metaphors from which these concepts originate and the articulation of these to make sense of specific phenomena. Such a framework is required to enable each individual to be comprehended and appreciated as an individual in all its uniqueness in relation to everything else in the world, and in relation to the world as a whole.

The articulation of metaphors into conceptual frameworks and the elaboration of concepts is a long, laborious process, essentially social in its nature. While the formulation of a particular concept (such as 'acid') might begin with a characterization in terms of immediately observable properties, science only advances through the development of theoretical concepts, originating in metaphors, then refined through being defined in relation to each other and in their application in efforts to understand the world. The nature of such formulation and elaboration can be seen in the development of the notion of field on which much of modern physics is based - from Leibniz's notion of the active monad (based on the analogy of mind), to Boscovitch's notion of point centres of power to attract and repel, to Priestley's rejection of points and reformulation of the notion of attraction and repulsion in terms of regions, to Faraday's application of this notion in his efforts to comprehend electricity and magnetism. In Faraday's work, the notion of field was still very vague, and was described in terms of mechanical metaphors and through the metaphorical use of such terms as 'tension', 'power' and 'force', and was frequently buttressed by the notion of a mechanical aether. The concept of field became increasingly better defined as the relationships between the electric field and other electromagnetic properties were more clearly specified. Then when Maxwell introduced the concept of the displacement current, it became possible to conceive light in terms of fields and to formulate Faraday's ideas in terms of a set of mathematical equations. While Maxwell himself was committed to explaining fields in terms of an aether, the recognition by Lorenz and Herz that the aether was not required by Maxwell's equations led to further fruitful lines of research, culminating in the development by Einstein of the general theory of relativity.

Granting a central place to metaphors and concepts still does not fully overcome the Platonic idea of knowledge as a perspectiveless eternal order transcending people's involvement within the world. The idea of using metaphors and concepts to make the world intelligible still implies an isolated, transcendent subject which acts like an executive in deploying abstract concepts to

---

1979, pp.215-271, p.218ff. For a defence of 'concepts' against criticisms from Quine, Witgenstein, Putnam and Kripke, see Tyler Burge, 'Concepts, Definitions, and Meaning", *Metaphilosophy*, Vol.24, No.4, 1993, pp.309-25.

39. As Schelling argued (echoing Aristotle in Zeta and Eta of *Metaphysics*) there is an *'unvordenkliches Sein'* (unprethinkable Being) - before all thought and presupposed by all thought. This view is shared by the 'existentialists' from Kierkegaard onwards, and by Whitehead, Heidegger and Merleau-Ponty.

40. See R. Harré and E.H. Madden, *Causal Powers*, Oxford, Basil Blackwell, 1975, Ch.1 for a defence of this view.

comprehend a previously uninterpreted world.[41] But the use of metaphors and the development of concepts and the associated conceptualization of the world take place through active, bodily engagement within a socially shared world which is itself active, which is always already partially understood, and of which people are part. It is only against the background of a pre-predicative experience of the world in relation to our bodily engagement in it, and through the generalization of body schemas, that it is possible to explicitly deploy metaphors and concepts.[42] The development of the capacity to constitute the world in new ways through new metaphors and concepts is at the same time the development of a mode of being in the world and of relating to others, a development of embodied subjects as the situated, social beings through whom the world is becoming conscious of itself. And it is generally associated with the development of the technology of experimentation, informed by theoretical concepts, which mediates experience of the world. Thus, knowledge is always situated and provisional. As Merleau-Ponty argued:

> As long as I cling to the ideal of an absolute spectator, of knowledge with no point of view, I can see my situation as nothing but a source of error. But once I have recognized that through it I am geared to all action and all knowledge which can have a meaning for me, and that it is gradually filled with everything which can exist for me, then my contact with the social in the finitude of my situation is revealed to me as the point of origin of all truth, including scientific truth. And since we have an idea of truth, since we are in truth and cannot escape it, all I can do is define a truth in the situation.[43]

Through his scarcely acknowledged influence on Michael Polanyi, and Polanyi's scarcely acknowledged influence on Thomas Kuhn, Merleau-Ponty was a major inspiration for the reorientation in the philosophy of science which led to the acceptance of this conception of knowledge. Polanyi argued that while scientific knowledge involves continual transcendence of the limitations of individual perspectives, it is still irreducibly personal; that it involves an 'indwelling' in the world such that each particular is perceived or known explicitly in terms of a background which is known tacitly so that there is a 'from-to' relation in all perception and all knowledge - we attend from the background to what interests us.[44] He compared such knowledge to what is involved in using an instrument such as a rake, and to understanding a sentence. Using a rake we 'indwell' in the rake so that it becomes an extension of our bodies, and our attention comes to be focused not on our hands manipulating the rake, but on the end of the rake. We attend from our hands and bodies and the perceptual background to the end of the rake and its relation to the task at hand. In understanding a sentence we have to 'indwell' in the meaning of each word and attend from these to focus on the meaning of the whole sentence, while to focus our attention on the meaning of any individual word in a sentence, we must 'indwell' in and attend from the meaning of the whole sentence. Extending this to science, the physiologist studying a body must 'indwell' in or

41. Only if conceptual frameworks are understood in this way is the argument of Donald Davidson valid (in 'On the Very Idea of a Conceptual Scheme', *Inquiries into Truth & Interpretation*, Oxford: Clarendon Press, 1984, pp.183-198). On this see Alasdair MacIntyre, 'Relativism, Power and Philosophy' in *After Philosophy*, ed. Kenneth Baynes et.al., Cambridge, Mass.: MIT Press, 1987, pp.385-411, esp. p.387.

42. The nature of this has been described by Mark Johnson in *The Body In The Mind: The Bodily Basis of Reason and Imagination*, Chicago: University of Chicago Press, 1987.

43. Maurice Merleau-Ponty, 'The Philosopher and Sociology' in *Phenomenology, Language & Sociology*, ed. John O'Neill, London: Heinemann, 1974, p.106; translation modified. A very similar view was defended by Whitehead, and later by George Herbert Mead in *The Philosophy of the Act*, ed. Charles W. Morris, Chicago: University of Chicago Press, 1938.

44. Michael Polanyi, *Personal Knowledge*, Routledge & Kegan Paul, London, 1958. See also Polanyi, *Knowing and Being: Essays by Michael Polanyi*, ed. Marjorie Grene, University of Chicago Press, Chicago, 1969, esp. 'The Logic of Tacit Inference.'

attend from physiological theory as a means to 'indwell' in or attend from the organism to comprehend any part of the organism. It is only because of such indwelling that the physiologist dissecting an organism is able to make sense of what he is focussing on. But similarly we 'indwell' in or attend from the parts of the organism to focus on it as a functioning whole. Polanyi argued that such indwelling and tacit knowing is involved to some degree in understanding all phenomena in nature. The importance of this is implicitly recognized in the work of Thomas Kuhn who emphasised the role of mastering exemplars - concrete problem-solutions, in scientific education.

In this context, the explicit formulation of propositions, including facts, must be seen as part of the process through which people develop their understanding of the world, to identify what observations are worth making, to communicate what they perceive and to dispute or negotiate with rival claims to knowledge or understanding - rather than as an end in itself, and the status of propositions can only be properly understood in terms of the intentions of those who present them. A proposition is, as the term suggests, a proposal, a bringing forward for consideration and exploration, and that a conjectured state of affairs be supposed to be the case. And facts are not what is in the world to which propositions conform if they are true, but propositions for which overwhelmingly compelling reasons have been produced for what they propose be supposed to be the case (at least for the time being), thus disposing people to think and act in new ways.[45] As the etymological root of the word fact (from *facere* - to make) suggests, a fact is something made, and its significance and status as such can only be judged by knowing the purpose for which it was made.[46] An assertion by Aristotle purporting to be a fact cannot be evaluated simply in terms of its accordance with what is held to be fact by modern science. It must be evaluated in terms of Aristotle's intentions in formulating it, his conflicts with other philosophers, and in terms of its role in the research and dialogue through which the potentialities and limits of Aristotle's research programme were revealed. And 'true' having originally meant 'trustworthy' would suggest that truth in science should be identified with those propositions and ways of understanding the world we can rely upon.

## *Mathematics, Logic and Language*

With this conception of knowledge it is necessary to reconceive the nature and role of mathematics. In the seventeenth century the applicability of mathematics to nature was seen in Pythagorean or Platonistic terms with mathematics understood as a transcendent realm of eternal truths and nature seen as being a mathematical order, while for logical empiricists, mathematics came to be seen as a system of tautologies useful for ordering knowledge to facilitate its storage and recovery and to make predictions. But the view of mathematics required to sustain these conceptions of the relationship between mathematics and the world has been undermined. Frege's effort to ground arithmetic in logic which inspired later efforts to conceive all knowledge as a logical structure was shown by Russell to be flawed. It implied that there is a class of all classes which are not members of themselves, which if it is not a member of itself, must be, and if it is a member of itself, cannot be. This is Russell's paradox. Gödel then showed that a non-trivial formalised system necessarily includes propositions the truth of which cannot be demonstrated in terms of the system, and that it is impossible to demonstrate the non-contradictory nature of such a system within the terms of that same system. The efforts to demonstrate the logical coherence of

45. For an overwhelmingly convincing argument that facts are nothing but 'true' propositions, see David Mitchell, *An Introduction to Logic*, 2nd ed., London: Hutchinson, 1964, pp.109-15.

46. That idea that facts are constructed by the scientific community has recently become very popular, but it was cogently argued for by Ludwik Fleck in *Genesis and Development of Scientific Fact* (tr. F. Bradley & T.J. Trenn, Chicago, Uni. of Chicago Press, 1979) which was first published in 1935.

mathematics revealed the impossibility of such demonstrations, undermining the certainty of mathematics.[47]

The nature of mathematics and its efficacy for science can be better appreciated by examining the way it was generated and has been developed. Saunders Mac Lane argues that different mathematical structures are grounded in and are elaborations of different basic human activities: counting, measuring, shaping, forming, estimating, moving, calculating, proving, puzzling and grouping.[48] This is why mathematics is applicable to the world. While this does capture both the early history of mathematics and how children begin to enter the mathematical realm, it does not account for how mathematics has developed beyond its elementary stages. To understand this it is necessary to turn to the work of Imre Lakatos on the history of mathematics.[49] Through his study of the development of Euler's theorem on polyhedra, Lakatos showed that far from being a discovery of eternal truths, the development of mathematics is itself a dialectical process of conjecture and attempted refutation, requiring much work to search for counter-examples, to elaborate concepts and proofs, to integrate these into a coherent system, and then to modify this system to deal with counter-examples. This process is essentially social, ideas being developed through an on-going dialogue between mathematicians in which theorems are proposed, and then definitions, proofs and refutations are proposed, revised and modified. In this process mathematical concepts are developed through negotiation and renegotiation, evolving to transcend and constrain each individual who participates in the development of mathematics.

The greatest advances in mathematics are achieved by utilizing the concepts developed in one branch to interpret another. For instance the major innovation of the Ancient Greeks was to interpret arithmetic through geometry. Descartes opened a new era in mathematics by interpreting geometry through algebra, which facilitated the development of calculus. From Cauchy to Weierstrass mathematicians concentrated on reinterpreting all branches of mathematics through arithmetic, which was then followed by the reinterpretation of all branches of mathematics through logic and set theory. In short, the realm of mathematics is a social construction based on the utilization of forms of cognition developed in practical experience as metaphors, articulated through negotiation and renegotiation into coherent frameworks of concepts, theorems, lemmas, proofs and refutations, and developed through a spiralling process of successively utilizing one branch of mathematics to interpret others. Like all metaphors, mathematical ideas are enduring structures of potential operations generated by and then constraining mental activity, rather than a set of eternal Platonic truths. And it is not as though mathematics is developing towards a fixed, eternal, logically coherent system which could guarantee certainty of logical deduction, the essential requirement of mathematics to fill the role prescribed for it by the logical empiricists.

Abandoning the fixation on eternal truths and focussing instead on mathematics as a social activity overcomes Russell's paradox. Once a class is seen as a theoretical object formed by a process of collecting, reflexivity becomes no more paradoxical than it is for the proverbial barber who has been told to shave all people who do not shave themselves. The barber shaves those who have not shaved themselves, and then at that time being one of those who have not shaved themselves, shaves himself. Similarly when conceiving a class as formed by collecting all classes which are not members of themselves - up until the time that the last such class other than itself is collected it is not collected, so it then collects itself.

Within the scheme of the new philosophy of science, branches of mathematics as systems of integrated concepts with relatively clearly defined principles of operation and transformation can be understood as important means to supplement and refine the non-mathematical metaphors and

---

47. See Morris Kline, *Mathematics: The Loss of Certainty*, Oxford: O.U.P., 1980, esp. Ch. XII.

48. Saunders Mac Lane, 'Mathematical Models: A Sketch for the Philosophy of Mathematics', *American Mathematical Monthly*, Aug. Sept. pp. 462-472. See also George Lakoff, *Women, Fire and Dangerous Things*, Chicago and London: The University of Chicago Press, 1987, Ch.20.

49. Imre Lakatos, *Proofs and Refutations*, Cambridge: Cambridge University Press, 1981.

conceptual frameworks of theories, thereby deepening understanding and facilitating the drawing of necessary conclusions and thereby the making of predictions.[50] For instance the development of the Cartesian coordinates, and following this, the development of the calculus, provided a way of refining the mechanistic concepts of matter and motion - particularly acceleration and revealing all the implications of these concepts. Similarly, Maxwell was able to develop mathematically the concepts of electric and magnetic fields of the mathematically illiterate Faraday, and thereby was able to go beyond him and demonstrate that electromagnetic fields could generate waves which would travel at 186,000 miles per second, the velocity which had already been measured as the velocity of light, and to postulate the existence of radio waves. In the twentieth century C.H. Waddington's work on epigenesis leading to notions of epigenetic landscapes, time paths (chreods), self-stabilization along such paths (homeorhesis) and switches, inspired René Thom to develop his catastrophe theory in terms of which such notions have been refined and applied to new areas. More recently, Bohm, Hilley and others attempted to clear up the chaos in quantum mechanics by using holography as an analogy to develop an intuitive notion of non-localizable order, and then developing this notion in terms of algebraic topology.

What then can be said about the apodictic logic which has been at the core of the logical empiricists' research programme? To begin with it must be acknowledged that considerable advances have been made in formal logic. In the twentieth century the ideal of formality and preoccupation with the procedures for making deductions have been pushed to the extreme, producing structures of propositions floating above the material world precariously anchored to it by a few rigid designators, occasionally breaking away in vast self-enclosed nets to become the whole of reality for the desiccated minds of their creators. But successes in the development of logic has been almost entirely in formalizing of valid deductive inferences and analysing and interpreting the nature of this formalization, and even these successes must be qualified by the limited success of logic in dealing effectively with probabilities, causal relations, psychological attitudes, mass terms (such as 'fire' or 'snow'), verbs of action and adverbs. Efforts to formalize inductive inferences have proved unsuccessful,[51] and deductive logic is not creative. It helps us present thoughts already thought out; it does not help us think up thoughts.[52] In fact by presenting old ideas in a forbiddingly formalistic manner, logicians have frequently inhibited the development of new ideas. Even its contribution to mathematics is questionable, and Lakatos has criticised the axiomatization of mathematics for disguising its creativity. It should be borne in mind that it was only by overthrowing the intellectual reign of the logicians that people such as Kepler, Galileo, Descartes and Newton were able to establish the new world-orientation of mechanistic materialism.[53]

Developments within logic itself have forced a recognition of the impossibility of the project of reducing knowledge of the world to a timeless set of logical relations between true propositions.[54]

---

50. This conception of the role of mathematics in science has been defended by Mary Hesse, by Cliff Hooker and by David Bohm. See David Bohm, 'Quantum Theory as an Indication of a New Order in Physics' in Bohm, *Wholeness and Implicate Order*, London: Routledge & Kegan Paul, 1980, pp.111-171.

51. This is by no means a universal conclusion. See for instance Brian Skyrmes, *Choice and Chance: An Introduction to Inductive Logic*, 2nd ed. Chicago Circle: University of Illinois Press, 1975.

52. This is Aristotle's view of logic. See *Sophistical Refutations*, 165a38-165b11. On the place given by Aristotle to syllogistic logic, see Jonathon Barnes, 'Aristotle's Theory of Demonstration' in *Articles on Aristotle*, ed. Jonathon Barnes et.al., London: Duckworth, pp.65-87. See also John Herman Randall Jr., *Aristotle*, N.Y.: Columbia University Press, 1960, p.51ff.

53. Thus it was noted by William and Martha Kneale in *The Development of Logic*, Oxford: O.U.P., 1962, pp.298-378 that there was a veritable Dark Age in logic from the fifteenth century to the beginning of the nineteenth century.

54. On this see David Bloor, *Wittgenstein: A Social Theory of Knowledge*, London: Macmillan, 1983, Ch.6; and Cornelius Castoriadis, *Crossroads in the Labyrinth*, [1978] tr. Kate Soper and Martin H. Ryle, Brighton: Harvester, 1984, pp.208-220.

These developments suggest that logicians are not discovering the universal structure of relations between propositions which reflect the world, but are making explicit and clarifying the forms of implication associated with different ways of conceiving the world - beginning with the way of conceiving the world presupposed in the culture in which formal logic is being developed. The applicability of different logics is dependent upon prior metaphysical commitments which provide the impetus for their development and the means for their interpretation,[55] although developments in logic can free us from old metaphysical assumptions or elucidate metaphysical positions, and problems in logic associated with particular ontological commitments can be taken as evidence against them. Thus Whitehead's development of classical logic was an attempt to transcend the substantialism implied by the medieval rendering of Aristotle's logic. Prior's tensed logic is applicable in a Newtonian world, Quine's extensional logic in a world conceived of as a space-time plenum, Vaughan Pratt's dynamic logic, dealing with the successive realization of chains of possible worlds, is appropriate to a world conceived of as consisting of discrete processes, and Routley's/Sylvan's intensional, relevance logic to an anti-reductionist conception of being in which individuals only exist in relation to their environment and constituents, but are irreducible to either of these.[56]

The limitations of formal logic and the liberating potential of advances in the field were obscured until recently by developments in the philosophy of language, particularly in USA In response to the failure of logical positivists to give an objectivist account of scientific knowledge, philosophers of language strove, in accordance with the tradition of Platonism, to describe the relationship between logic, language and the world and to characterize meaning and reference so as to exclude all 'subjective' elements.[57] It was argued that meaning is based on, or reducible to, reference and truth, that there is an objectively correct way to associate terms represented by arbitrarily defined signs with things, and that truth consists in a correspondence between propositions or sentences and states of affairs in the world.[58] But the proponents of these ideas have been blind to the freedom of language and thought from reference,[59] and to the background knowledge and understanding involved in the use of language, even when no more is involved than referring and making inferences. More fundamentally, they have been blind to the role played by the body, image schemas, metaphors, metonymy and imaginative projection and to the importance of focusing, scanning, superimposition, figure-ground reversal and reflexivity in the development

55. As Whitehead argued: '... logic presupposes metaphysics.' *Modes of Thought*, p.107. At the same time Whitehead suggests that the abstractions of logic will never be adequate for the complexity of the world.

56. On the relationship between logics and theories of being see Susan Haack, *Philosophy of Logics*, Cambridge: Cambridge University Press, 1978, Ch. 9. Vaughn Pratt's dynamic logic is described in V.R. Pratt, 'Logic of Processes', manuscript dated November 29, 1977 and Krister Segerberg, 'Applying Modal Logic', *Studia Logica*, Vol.XXXIX, 2/3, pp.275-295. Richard Sylvan and Val Plumwood have described the relationship between their intensional logic and ontology in Val and Richard Routley, *Social Theories, Self Management, and Environmental Problems*, Don Mannison et.al. eds, *Environmental Philosophy*, RSSS, A.N.U. Press, 1980, pp.217-332, esp.239-250. Richard Sylvan has addressed the issue of process more recently in an unpublished paper, 'Process and Action: Relevant Theory and Logic', 1992.

57. Hilary Putnam's, 'Reference and Truth' in *Realism and Reason: Philosophical Papers, Volume 3*, Cambridge: C.U.P., 1983, pp.69-86, gives a good overview of mainstream philosophy of language. This tradition has also been briefly characterized, and severely criticised by Richard Rorty in *The Mirror of Nature*.

58. The most influential doctrines of such objectivist semantics have been the causal theory of reference where the meaning of terms are seen to be given by an act of naming, and the truth conditional theory of meaning of sentences. The truth conditional theory of meaning was first proposed by Carnap who argued: 'To know the meaning of a sentence is to know in which of the possible cases it would be true and in which not.' (Rudolf Carnap, *Meaning and Necessity*, Chicago: University of Chicago Press, 1947, p.10.)

59. For a sustained attack on the reference theory of truth and meaning, see Richard Routley, *Exploring Meinong's Jungle*, Canberra: Philosophy Dept., R.S.S.S., A.N.U.

of cognition and in using language. And it has led them to dogmatic assumptions about the nature of the world. In order to fit the world into their dessicated philosophy of language, the world is assumed to consist of entities with fixed properties and relations holding among them at any instant, and to be divided up into natural kinds consisting of sets defined by the essential properties shared by their members. Complex properties of entities are assumed to be logical combinations of primitive properties.

Work in the philosophy of language, particularly in the area of cognitive semantics, is forcing philosophers to recognize the centrality of metaphor and metonymy in language, the importance of background understanding, and it is forcing them to question such assumptions about the world. In doing so it is contributing further to the development of a dialectical theory of knowledge. George Lakoff has summed up the findings of cognitive semantics:

> Meaning is based on the understanding of experience. Truth is based on understanding and meaning. Innate sensory-motor mechanisms provide a structuring of experience at two levels: the basic level and the image-schematic level. Image-schematic concepts and basic-level concepts for physical objects, actions, and states are understood directly in terms of the structuring of experience. Very general innate imaginative capacities (for schematization, categorisation, metaphor, metonymy, etc.) characterize abstract concepts by linking them to image-schematic and basic-level physical concepts. Cognitive models are built up by these imaginative processes. Mental spaces provide a medium for reasoning using cognitive models.[60]

Different logics are themselves founded on, can only be made sense of, and must be evaluated in terms of the experience of embodied engagement in the world, of body schema, of imagination and of metaphors. Traditional and classical logic are elaborations of the metaphor of spatial containment and exclusion, while modal logic (dealing with necessity, impossibility and possibility) adds an extra dimension through the metaphor of force and barriers or absence of barriers to it.[61] After studying the role of metaphor in the language of science, Richard Boyd rejected prevailing theories of reference, arguing that it is: 'essential that one adopt a *dynamic* and *dialectical* conception of reference, in contrast to conceptions of reference which present *synchronic*, *piecemeal*, and *nondialectical* idealization of the relation between individual words and features of the world.'[62] So called 'literal' meanings are not simply denotations but are frozen metaphors. Scientists who are advancing science are always struggling to free people from the assumption that terms simply refer to what there is in order to extend the limits of prevailing language.

Showing the 'relativity' of theories, the perceptual world, concepts, experimental design and technology, facts, mathematics and logic to each other revealed the incommensurability of theories from the point of view of logical empiricists; that is, the impossibility of comparing opposing theories point by point or through an ideal, theory neutral language based on symbolic logic supposedly representing states of affairs in the world. But this does not mean that theories cannot be compared. Once the creative potential of language is acknowledged, it can be seen that the barriers to communication assumed by logical positivists simply do not exist. As Paul Feyerabend pointed out: 'Philosophers insist on stability of meaning throughout an argument while scientists, being aware that speaking a language or explaining a situation means both *following* rules and *changing* them, are experts in the art of arguing across lines which philosophers regard as

---

60. George Lakoff, 'Meaning and Mental Representations', in Umberto Eco, Marco Santambrogio and Patrizia Viola eds, *Meaning and Mental Representations*, Bloomington and Indianapolis: Indiana University Press, 1988, pp.119-154, p.150.

61. These metaphors have been explicated by Johnson, *The Body in the Mind*, pp.'s 38ff. and 63f.

62. Boyd, 'Metaphor and Theory Change', p.381.

insuperable boundaries of discourse.'[63] The rationality of science can only be properly understood in socio-historical terms in relation to the struggles between proponents of competing theories and research programmes to establish a 'ratio' between different domains of experience, experiments, metaphors, concepts, insights, forms of thinking and opposing theories. It requires competing research programmes for the inadequacies and limitations of each research programme to be revealed.[64] And in the process of proposing, developing and comparing research programmes, the criteria of valid inference itself change.[65] In other words, the rationality of scientific progress is dialectical.

## *Dialectics*

Dialectics first and foremost implies dialogue, although it also implies other things, notably the absence of any element of experience, knowledge or reasoning which can be taken as the absolute foundation on which all knowledge is built or in terms of which it can be judged. It implies that the advance of understanding can only be achieved through the critical examination, confrontation and appreciation of the different points of view of people who are engaged in the world trying to make sense of it without any absolute reference points. Dialectics is opposed to both the attempt to reduce the development of knowledge to the mechanical application of a method and to relativism, since both of these exclude dialogue - methodologism by denying the assumptions underlying any method, and relativism by denying the possibility of mediating between ways of thinking and living based on different assumptions. Dialogue is essential to expose and comprehend the assumptions underlying all claims to knowledge, to reveal differences in assumptions and to open the possibility of replacing these assumptions, of developing radically new starting points to transcend old problems; and also, at the same time, to appreciate diverse points of view. The participants in such dialogue are embodied subjects, and they participate from the stand-point provided by their socio-historical situation. Scientific knowledge is essentially social not in the sense that what the majority accepts is true, but in the sense that individuals only make judgments as participants in forms of life, usually embedded in material transformations of the world, in which there is some degree of fusion of horizons between members. Rather than theories, concepts, mathematics, methods, experiments and facts rigidly and logically implying or excluding each other, what we have is people working as theoreticians, conceptual analysts, mathematicians, methodologists, experimenters and logicians all aware of and guided by the activities, endeavours, achievements and conflicts between others, and striving in their own particular work to throw light on these problems and controversies and thereby to make their own distinctive contribution to understanding the world. Only insofar as individuals understand to at least some extent the work of others and the proponents of opposing ideas are they in a position to judge some ideas as superior to others and to contribute to research. To talk about progress in scientific knowledge from a stand-point outside such common understanding is meaningless, and when such fusion of horizons breaks down, as it is arguably doing at present in many areas of science,[66] the notion of scientific progress is an insupportable myth.

Dialectical rationality is relational to begin with in the sense that the meanings of concepts are understood in relation to and in opposition to each other, as Plato argued, and in the Hegelian sense that advances in knowledge can only be understood by defining their achievements in relation to the ideas transcended. Knowledge advances not by moving towards a full and final truth which pre-

---

63. Paul Feyerabend, *Farewell to Reason*, London: Verso, 1988, p.272.

64. Paul Feyerabend in *Against Method*, London, Verso, 1978, Ch.3, argued that without a diversity of competing theories, the limitations of theories are invisible.

65. On this, see Stephen Toulmin, *The Uses of Argument*, Cambridge: Cambridge University Press, 1968, esp. p.257.

66. This is argued by David Bohm and F. David Peat in *Science, Order and Creativity*, Toronto: Bantam Books, Ch.2.

exists all enquiry, but by revealing and overcoming the failures and limitations of old ways of thinking and conceiving the world. As Kuhn argued:

> Can we not account for both science's existence and its success in terms of evolution from the community's state of knowledge at any given time? Does it really help to imagine that there is some one full, objective, true account of nature and that the proper measure of scientific achievement is the extent to which it brings us closer to that ultimate goal? If we can learn to substitute evolution-from-what-we-do-know for evolution-towards-what-we-wish-to-know, a number of vexing problems may vanish in the process.[67]

The reasons why such theories must be regarded as advances can only be fully comprehended in the context of the particular situations in which new theories are proposed.[68] It is impossible to evaluate them in terms of some absolute criteria because major advances in knowledge transcend old assumptions and create new ways of arguing, changing the standards of relevance and proof. They advance our understanding of understanding and what is involved in achieving it. The superiority of the new theories is only revealed by the comprehension they facilitate of the achievements and limitations of the theories transcended. As Alasdair MacIntyre pointed out:

> Wherein lies the superiority of Galileo to his predecessors? The answer is that he, for the first time, enables the work of all his predecessors to be evaluated by a common set of standards. The contributions of Plato, Aristotle, the scholars at Merton College, Oxford and Padua, the work of Copernicus himself at last all fall into place. Or to put matters in another and equivalent way: the history of late medieval science can finally be cast into a coherent narrative.... What the scientific genius, such as Galileo, achieves in his transitions, then, is not only a new way of understanding nature, but also and inseparably a new way of understanding the old sciences way of understanding... It is from the stand-point of the new science that the continuities of narrative history are reestablished.[69]

Dialectical rationality is also relational in that the meaning of the enterprise of striving for knowledge and understanding only makes sense in relation to social practices of particular forms of life, which in turn only make sense in terms of broader social and cultural contexts of which they are part. Wittgenstein made this point when he argued:

> "So you are saying that human agreement decides what is true and what is false?" - It is what human beings say that is true and that is false; and they agree in the *language* they use. That is no agreement in opinions, but in forms of life.[70]

But this is only the most basic agreement constituting the enquiring community as a form of life. The scientific community as a whole is underpinned by common assumptions about what science is, what are the goals of science, about what have been its major achievements, about what place science has in society, and about the nature of the world in general. Such assumptions are not only institutionalized; they are embodied in the transformations of the material world - in buildings, laboratories, technology and experimental apparatuses. This community is in turn subdivided into a multiplicity of disciplines and sub-disciplines constituted by more specific shared commitments,

67. Thomas Kuhn, *The Structure of Scientific Revolutions*, 2nd ed., Chicago: University of Chicago Press, 1970, p.171.

68. Support for this contention is provided by the detailed analyses of scientific reasoning made by Dudley Shapere in *Reason and the Search for Knowledge*, Dordrecht: Reidel, 1984, Part III.

69. Alasdair MacIntyre, 'Epistemological Crises, Dramatic Narrative and the Philosophy of Science' *Monist*, Vol. 60, 1977, pp.453-472, pp.459-60 & 467.

70. Ludwig Wittgenstein, *Philosophical Investigations*, tr. G.E. Anscombe, 3rd ed., Oxford: Basil Blackwell, 1968, 241.

including technologies, symbolic generalizations, models (analogies and ontologies) and exemplars: the concrete problem solutions accepted by these communities as paradigmatic.[71] It is the condition for the possibility of science that people are socialized, through education and apprenticeships, into such forms of life.

The precedence given to certain discourses, organizations and individuals to adjudicate truth claims is an essential constituent of the order of power within society. This is the point made by Foucault who argued:

> Each society has its regime of truth, its 'general politics' of truth: that is, the types of discourse which it accepts and makes function as true; the mechanisms and instances which enable one to distinguish true and false statements, the means by which each is sanctioned; the techniques and procedures accorded value in the acquisition of truth; the status of those who are charged with saying what counts as true.[72]

In the forms of life in which scientific ideas are formulated, communicated and legitimated, the relationship between power and knowledge is indissociable.[73] To begin with, there are the power structures and struggles within scientific laboratories, within research institutions, within cultural fields and discursive formations and within the organizations which sustain these. People struggle for power within disciplines for teaching positions, the means for research, to choose what research to do, for positions in different research establishments, for the brightest students and research assistants, and for the means to disseminate ideas and to ensure that they are seriously considered. This includes the struggle for the appointment of former students to teaching institutions and for the editorships of the most respected journals. There are also struggles over the power structures within disciplines, between existing disciplines over status and finance, and to establish and legitimate new disciplines, and struggles within and between institutions of learning and research. Such struggles involve complicated interpersonal and institutional manoeuvring, the formation of alliances, the accumulation and deployment of symbolic capital, and the construction of mythologies (presented as histories) to legitimate the claims to authority, and thereby the power of different groups of researchers to carry out research and promote their ideas.[74] These structures of power and power struggles are then intimately related to the broader political and economic contexts which constrain what sort of research and teaching institutions can be legitimated in the eyes of those who ultimately control or supply finance. Finally there are the broader cultural processes, from the ideological power struggles within and between discursive formations such as those studied by Foucault, to the ideological struggles affecting whole societies and civilizations focussed on by Hegelian and Marxist historians of science, which limit what will be tolerated or even understood by anyone striving for legitimacy.[75]

---

71. See Thomas S. Kuhn, *The Essential Tension*, Chicago: University of Chicago Press, 1977, p.297ff.

72. Michel Foucault, *Power/Knowledge*, ed. Colin Gordon, Brighton: The Harvester Press, 1980, p.131.

73. On this see Pierre Bourdieu, 'The specificity of the scientific field and the social conditions of the progress of reason' in *Social Science Information*, Vol.14, No.6, 1975, pp.19-47; Bruno Latour, *Science in Action*, London: Open University Press, 1987; Bruno Latour and S. Woolgar, *Laboratory Life: The Social Construction of Scientific Facts*, London and Beverly Hills: Sage, 1979; and K.D. Knorr-Cetina, *The Manufacture of Knowledge: An Essay on the Constructivist and Contextual Nature of Science*, Oxford: Pergamon Press, 1981.

74. For an excellent study of this see Pnina Abir-Am, 'Themes, Genres and Orders of Legitimation in the Consolidation of New Scientific Disciplines: Deconstructing the Historiography of Molecular Biology' *History of Science*, Vol.23, Part 1, No.59, 1985, pp.73-117 and 'Essay Review: How Scientists View Their Heroes: Some Remarks on the Mechanism of Myth Construction' in *Journal of the History of Biology*, Vol.15, no.2. Summer 1982, pp.281-315.

75. One of the best theoretical - and polemical statements on this project, see Bob Young, 'Science *is* Social Relations' in *Radical Science Journal*, Vol.5, 1977, pp.65-118. For a general survey of such analyses see Michael Mulkay, *Science and*

The pervasive nature of these power struggles and their social contexts has given rise to the problem of the relationship between the internal history of science - the development of ideas themselves, and the external history of science - the history of the external conditions which have led to the production of scientific ideas. In general it appears that certain external conditions are conducive to major intellectual advances: the existence of a diversity of competing intellectual centres with a major centre but without centralized control, tied together into a single network - as occurred for instance in Ancient Greece, Renaissance Italy, eighteenth century France and nineteenth century Germany.[76] But these conditions do not account for the nature of the intellectual advances. These can only be accounted for through the dialectics of internal and external conditions. In terms of dialectics post-Hegelian dialectics, power relations and broader social dynamics are not merely non-rational influences on the creation and legitimation of ideas. They also have a rationality, closely associated with the rationality of the development of explicit ideas, which can be investigated and evaluated. The forms of life of a society embody world-orientations incorporating metaphors, generally elaborated by using social relations as a metaphor for nature then using nature as a metaphor for understanding society, thereby legitimating its institutions, organizations and social movements. Forms of life can be evaluated in terms of the success or failure of the explicitly developed ideas which are engendered by and produced within them to make the world intelligible and as means to confront, mobilize people and resolve the problems of these forms of life. Where it becomes impossible to develop the ideas required to properly comprehend the world and its problems within the forms of life of a society, then the limits of these forms of life are revealed, and this must be faced up to and society transformed accordingly. In attempting to advance beyond a particular set of ideas it is therefore not only important to cast past ideas into an historical narrative, but also the forms of life - the institutions and socio-economic formations which have produced these ideas, and the successes, problems and failures of these forms of life. This project only becomes fully intelligible in relation to a philosophy of history, a conception of humanity, and ultimately, as part of metaphysics.

## *Dialectics, Metaphysics and Science*

Radically opposed to the conceptions of knowledge based on the classical logic of Bertrand Russell, dialectical rationality is oriented towards achieving a comprehension of the whole. This relational conception of knowledge oriented towards the totality is also associated with a far greater concern with contradictions between diverse knowledge claims and between theories, experiments and social practices than is the case with theories of knowledge centred on formal deduction. The dialectic of understanding involves both a struggle to grasp each individual in its uniqueness and a struggle to attain a comprehensive perspective, a process which by its very nature can never be complete. Individuals as participants in the struggle to understand the world can only make provisional commitments to particular ways of conceiving the world in the struggle to deepen understanding. Lucien Goldmann pointed out the significance of this:

> Both rationalism and empiricism are ... opposed to dialectical thought, for this affirms that there are never any absolutely valid starting points, no problems which are finally and definitely solved, and that consequently thought never moves forward in a straight line, since each individual fact or idea assumes its significance only when it takes up its place in the whole, in the same way as the whole can be understood only by our increased knowledge of

---

*the Sociology of Knowledge*, London: George Allen & Unwin, 1979. For some of the best works in this area see *Radical Science Journal* which has now been renamed *Science as Culture*.

> the partial and incomplete facts which constitute it. The advance of knowledge is thus to be considered as a perpetual movement to and fro, from the whole to the parts and from the parts back to the whole again, a movement in the course of which the whole and the parts throw light on each other. [77]

This movement between wholes and parts is characteristic of both efforts to understand particular situations and to understand the world as whole. What counts as a part is determined from the perspective of the whole, while the whole must be defined as such from the perspective of the parts which compose it. Thus science is articulated into various domains, each defined by some problem (the inexplicable existence of some kind of order which needs to be accounted for, for example) which theories are required to solve together with the information relevant to the effort to solve this problem, or by a theory with its associated objects and relevant information entailing a research programme to elaborate it.[78] But what counts as problematic in the first place is largely determined by other domains and their relationships, which also determine which theories can be plausibly entertained. The endeavour to grasp the relationships between and to put in perspective all domains, to achieve a comprehensive understanding of the world as a whole is speculative philosophy, and what speculative philosophy elaborates is a metaphysics.

Speculative metaphysics, by elaborating categorial schemes, strives to make intelligible the relationship between each and every entity, component and aspect of the world by defining the generic features of the primary being or beings of the world, as opposed to the characteristics of what is merely an aspect or part of something else. The problem is to obtain a unity of understanding through a theory of being or beings which puts all particular domains into a coherent perspective, and thereby provides science with a grand research programme. Metaphysics must define the basic characteristics of the beings which particular sciences are to investigate, enabling each science to define its domain in relation to other domains in terms of the kinds of being it is investigating. In doing this, it must also show that it is possible for these 'objects' to be understood, provide a general characterization of what it means for them to be understood, and a general direction for attaining this understanding. After the question: What is being? the most significant questions for attaining such a general perspective are: What is the cosmos? What is life? and What is humanity? In terms of the notion of humanity, it is then necessary not only to provide answers to such questions as what is worth striving for, how should we live, and how should society be organized, but also to account for the possibility of humans attaining an understanding of the nature of being, of the cosmos, of life, and of themselves and the point of their existence. Any metaphysical system which cannot account for the comprehensibility of the world and the existence of beings who can comprehend it (which is the case with both mechanistic materialism and field theory) is self-contradictory. The importance of any particular research can be judged by how basic are the questions which it illuminates. The development of research programmes, whether dealing with the nature of being as such or with more specific issues, generates new problems, and thereby opens up new domains, inspires the development of new theories, and thereby leads to the development of new research programmes.

Virtually all the most significant advances in science have been engendered by the struggle to attain a coherent conception of the nature of the world. Where enquiry has been divorced from concern with broader questions and been reduced to a means to develop technology, as in the

---

76. On this, see Joseph Ben-David, *The Scientist's Role in Society*, Englewood Cliffs: Prentice-Hall Inc., 1971, and RAndall Collins, 'On the Sociology of Intellectual Stagnation', *Cultural Theory and Cultural Change*, ed. Mike Featherstone, London: Sage, 1992.

77. Lucien Goldmann, *The Hidden God: A Study of Tragic Vision* tr. Philip Thody, London: Routledge & Kegan Paul, 1964, p.4f.

78. The nature of domains and their formation has been studied by Dudley Shapere in 'Scientific Theories and Their Domains' in Suppe ed. *The Structure of Scientific Theories*, pp.518-565.

medieval Arab world, in France after the French revolution and in Stalinist Russia, or ghettoised into separate domains as in late medieval scholasticism, understanding has stagnated or regressed. The advances in knowledge achieved during antiquity and in the medieval world were only possible because the Greeks had articulated coherent conceptions of being which could serve as the foundations for research programmes to attempt to understand all aspects of the world. Without such conceptions of being there would have been no way to begin enquiry, no way to work out the important questions to put to nature. The revolution in the seventeenth century was first and foremost a metaphysical revolution, and Galileo for one claimed that he had spent as many years thinking about philosophy as months thinking about mathematics. The development of science since then has only been possible because of the coherent metaphysics which was articulated at that time.[79] Chemistry, biology and psychology have successively been advanced on the basis of this theory of being. But while the advance of science has generally involved the transcendence of all concepts deriving from Aristotle: phlogiston, entelechies and so on, the mechanistic ontology has been largely undermined in physics by the alternative theory of being - field theory - which has its roots in the ideas of Leibniz. And as I will go on to argue, both these theories of being are now under attack from a science based on the process view of the world. Thus, as opposed to logical empiricists such as Carnap and Ayer who defined science in opposition to metaphysics, the historical evidence suggests that it is the effort to investigate and explain the world in terms of a coherent metaphysics which defines them as scientific, and the present balkanization of disciplines and fragmentation of discourse in which the underlying metaphysical assumptions are being lost sight of and confused, must be regarded as a corruption and degeneration of science.

While hack scientists can ignore their taken for granted assumptions, metaphysics is vital to more creative scientists. This was clearly revealed at the biological conferences at Belagio organized by C.H. Waddington. The physicist David Bohm summed up the conclusions of one of these conferences:

> I think the most important aspect of the interchange is the emergence of a common realization that metaphysics is fundamental to every branch of science. Metaphysics is ... something that pervades every field, that conditions each person's thinking in varied and subtle ways, of which we are not conscious. Metaphysics is a set of assumptions about the general order and structures of existence ... It seems clear that everybody has got some kind of metaphysics, even if he thinks he hasn't got any.

He then went on to point out the implications of this:

> ... the practical 'hard-headed' individual has a very dangerous kind of metaphysics, i.e. the kind of which he is unaware... Such metaphysics is dangerous because, in it, assumptions and inferences are being mistaken for directly observed facts, with the result that they are effectively riveted in an almost unchangeable way into the structure of thought... [W]hat is needed is a the conscious criticism of one's own metaphysics, leading to changes where appropriate and, ultimately, to the continual creation of new and different kinds. In this way, metaphysics ceases to be the master of a human being and becomes his servant, helping to give an ever changing and evolving order to his overall thinking.[80]

---

79. As clearly demonstrated by Ivor Leclerc in *The Nature of Physical Existence*, London: George Allen & Unwin, 1972. See also the work of Alexandre Koyré.

80. David Bohm, 'Further Remarks on Order' in C.H. Waddington ed. *Towards a Theoretical Biology: 2 Sketches*, Edinburgh: Edinburgh University Press, 1969, p.41f.

The relationship between science and metaphysics is clarified by the Robin Collingwood's logic of question and answer.[81] Developing ideas from both Plato and Hegel, Collingwood elaborated this logic in opposition to the logic of Russell and Whitehead, arguing that this was a valid characterization of the rationality of both scientific and historical investigation, and using it to reveal the different levels of assumptions dominating historical eras. To defend this logic, Collingwood argued that the validity of any proposition can only be understood and judged when the question it is attempting to answer is understood. Each question presupposes a set of assumptions which in turn are answers to other questions. For instance, the search for the type of virus making someone ill presupposes that types of illness are due to viruses, which in turn is a theory based on other assumptions about the nature of organisms and their normal functioning. The ultimate assumption underlying this research is that all events, in this case, becoming ill, have some cause. This is a metaphysical assumption.

Such metaphysical assumptions cohere as categorial schemes which are held together by what were referred to by the eighteenth century German philosopher Lichtenberg as *paradeigma*.[82] Lichtenberg argued that in physics puzzling phenomena are made intelligible by relating them to some standard form or process which we must accept as self-explanatory.[83] Since a theory of all that is must ultimately account for the world in its own terms rather than in terms of something else, such *paradeigma* are an unavoidable part of metaphysics. However with metaphysical revolutions, these *paradeigma* are brought into question and replaced. For instance, Aristotle's metaphysics, being based on the analogy of organisms, took organic growth and a stationary state in relation to the earth as paradigmatic, not in need of explanation, and the starting point for explaining everything else. It was for this reason that it was assumed that base metals were evolving into some higher form, and that all that is necessary to transmute base metals into gold is to find the conditions which would hasten this development, while the motion of a thrown object after it had lost contact with its mover was seen as something which had to be explained. However with the metaphysical revolution of the seventeenth century with the elaboration of the analogy of a machine, inert matter located in space and moving according to fixed laws of motion through time came to be taken as paradigmatic. Uniform motion in a straight line was then not something to be explained, but the starting point for explaining everything else, while organic growth came to be seen as something which had to be explained in terms of the arrangement and motion of bits of inert matter and the forces of attraction and repulsion between them. However in terms of mechanistic materialism, such forces of attraction and repulsion were inexplicable. With the adoption of field theory, fields of force became paradigmatic, and the real problem came to be accounting for the existence of particle-like centres of force within the fields. Such *paradeigma*, founded on basic metaphors and encoded in the basic categories of cultures, are assumed by whole eras. No investigation of any sort can escape these metaphysical assumptions since they are presupposed by all questions.

## *Metaphysics and Society*

The acceptance of *paradeigma* plays a major part in the process of embodiment of categories into the social practices of society. The transformation from a view of the world in which all

81. See R.G. Collingwood, *An Autobiography*, Oxford: Oxford University Press, 1939, Ch.5 and *An Essay on Metaphysics*, Oxford: Oxford University Press, 1940, p.23ff.

82. This concept has been developed by Stephen Toulmin in *Foresight and Understanding*, London: Hutchinson, 1961 and *Human Understanding* Vol.1, Oxford: Oxford University Press, 1972, p.106f.

83. These are closely related to what Imre Lakatos regarded as the 'hard core' of scientific theories, and also to what Gerald Holton called 'themata'. See Lakatos, *The Methodology of Scientific Research Programmes*, p.48 and Gerald Holton, *The Scientific Imagination: Case Studies*, Cambridge: C.U.P., Ch.1.

entities are growing to a world in which all entities are naturally in uniform motion unless acted on by an external force, and in which every event has an identifiable cause, was associated with the transformation of society and the development of practices devoted to the total control of nature and people. The possibility of total control required such a *paradeigma*. However with the world seen as composed of bits of inert matter there were still limits to such control. In Newtonian physics, atoms are immutable and can only be rearranged. To hold out the possibility of absolute control it was necessary to reconceive the world in a way which would enable these bits of matter to be seen as derivative - the view of the world defended by the field theorists. As I have shown in earlier chapters, such assumptions about the physical world are inseparable from assumptions about people, social relations and ideals of social order. The categories of mechanistic materialism have developed as constituents of and as constitutive of social life in capitalist societies, and are presupposed not only in inquiry, but also in decision-making, in deliberate action and in the language of justification.

While Collingwood gave no place to the actual development of metaphysical systems, there is no reason why this logic of question and answer should not be extended to asking and answering questions about the nature of primary beings in answer to developments and conflicts in and between different domains of science and with other domains of social life. The dialogue associated with the development of understanding must take place at a number of different levels, ranging from those associated with highly specific questions within particular research programmes or domains, through those associated with assumptions specific to such programmes and domains, to the epistemological and metaphysical assumptions underlying the entire scientific endeavour, and finally to the metaphysical schemes underlying ethics, politics, the social order and civilization itself. The form of rationality involved in metaphysics is no different from, and is inseparable from, the form of rationality involved in particular sciences. In each case, comprehension is developed through the articulation into conceptual frameworks and elaboration of metaphors in competition with or in relationship to other metaphors in the struggle to understand the world and its anomalies.[84] But there are unique problems in the development and justification of such metaphysical schemes, because as the foundation for total conceptions of the world, nothing can be simply assumed. As Hegel succinctly summarized the problem:

> Philosophy misses an advantage enjoyed by the other sciences. It cannot like them rest the existence of its objects on the natural admissions of consciousness, nor can it assume that its method of cognition, either for starting or for continuing, is one already accepted... We can assume nothing dogmatically; nor can we accept the assertions and assumptions of others. And yet we must make a beginning: and a beginning, as primary and underived, makes an assumption, or rather is an assumption. It seems as if it were impossible to make a beginning at all.[85]

The solution to this problem is to justify the assumptions on which the starting point is based by the system which is developed from it. As Hegel put it:

> The very point of view, which originally is taken on its own evidence only, must in the course of the science be converted to a result - the ultimate result in which philosophy returns into itself and reaches the point with which it began. In this manner philosophy exhibits the

---

84. On the nature of metaphysical thinking, showing the central role of analogies, see Dorothy Emmet, *The Nature of Metaphysical Thinking*, London: Macmillan, 1966. See also Whitehead, *Process and Reality*, Ch.1. Whitehead refers to 'descriptive generalization' rather than the elaboration of analogies.

85. G.W.F. Hegel, *Hegel's Logic*, tr. William Wallace, 3rd ed. Oxford: Clarendon, 1975, § 1, p.3.

appearance of a circle which closes with itself, and has no beginning in the same way as other sciences have.[86]

But being a closed, internally consistent circle is not enough by itself. The statement: 'All statements but this are absurd' starts and finishes with itself in an entirely consistent way, but gets nowhere. As noted previously, while there may be no Archimedean point on which knowledge claims can be built, the quest for understanding inevitably reveals the limits of such a quest, that as Schelling argued against Hegel, there is an '*unvordenkliches Sein*' (unprethinkable Being) before all thought, presupposed by all thought, and ultimately beyond the full grasp of thought. Before any enquiry we are always already engaged in a world shared with others which is already partially understood, and such partial understanding of this world is presupposed by such enquiry. Proponents of metaphysical systems cannot avoid relating their speculations to their prior understanding of the world as they have previously engaged in it. They are impelled to be reflexive towards their erstwhile assumptions and to acknowledge that they themselves are part of an on-going struggle with others to make sense of this world. This means that metaphysicians must come to terms with rival efforts to advance understanding, including rival metaphysical systems and the claims made by them. To retain their plausibility a metaphysical system must account for the achievements and reveal the limitations of these rivals, or at least provide a research programme for doing so. Furthermore, to escape the charge that philosophy itself is idiosyncratic, such encompassing cannot stop at the productions of philosophers. Metaphysicians must also come to terms with scientific ideas, with conceptions of the world embodied in social relations, in forms of life and institutions, and in all other symbolic productions, including art, literature, history, ceremonies, religious practices, and so on.

To avoid the idiosyncrasy of conceiving the world from the point of view of one culture or from one geo-socio-historical situation, metaphysicians must accept the task of achieving a critical perspective able to comprehend the achievements and limitations of all other cultures, both those co-existing and those of past societies. While obviously this task could never be complete, a contribution to this has been attempted in this work by showing the role played by metaphysical assumptions in the evolution of European civilization and the influence of Neoplatonist metaphysics in the dynamics of Russian society; and from a very similar perspective it is what Joseph Needham has accomplished in far greater depth for China in his monumental study, *Science and Civilisation in China*. But most importantly, a metaphysical system must come to terms with the way the world is presently understood. It should provide a critical perspective on the present era and the metaphysical assumptions which underlie it. Through engaging with conceptions of the world embodied in current social practices and institutions and offering modifications of or alternatives to these, metaphysical schemes can then become more than simply theories or grand research programmes; they can become world-orientations which can challenge, and if successful, replace the foundations of civilizations. By providing concepts which can mediate people's relations in practices and institutions in new ways they have the potential to become incorporated as new forms of life, as the foundations for new social formations, and ultimately civilizations, with dynamics of their own.

It follows from the historical analyses of civilizations in this work and its precursor that there are two further ways in which it is necessary to dissent from Hegel's characterization of metaphysics. Firstly Hegel's assumption (which he himself occasionally questioned) that there is an end point to philosophy, a final system capable of a complete vision of the world which can be captured in a system of logic, and such that all earlier philosophies can be seen as mere stages on the path to this end point, must be rejected. There is no justification for such an assumption. One can only hope to achieve a way of understanding the world to which a provisional commitment can be made on the basis of its demonstrated or promised superiority over all known rivals. But earlier

86. Ibid. § 17, p.23.

ideas are not just stage-posts on the way to one's own conception of the world, even if one does successfully transcend their limitations. Although it is important for the justification of a philosophy that it be shown to provide a perspective on the past, each philosophy is a more or less successful effort to come to grips with its age, and must be regarded as an end in itself in this regard. Furthermore while it might be possible to provide a philosophy fully adequate to the present, it can be expected that such ideas will also be shown to be limited and will be transcended in the future. Thus rather than conceiving of a metaphysical system as a circle which closes on itself, a metaphysical system should be regarded as a spiral which begins with a set of assumptions in terms of which the world, including the history of philosophy (and the history of science), is investigated, and which eventually explicates and validates these, but which at the same time reveals their provisional nature, thereby providing the point of departure for new efforts to achieve a comprehensive understanding of the world.

Secondly, while Hegel has acknowledged that metaphysical systems are not simply ideas entertained about the world but are embodied in the institutions of societies, and that consequently the ideas of metaphysical systems are intimately related to the way society is organized, he has maintained a division between theory and practice by arguing that metaphysics is simply the bringing to full consciousness of forms of thinking which have already been developed within practices and partly brought to consciousness within art and then in religion. It is the final coat of icing on the cake. It can therefore never be a guide to action. As he put it:

> ...it is only when actuality is mature that the ideal first appears over against the real and that the ideal apprehends this same real world in its substance and builds it up for itself in to the shape of an intellectual realm. When philosophy paints its grey in grey, then a shape of life has grown old. By philosophy's grey in grey it cannot be rejuvenated but only understood. The owl of Minerva spreads its wings only with the falling of the dusk.[87]

As opposed to this, and also to the dialectics of orthodox Marxists who underplay the importance of theory, I have argued here for the indissociability of theory and praxis, and for the capacity of metaphysics to go beyond prevailing forms of thought and praxis and thereby to reveal the limitations of the metaphysical assumptions which dominate them. As well as serving to make the world intelligible, a metaphysical system must articulate the problems and aspirations of people and reveal to them how such problems can be overcome and how their aspirations can be realized. In earlier chapters of this work the nature of this dialect between metaphysics and action has been shown: how in the early Middle Ages a version of Neoplatonic Christianity served to unify society and then to provide the means whereby the church was able to achieve ascendancy over secular rulers, how in the seventeenth century mechanistic materialism was able to provide a coherent perspective on both the social and natural world to provide the rising bourgeoisie with a new basis for interpreting the past and legitimating their struggle for political power, and how Neoplatonic Marxism provided the ideological means for the radical intelligentsia and the proletariat to gain and maintain power in the Soviet Union. In opposition to Hegel and vulgar Marxists it has been argued that the picture is closer to that drawn by Whitehead:

> [Metaphysics] is the most effective of all the intellectual pursuits.... It is the architect of the buildings of the spirit, and it is also their solvent:- and the spiritual precedes the material.

87. G.W.F. Hegel, *Hegel's Philosophy of Right* tr. T.M. Knox, Oxford: Clarendon, 1952, p.12. This conservatism also derives from the assumption of the identity between being and thought, since this allows no gap between the real and the rational.

> Philosophy works slowly. Thoughts lie dormant for ages; and then, almost suddenly as it were, mankind finds that they have embodied themselves in institutions.[88]

To be developing an alternative metaphysical system is to be challenging the existing power relations and forms of legitimation in society. It is not simply to be developing a set of ideas but to be developing a mode of being and engaging in the world. To comprehend such a system is to at least be open to the possibility of changing one's mode of being in the world and thereby of radically changing oneself. To change one's mode of being in the world is to see different possibilities, to evaluate the world differently, and to realign oneself in relation to the different tendencies within society and nature. A system opposed to the dominant metaphysics and the social order based upon it must also provide the conceptual foundations, at least in crude form, for a new society. This is what Neoplatonic Christianity did at the end of the Dark Ages, what mechanistic materialism did in seventeenth century Britain and what Neoplatonic Marxism did in twentieth century Russia. And so a metaphysical system must ultimately be evaluated also as an orientation for action, in terms of its success in mobilizing people for action and in terms of the success of their actions, as the constituents for new relations between people and between humans and nature, and in terms of how successful the socio-economic order based on these relations is. It is only when the new social order incorporating the world-orientation of a metaphysical scheme is established that the potentialities and limitations of this scheme will be fully revealed, and this will then provide a point of departure for the development of a new metaphysical scheme, a new comprehensive conception of the world.

## *The Present Work as a Metaphysical System*

Against this background it is now possible to explicate the systematic structure underlying the present work. From the beginning of this work a conception of humans as embodied subjects participating in the creative becoming of their society, of humanity and of nature has been assumed. To bring into focus the opposition between this set of assumptions and those which now dominate the modern world, I have focused on the environmental crisis. This not only is the most important practical problem confronting humanity, but it highlights the most significant features of modern civilization: its blindness to the environmental conditions of its existence, and what must be regarded as its most acute cultural and philosophical problem - its nihilism. If philosophy cannot provide compelling reasons for people to confront environmental problems, to do something about the ten to fifteen million people who die each year from starvation, to concern themselves with the whole future of humanity, then it can provide no reason for anything. Yet mainstream philosophy was shown to be impotent in the face of such questions, and it was this which justified a thorough investigation into the formation of European culture. This investigation revealed the roots of both the destructive orientation to the world and the nihilism of Western civilization in metaphysical assumptions of mechanistic materialism, assumptions which evolved out of Platonism and which have culminated in Darwinian evolutionary theory and information theory, and which have been incorporated into institutions and into the very bodies of people as modes of being-in-the-world. The form of Marxism which triumphed in the Soviet Union (at the expense of the 'process Marxism' of Bogdanov and the *Proletkul't* movement), did not provide an alternative to this culture, or a solution to the problems confronting humanity. The full development and defence of Bogdanov's process conception of the world is required if what has proved fruitful within Marxism is to be salvaged.

All these analyses have assumed a conception of the world as a process of creative becoming, and the study of the development of Western and Eastern European civilizations has not only been

---

88. Alfred North Whitehead, *Science and the Modern World*, [1925], New York: Mentor 1964, p.viif.

designed to reveal the need for a metaphysical revolution, but has been an attempt to develop this alternative. The remainder of this work will be an explicit formulation and defence of process philosophy. A series of categories will be outlined and attempt made to show the validity of the process conception of being through an examination of developments within the natural sciences. It will then be show how humanity can be understood on the basis of this new science in a way which transcends the problems which have plagued philosophy for the last three centuries: specifically, the relationships between mind and body, free will and determinism, knowledge and reality, subjectivity and objectivity, facts and values. This conception of humanity will be used as the foundation for a new ethics, political philosophy and the sciences of humanity. This new vision will not only make the environment a central theme of life in the context of humanity's process of self-creation, but will overcome the nihilism of the modern world. In the last chapter the problem of action and how people, both individually and collectively, can act to change the world and to establish an environmentally sustainable, post-nihilistic civilization, will be considered explicitly. However metaphysics cannot end with a discussion of action; ultimately, process philosophy must be developed and validated in action - as an orientation for action against the present order and as the basis for new forms of relationships between people, between individuals and society, and between humanity and the rest of nature.

# PROCESS PHILOSOPHY AS METAPHYSICS

Process philosophy originated in the West with Heraclitus. Heraclitus argued that: 'All things are passing and nothing abides'; and that: 'Nothing is, everything is becoming.'[1] The world is in flux, a process of becoming in which whatever is, is an enduring pattern of activity, an island of stability within the flux which can only maintain itself through constant interaction with the background flux and other patterns of activity. A conception of the world similar to this has been common in China and is virtually embodied in Chinese language. For example the idea of process is implicit in the commonly used term *ch'i* which means 'the directed and structured expression of movement', a notion difficult to express in English. The notion of endurance within flux is beautifully conveyed by Li Po, *one of China's greatest poets*:[2]

Petals are on the gone waters and on the going,
And on the back-swirling eddies,
But today's men are not the men of the old days,
Though they hang in the same way over the bridge-rail.

In the West, by contrast, this conception of the world has been held only among those rebelling against the dominant order; for instance among the peasants in the Middle Ages who celebrated the carnival, laughter and the cycle of birth, death and regeneration in defiance of the petrified seriousness of the Church. The conception of the world as a process of creative becoming was also an important, if subordinate theme, among the radical Neoplatonists. In the sixteenth and seventeenth centuries the Hermetic philosophers or Nature Enthusiasts were concerned to develop such a conception of the world to justify their view that humans are capable of transforming society to create a new harmony between people and with nature. In eighteenth century France the conception of nature as active was defended by Diderot for similar reasons. Then in Germany, Herder, Goethe, Schelling and von Humboldt conceived nature, individuals and societies as processes of becoming, though they tended to see such becoming as actualizing predetermined ends. Much of Hegel's philosophy accords with a process view of the world, although these aspects are ultimately subordinated to the eternal logical structure of the Absolute. Both Marx and Nietzsche defended the primacy of becoming in their very different attacks on reified abstractions. In the twentieth century the most significant proponents of process philosophy in the West have been Bergson, Alexander, Whitehead, Collingwood, Hartshorne, Lawrence, Pols, Cobb, Capek and Leclerc, while in the Soviet Union Mikhail Bakhtin and his circle were outstanding exponents of the primacy of becoming. There have also been a number of philosophers whose ideas accord with process philosophy, notably the early pragmatists: Peirce, James, Dewey and Mead, the Monists in Germany, Bogdanov in Russia, anti-mechanistic systems theorists such as von Bertalanffy and Ervin Laszlo, some phenomenologists, most notably Heidegger, Merleau-Ponty, and Merleau-Ponty's former student, the political philosopher Castoriadis. Some of the work of the post-structuralists can be seen as a struggle to come to terms with the idea that the subject is not a substance but a process. Perhaps most importantly, a number of scientists and mathematicians have embraced and developed the categories of process philosophy in their work, the most well known of these being David Bohm, Ilya Prigogine, C.H. Waddington, Charles Birch, Roger Sperry, Brian

1. Plato *Cratylus* 402a8 and *Theaitetus* 152e1.
2. From *'Poem by the Bridge at Ten-Shen'*, *Ezra Pound Translations*, [1926] N.Y.: New Directions, 1963, p.193.

Goodwin and René Thom. There are also a number of anti-reductionist scientists closely aligned with process philosophy, including some dialectical materialists, both in the West and in the former Soviet Union. In most of these cases the thinkers involved in the promotion of these ideas have been concerned to oppose the nihilism deriving from the mechanistic view of the world. Process philosophy can thus best be understood as the development of that tradition of thought which has exalted life in opposition to the mainstream of Western culture. It is the tradition which has refused to accept either the victory of mechanistic thinking or the social order based upon it. But as such it has been a tradition with little influence.

## *The Categories of Process Philosophy*

To develop process philosophy it is necessary to elaborate and defend a categorial scheme to oppose the categories which dominate people's present thinking. 'Categories' are to be understood as the most fundamental concepts for understanding the world, or equivalently, as Whitehead put it, 'tentative formulations of the ultimate generalities.'[3] They are the concepts which define the nature and generic characteristics of primary being (or beings), in terms of which (or at least in relation to which) all other concepts must ultimately be understood, and which are presupposed by every proposition.[4] This presents the problem of how categories themselves are to be defined.

The problem of defining categories has been avoided by most philosophers - who have merely striven to eliminate inconsistencies and to refine and reconstruct the relationships between categories already dominating thought. Kant in his later work and Neo-Kantians typify this approach, as do the 'analytic metaphysicians' of recent Anglo-American philosophy. Most analytic philosophers, under the influence of Frank Ramsey, attempt to reduce the number of categories by showing how some can be reduced to others. They ignore the problem of how the more fundamental categories are to be understood, or reduce this to a problem of the survival of forms of life based upon them.

The philosophers who have most squarely confronted the problem of defining categories are the Neoplatonist thinkers, from Plotinus to Hegel. It was the early Neoplatonists following Iamblichus, a student of Plotinus, who argued that since forms can only be defined in relation to each other, the ultimate, identified with God, is unknowable except by negative definition. Hegel, under the influence of Fichte's effort to deduce the categories of Kant's philosophy, attempted to solve this problem by deducing a categorial scheme 'dialectically', beginning with the most empty category (Being), and then by revealing the limitations of each category in turn, generating a series of categories to eventually arrive at the ultimate category, the Absolute Idea which contains all previous determinations, and includes our consciousness of it: the ideal union of objective reality in its essential features with the human world of thought. He conceived this dialectical deduction of categories to be possible only after they had already been revealed or developed through the evolution of society and of science. It was designed to exhibit the conceptual structure familiar to us, and to be constructive only to the extent of filling in the gaps of this structure. This approach not only freezes our understanding of the world at its present state of development (after making a few refinements), but it makes the necessity involved in this dialectics very ambiguous, and few people have been convinced by this aspect of Hegel's philosophy.[5]

---

3. Alfred North Whitehead, *Process and Reality*, [1929], corrected edition, David Ray Griffin and Donald W. Sherburne eds., N.Y.: Free Press, 1978, p.8.

4. This is different from, but is also a development from Aristotle's, Kant's and Hegel's understanding of categories. On the history of the concept of categories culminating in the work of Whitehead, see A.Z. Bar-On, *The Categories and the Principles of Coherence*, Dordrecht: Martinus Nijhoff, 1987, esp. the 'Extensive Summary of the Exposition', pp.5-17.

5. Uwe Petersen is attempting to revive and carry through Hegel's project. His work is not yet published.

The solution to the problem proposed here is based on the dialectical epistemology outlined in the previous chapter and has a number of dimensions. Firstly, it is necessary to acknowledge that new categories are developed from within the culture of an already functioning community. To borrow and build upon an analogy from Otto Neurath, if developing our knowledge is like repairing a boat at sea, then developing a new categorial scheme is like repairing the keel of the boat. It is much more difficult, but it does not require a standpoint completely outside one's culture. Rather, it involves confronting the problems of one's culture by drawing on its resources, giving old terms new meanings which can be at least partially defined through existing language. Secondly, it is possible to generate these new meanings through the elaboration of an analogy (what Whitehead called 'descriptive generalization'). This involves applying forms of cognition which have developed in a domain with some autonomy within the prevailing culture to domains from which in the past these forms of cognition have been excluded. By counterposing a new analogy in this way to the analogy underlying the dominant categories it becomes possible to transcend these categories (though achieving this generally requires an historical study of the way the prevailing categories were originally articulated and why they came to be adopted). Thirdly, it is possible to further refine these categories by defining them in opposition to, and through a critical analysis of, the categories they have been designed to replace or transcend and to the categories of rival categorial schemes. Ultimately this should involve casting past and rival schemes into an historical narrative from the point of view of the new categories. Fourthly, categories can be further developed through their application and associated elaboration in the comprehension of particular situations and by their incorporation into practices. Rather than thinking of concepts as fully definable through other concepts, concepts should be seen as being ultimately grounded in discourses and the practices associated with them. Meaning should not be seen as finally fixed but as forever open to further development (and possibly, replacement) both through reflection and through practices in the struggle to come to grips with the world. Finally, in relation to process philosophy, formulating a categorial scheme does not involve a completely new beginning. Bergson, Whitehead and those influenced by them have already done much to conceptualize the world as a process of creative becoming, and they have strongly influenced the sciences. Concepts proposed by these philosophers have been selected and refined through their applications within science. Process philosophers today are participating in the on-going development and refinement of concepts which have already proved themselves in a number of areas.

It follows from this that it is not possible to precisely specify and delimit which concepts are fundamental and which are derivative. What is more important is to define only a sufficient number of concepts as can be easily grasped, kept in mind, and then deployed in any situation to displace those concepts which are at present dominating people's thinking.[6] The most important concepts to displace, those inherited from the seventeenth century revolution in thought, are space (the receptacle of matter), time (during which matter changes place in space), matter (identified with body and the occupancy of space), and motion (identified with locomotion of matter through space over time). The categories which are proposed to define the nature of the cosmos as a process of creative becoming consisting of a multiplicity of emergent processes, each being in a complex relation to other co-existing processes and having some degree of autonomy from all others, and to define the nature of these emergent processes, are: activity, order and potentiality; process, structure and event; cause; and spatio-temporal position.[7]

6. Whitehead lists forty-five categories, something for which he has been often criticised. See Andrew J. Reck, 'Process Philosophy, A Categorial Analysis', *Tulane Studies in Philosophy,* Vol. XXIV, 1975, pp.58-91, p.63f.

7. In my efforts to develop these categories I have been most influenced by Aristotle's *Metaphysics*, by Whitehead's *Process and Reality*, by Ivor Leclerc's *The Nature of Physical Existence*, London: George Allen & Unwin, 1972 and *The Philosophy of Nature*, Washington: Catholic University of America Press, 1986, and by the works of David Bohm, Milic Capek and Edward Pols. It should be noted here that while influenced by Whitehead, I have aligned myself with Pols and Leclerc (and also with the Bergsonian philosopher, Capek) against orthodox Whiteheadians in taking composite entities as

The most important problem for process philosophy is: How can 'becoming' be made sense of? As Nietzsche noted of European languages, 'Linguistic means of expression are useless for expressing "becoming".'[8] The European originator of process philosophy, Heraclitus, used the analogies of both fire and flowing water to elaborate his conception of the world. However to free our thinking from the Parmenidean notions which subsequently came to dominate Western culture, Bergson suggested that we must transcend visual analogies altogether and think in terms of auditory analogies. Only in this way is it possible to fully comprehend the nature of becoming (always characterized by duration), of change which is more than changing relationships between elements, of creative emergence constrained by the conditions of this emergence without being determined by them, of individuality within continuity, of order which is prior to the existence of space, and of space as emerging through the ordering of change. Bergson's arguments have been further developed by Milic Capek, and the auditory analogy has been explored in great depth by Victor Zuckerkandl.[9] In elaborating these categories this analogy will be assumed rather than Whitehead's 'mind' analogy.

The first categories to be defined (the categories of the ultimate - into which all primary beings can be analysed) are activity, order and potentiality. These are required to define the other categories without being presupposed by them. As such they are the most difficult categories to define. The second categories (the categories of existence), process, structure and event, characterize what exists as primary beings in the world. The third categories (the categories of explanation), of causation, pertain to the explanation of all that has existed, does exist and could exist, while the fourth categories (the categories of ultimate potentiality), of spatio-temporal position (where space and time are shown to be inseparable from each other and ontologically derivative), are the most fundamental concepts defining potential relationships between actual or potential existents.

## *The Categories of the Ultimate*

'Activity' (corresponding to 'creativity' in Whitehead's philosophy - 'the ultimate behind all forms, inexplicable by forms, and conditioned by its creatures') can be equated with the 'energy' of modern science, while giving this concept a new meaning - or at least a more definite meaning since the notion of energy has never had a clear meaning in physics.[10] In terms of the auditory analogy, the very being of sound is activity, and no unchanging substratum of activity need be supposed. It is in this sense that 'activity' must be understood, and then identified with 'energy'. The concept of 'energy' derives from the Greek *energeia*. Aristotle, who gave *energeia* its technical meaning, meant by it: 'enacting of form'. As such this concept is closer to what I define as 'process' than

---

primary beings, in rejecting Whitehead's panpsychism (rejecting the concept of 'experience' as a category), and rejecting the characterization of potentialities as 'eternal objects' which 'ingress' in actual occasions. Following Pols I have also rejected the complete replacement of efficient causation by 'prehensions'. My aim has been to encompass everything which Whitehead recognized as important while avoiding the problematic aspects of his categorial scheme.

8. Friedrich Nietzsche, *The Will to Power*, tr. Walter Kaufman, N.Y.: Vintage, 1978, § 715, p.380.

9. See Milic Capek, *Bergson and Modern Physics*, Dordrecht: Reidel, 1971 and Victor Zuckerkandl, *Sound and Symbol: Music and the External World*, tr. Willard R. Trask, Princeton: P.U.P., 1969.

10. Whitehead, (*Process and Reality*, p.20). On Whitehead's concept of creativity, showing its relationship to the concept of energy, see Dorothy Emmet, 'Creativity and the Passage of Nature' and Friedrich Rapp, 'Whitehead's Concept of Creativity and Modern Science' in *Whitehead's Metaphysics of Creativity*, ed. Friedrich Rapp and Reiner Wiehl, N.Y.: S.U.N.Y. Press, 1990. For a critique of this concept, along with the concept of 'eternal object', in each case for implying some reality more basic than 'actual occasions', see Edward Pols, *Whitehead's Metaphysics*, Carbondale: Southern Illinois University Press, 1967. For a history of the concept of energy, see Yehudi Elkana, *The Discovery of the Conservation of Energy*, London: Hutchinson, 1974.

what I refer to as 'activity'. 'Activity' corresponds more closely to the concept of *kinesis* as it was used by the early Presocratic philosophers, meaning the eternal motion pervading everything, without this motion being understood, as it came to be after Parmenides, as something requiring an unchanging being which is active. Aristotle redefined the original concept of *kinesis*, bringing it closer to its original meaning by allowing entities to be self-moving, but this concept is still different from the one being defended here. Aristotle defined *kinesis* as incomplete process towards some goal which ceases when the goal is reached, in opposition to *energeia* which is a completed act.[11] This ties the notion to an end in a way which I wish to avoid. As the term is to be understood, activity has more affinities with Aristotle's concept of matter or *hyle*: that which is formed, which is the potency to be reformed and which is the principle of individuation of forms, but which is unknowable in itself (as distinct from the way matter came to be understood in the Renaissance and after). In fact 'activity' can be understood as an identification or conflation of the Milesian (and Heraclitean) concept of *kinesis* and Aristotle's notion of *hyle*.

Order is perhaps the most difficult category to define. Whenever anyone thinks about anything they are assuming order - its quantity, its quality, its endurance, its composition and its context, and of the spatio-temporal order in which it is located. To gain some idea of the notion of order it is necessary to free oneself from such assumptions of order, to imagine each type of order which is normally assumed - ceasing to exist, including the endurance of things and space-time itself. Such complete absence of order can be defined as flux. Any order in this flux can then be seen as some type of constraint which differentiates it, and in doing so makes possible other types of order. And in fact, when this starting point is taken, it becomes evident that all order is facilitating constraint. For instance in thinking of sound and the different types of constraining which occur with the becoming of a piece of music, the ordering of activity can be seen as any constraining or modulation of the sound which differentiates it into identifiably similar aspects, or which constrains such differentiated aspects into similarly different aspects.[12] Through the constraining of sound notes emerge, which makes possible their ordering into melodies, and so on. This notion of order corresponds in some ways to the notion of *eidos* ('idea' or 'form') in Greek philosophy, though it is narrower in meaning and is defined in such a way as to facilitate analysis. As in the philosophy of Aristotle, it must be conceived as immanent within the world, as its 'definiteness'. It is more general than the basic concepts of mechanistic materialism and field theory, the proponents of which must be regarded as attempting to explain all order in the world in terms of particular types of order: the motion of unchanging matter, or extensive force fields.

The way such a conception of order provides a basis for analysis can be seen by considering what is involved in the generation of extension. Extension can be understood as the order generated through the emergence and transformation of potentialities for independence and interaction, and locomotion as change of position can be seen as a particular type of ordering whereby potential relations for independence and interaction are changed in an orderly way. A line can be then understood as a similar difference in point positions. The generation of a circle can be understood as similar differences between similar differences in point positions, and a spiral, with three dimensions, as simultaneously three separate similar differences of point positions.[13] In each case, the order makes possible further ordering to generate new types of order.

Potentiality can be understood as the possibilities - the powers and liabilities for ordering and being ordered which are produced and which can be realized or undermined in the becoming of the world. Referring back to the auditory analogy, it is the 'oriented tension' or 'directed anticipation'

---

11. See *Metaphysics* 1048b28-35.

12. For an elaboration of a conception of order in terms of similarities and differences see David Bohm, *Wholeness and the Implicate Order*, London, Routledge & Kegan Paul, 1980, p.115ff. and David Bohm and F. David Peat, *Science, Order, and Creativity*, Toronto: Bantam, 1987, pp.104-191. Whitehead has defined order differently, and I have not used his concept.

13. See Bohm, *Wholeness and the Implicate Order*, p.116f.

in a piece of music which constrains without determining its becoming. This notion of potentiality corresponds to the Greek notion of *dynamis* - in the more traditional sense of a power only inherent in something without being manifest and in Aristotle's more specific sense of potentiality only illustrated in its realization - the concept which was virtually eliminated in the deterministic mechanical conception of the world and by field theory.[14]

## *The Categories of Existence*

The category of process is meant to characterize primary being or an actual entity, *ousia* - that which exists in the full sense rather than through analysis or derivatively. A process can be defined as an ordering activity which is to some extent (although never entirely) an immanent cause of its own being (or more accurately, becoming), a self-ordering activity in which activity constrains itself and reproduces these constraints.[15] So, to be in the primary sense is to be a process (although it is also possible for there to be a background of unordered activity unknowable in itself, but knowable as the condition for the emergence and continued existence of processes - for example, the vacuum in quantum field theory), and everything else must be understood as a part of or as an aspect of some process or processes, or an aspect of the relationship between processes. As such 'process' corresponds to the place given to 'substantial form' in medieval Aristotelian metaphysics (conceived as the outcome of a process),[16] and to 'actual occasion' or 'actual entity' in Whitehead's metaphysics. This notion of process is designed to replace the post-Renaissance category of self-subsistent matter or body conceived of as essentially inert, along with the associated categories of space, time and motion which have also been conceived as primary beings within the mechanical view of the world. A process is that which in Aristotle's terminology has in it its own source of movement, or in Whitehead's terminology, that 'which constitutes its own becoming'[17]. Assuming the underlying auditory analogy, along with Whitehead I wish to stress both the durational nature of this becoming and interdependence of primary beings. But in opposition to Aristotle and Whitehead, the idea that primary beings must be actualized in some completed end is rejected. Rather, primary beings are identified with processes of becoming, whether such becoming completes itself in some definite end, or endures indefinitely, as protons well might.

By focussing on ordering as an activity, the relationships between different processes and the emergence of new composite processes as primary beings become comprehensible. What is involved

---

14. 'Potentiality' is offered here as the basis for a partial alternative to the 'eternal objects' of Whitehead's categorial scheme, an alternative which avoids the dualism to which Whitehead's philosophy is prone. On the debate surrounding Whitehead's category of 'eternal objects', see Bart F. Kennedy, 'Whitehead's Doctrine of Eternal Objects and Its Interpretations', *Tulane Studies in Philosophy*, Vol.XXIII, 1974, pp.60-86. This debate is not reviewed here, but the position supported is that which is against Whitehead's Platonism. Most of Whitehead's 'eternal objects' are here equated with 'structures'.

15. I have used the term 'process' to emphasise my rejection of Whitehead's temporal atomism in favour of Capek's notion that becoming is 'pulsational' (also argued for by Leclerc), and my acceptance of Pols' and Leclerc's inclusion of compound actualities as primary beings (in accordance with Aristotle's metaphysics). On the importance of recognizing composite entities as primary beings, and the intellectually disastrous consequences of failing to do so, see Leclerc, *The Philosophy of Nature*, Ch.10, 'The Physical Existent as Compound Actuality', pp.130-138.

16. It is possible to interpret Aristotle to accord with the position being defended here by construing what he meant by actualized form as forming activity. See Emerson Buchanan, *Aristotle's Theory of Being*, Cambridge, Mass.: University, Mississippi, 1962 and John Herman Randall, *Aristotle*, N.Y. and London: Columbia University Press, 1960, p.129ff. See also L.A. Kosman, 'Substance, Being, and *Energeia*' in *Oxford Studies in Ancient Philosophy*, Volume II, Oxford: Clarendon Press, 1984, pp.121-149.

17. Whitehead, *Process and Reality*, p.23. What I have called 'process' corresponds roughly to what Whitehead called 'the concrescence of an actual entity or occasion'.

in any causal relationship is always additional constraining of activity so that processes relating to other processes are different than processes not relating to these processes. Emergence and hierarchical ordering can then be seen as self-ordering activities coming to be or being involved in further ordering, that is, being further constrained, as parts of higher level processes which are the ordering activity creating and reproducing these, and other, constraints. Under these circumstances the constituent processes of the supervening process are changed so that they act to constitute the emergent activity and thereby to produce and reproduce these constraints.[18] It is possible for supervening processes to emerge which are the ordering of the emergence of a sequence of such emergent processes, or involving the ordering of even more complex relationships between processes.

Ordered potentialities for ordering produced and maintained by processes are 'structures'. The term 'structure' being a verb as well as a noun (deriving from the Latin '*struere*' - to build) points to 'structures' being derivative from ordering activity or processes as something produced, but they are also derivative from processes which might actualize these potentialities. Thus 'structure' also corresponds to what Whitehead designated as the 'potentiality [of an actual entity] for "objectification" in the becoming of other actual entities', where '"objectification" refers to the particular mode in which the potentiality of one actual entity is realized in another actual entity.'[19] In other words, as ordered potentialities for ordering, structures cannot be understood only in terms of being maintained and produced by processes. They must also be understood as such in relation to processes which could realize these potentialities through their becoming - frequently, but not always, involving their 'objectification'. However no distinction is made here between whether the ordered potentialities for ordering generated by a process are potentialities which could be realized by processes other than the generating process, or whether the potentialities could be realized by the process which generated them. As such I take 'structures' to be equivalent for the most part to Whitehead's 'eternal objects'.[20] It is through identifying such potentialities and their relationships that processes of becoming can be analysed - and also evaluated. What are described as forms by Georg Simmel in sociology, and by D'Arcy Thompson in biology, and following him, by Brian Goodwin, are structures as defined here. The notion of 'formation' understood as referring simultaneously to a structure and to the process which maintains it (as is implied in the concept of 'socio-economic formation', for example) is a useful concept, but as such it should not be thought of as a basic category.

Most of what people identify in the world as existing 'things' are 'structures'.[21] For example, a tree must be regarded as a process of becoming which is durational. What we normally identify as a tree at a particular time as a 'thing' is its structure, the ordered potentialities produced and maintained by this process of becoming: the potential to maintain shape, impenetrability, opacity etc. - which then are realized as such in the process of becoming of not only other processes, but

18. This makes sense of what Whitehead had claimed in *Science and the Modern World* that 'the plan of the *whole* influences the very characters of the various subordinate organisms which enter into it.' *Science and the Modern World*, [1925] N.Y.: Mentor, 1964, p.76. On this, see also Whitehead's *Adventures of Ideas*, [1933], N.Y.: Free Press, 1967, p.199. This notion of emergence contrasts with that ultimately defended by David Blitz in his history of the notion of emergence (which strangely excludes Whitehead's contribution), *Emergent Evolution*, Dordrecht: Kluwer, 1992.

19. Whitehead, *Process and Reality*, p.23.

20. Roy Sellars argued that eternal objects "are, then really expressions of operations and discriminations made possible by the similarity of things." (Roy Sellars, 'Philosophy of Organism and Physical Realism', *The Philosophy of Alfred North Whitehead*, ed. Paul Arthur Schilpp, 2nd ed., La Salle: Open Court, 1951, p.432). I am accepting this, but holding that such operations and discriminations, and the consequent 'universals', are aspects of processes and part of reality rather than merely features of cognition, and that there are also unrealized possible, and even impossible operations and discriminations.

21. As Whitehead argued: '... "potentiality for process" is the meaning of the more general term "entity" or "thing"...' (*Process and Reality*, p.41).

also by the tree itself as the necessary condition of its becoming. However not all structures are 'things', namely those which are not objectified. Examples of unobjectified structures are the cognitive structures referred to by Piaget which can be regarded as ordered potentialities to order action and experience produced, maintained and developed by organisms in interaction with their environments, and social structures which are the potentialities maintained by social processes for various types of interaction between people and organizations.

Along with processes and structures there are also events, such as the coming into being or the destruction of structures and processes, 'decisions' by processes to take one path of development rather than another, significant changes within or differentiated activities of processes, and contingent interactions between processes. Events must always be understood in relation to structures and processes, and it is not possible to completely analyse processes into events. Regularities in the relationship between events should be seen in relation to structures and as manifestations of processes.

An event presupposes Process : event { a constructive/destructive happening ⇒ a creative/discreative act involving decision or choice (or composite compound thereof) ⇒ an instance of Process

## *The Categories of Explanation*

To explain something is to identify its causes. The notion of cause has a long and complex history.[22] The term derives from the Latin *causa* which was a translation of the Greek *aiton* or *aita*. This term referred to the voluntary action of an agent for which he or she could be held responsible. It was originally applied in legal contexts but was generalized to refer to any action designed to bring about an event or state of affairs. This was then applied by analogy to nature, first to events produced by people designed to get nature to do things for them (for instance, lighting a fire to cook food), and then as a simple explanatory principle as when lightning is seen as the cause of fire. It was this notion of causation which was developed systematically by Aristotle who analysed it into four aspects: the material cause, the efficient cause, the formal cause and the final cause; the material cause being the matter involved in the process, the efficient cause the exercise of power, the formal cause the form aimed at by this action and the final cause the reason for aiming at this form. In describing biological growth efficient, formal and final causation tended to be conflated.

However with the birth of modern science there was a radical break with Aristotelian concepts. It is widely assumed that final causes were excluded from scientific explanation. However the Pythagorean Platonism of the major proponents of the 'new philosophy' in the seventeenth century excluded not only final causes but also efficient causes. The notion that power was exercised in causation was replaced by the notion that inert matter moves according to formal principles or laws. The failure to grasp this change in thinking led to a conflation by many of formal and efficient causation. Thus, cause was defined by Hobbes as 'the aggregate of all the accidents both of the agents how many so ever they be, and of the patient, put together; which when they are all supposed to be present, it cannot be understood but that the effect is produced at the same instant…'.[23] But this would imply that each cause and effect, and therefore all causes and effects, must occur instantaneously. There can only be one instant. The incoherence of treating the exercise of power or natural necessity as a logical necessity in this way paved the way for David Hume to argue that the world consists of atomic events without any relation between them but that they follow each other in a regular way. The laws of nature are then conceived as simply the means for making predictions from one event to another.

Where the world is conceived of as a multiplicity of semi-autonomous self-producing processes, causation can best be seen to consist of, firstly, immanent causation (that is, self-creation)

---

22. In focussing on cause here I am departing radically from Whitehead, and from also from his own analysis of cause in terms of prehensions.

23. *The English Works of Thomas Hobbes* ed. Sir William Molesworth, London: John Bohn, 1939, Vol. 1, 'Elements of Philosophy Concerning Body' Ch.IX, p.121.

The force that through the green fuse drives the flower is not in the seed, nor the stem nor the bud, but in the whole plant (which is the material, efficient, formal and final cause in one) consisting of supervening causation whereby constituent processes or activities are constrained, and efficient causation or action on the rest of the world, and secondly, conditional causation (the production of the conditions of any process's existence) which on the emergence of a process differentiates into environmental causation, the environmental conditions of a process - ultimately extending to the entire past of the universe, and material causation, the maintenance of the constituents of the process (although these are not always entirely separable). The notions of immanent and conditional causation are complementary, with each instance of causation being characterizable as either an immanent or a conditional cause depending on from which individual they are being defined in relation to. Structures and events should be seen as causes or effects only insofar as these are understood in relation to processes and their immanent and conditional causation. Structural causation, the more basic or these, is an aspect of conditional causation, the production of potentialities by processes which are causes insofar as they are utilized by one or more processes in their self-formation. Causation cannot be understood as the production by a process, structure or event of an effect, since the effect must be seen as itself an active response of or appropriation by a process,[24] as part of its coming into being or of its becoming. A causal relation between events must be seen as first, presupposing the existence of structures, and more basically, processes which produce and utilize these structures in their becoming.

This complex notion of causation reintroduces and extends the notion of causation as activity realizing potentialities. It emphasises that the very existence of anything must be self-creating activity. In this sense it is closest to Aristotle's notion of causation as applied to the growth of organisms, in which material, formal and final causation are fused. It differs in its emphasis on the actual process of becoming as distinct from the realization of an end (this is accentuated by the distinction between ordering and structure), and through the introduction of the notion of environmental causation pertaining to the relation between any activity to its context, and ultimately, thereby, to the entire past of the universe. Where change of position is concerned, that is, locomotion, this should not be regarded as a state of inactivity as in Newtonian physics, but as an aspect of acting, of immanent causation.

Also in accordance with Aristotle's notion of causation, causation must always be understood as durational.[25] This view has been elaborated by Edward Pols:

> The power is exerted in and through a time-unit, and it cannot therefore be isolated as an exercise of power unless we take the whole time-unit into consideration. Any present moment of that time-unit is like a Bergsonian durée, carrying with it its past as qualifying it, and carrying it with it as a means to its own completion... The end of the action is already present in the beginning, and as the action develops, its beginning and all its past phases are carried with it. What exists at any moment of the action - any temporal 'point' in it - is an abstraction, for the time of the entity's action is not composed of discrete instants. And what exists in any period of the action short of the totality leaves us equally unable to isolate the action.[26]

While such causation is potentially divisible in that it is possible to divide it, it is actually indivisible in the sense that in a shorter period than this duration the 'exercise of power' does not exist.[27] This also is fully intelligible if referred back to the analogy of a piece of music which ceases

---

24. What Whitehead captured with his conception of actual occasions as consisting of 'prehensions' (*Process and Reality*, p.19).

25. Aristotle, *Nicomachean Ethics*, 1174a20. As Collingwood pointed out in *The Idea of Nature*, [1945] Oxford: Oxford University Press, 1960, p.20ff., this resolves Zeno's paradoxes.

26. Edward Pols, 'Power and Agency', *International Philosophical Quarterly*, Vol.11, 1971, pp.293-313, p.297.

27. As Whitehead put it, 'in every act of becoming there is the becoming of something with temporal extension; but ... the act itself is not extensive, in the sense that it is divisible into earlier and later acts of becoming which correspond to the extensive divisibility of what has become.' (*Process and Reality*, p.69.)

to be that piece when it is divided; and this leads to another point. Different causal activities require different durations. This is extremely important in hierarchical relations where, as in a melody where the ordering of notes must be of a longer duration than the individual notes, the immanent causation of higher level processes must have longer durations than the processes ordered by them.

## *The Categories of Ultimate Potentiality*[28]

With these categories of activity, order, potentiality, process, structure, event and cause defined in a preliminary way, it is now possible to redefine the concept of position, rejecting the conception of space and time as the self-subsistent, continuous receptacles of all other entities in the universe and defining space-time in relational terms. That is, 'position' can be defined as the set of potential causal relations of an entity to all other entities, while 'space-time' can be conceived as emerging or becoming as an order of potential (and actual) relations between such positions. This idea is difficult to comprehend when conceived in terms of visual analogies but follows automatically once the world is conceived in terms of auditory analogies.[29] Listening to music the differentiation of sound produces from the indefinite a directionality and extension. In this, becoming is basic and extension derivative, the opposite of the mechanistic conception of space and time where time is reduced to a dimension of space. The past is that which has been formed, while the future is open and yet to be established. And rather than space being an order of places external to each other, this is an order from which extended places emerge from a dynamic, flowing continuum as emergent processes differentiate themselves and achieve some degree of autonomy. Space-time can then be defined as an order of potentialities for independence (space) and interaction (time) which becomes or emerges from a process of becoming with the emergence of semi-autonomous sub-processes. It is the order of potentialities for co-existence and interaction between these emergent processes, or the structures and events associated with them, and is continually produced and reproduced with the becoming of both the supervening and the emergent processes. The past is what a process, structure or event can be causally influenced by and the future is the realm of what it can causally affect. There is no reason to assume that space-time as an order of potentialities must be of any particular dimension, and three-dimensionality must be seen as a particular constraining of activity.

Conceiving of space-time in this way opens the possibility of there evolving a number of space-time orders. Since all processes, and the space-time orders they produce, are locatable within the space-time of the universe as a whole, it is necessary to acknowledge this as the most basic space-time. However it would be a mistake to disregard the reality of the sub-orders of space-time which have emerged and continue to emerge. The potentials for interaction between various levels of sub-processes cannot be adequately understood without taking into account the limited divisibility of the extensive becoming of any process, and the relationship between different orders of divisibility. It is necessary to acknowledge that space-time has been articulated in a number of ways and to pay due regard to this articulation. Thus while galaxies must be seen as co-existing and interacting within cosmic space-time, the nature and co-existence of stars can only be fully understood in terms of galactic space-time, geological processes in planetary space-time, life processes in ecological space-time, organic processes in life space-time, personal life in terms of a complex of social space-times, perception and action in personal or inter-personal space-time, and so on.[30]

---

28. This division of categories is offered as an alternative to Whithead's 'Categoreal Obligations'.

29. The fullest elaboration of time and space in terms of an auditory analogy is provided by Victor Zuckerkandl, *Sound and Symbol*, Princeton: Princeton University Press, 1969, Ch.'s XII-XVIII.

30. An argument somewhat along these lines has been made in relation to time by J.T. Fraser in *The Genesis and Evolution of Time*, Amherst: The University of Massachusetts Press, 1982.

## *Process Philosophy as a Grand Research Programme*

Adopting the categories described above involves overthrowing the reigning paradeigmas and replacing them with a more abstract paradeigma. The notion that there are things conceivable as primary beings characterized by a simple location in space and time, the central assumption of mechanistic materialism, is abandoned. But even more fundamentally, process philosophy rejects the assumption that there is an ultimate order of positions external to each other, a notion which is still assumed by field theory. It is necessary to assume as a starting point (that is, all that is taken to be not in need of any explanation) a complete absence of order, with both dependence and independence of different parts of the universe being taken as problematic. This means that order emerging within the universe, since it cannot be explained entirely by pre-existing order, must be explained at least in part as self-causing. Along with defining the basic objects particular sciences must concern themselves with, providing the basic concepts in terms of which explanations must be formulated, and prescribing the basic forms such explanations must take, the categories of a process philosophy must also provide a very general direction for particular research endeavours. This programme should encompass the grand research programmes of mechanistic materialism, field theory and formism, accounting for their successes, but going beyond all these to account for their failures and to open up new dimensions of the world for investigation.

To begin with, the categories defined provide a way of characterizing the basic 'theoretical objects' to be investigated. These are not 'things', nor force fields, nor forms, although each of these have a derivative place. Ultimately, all systematic enquiry must be seen to be concerned with the nature of and the relationships between processes. There are some difficulties here, since to begin with, the world is known through actualizing its potentialities, and it is as actualized potentialities, that is, through events and structures, that processes are first known. To penetrate beyond apparent reality, beyond events and structures to the reality of 'primary beings', it is necessary to explicitly identify which entities are processes, that is, which entities are to some degree self-creative, and to identify the relationships of dependence and independence between them; that is, their spatio-temporal positions and their causal relations - in particular, the conditional causes of their existence - their environmental causes and their material causes. In mechanistic thought, the environmental conditions tend to be simply assumed, while in field theory, the environmental cause is taken as the whole of reality. Once the conditional causes are understood, the next thing to focus on is efficient and supervening causation, that is, the powers and liabilities of processes both in relation to their environments and to their constituents given specified conditional causes. Field theorists tends to focus only on potentialities for supervening causation and so represent the world as consisting of force fields, while mechanistic materialists tend to focus only on potentialities for efficient causation and so represent the world as consisting of things.

The role of experiments in science needs to be reconceived accordingly. Experiments are designed to actualize potentialities as events through various forms of causal intervention (which deluded logical empiricists into believing that science is about discovering predictable relationships between observed events), and thereby to reveal the full range of potentialities or structures of enduring entities. Sometimes experiments are designed at the same time to reveal the relationships between the constituent or environmental structures which make revealed potentialities or structures possible. However in terms of the categories outlined above even this is not enough. Such investigation must be taken as the means to determine the nature of the ordering activity which maintains and accounts for the existence of structures, sub-structures and environmental structures. But trying to characterize the nature of each self-ordering activity raises a number of problems. It is usually thought that intelligibility requires either an account of an entity in terms of its constituents or an account of the effect of a whole on its parts. Each of these efforts generates problems. In the first case, the constituents themselves would have to be comprehended in terms of their constituents, and so on either *ad infinitum*, or until the ultimate constituents are discovered. In either case comprehension must ultimately be based on constituents which are themselves not comprehended, which would seem to throw into doubt any intelligibility attained. In the second case, the whole in terms of which any particular differentiation is understood must itself be accounted for in terms of some larger whole in order to be intelligible. This must go on *ad infinitum*, or there must be some ultimate whole; and the same problem arises.

To some extent this problem is avoided by proposing a number of different elementary entities, either particles or fields, which can then be defined in relation to each other; for instance defining the electrical charge of quarks as a proportion of the electrical charge of leptons, or the strength of a field as a proportion of the strength of another field. However this means that the existence of such ratios must themselves be unintelligible. Another possibility is to allow both means of attaining intelligibility, so that ultimate entities are made intelligible as an effect of some whole, while the ultimate whole is made intelligible in terms of its constituents. However this merely hides the problem unless wholes are more than the effects of their constituents and constituents are more than manifestations of wholes. The problem then is to specify the existence of individuals over and above both relations to wholes and relationships between constituents, although not being completely independent of either of these. But if such individuals exist, they cannot be made entirely intelligible in terms of either wholes or constituents. How then are individuals to be understood?

This problem can be overcome by recognizing that there are two other aspects to understanding ordering. Firstly, ordering is not merely a relation between wholes and parts, but is a temporal activity. The notion of an individual which is both more than the effects of its constituents and the wholes of which it is part, yet which is not independent of either of these, can be made sense of when durations are considered. An individual can then be seen as a semi-autonomous pattern of differentiating activity through which the structures of constituent and environmental processes are constituted and realized as such over a duration. Explanations of individuals in terms of constituents and environments is essentially an account in terms of the potentialities or structures produced by constituent and superordinate processes which are the conditions for the individual. The individual cannot be conceived separately from these; but then neither can potentialities or structures be conceived independently of the individuals for which they are potentialities and through which these potentialities are realized. The individual itself also generates potentialities or structures which are realized by both itself and its constituent and environmental processes, and these processes are constrained in their becoming by what potentialities are produced by the individual.

How then is this differentiating activity of an individual comprehended? As I have pointed out, it is only as actualized potentialities, that is, through the objectification of processes, that they can be identified. The relationship between the knowledge of objects and knowledge of durational activity is such a problem that Bergson accepted a dichotomy between two forms of knowledge without any possible reconciliation between them. Duration was seen to be accessible only through intuition. Avoiding this solution brings us to the second aspect of understanding ordering. That is, it is necessary to recognize the role of 'indwelling' in attaining intelligibility. The development of understanding of each individual involves 'indwelling' in the process itself, so that while focal awareness is directed at environmental and constituent structures, there is a development of subsidiary awareness of the ordering activity of the individual through which these potentialities are actualized. In this way the objectified potentialities of constituents or the environment can be recognized as just that, as potentialities being actualized by processes in their becoming. It must be the goal of science to facilitate such indwelling and to overcome the fixation on events and 'objects'.

Having allowed for these two aspects to understanding of ordering, it is then possible to consider another two aspects of ordering. It is not only changing relationships between unchanging constituents of an individual, and between an unchanging individual to a changing whole of which it is a part which are significant for understanding an individual, but also the changing of the individual's constituents and the changing of the individual through participating in a whole which are important. In fact, such changing is likely to be a more significant aspect of any individual than those forms of change focussed on by atomists and field theorists. By thinking in terms of auditory analogies and by allowing for the role of indwelling whereby parts of durations are immediately grasped in terms of the unfinished becoming of the whole of which they are parts, such durational changes can be understood as possible prior to spatial differentiation.

The successful understanding of the nature of particular processes in such terms (including their durational aspects) should then provide the basis for comprehending the complex inter-relations between types of processes. However the research programme of process philosophy cannot aspire to total understanding of the world, as the world is acknowledged from the beginning to be both irreducibly complex and creative. The development of understanding involves identifying, characterizing and analyzing the different islands of stability within the flux, and can only provide predictions in limited contexts, and except in rare or artificially constructed cases, it can reveal only trends and tendencies.

## *Scientific Laws, Mathematics and Reality*

This must change the way laws of science are understood and the place of mathematics in science. Paul Davies has noted that 'the [post-classical] physicist's image of reality is rooted in a sort of meta-universe of mathematical objects and relationships that are concrete, eternal and totally dependable, while the Universe is nebulous, shifting and unpredictable.'[31] So it is the mathematically expressible laws of physics which are taken as real, while the Universe itself is granted only a shadowy existence. What is being suggested here is that the 'laws' of nature in science be seen as mappings of the potentialities and their transformations, while reality itself as a process of becoming should be recognized as indefinite - gaining definition only in becoming the potential for processes or activities (measurement being one case of becoming such a potential).[32] Laws must be recognized as abstractions which take for granted the existence of processes and their environmental and material conditions, and the potentialities which they map are usually only fully realized in situations created by carefully constructed experiments in which initial conditions, which cannot be accounted for in terms of the laws, can be controlled. The fundamental laws of science are the mappings of what are the most universal potentialities of being, and there is no reason to suppose that more specific laws (such as laws in chemistry or biology), as specifications of the transformations of potentialities of emergent processes which constrain these universal potentialities, will be deducible from fundamental laws. The laws of science are 'eternal' or 'transcendent' because they pertain to potentialities - whatever has come to exist must eternally have been a potentiality in some sense. However granting eternal status to the laws of science even in this sense is somewhat misleading since potentialities are only such for the becoming of the processes themselves or for other processes which utilize them. The laws of nature must be seen as having emerged with the becoming of the universe - the fundamental laws with its origin, more specific laws with processes which emerged later, as processes emerged for which the universe itself and then these emergent processes were utilizable for other processes. And in some cases,

---

31. Paul Davies, 'Law and Order in the Universe,' *New Scientist*, 1634 (Oct. 15th, 1988): 58-60, p.60.

32. The different theories of the laws of nature are described and analysed by Alfred North Whitehead in *Adventures of Ideas*, [1933], London: Free Press, 1967, pp.103-139. Whitehead distinguishes between the positivistic conception of law understood as relating observations to each other, the Platonic according to which mathematical laws are part of reality (assumed by Descartes and Newton), the Aristotelian according to which laws describe causal processes (which has recently been defended by Harré and Bhaskar) and the conventional. For a defence and elaboration of Whitehead's notion of immanent scientific law (a development of Aristotle's) see Ann Plamondon, 'Whitehead and the Philosophy of Science' in John B. Cobb Jr. and David Ray Griffin eds, *Mind in Nature*, Washington: University Press of America, 1978, pp.109-119.

where the potentialities revealed by laws are only utilizable by humans after they have been revealed as potentialities by science, the formulation of these laws must be seen as having partially created these potentialities.

In this scheme mathematics (associated with the 'quantitative' aspect of reality) must be seen differently than within the framework of other ontologies. As Cantor showed, all mathematics can be characterized in terms of set-theoretical logic, and presupposes that: '"what is - what can be thought" is capable and must always be capable of being fully and distinctly defined, composable and decomposable into totalities definable by universal properties and comprising parts defined by particular properties.'[33] No matter how far afield one ventures in mathematics, this same logic presides, and nothing would change by switching to multivalued logics or fuzzy sets. Mathematics deals with what is definite. Genuine becoming - order in the process of emerging out of disorder, the emergence of new types of partially autonomous individuals - essential features of a world of activity and of processes, cannot be fully captured by mathematics. So, the success of mathematics must be seen in a new light. Rather than seeing mathematics as defining and describing the nature of primary beings, mathematics must be seen as defining and mapping potentialities or structures and their transformations which are created, sustained and transformed by processes. As Whitehead argued, 'mathematics is concerned with certain forms of process issuing into forms which are components for further process.'[34] Mathematics is important to science 'as the search for infinitely rich and diverse patterns of order...'[35], that is, for the search for and analysis of structures. The movement in the twentieth century towards seeing mathematics as the study of structures,[36] where 'structures' are understood as both part of reality and as abstractions from reality dealing with relations beyond any exemplification, can be interpreted to support this contention.

Claiming that mathematics and mathematically expressed laws pertain to potentialities does not mean that mathematics cannot illuminate processes of becoming; but it does imply that it can only do so indirectly - by mapping out existing potentialities and tendencies of processes and what potentialities will be realized in different circumstances. Furthermore it means that mathematical analysis and description cannot take the place of causal analysis and description, and ultimately, 'indwelling' in the processes of becoming of the world as the starting point and ultimate goal of science. The application of mathematics, to be successful, always presupposes indwelling by means of non-mathematical causal theories, and as a goal, mathematical prediction must always be subordinated to the goal of achieving a better understanding of the world as a process of creative becoming in this sense.[37]

---

33. Cornelius Castoriadis, *Crossroads in the Labyrinth*, [1978] tr. Kate Soper and Martin H. Ryde, Brighton: Harvester Press, 1984, p.210.

34. Alfred North Whitehead, *Modes of Thought*, p.92. Whitehead argues that when we say 'twice three is six' we are not uttering a tautology, but are describing the process which issues from two forms of three in the form of 'six', which is then a potential for other processes beyond itself. On Whitehead's view of mathematics see Murray Code, *Order & Organism*, N.Y.: State University of New York Press, 1985.

35 As Ralph V. Norman, Jr. put it in his explication of Whitehead in 'Whitehead and "Mathematicism"', *Alfred North Whitehead: Essays on his Philosophy*, ed. George L.Kline, Englewood Cliffs: Prentice-Hall, 1963, pp.33-40, p.34.

36. On the concept of structure in mathematics, see Jean Dieudonné, 'The Difficult Birth of Mathematical Structures (1840-1940)' in *Scientific Culture in the Contemporary World*, V. Mathieu and P. Rossi (eds), Milano: Scientia, 1979. For a view of the relationship between mathematical structures and physical reality similar to that proposed here (but with the dynamic aspect omitted), see James Franklin, 'Mathematics, Necessity and Reality', *Australasian Journal of Philosophy*, Vol. 67 (1989), pp.286-94.

37. The different place of causal accounts and scientific laws in science, and the primacy of causal accounts for understanding reality, see René Thom, *Structural Stability and Morphogenesis*, tr. C.H. Waddington, Reading, Mass.: W.A. Benjamin, 1972, p.5, and Nancy Cartwright, *Nature's Capacities and Their Measurement*, Oxford: O.U.P., 1989.

# 14

# PROCESS PHILOSOPHY AND THE NATURAL SCIENCES

Modern science originated in the seventeenth century with the mechanistic view of the world, and while it is acknowledged that it breaks down at the extreme macro and extreme micro levels, it is still Newton's mechanics which is the ultimate point of reference for all science. Theories not comprehensible as developments of Newton's mechanics are presented as though only the mathematical formalism and the predictions facilitated by it are of significance.[1] But advances in the natural sciences over the last hundred years, particularly in physics itself, have invalidated the metaphysical assumptions of classical science. This has led to the odd situation described by David Bohm where:

> ... just when physics is moving away from mechanism, biology and psychology are moving closer to it. If this trend continues it may well be that scientists will be regarding living and intelligent beings as mechanical, while they suppose that inanimate matter is too complex and subtle to fit into the limited categories of mechanism.[2]

It is these developments in the physical sciences which led Ilya Prigogine to claim that 'we are in a period of revolution - one in which the very position and meaning of the scientific approach are undergoing reappraisal - a period not unlike the birth of the scientific approach in ancient Greece or of its renaissance in the time of Galileo.'[3]

The significance of the present state of science has been disguised until very recently, not only by its positivistic interpretation, but also by what can only be described as the corruption of science by poor pedagogy, over-specialization and over-industrialization.[4] To begin with, science is presented to students as a body of knowledge which is merely being added to by practicing scientists. So while most theories in physics have replaced particles as the fundamental material entities of the universe with fields, these are still treated in accordance with the Newtonian mechanics as being determined by laws of motion plus initial conditions. Consequently for most physicists the goal remains, as Leon Lederman, director of the Fermi National Accelerator near Chicago, put it: 'to explain the entire universe in a single, simple formula that you can wear on your T-shirt.'[5] Overspecialization has further blinded scientists. While the cutting edge of each domain of science has broken out of the mechanistic framework of concepts and the ideals of explanation associated with it, scientists presuppose ideas from related disciplines which have been superseded decades ago. Most scientists are therefore ignorant of how all the specific developments within different domains cohere and undermine the prevailing reductionist ideal of science. Consequently the domination of everyday life by mechanistic categories of thought in terms of which recent advances in science make no sense continues almost unquestioned. Instead of seeing science as invalidating these categories, science is presented as moving away from concrete

1. On this, see Jean-Marc Levy-Lebond in 'Ideology of/in Contemporary Physics' in Hilary Rose and Steven Rose, *The Radicalisation of Science*, London: Macmillan, 1976, pp.136-175.

2. David Bohm, 'Some Remarks on the Notion of Order' in *Towards a Theoretical Biology, 2 Sketches*, ed. C.H Waddington, Edinburgh: Edinburgh University Press, 1969, p.34.

3. Ilya Prigogine, *From Being to Becoming*, San Francisco: Freeman, 1980, p.xiif.

4. For an account of this corruption see Mark Chargaff, *Heraclitean Fire*, N.Y.: Rockefeller University Press, 1978; and Jerome Ravetz, *Scientific Knowledge and its Problems*, Harmondsworth: Penguin, 1973, esp. Ch.2.

5. Cited by Paul Davies, *The Cosmic Blueprint*, London: Heinemann, 1987, p.13.

experience towards levels of abstraction which are the exclusive province of the scientific elites. Finally, the harnessing of science to industry has reinforced poor pedagogy and overspecialization and discouraged efforts to think of science as anything but a means to advance technology. Science is now almost completely dominated by administrators concerned to ensure that what scientists produce is economically profitable.[6] My contention is that once mechanistic categories are brought into question, modern science becomes comprehensible as a revolution in progress, a revolution in which our conception of the world and our place within it are coming to be understood in terms of a metaphysics of process.[7]

Within physics the major advances beyond mechanistic categories are based on three basic theories: relativity theory, quantum theory and thermodynamics. What is most significant about these theories is that while they were developed independently of each other to deal with different problems, as far as predictions go they dovetail together without conflict in the explanation of a vast variety of phenomena. This harmony between the theories is frequently not obvious and is only revealed by careful analysis which invariably reveals how the validity of each theory must be accepted to defend the validity of the others. For instance a thought experiment proposed by Einstein to invalidate quantum theory was shown by Bohr to be invalid because it had not taken into account the implications of relativity theory. All efforts to invalidate the second law of thermodynamics by working out the more arcane implications of the general theory of relativity have been shown to have overlooked some feature of the general theory itself. Yet conceptually these theories are not easily reconcilable with each other, even in the ontological status ascribed to the objects of the theories. What I hope to show is that if relativity theory, quantum theory and thermodynamics are interpreted through process philosophy, there is hope that they can be conceptually reconciled. This will require of scientists that they acknowledge the primacy of becoming, the irreducibility of complexity, and that humans as conscious agents are part of the world.

## *Relativity Theory*

Relativity theories, that is, the special theory of relativity developed from the theory of electro-magnetism and the general theory of relativity designed to explain gravity, are essentially developments of field theory, the conception of being according to which the world does not consist of discrete bits of matter but of continuous force fields.[8] While the concept of force field has much in common with notions of Stoic physics, the modern concept evolved through Leibniz's criticism of Newton's notion of a duality between force and matter, Boscovitch's dynamism (an attempt to reconcile Leibniz and Newton) in which explanation is ultimately in terms of point centres of power, Priestley's rejection of point centres and his description of nature in terms of active forces alone, Faraday's elaboration of Priestley's ideas to describe electrical and magnetic phenomena, Maxwell's mathematical treatment of Faraday's ideas, and the jettisoning by Herz and Lorenz of the notion of ether by which Maxwell had tried to give a mechanical explanation of force fields. While field theory, like mechanistic materialism, is deterministic and ultimately leads to the Parmenidean conception of the universe as an 'iron block', its development in the theories of relativity decisively undermines many other central features of the mechanistic conception of the

6. On this, see David Dickson, *The New Politics of Science*, Chicago: University of Chicago Press, 1988.

7. That this is the case has been argued in a number of works, most notably Prigogine *From Being to Becoming* and Ilya Prigogine and Isabelle Stengers *Order Out of Chaos*, Toronto: Bantam, 1984; David Bohm *Wholeness and the Implicate Order*; Erich Jantsch *The Self-Organizing Universe*, Oxford: Pergamon Press, 1980; and Paul Davies, *The Cosmic Blueprint*. See also the anthology edited by John B. Cobb, Jr. and David Ray Griffin, *Mind in Nature: Essays on the Interface of Science and Philosophy*, Washington D.C.: University Press of America, 1978.

8. For a history of the theory of fields see Mary Hesse, *Forces and Fields*, London: Sheed and Ward, 1963.

If mass is one aspect of primordial activity, then mass-effects are more or less significant according as the activity is more or less energetic — hence the connection between mass & energy.



E is proportional to m, not equivalent to m.

world. To begin with, the special theory of relativity emancipates the theory of electro-magnetism from classical physics by invalidating the idea of an underlying ether supporting wave motion, conceiving mass as a function of velocity, thereby revealing the equivalence of energy and mass (given in the famous formula $E=mc^2$), eliminating the concept of a rigid body, while defining simultaneity, space and time in terms of interactions at the speed of light c (held to be constant in all inertial reference systems).

The special theory of relativity is generally taken to support a field conception of being in which the laws of nature will be continuous field variables defining points in the field in terms of the whole. This Parmenidean view of the world appears to be reinforced by the geometrical representation of relativity theory in the Minkowski diagram, since this seems to imply that what is taken to be future and what past is relative to what reference system happens to be chosen. However the special theory of relativity can also be interpreted in terms of, and thereby be shown to provide support for, a process conception of the world.[9] Without going into all the arguments for and against the different interpretations, there are three ways in which understanding of the world can be deepened when it is interpreted in this way.

Firstly, the theory of relativity reveals how all knowledge of the world is situated within a process of becoming. In the Minkowski diagram an inertial frame of reference t is represented by a world-line or world-tube with the light cone ABC representing the future and the light cone DBE the past. ABD and CBE represent the 'elsewhere', the region which cannot interact with B in any way.

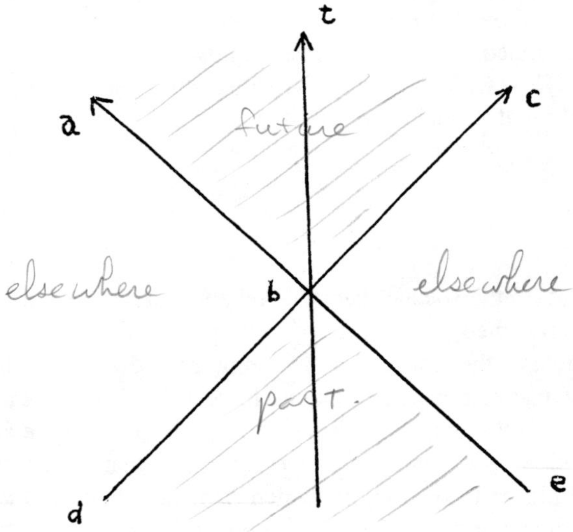


While this diagram is usually interpreted to mean that time is nothing but a dimension of space, it actually reveals the primacy of becoming and the relativity in the concept of space. In all frames of reference the order of causal succession is absolute, while in any frame of reference the future, that which can be causally influenced from a situation, is separated from the past, that which can be known about or which can causally influence the situation, by the four dimensional wedge of the 'elsewhere'. It is no longer possible to define the present as a simultaneous juxtaposition of points, and this virtually destroys the traditional notion of space as a timeless order spread out under events. Further reinforcing the primacy of the notion of becoming, the diagram reveals the

9. The special theory of relativity was interpreted in these terms by Whitehead, and more recently by David Bohm in *The Special Theory of Relativity*, N.Y.: W.A. Benjamin, 1965 and by Milic Capek in *The Philosophical Impact of Contemporary Physics*, Princeton: Van Nostrand, 1961 and 'The Myth of Frozen Passage: The Status of Becoming in the Physical World' in *Boston Studies in the Philosophy of Science*, Vol.2, N.Y.: Humanities Press, 1965, pp.441-463.

impossibility of completely predicting the future. The region in the light cone ABC will be influenced by what happens 'elsewhere', and it is impossible to work out completely what will happen elsewhere from what happened in the past.

Secondly, the special theory of relativity suggests how a spatio-temporal order emerges. According to the theory there is a Lorenz contraction in the direction of relative motion in the ratio of $\sqrt{1-v^2/c^2}$ (where v is the velocity of relative motion and c is the velocity of light) which implies that at the speed of light there is no spatial separation between the emission and absorption of a quantum of light, and no passing of time. This implies that if all activity were unordered, space and time would have no meaning. Spatial and temporal features emerge with the ordering of activity, as is revealed by the following Minkowski diagram.

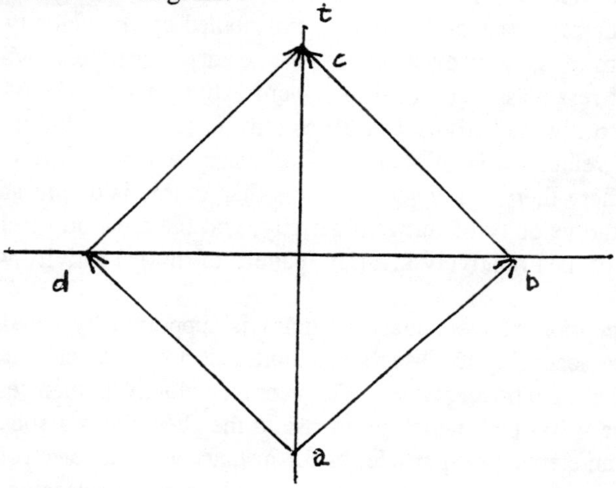


In this, AB and AD represent light rays in primary contact, while ABC and ADC represent light rays in secondary contact. Two primary contacts in the same direction combine to give a time-like interval or duration AC while two oppositely directed contacts such as AD and AB give rise to a space-like interval or extension DB. However before it is possible to talk of time and space as such, it is necessary to refer to relations between ordering activities which are such extensive durations. Space-time can then be thought of as the order of the potentials for interaction between such entities. The Minkowski diagram should then be seen not as a representation of a space-time plenum but as a map of these potentialities. What is past is that which can in principle be known about, while what is in the future is that which can in principle be causally influenced. It is in relation to such potentialities and actual interactions that the notion of extensive duration takes on its full meaning.

Since it is the ordering of activity into patterns of relations which produces a spatio-temporal order, it is no longer possible to conceive of these relations in terms of locations specified in terms of a set of continuous Cartesian co-ordinates. Rather than being an external system in terms of which things in the world can be measured, spatio-temporal relationships must be understood topologically, with things defined in relationship to each other. A most fundamental relationship of this kind is containment where one process is a constituent of another and is therefore contained within the extensive becoming of the superordinate process. A number of constituent processes ordered into such a process can then be understood as 'in' the spatio-temporal order of the whole process, but the spatial and temporal order in which they are 'in' is created by their ordered interaction. In the incomplete process of becoming of the superordinate process, space-time is the potential for interaction between semi-autonomous sub-processes. The continuous space-time order of the universe comes into being through the articulation of the world into a multiplicity of hierarchically ordered processes. Measurement can only be the establishment of ratios between structures produced by processes.

Thirdly it is possible to develop a better understanding of energy as activity and its relation to matter and mass. To clarify this notion it is necessary to draw a distinction between outward activity as in the change in position of a body and inward activity such as the thermal motion of constituent molecules which cancel each other out on a large scale. This distinction is relational since what is outward activity at one level is inward activity at a higher level. Mass can be understood as the sum of both inward and outward activity. The outward activity increases mass in a straightforward way which is easily calculated from Einstein's equations. But the rest mass can also be seen as due to the velocity of movement; of inward activity. For instance a major component of the rest mass of an atom is contributed by the velocity of the electrons. This leaves only the problem of the rest mass of such elementary constituents which can be neither points nor rigid objects. If rest mass is defined as inward activity then it follows that where there is no rest mass, there is no inward activity and all activity is outward. This is true of all forms of radiation which travel at velocity c in all frames of reference, and so can never be considered at rest. This suggests that where there is a velocity of less than c, this is due to an inward reflecting of activity cancelling out the velocity of outward activity, and the creation of elementary particles or entities can be thought of as a relatively invariant pattern of inward activity with no substratum apart from this activity.

Such a conception of elementary particles is supported by relativistic quantum mechanics of electrons where, according to Dirac's equations, electrons travel at the speed of light in trembling movements called *Zitterbewegungen*. The average velocity is then less than the speed of light and corresponds to a spiral path which gives rise to the phenomena associated with electron 'spin'. The annihilation of an elementary particle, as for instance when an electron collides with a positron, can be thought of as a breaking down of the inward ordering releasing activity in a purely outward form as radiation. The nature of the rest mass of such an entity can be clarified by means of one of Einstein's thought experiments. A box of radiant energy in thermodynamic equilibrium produces a radiation pressure on the walls. If it is accelerated, the radiation on the rear wall will gain more momentum than the radiation which reflects off the front wall will lose, producing a resistance to acceleration which is the characteristic manifestation of what we call mass. An elementary entity conceived as the ordering of outward energy inward is analogous to such a box. A 'state of motion' due to the inertia of such a body can then be seen as an aspect of the activity of this body relating both to itself and to the rest of the universe.

With the distinction between inward and outward activity it is possible to define potential energy as inward activity which can be converted to outward activity. In terms of the above analysis of the nature of elementary entities and the nature of their rest mass, the diminishing potential energy of a body being accelerated in a gravitational field, that is, a falling body, corresponds to decreasing inward activity in the body as defined from an inertial system which is manifest in a lowering of its rest mass as it falls.

The general theory of relativity was developed to deal with accelerating frames of reference, and thereby to deal with the relationship between inertial and gravitational mass. This was achieved by replacing the Euclidean space-time of the Minkowski diagram with Reimannian curved space-time. In terms of this, Einstein represented gravitation, and ultimately hoped to represent matter itself, as a curvature of space-time. The general theory of relativity has become central to the development of theories of the cosmos, especially with the realization that an adequate formulation of general relativity requires the universe to be seen as expanding and that gravity has played a central role in differentiating the universe into stars and planets, galaxies, clusters of galaxies and clusters of clusters of galaxies throughout its expansion. While the general theory has served as the starting point for efforts to unify science in terms of field theory,[10] the theory can equally be interpreted in terms of process philosophy with space-time being conceived as derivative from causation rather than as a representation of a Parmenidean plenum.[11] The idea that the universe had a beginning and developed through a process of differentiation and integration is fundamentally in accord with the view that the universe is a process of creative becoming such that the future is not contained in the past. This view could only be invalidated by a theory able to predict the nature and time of each and every differentiation which has occurred in this becoming.

Classical field theory with its Parmenidean implications foundered on the discrete nature of energy, the phenomenon around which quantum theory is built. Since any satisfactory reunification of the theory of gravitational fields with the rest of physics must be quantized, it is to quantum theory that we must now turn.

discrete nature of energy = atomistic (quantized) nature of primordial activities.

## *Quantum Theory*

Quantum theory had its origins in problems of radiation. Its most revolutionary feature was its postulation of a fundamental discreteness in the world in opposition to the assumption of continuity in change in both classical mechanics and classical field theories. The theory has since been extended as the theory of the micro-world: the structure of atoms and the nature of chemical bonding, the interaction between radiation and matter, and the nature of the elementary constituents of the universe; and it is now playing a central role in the development of theories of cosmology.[12] The major efforts in theoretical physics over the last fifty years have been devoted to developing theories which unite quantum and relativity theories: as relativistic quantum theories and as quantized field theories. The most important achievement in this regard was the development of quantum-electrodynamics (QED), beginning with Dirac's equations for a relativistic quantum theory and culminating in Feynman's formulation of it in terms of path integrals. This approach has been successfully developed by Glashow, Weinberg and Salam to include weak forces, revealing these to be manifestations of electro-magnetism. There is also a quantum field theory of strong forces: quantum chromodynamics, formulated on the model of QED as a gauge field theory, and efforts are being made to develop a grand unified theory which will relate the strong force to the electro-weak force. The ultimate aim is presented as the development of a unified theory which will at the same time unite all the forces, including gravity, and be a theory of the elementary entities of the universe. These are the supergravity theories. However a new contender to unify physics has emerged with the superstring theories, which reject the idea of elementary particles and replace them with strings.

While it is generally accepted that quantum theory has been a remarkably productive research programme, there is little consensus on how it should be understood. While this is partly due to the nature of the formalism, it is also due to its inconsistent formulations. As Ted Bastin wrote, scientists 'habitually work with a jumble of elements taken from a variety of different conceptual frameworks none of which, singly, is adequate to present the facts that are known, and each of

---

10. For a defence of this, see A. Mercier, H.J. Treder and W. Yourgrau, *On General Relativity Theory*, Berlin: Akademie - Verlag, 1979.

11. The beginnings of an argument along these lines was made by John A. Winnie, 'The Causal Theory of Space-Time' in J.S. Earman et. al. eds, *Minnesota Studies in the Philosophy of Science*, Vol. VIII, 1977, pp.134-203. These arguments have been extended in a paper by Adrian Smith, 'The Causal Structure of Space-Time' circulated at Australian Philosophy Conference at Newcastle in August, 1981.

12. For the history of the beginnings of quantum theory see Max Jammer, *The Conceptual Development of Quantum Mechanics*, N.Y.: John Wiley and Sons, 1974. For a brief account of the development of modern particle physics see J.E. Dodd, *The Ideas of Particle Physics*, Cambridge: Cambridge University Press, 1984.

which is partly or even largely incompatible with the rest.'[13] For instance while the Born interpretation of the wave or *psi* function, according to which waves are seen as referring to the probabilities of finding particles in particular places, can be adopted to interpret scattering experiments, interference experiments require the wave to be seen as a physical phenomenon. And while the entities associated with quantum theory are usually spoken of as particles, with quarks and leptons represented as the ultimate building blocks of the universe, the notion of a particle is incomprehensible in terms of quantum field theories. In these theories, the so-called ultimate building blocks of matter are treated as points in the fields. Such contradictions are avoided by representing the formalism of quantum theory positivistically as simply the means for making correct predictions. But the kind of mathematics used in a scientific theory already implies a way of conceiving the world, and the confusion in quantum theory also exists its mathematical formalism - as C.A. Hooker has pointed out.[14] Furthermore, when it comes to speculations and experiments on elementary entities and their interactions or to speculations on the origins of the universe based on quantum theory, scientists immediately become realists. In the face of this situation, Feynman could do no more in his famous lectures than comment 'we must be careful not to attribute too much reality to the waves in space. They are useful for certain problems, but not for all.'[15]

To overcome this confusion it will be necessary for any proposed interpretation of quantum theory to confront six unique implications of the theory. The first is that, as in relativity theory, the observational situation has to be taken into account, and no independent 'reality' can be abstracted from it. The properties of 'objects' exist in a twilight state of 'superposition' until they are measured. Second, the quantum of action is indivisible. Transitions between stationary states are discrete, with systems moving from one state to another without passing through intermediary states. Third, matter has a wave-particle duality, behaving in some cases more like a wave, at others more like a particle, but always in certain ways like both together. Fourth, it is impossible to predict in detail what will happen in each individual observation, implying some degree of indeterminacy in the world. Fifth, a particle travelling between two points travels through all possible paths between them simultaneously. Sixth, particles that are millions of miles apart can affect each other instantaneously.

Given the prevailing formalism, the most coherent interpretation of quantum mechanics is Niels Bohr's complementary theory. It is associated with a Neo-Kantian (Wittgensteinian) position according to which science is only concerned with what we can say about the world, but in addition it is argued that what we can investigate and describe cannot be combined into a coherent picture. Among other things, it is necessary to use concepts from both mechanistic materialism (the particle) and field theory (the wave) in a complementary way to investigate and interpret the quantum domain. As Bohr put it:

> ... the impossibility of combining phenomena observed under different experimental arrangements into a single classical picture implies that such apparently contradictory phenomena must be regarded as complementary in the sense that, taken together, they exhaust all well-defined knowledge about atomic objects.[16]

In this it is recognized that humans are actors within the world striving to make it intelligible. Science is then not a description of reality itself, but reality in particular experimental situations in

---

13. Ted Bastin, 'Introduction' to Ted Bastin ed., *Quantum Theory and Beyond*, Cambridge: Cambridge University Press, 1971, p.8f.

14. C.A. Hooker 'Metaphysics and Modern Science' in C.A. Hooker ed. *Contemporary Research in the Foundations and Philosophy of Quantum Theory*, Dordrecht: Reidel, 1973, pp.174-304.

15. Richard P. Feynman, Robert B. Leighton and Matthew Sands, *The Feynman Lectures on Physics: Quantum Mechanics*, Reading: Addison-Wesley, 1965, 3-9.

16. Niels Bohr, *Essays 1958/1962 On Atomic Physics and Human Knowledge* N.Y.: John Wiley, 1963, p.25.

which the experimental situation must be treated as a whole. The idea of abstracting the experimental object from the experimental apparatus is rejected as irrelevant. This means that quantum theory cannot be held to describe a reality independent of experimental situations, and the rejection of this form of realism is essential to account for situations in which quantum theory implies a violation of the principle that inter-actions cannot occur at faster than the speed of light (the Einstein-Rosen-Podolsky paradox).[17] J.S. Bell revealed the theoretical possibility of experimentally testing whether quantum theory was in fact valid in such situations, and the predictions of quantum theory were experimentally validated by A. Aspect and his colleagues in 1982.[18]

However Bohr's arguments that the phenomena together exhaust well-defined knowledge is misleading. While each of these phenomena are understood deterministically, it is also well-defined knowledge that using the two forms of description together implies an indeterminacy in the world, that the world is in some sense genuinely creative, and that the experimenter participates in this creativity. And Bohr's arguments that it is impossible to go beyond the concepts of classical physics to take this into account are invalid. If two deterministic theories of being turn out to be unsatisfactory in isolation but usable when treated as complementary to each other, it is oddly conservative to believe that a non-deterministic theory of being could not be developed which would account for the distinguishing features of the quantum domain, including the extent to which the domain is predictable. Such a belief suggests a poverty of imagination. Furthermore the makeshift way of combining the two theories of being, pressing fragments of the particle and the field schemes into service blindly as the situation demands, has manifest itself in problems and limitations in the developments of quantum theory. Since there is no way to introduce extended structures into relativistic quantum theory, particles are treated as points. But this leads to infinite energies in calculations which can only be removed by a mathematically and physically *ad hoc* 'renormalization' procedure. And while theorists speak glibly of quantisation procedures to represent the change from continuity to discreteness, there is no comprehension of why this should occur or what are its ramifications. In grand unified and super-gravity theories the existence of infinities have not yet been shown to be renormalizable, and as quantum theory advances, problems are increasingly being left unaddressed. As Christine Sutton complained in a popular study of elementary particle physics:

> Why ... is electric charge quantized, with the proton's charge the same size (but opposite sign) as the electron's? This comes down to asking why the quarks have charges of 2/3 and 1/3, and leptons have charges 0 and 1 in units of e, the charge of an electron. Electroweak theory does not say what these charges should be; they have in effect to be inserted 'by hand'. Moreover the masses of all the quarks and leptons are quite arbitrary, as are the strengths of the interactions...[19]

After revealing the confusion of the mathematics of quantum theory, C. Hooker concluded that:

> ... quantum mechanics demands either a new conceptual-ontological scheme (a revision of the two conceptual schemes more thoroughgoing even than their logic) or the abandonment of quantum mechanics as a hopelessly bastard offspring of an attempted marriage of the two great

---

17. See Bernard d'Espagnat, *In Search of Reality*, N.Y.: Springer-Verlag, 1983, esp. Ch.4.
18. A. Aspect, P. Grangier, G. Roger, *Physical Review Letters*, Vol.49, Dec. 1982, p.'s 91 and 1804.
19. Christine Sutton, *The Particle Connection*, N.Y.: Simon and Schuster, 1984, p.163.

> classical theoretical structures, doomed forever to a jerrymandered interpretation in terms of one of them.[20]

In the light of this state of affairs it is to approaches which break with traditional ontologies which must be looked at.

One of the most important developments in this respect is the attempt by the theorists at Birkbeck College, London University, originally under the leadership of David Bohm, to develop a non-localizable hidden variable theory. Non-localizable hidden-variable theory has been explicitly formulated in terms of process philosophy. It is a theory of the quantum domain which accepts the role of the scientist in the world and the indivisibility of the experimental situation, but takes these features to be characteristics of the world which must be explained. In this theory, the quantum mechanical wave-function becomes an actual field, but with unusual properties. The particles and waves acknowledged in present quantum theory are seen as manifestations, 'relevated' by particular experimental arrangements, of the more basic non-local order of the quantum field. This provides a coherent conception of the quantum domain in which what are generally taken to be 'things' are seen as emergent processes within the becoming of the universe. As Bohm described his theory:

> What we are suggesting ... is that all matter is to be understood as a relatively autonomous and constant set of forms built on and carried by the universal and indivisible flux ... Such material forms have a certain subsistence, in the sense that under appropriate conditions they can continue with a certain limited possibility for stable existence. However they are not to be regarded as substance, which would be completely stable, permanent and not dependent on something deeper for their continued existence.[21]

The universal flux, the 'holomovement' as it is elsewhere described by Bohm, is an undivided whole, not in the sense that it is indivisible, but in the sense that division has no meaning in relation to it. Associated with this, Bohm developed his concept of a new type of order, the non-local 'implicate' order, by using the hologram as an analogy.[22] Holograms are such that if a photographic plate is illuminated by a laser beam, the eye will see from a range of possible viewing points a three dimensional structure as though looking through a window. But the order in the photographic plate is not localized. If only a small part of the plate is illuminated the viewer will still see the whole structure, but with less sharply defined detail and with less possible points of view, as though looking through a smaller window. There is an order 'implicated' non-locally in the whole plate which is 'explicated' by illuminating it. But this analogy is slightly misleading because it is static rather than dynamic, and to emphasise the dynamic nature of the becoming of the holomovement Bohm and his colleagues have used an auditory analogy. In doing so they have tried to show how the causation involved in this becoming cannot be comprehended in terms of a chain of events, but must be understood as a 'formal' cause, corresponding to what I have described as immanent causation. This generates localizable particle-like phenomena. As they described this:

> Let us begin by considering a musical theme. The order of successive notes in such a theme evidently cannot be understood as dynamically determined. Rather the entire theme is a single whole form, which is perceived directly as such. One theme may then be followed by another in a developing structure, which in turn constitutes a higher order form, and this sort of

---

20. C.A. Hooker 'Metaphysics and Modern Science' in C.A. Hooker ed. *Contemporary Research in the Foundations and Philosophy of Quantum Theory*, Dordrecht: Reidel, 1973, pp.174-304, p.270.

21. David Bohm, 'The Implicate of Enfolded Order: A New Order for Physics' in *Mind in Nature: Essays on the Interface of Science and Philosophy*, eds. John B. Cobb Jr. and David Ray Griffin, Washington: University Press of America, 1978, p.40.

> development can go on further to indefinitely higher levels. ... The development of themes in successive stages is then like a particle which is first in one quantum state and then in another etc. As there is no dynamical cause of successive quantum themes, so there is no dynamical cause of successive quantum states. Rather the whole order and form of the development is the cause. ... We compare the many-particle system to an orchestra (each particle to an instrument). When the whole orchestra is playing one theme all the instruments are related in an essential way... We thus obtain an analogy to the nonlocal correlation implied by the many-body wave function... A new process can now be envisaged in which the orchestra is playing together as a whole (i.e. in nonlocal relationship) begins suddenly, as part of the whole structure of the composition, to break up so that each instrument plays independently (i.e. solo) in a way that is not related to how the others are playing. This is our analogy for the spontaneous process of localization of states.[23]

This conception of the quantum reality has been formalized using algebraic topology.[24]

Ideas complementary to this which also accord with the categories of the metaphysics of process have been developed by Geoffrey Chew. Chew has proposed what he calls a 'bootstrap model' to explain hadrons (strongly interacting particles such as the proton and neutron).[25] Rejecting the idea that nature can be analysed into fundamental entities, Chew has argued that hadrons are temporarily stable configurations which result from the interaction of processes. These may transform themselves into each other, help other hadrons in their transformations, appear as composite particles, constituents of other particles, or binding forces. While the actually unfolding process chains and the resulting process webs are unpredictable, they obey certain rules based on the single principle of self-consistency. Whatever comes into being has to be consistent with itself and with everything else, so that the set of hadrons 'pulls itself up by its own bootstrap'. Recently Chew and his colleagues have been able to obtain results consistent with the achievements of its main rival research programme, Gell-Mann's quark model of hadrons, again through the use of algebraic topology.

The most recent fashion in theoretical physics is superstring theory.[26] Rejecting both the conception of particles and of fields as the fundamental entities of the universe, superstring theorists are trying to unite the general theory of relativity or gravity theory with quantum theory by conceiving of the universe as composed of spatio-temporally oscillating and vibrating 'strings'. The 'elementary particles' can then be thought of as different modes of oscillation or vibration so that electrons, gravitons, photons, neutrinos etc. can be seen as different harmonics (like different musical notes) of a fundamental string.[27] As yet this theory is in the early stage of its development. There is little conceptual understanding of what the mathematics is about while the mathematics has not been sufficiently mastered to provide definitive tests of the theory. However, along with non-local hidden variable theories of quantum mechanics and bootstrap theories, work on superstring theories clearly manifests the growing dissatisfaction with the dominant theories of being and the struggle to develop alternatives in which the world is seen as consisting of patterns of activity.

---

22. See David Bohm, *Wholeness and the Implicate Order*, London: Routledge & Kegan Paul, 1980, Ch.'s 5 & 6.
23. A. Baracca, D.J. Bohm, B.J. Hiley and A.E.G. Stuart, 'On Some New Notions Concerning Locality and Nonlocality in the Quantum Theory' in *Il Nuovo Cimento*, Vol.28 B, No.2, 11 Aug. 1975, pp.453-465, p.458.
24. The best description of what is involved in this formalization is B.J. Hiley, 'Towards an Algebraic Description of Reality' in *Annales de la Foundation Louis de Broglie*, Vol.5, 2, 1980, pp.75-103.
25. See *Zygon*, Vol.20, No.2, June 1985 on this, particularly the article by Chew.
26. On this see P.C.W. Davies and J. Brown eds, *Superstrings: A Theory of Everything?* Cambridge: Cambridge University Press, 1988; and E. David Peat, *Superstrings and the Search for The Theory of Everything*, Chicago: Contemporary Books, 1988.
27. As Edward Witten and John Ellis have put it, *Superstrings: A Theory of Everything?* p.93 & p.153.

## *Thermodynamics*

Thermodynamics originated in Jean-Joseph Fourier's mathematical description of heat flow in solids in 1811.[28] Here a physical theory had been created which was just as mathematically rigourous as the mechanical laws of motion, yet remained completely alien to the Newtonian conception of the world. In 1824 Carnot, who was concerned with the efficient use of fuel in engines, formulated the principle of irreversibility: that fuel once used, disappears as fuel forever. Mayer (1842) and Helmholtz (1847), influenced by the *Naturphilosophen* who, under the influence of Leibniz, had postulated the existence of a universal *vis viva*,[29] proposed that the various sciences of heat, mechanics, chemistry, electricity and biology could be united by the principle of energy conservation. According to this scheme, 'energy' is merely transformed by various physical, chemical and biological systems. Then in 1850 Clausius formulated Carnot's principle from the new perspective provided by the conservation of energy, and the science of thermodynamics came into being. In 1865 in the process of generalizing the principles of irreversibility from technology to cosmology, Clausius coined the term 'entropy' and explicitly formulated the first two laws of thermodynamics. In opposition to classical mechanics which was still thought to govern the behaviour of the elementary constituents of the world, the new science of thermodynamics dealing with large aggregations of atoms or molecules implied an asymmetry in the relationship between the present and the future and the present and the past. The universe was seen to be running down to a 'heat death' in which all energy would be uniformly distributed throughout the universe.

Later in the century, Bolzmann attempted to reconcile thermodynamics with mechanics by explaining the thermodynamic properties of gases in terms of the behaviour of atoms or molecules.[30] Although he was only concerned with systems moving towards equilibrium, and he himself acknowledged that he had not reconciled thermodynamic systems to mechanics, his research project was in accordance with the reductionist tendencies of the mechanistic conception of the world. As a consequence, the phenomena of thermodynamics have been widely held to be epiphenomena produced by the mechanical laws governing the elementary constituents of the universe, and of significance only because of our ignorance of individual constituents. Thus a recent textbook on thermodynamics defined a thermodynamic system as 'a system in which there *are so many relevant degrees of freedom that we cannot possibly keep track of all of them.*'[31]

However the whole research programme of mechanistic reductionism was revealed to be impossible by Bruns and Poincaré. They showed the so-called 'many-body problem' or 'three-body problem' to be insoluble; that is, that it is impossible to analyse a system containing more than two bodies in terms of deterministic equations of motion describing each body (in terms of co-ordinates and momenta) in the system.[32] The interactions between all the bodies were shown to be more than the sum of the interactions between each of them. As Prigogine and Stengers put it: 'Nature as an evolving, interactive multiplicity thus resisted its reduction to a timeless and universal scheme.'[33] Later developments of thermodynamics have brought this home, forcing people to recognize that thermodynamic phenomena are genuinely emergent features of the universe. In 1931 Onsager formulated the first general relations in non-equilibrium thermodynamics. On this foundation

---

28. For an account of the history and state of thermodynamics see Prigogine *From Being to Becoming* and Prigogine and Stengers *Order Out of Chaos*.

29. See Thomas S. Kuhn, 'Energy Conservation as an Example of Simultaneous Discovery' in *The Essential Tension*, Chicago, University of Chicago Press, 1977, pp.66-104.

30. For the history of this see Yehuda Elkana, 'Bolzmann's Scientific Research Program and its Alternatives' in Y. Elkana ed. *The Interaction Between Science and Philosophy*, Atlantic Highlands, N.J.: Humanities Press, 1974, pp.243-279.

31. J.R. Waldram, *The Theory of Thermodynamics*, Cambridge: Cambridge University Press, 1985, p.2.

32. This issue has been lucidly described by Prigogine and Stengers in *Order out of Chaos*, p.70ff. Poincaré's proof is recounted in Ivar Ekeland, *Mathematics and the Unexpected*, Chicago: University of Chicago Press, 1988, Appendix 1.

33. Prigogine and Stengers, *Order Out of Chaos*, p.72f.

Prigogine and his colleagues formulated principles to describe far from equilibrium states, inaugurating a new era in thermodynamics. The central concern of non-equilibrium thermodynamics is the study of the generation of new order in thermodynamically far from equilibrium systems: the dissipative structures which maintain continuous entropy production and dissipate the accruing entropy.[34]

Far from equilibrium thermodynamics most clearly reveals the necessity of conceiving the world as a process of creative becoming. It deals with the emergence of types of ordering through the amplification of fluctuations which, once established, have a dynamics of their own beyond the conditions of their emergence which then constrain the rate and way in which negative entropy is dissipated. Typical examples of this are the turbulence which develops in a laminar flow of liquid (as for instance when a tap is turned on until turbulence develops in the flow of water) and the cellular convection patterns which develop when a liquid is being heated at one end and cooled at the other. Each of these increase the rate of creation of entropy. Dissipative structures have also been revealed in chemical reactions which exchange energy and matter with the environment and are auto- or cross-catalytic. In these there can be a multiplicity of types of order: temporal organization as in a limit cycle, stationary inhomogeneous structures, spatio-temporal organization as in a wave form, and localized structures. In all these cases a large number of molecules manifest a coherent order over a large region and period of time. Unlike equilibrium structures which are uniquely determined by their environmental parameters, dissipative structures are involved in cycles of activities in which, if the systems are large enough, they establish their own boundaries and undergo state transitions autonomously. In terms of the categorial framework outlined in the previous chapter such 'dissipative structures' are processes which have structures; it is 'processes' as self-ordering patterns of activities which 'do' things.

While the work of Prigogine and his colleagues has been concerned with the emergence of order from disorder, this has been complemented by studies of how determinate systems generate indeterminacies, a field which has become widely known through the development of chaos theory.[35] Chaos theory enables systems which were once only describable through statistics to be conceptualized by a form of mathematics which reveals why determinate systems develop unpredictably. When the notions of dissipative structure and chaos are combined, a picture emerges of a world consisting of both indeterminate and determinate processes, with neither being more basic than the other. Any appearance of determinate order must be seen as emerging from an indeterminate order (or disorder) at one level while generating unpredictable outcomes at another level.

Combined with the breakdown of the reductionist project with the developments of relativity theories and quantum mechanics, these developments in thermodynamics have inaugurated a new era in science concerned with the emergence of new levels of order, the relationship between microscopic and macroscopic order, and with complexity.[36] Such notions provide a bridge between the science of the animate and the inanimate world. Life forms can be conceived as complexes of dissipative structures emerging from indeterminate physical and chemical processes and generating in turn indeterminate biological processes. While entropy initially appeared to be an anthropocentric concept, defined only in terms of potentiality for human purposes, the concept of

---

34. See G. Nicolis and I. Prigogine, *Self-Organization in Non-Equilibrium Systems* New York: Wiley, 1977. For a summary of the work on this subject see also Ilya Prigogine, 'Order Through Fluctuation: Self-Organization and Social System' in Erich Jantsch and Conrad H. Waddington eds, *Evolution and Consciousness,* Reading Mass.: Addison Wesley, 1976, pp.93-133.

35 On this see Nina Hall ed. *The New Scientist Guide to Chaos*, Harmondsworth: Penguin, 1991.

36. See for example H. Harken, *Synergetics: An Introduction*, 3rd ed., Berlin: Springer, 1983; E. Frehland ed. *Synergetics - From Macroscopic to Microscopic Order*, Berlin: Springer-Verlag, 1984; Giorgio Careri, *Order and Disorder in Matter*, tr. Kristin Jarratt, Reading: Benjamin/Cummings, 1984, and Davies, *The Cosmic Blueprint*. On the science of complexity see John L. Casti, *Complexification*, London: Abacus, 1994.

dissipative structures, themselves defined in terms of the transformation of negative entropy into entropy, enables negative entropy and entropy to be defined in terms of potentiality for dissipative structures, of which humans can then be seen as a kind. Humans, cognizing, analysing, experimenting on, and engaging with or utilizing negative entropy, must be seen as themselves ordering activity within nature in relation to which potentialities and the processes which generate and maintain them must be defined. This finally invalidates all efforts to reduce thermodynamics to mechanics. As Prigogine and Stengers wrote:

> ... irreversible processes have an immense constructive importance: life would not be possible without them. The subjective interpretation [of thermodynamics] is therefore highly questionable. Are we ourselves merely the result of our ignorance, of the fact that we only observe macroscopic states?[37]

## *Process Philosophy and the Life Sciences*

Mechanistic materialism is even more firmly entrenched in the life sciences than in the physical sciences. This follows a long history of struggle into the twentieth century by mechanists against the surviving concepts of Aristotelian biology as espoused by the vitalists. Opposing notions such as those of Claude Bernard who argued: 'As long as a living being persists, it remains under the influence of ... [a] creative force, and death comes when it can no longer express itself'[38] the German biologist Virchow argued in 1845 that: 'The new medicine ... has shown that life is nothing more than the sum of the phenomena which proceed from general physical and chemical (that is to say mechanical) laws. It denies the existence of an autocratic Life or Healing Force.'[39] The most important advance of the mechanistic approach was the development of Darwin's theory of evolution, for this offered an explanation for the appearance of complex order in the world in purely mechanistic terms.[40] Evolutionary theory was bolstered by Mendelian genetics, population biology and then by the development of molecular biology which described the mechanism of inheritance chemically in terms of the replication of DNA. It has been reformulated through these as the 'synthetic theory' of evolution.

The essence of the synthetic theory of evolution is the notion that more complex organisms have descended from less complex organisms, and that this process is explained by the way populations produce more descendants than will survive, by the variability of these descendants, and by this variability affecting their chances of survival. Following Weissman (1885), inheritance and variation are seen to derive from the germ plasm which is held to be continuous from generation to generation, unaffected by the body or environment of the organism. This germ plasm, which has since been described in terms of genes and DNA, is taken as the sufficient cause of biological form. The adult organism, the phenotype, is represented as a complex of discrete traits produced by the genes and the environment. This led evolutionary theory to focus on populations of genes and fitness of genotypes. Sewell Wright was the most important instigator of this approach.

37. Prigogine and Stengers, *Order Out of Chaos*, p.125.

38. Claude Bernard, *Introduction to the Study of Experimental Medicine*, cited by Stephen Toulmin and June Goodfield, *The Architecture of Matter*, London: Hutchinson, 1962, p.335.

39. Cited by Everett Mendelsohn in 'Revolution and Reduction', Y. Elkana ed., *The Interaction Between Science and Philosophy*, Atlantic Highlands, Humanities Press, 1974, p.415.

40. As argued by G. Montanelli in 'From Aristotle to Democritus via Darwin', Francisco Jose Ayala and Theodosius Dobzhansky eds, *Studies in the Philosophy of Biology: Reduction and Related Problems*, London, Macmillan, 1974, pp.3-19. The major orthodox history of biology and the development of evolutionary theory is Ernst Mayr's *The Growth of Biological Thought: Diversity, Evolution and Inheritance*, Cambridge, Mass.: Harvard University Press, 1982.

Dobzhansky then redirected attention from the fitness of individual genotypes to the fitness of populations of genotypes, but genes remained at the core of the theory. This is evident in, and basic to, the work of the sociobiologists. For instance E.O. Wilson wrote:

> Natural selection is the process whereby certain genes gain representation in the following generations superior to that of other genes located at the same chromosome positions. When new sex cells are manufactured in each generation, the winning genes are pulled apart and reassembled to manufacture new organisms that, on the average, contain a higher proportion of the same genes. But the individual organism is only their vehicle, part of an elaborate device to preserve and spread them with the least possible biochemical perturbation. Samuel Butler's famous aphorism, that the chicken is only the egg's way of making another egg, has been modernized: the organism is only DNA's way of making more DNA.[41]

Developments in bio-chemistry and molecular biology have further supported the reductionist programme in biology. The general view of modern biologists was summed up by the Nobel laureate J. Lederberg in 1970:

> A few eccentrics aside, the whole community of contemporary science shares the view that the laws of nature apply to nonliving and living matter alike. All of us who investigate the chemistry and physics of living organisms pursue our work as if organisms were complex machines, and we find man to exhibit no tissues or functions that would except him from this way of analysing human nature.[42]

However there are alternatives to the mechanistic view of life which are not vitalist. The most important of these has been inspired by process philosophy. In 1931 a group was formed in Cambridge centred around Waddington, Needham, Wrinch, Bernal and Woodger.[43] Waddington and Needham in particular had been strongly influenced by both Whitehead and D'Arcy Thompson, and all had been somewhat influenced by a lecture given by Bukharin in England in 1931 defending Engels' anti-reductionist philosophy of science. Inspired also by the recent advances in physics (especially in quantum theory) these scientists formulated a research programme which was physicalist but anti-reductionist, an approach they called 'mathematico-physico-chemical morphology.' While the Rockefeller Foundation was willing to finance this programme, they were unable to attain the support of Cambridge University and the group disintegrated. However Waddington continued his research in biology, specializing in the field of epigenesis - the study of the genesis of form and the differentiation of cells. Forced out of Cambridge he established himself at Edinburgh, and before he died he organized four major symposiums at Belagio, the contributions to which he edited and published in four volumes between 1968 and 1972 as *Towards a Theoretical Biology*. Most of the participants at these symposiums together with a number of other biologists have continued to develop ideas implicitly or explicitly in accordance with the process view of the world, and these provide a framework for interpreting other unorthodox developments in biology. This framework is far more promising as a basis for understanding the phenomena of life than the orthodox molecular biology and population biology based on reductionist materialism, and it also provides the basis for a critique of this

41. Edward O. Wilson, *Sociobiology: The New Synthesis*, Cambridge Mass.: Harvard University Press, 1975, p.3.

42. J. Lederberg 'The Perfection of Man' in *Nobel Symposium 14: The Place of Value in a World of Facts*, Stockholm: Almquist and Wiksell, 1970, p.55.

43. For a study of this group see Pnina G. Abir-Am, *The Biotheoretical Gathering in England, 1932-38 and the Origins of Molecular Biology*, Ph.D. thesis, Universite de Montreal, 1983 and 'Recasting the Disciplinary Order in Science' in *Humanity and Society*, Vol.9, No.4, November 1985, pp.388-427.

orthodox reductionist position. My contention is that this framework can be clarified by the categories of process metaphysics outlined above.

The first reason why the process conception of life should be considered as the foundation for biology is the increasing evidence of failure in the mechanistic reductionist research programme. In particular, and most fundamental for biology, evolutionary theory in its orthodox formulation is showing increasing signs of breakdown.[44] The most obvious problem raised by it is that if the organism as a functioning whole is reduced to a mere epiphenomenon of the genes or DNA, then what is it that is being explained? Evolutionary theory becomes a mere tautology in which the fitness of a gene or sequence of DNA, defined by the fact that it survives, is used to explain its survival. Any attempt to overcome this tautology must return to the issue of what evolutionary theory is about - the existence of complex forms of life. If such complex forms are mere collections of chemicals no more significant than any other, then to talk of evolution is meaningless. If they are more than this, what is the relationship between the genotype and the phenotypic forms? As it stands, orthodox evolutionary theory explains biological form solely by tracing genealogies. This is like explaining why the earth is following an elliptical trajectory around the sun by the fact that it did so last year.

If some minimal status is granted to the phenotypes, and fitness is defined in terms of the propensity for survival of the phenotypes or their traits, then the orthodox theory does become a testable hypothesis. But such a theory would lead one to expect evolution to occur gradually. This has not been born out by the evidence. S.J. Gould in particular has argued that the palaeontological evidence points to a punctuated equilibrium in which periods of rapid evolution are followed by long periods of stability.[45] There is no way for the orthodox theory to account for this.

Giving meaning to evolutionary theory and trying to account for such observations has produced a far more complex picture of life. And this picture undermines the mechanistic reductionism of the theory. It involves recognition of the holistic dynamics of living processes, from the DNA to species and eco-systems. These dynamics are dependent upon their environments and constituents, but not reducible to them. This does not involve an extra force, a 'life force' for instance, over and above physical processes. In accordance with the conception of process as self-ordering activity, this involves immanent causation which is simply a matter of additional constraints. As Howard Pattee wrote: 'the constraint is simply some additional regularity or order which is not explicitly found in the initial conditions.'[46]

There are a number of issues involved in understanding the place of the organism or phenotype in biological evolution. The first is the problem of epigenesis - the process of differentiation and the genesis of form or structure. The complex nature of this undermines the synthetic theory and suggests a far more complex relationship between the genotype and the phenotype in evolution. The second pertains to the nature of the on-going organization of organisms. This appears to involve hierarchical ordering based on entrainment of oscillations, and involves features irreducible to molecular biology. The third is the nature of teleology and subjectivity. Ideas in theoretical biology, philosophical biology and ethology, when interpreted in terms of process philosophy, will

44. See B.C. Goodwin, N.J. Holder, C.C. Wylie eds, *Development and Evolution*, Cambridge: Cambridge University Press, 1983; M.W. Ho and P.T. Saunders eds *Beyond Neo-Darwinism*, London: Academic Press, 1984, and J.W. Pollard ed., *Evolutionary Theory*, Chichester: John Wiley & Sons, 1984. Elliot Sober has examined many of the conceptual issues in evolutionary theory in *The Nature of Selection*, Cambridge Mass.: MIT Press, 1984.

45. S.J. Gould and N. Elridge, 'Punctuated equilibria: the tempo and mode of evolution reconsidered', *Paleobiology*, 1977, Vol.3, pp.115-151 and S.J. Gould 'The meaning of punctuated equilibrium and its role in validating a hierarchical approach to macro-evolution', in R. Milkman ed. *Perspectives on Evolution*, Massachusetts: Sinauer Associates, 1982, pp.83-105.

46. H.H. Pattee 'Physical Theories of Biological Co-ordination' in Marjorie Grene and Everett Mendelsohn eds, *Topics in the Philosophy of Biology* Dordrecht: Reidel, 1976, p.161.

be seen to support each other, and suggest the impossibility of accounting for evolution without taking the conscious striving of individual organisms into account.

To begin with, experimental work has revealed the fallacy of Weismann's hypothesis that the genotype in any particular organism is inviolable and can only be changed over generations through selection. Barbara McClintock demonstrated the existence of moveable genetic elements in maize by their genetic effects. These could not be accounted for in terms of previous models of mutation. McClintock's thesis has been amplified by new experimental techniques, and this work reveals the genome to be itself a highly complex self-organizing system in interaction with the dynamics of the organism as a whole.[47] This may account for both the inheritance of some acquired characteristics and for rapid changes in DNA in particular circumstances, and makes it impossible to treat the phenotype as merely an expression of the genotype.

Since the DNA complement of each cell in a multicelled organism is the same, this raises the question of how differentiation of cells occurs, and in particular how this differentiation gives rise to coherent structures such as limbs, eyes, nerves, and so on. This obviously cannot be accounted for simply in terms of DNA. In actual fact DNA cannot divide or do anything except as part of the highly organized processes of life, even at the cellular level, and the linear scheme of DNA producing RNA which in turn produces protein is nothing but fiction. Even in protozoa the role DNA plays is more like that of a set of instructions read according to the requirements of the functioning organism than encoded information mechanically producing proteins to constitute the organism. The activation and de-activation of the different parts of DNA is dependent upon a complex of interacting feedback systems based on the production of enzymes which catalyse or inhibit the synthesis of different proteins. In metazoa differentiation of cells leads to the establishment of emergent dynamics through which cells and the part played by DNA are constrained in their development by their position within the total organism.

While such epigenesis is not yet fully understood, a number of facets have been revealed. Most importantly it has been shown how the development of the organism is canalized along different paths. These have been described by Waddington as 'chreods' (time-paths) and the self-stabilization along these paths as 'homeorhesis', corresponding to the notion of homeostasis as self-stabilization at a point. For instance the development of a piece of tissue is canalized to form a limb, and then canalized to form a fore-limb or hind-limb. If before the canalization to hind-limb tissue, tissue from the hind limb is grafted onto the fore-limb region, the disturbance will be buffered out and the tissue will develop into part of a normal fore-limb. If this transplantation is made after canalization to hind-limb tissue, it will develop as hind limb tissue, but in accordance with its position in the fore-limb. For instance if tissue from the thigh of a bird is transplanted to its wingtip, it will develop into toes and claws.

The question then arises of how morphogenetic fields operate, how do individual cells gain the positional information which enables them to develop in the appropriate manner, and how are individual cells able to respond to this positional information. There is no reason to think that there is only one means for achieving this, but there is evidence that a major role is played by fluctuations or oscillations. This could explain the differentiation into fore-limb and hind-limb tissue which cannot be entirely explained in terms of gene activation, since the behaviour of transplanted tissue rules out the existence of genes for the fore-limb or for the hind-limb. As C.H. Waddington wrote:

> We could not have a 'neural plate substance, a fore-limb substance, a hind-limb substance' etc. but neural plate, fore-limb or hind-limb oscillatory patterns, which could be regarded as analogous to musical themes or chord sequences. The later phases of differentiation into the

47. See Howard M. Termin and William Engels, 'Moveable genetic elements and evolution'; Christopher A. Cullis, 'Environmentally induced DNA changes'; and Edward J. Steele, Reginald M. Gorczynski and Jeffrey W. Pollard, 'The

various cartilages, bones, muscles, etc., must certainly involve the 'activation' of different structural genes controlling the proteins in these different sorts of cells; but we could interpret these changes as similar to the development of the initial themes according to the conventions of some school of classical music composition.[48]

When differentiation is conceived in this way then it is possible to account for the field effects which enable cells to determine their position in the organism. The neighbouring cells act as temporal templates which entrain the oscillations of the cells according to their position in the organism.[49]

Clarificatory evidence of the oscillatory ordering of epigenesis has been supplied by the study of the slime mould which transforms itself from a community of protozoa into a single, multi-celled organism. The isolated cells (between 10 and 100,000) which develop from spores exude the chemical acrasin at increasing rates as the food supply is depleted, while at the same time becoming more sensitive to this chemical. The increased production destabilises the homogeneous solution producing a far from thermodynamically equilibrium state which generates dissipative structures in the form of oscillations. A certain critical wavelength exists which determines the spatial distribution of the cells. The cells oriented by this wavelength then aggregate, eventually forming a structure in which some cells become rich in cellulose and develop into a foot or base while others rise above it and become rich in polysaccharides. The mass on top eventually develops as a fruit, producing a large number of spores. Predictions of the behaviour of individual cells based on the theory of dissipative structures have been verified by Keller and Segal.[50]

In more complex organisms there is a multiplicity of such patterning activities occurring simultaneously, with the different morphogenetic fields constraining each other. The constraints generated by the dynamic inter-relations between these fields have been investigated by Brian Goodwin, among others, and have been shown to account for many of the characteristic features of the structures of adult organisms.[51] On this basis, Goodwin argued: 'Organisms are not aggregates of elements, whether molecules, cells, organs, skeletal or other components, whose random variation results in an unconstrained variety of forms. They are self governed wholes governed by laws describing spatial and temporal organization such that processes of biological change involve constrained transformation, whether ontogenetic or phylogenetic.'[52] Such ordering precludes any simple relation between the genes and the phenotype. In such self-organizing activity the genes are, as Waddington argued, merely 'the pebbles in the concrete' and as such are 'almost irrelevant to the engineering of the bridge.'[53]

---

somatic selection of acquired characters' in Pollard, ed. *Evolutionary Theory*; and M.W. Ho and P.T. Saunders, 'Beyond Neo-Darwinism - An epigenetic approach to evolution,' *Journal of Theoretical Biology,* 1979, Vol.78, pp.573-591.

48. C.H. Waddington, 'Cellular Oscillations and Development' in C.H. Waddington ed. *Towards a Theoretical Biology 2 Sketches,* Edinburgh: Edinburgh University Press, 1969, pp.179-83, p.180f.

49. Since Waddington, such processes have come to be much better understood. On this, see Stuart A. Kauffman, *The Origins of Order: Self-Organizatin and Selection in Evolution*, N.Y. and Oxford: O.U.P., 1993.

50. See Ilya Prigogine, 'Order Through Fluctuation' in Jantsch and Waddington, eds, *Evolution and Consciousness*, p.107ff.

51. For an account of this work see B.C. Goodwin, *Analytical Physiology of Cells and Developing Organisms*, London, Academic Press, 1976; and Brian Goodwin, *How the Leopard Changed its Spots*, London: Weidenfeld and Nicolson, 1994. See also papers in *Development and Evolution* ed. B.C. Goodwin et.al., Cambridge: Cambridge University Press, 1983.

52. Brian C. Goodwin, 'Changing from an evolutionary to a generative paradigm in biology', Pollard ed. *Evolutionary Theory: Paths into the Future*, pp.99-120, p.113.

53. C.H. Waddington, 'The Theory of Evolution Today' in Arthur Koestler and J.R. Smythies eds, *Beyond Reductionism*, London: Hutchinson, 1969, pp.357-95, p.371f.

Acknowledging the existence of chreods and a more complex relation between the genotype and the phenotype gives another dimension to evolution, the possibility of genetic assimilation. Whether an organism develops along one chreod or another is dependent upon both the genes and the environment. A change in the environment can lead to an adaptation by some organisms so that development occurs along a different chreod. If this adaptation is beneficial, those organisms which are capable of switching chreods in response to environmental stress will be selected for, and there will be a concentration of genes in the population facilitating this switch. This can result in the development within individual offspring of the new chreod without the environmental stress. The stress produced phenotypic alteration becomes assimilated by the genotype and the acquired characteristic becomes hereditary. In this way the population of organisms is able to imitate Lamarckian evolution. Waddington has demonstrated such an effect with fruit-flies, many of which will develop shorter wings in higher than normal temperatures.[54]

Both while organisms are developing and after they have reached maturity they are engaged in a perpetual process of self-maintenance and self-realization directed by internally defined criteria of stability and organization. They are involved in self-creation or, as Maturana and Varela described it, 'autopoiesis'.[55] In the light of what has been revealed about the nature of epigenesis, this cannot be explained away mechanistically simply in terms of what order of material happens to have survived in the struggle of all against all. Epigenesis raises the question of the nature of this self-creation.

Self-creation in the organism has two fundamental dimensions. While it involves a struggle by the organism to maintain itself as a distinct unit, it must differentiate itself in order to meet requirements which cannot be met in the same place or simultaneously. For instance in a single cell, chromosome replication must involve temporal differentiation, and since ribosomes cannot occupy the same place as DNA, a nuclear zone is required involving a spatial differentiation. This means that the stability of self-creation cannot be the classical type in which a system is stable in relation to a point, but must be a dynamic stability in which there is a spatio-temporal differentiation. The central feature of this form of organization is that it involves hierarchical levels of constraints of a particular kind. For instance in the cycle of events by which organisms reproduce themselves there must be a supervening order to coordinate the temporal differentiation by providing phase information for the relative timing of such events as DNA replication and cell division.

In a crystal there is a structural hierarchy characterized by a permanent loss of degrees of freedom. This involves constraints too rigid to be important in biological coordination. On the other hand liquids and gases involve too few constraints. What is missing in both these cases is a recognizable 'function'. A function is, as Howard Pattee pointed out, 'a process in time, and for living systems the appearance of time-dependent functions is the essential characteristic of hierarchical organization.'[56] With this function the constraints must be variable and imposed on only select degrees of freedom of the constituent processes or entities. These are called 'non-holonomic' constraints because they can only be described by equations which relate coordinates to the trajectories, but cannot be derived from the ordinary equations of motion and the initial conditions of the system.

Such hierarchical ordering can be achieved on the basis of oscillations generated by states of far from thermodynamic equilibrium. Such oscillations allow for both hierarchical ordering and ordering through entrainment, as with the epigenetic ordering of morphogenetic fields described above. The central feature of hierarchical ordering is that 'levels of control must be distinguished

---

54. Ibid. p.374.

55. Humberto R. Maturana and Francisco J. Varela, *Autopoiesis and Cognition: The Realization of the Living*, Dordrecht: Reidel, 1980.

56. H.H. Pattee, 'The Problem of Biological Hierarchy' in Waddington ed. *Towards a Theoretical Biology: 3 Drafts*, Edinburgh: Edinburgh University Press, 1970, p.127.

by different time constants' (that is, the relaxation times or times required for the variables to reach a steady state after a 'small' disturbance).[57] If two systems have very different relaxation times, the variables of the faster system can be regarded as always being in a steady state relative to the time required for significant changes to occur in the slower system, while the variables of the slow system will enter into the equations of motion of the fast system as parameters of the environment rather than as variables. In this way the genetic system can be seen as constraining the epigenetic system, and the epigenetic system the metabolic system.[58] A.S. Iberall has shown entrained oscillations to be ubiquitous in organisms.[59] They include the bio-electric nervous cycle, the endocrine systems, the heat balance system, water cycles and so on. The time scales of these were shown by Iberall to vary greatly but to be such as to be able to be entrained in chains so that each oscillation comes to form a coherent part of a whole system. Research in this area has made rapid advances in recent years associated with the advances in non-linear thermodynamics, virtually transforming biology.[60] Such research suggests that it is oscillations which account for the distinctive characteristics of life, and life has been redefined accordingly by Iberall:

> Thus life is tentatively defined as any compact system containing a complex of sustaining non-linear limit cycle oscillators, and a similar system of algorithmic guiding mechanisms, that is capable of regulating its interior conditions for a considerable range of ambient environmental conditions so as to permit its own satisfactory preservative operation; that is capable of seeking out in the environment and transferring and receiving those fluxes of mass and energy that can be internally adapted to its own satisfactory preservative operation; that is capable of performing those preservative functions for a long period of time commensurate with the 'life' of its mechanical-physical-chemical elements...[61]

## *The Emergence of Awareness*

This conception of life provides the basis for reconciling the science of biology with philosophical biology and ethology, achieving an intelligible notion of what it means to be a purposefully acting subject.

The central notion underlying philosophical biology is that what distinguishes living from non-living beings is that they define their environments in terms of themselves, thus constituting these environments as fields of potentialities or worlds and themselves as subjects.[62] Thus the philosophical biologist Helmuth Plessner has defined life in terms of positionality. Whereas non-living things have a position, an organism takes its place in the environment, arises in it, is dependent upon it, and yet is opposed to it.[63] In a similar vein, Maurice Merleau-Ponty argued that:

---

57. Ted Bastin, 'A General Property of Hierarchies' in Waddington ed., *Towards a Theoretical Biology: 2 Sketches*, pp.252-265, p.256.

58. See B.C. Goodwin, *Temporal Organization of Cells: A Dynamical Theory of Cellular Control Processes*, London: Academic Press, 1963, p.20f; and *Analytical Physiology of Cells and Developing Organisms*, Chs 1,2 & 3 and p.190f.

59. A.S. Iberall, 'New Thoughts on Bio-Control' in Waddington ed. *Towards a Theoretical Biology: 2 Sketches*, pp.166-177.

60. For recent developments of this area see Leon Glass and Michael C. Mackey, *From Clocks to Chaos: The Rhythms of Life*, Princeton: Princeton University Press, 1988; Kauffman, *The Origins of Order*; Goodwin, *How the Leopard Changed its Spots*.

61. Iberall, 'New Thoughts on Bio-Control', p.168.

62. This is how George Herbert Mead defined teleological objects in *The Philosophy of the Act*, ed. Charles W. Morris, Chicago: University of Chicago Press, 1938, p.301ff.

63. Marjorie Grene, *Approaches to a Philosophical Biology*, N.Y.: Basic Books, 1969, p.74f.

> We speak of vital structures ... when equilibrium is obtained, not with respect to real and present conditions, but with respect to conditions which are only virtual and which the system itself brings into existence; when the structure, instead of procuring a release from the forces with which it is penetrated through the pressure of external ones, executes a work beyond its proper limits and constitutes a proper milieu for itself.[64]

Developing such ideas, Hans Jonas argued that life is characterized by three basic features. First, it is a metabolism with a double aspect, 'denoting on the side of freedom, a capacity ... to change its matter, ... [while] equally the irremissible necessity for it to do so.' Second, it must attain this matter from outside itself. It must thereby be 'turned outward and toward the world in a peculiar relatedness of dependence and possibility' thereby referring 'beyond its given material composition to foreign matter as needed and potentially its own.' Third, 'there is an inwardness or subjectivity involved in [this] transcendence, imbuing all the encounters occasioned in its horizon with the quality of felt selfhood, however faint its voice.'[65]

These descriptions of life would be little more than suggestive and their relationship to theoretical biology would remain vague so long as the world were understood in terms of the categories of mechanistic materialism. However, these descriptions become intelligible when the categories of process metaphysics are assumed, and can thereby be integrated with ideas developed in theoretical biology, ethology and neurophysiology. There have been two major obstacles standing in the way of making purpose and subjectivity intelligible - conceiving of causation in such a way that self-creation is incomprehensible, and conceiving of time in spatial terms, eliminating real becoming from the world. Both these obstacles are overcome by the categories of process metaphysics.

To begin with, primary beings are understood as processes, defined as self-ordering activities essentially durational in their nature. This means that whatever is identified as a primary being must be seen as an immanent cause of its own becoming. In the case of a living being, the constraints associated with its immanent causation are non-holonomic and apply to both constituent processes and to interchanges of the organism with its environment. And first through the evolution of species, then through the development of cognition in individual organisms, there is an ordered development of such constraints. Further, space-time must always be seen as becoming, with the future never being entirely determined by the past and potentiality thereby being a real part of the world, with spatio-temporality defined in terms of the potentialities maintained by superordinate processes for co-existence and interaction between actual or potential sub-processes. On this basis, the possibility of a multiplicity of spatio-temporal orders must be allowed for, with many being the condition for the existence of others. It has been noted that in biological organization, supervening causes are of long duration compared to constituents. As such they are of limited divisibility. To subdivide them durationally is to destroy them, as would the subdivision of a melody destroy it as a melody. The existence of supervening causes constraining the interaction between the organism and the environment generates a spatio-temporal order of potentialities for constituent sub-processes associated with this exchange. But it is in terms of the durational supervening causation of the higher level ordering of the organism that this space-time is defined. This implies that the ordering activity of this supervening causation transcends this space-time. It is this which allows it to be conceived of as a final cause, not through being seen as an event in the future affecting the present, but by forcing a reconception of the notions of event, the future and the present. It involves a causation which is indivisible in terms of the spatio-temporal order of the potentialities of the sub-processes associated with exchange between the organism and the environment which is defined in terms of it. In this scheme of things, the organism as an unfinished process of becoming consisting

64. Maurice Merleau-Ponty, *The Structure of Behaviour*, [1942], tr. Alden L. Fisher, Boston: Beacon, 1967, p.145f.

65. Hans Jonas, *The Phenomenon of Life: Towards a Philosophical Biology*, Chicago: University of Chicago Press, 1982, p.83f.

of such supervening causation must then be seen (or rather, understood, since appreciating reality as becoming requires 'indwelling') as constituting its environment as the field of its potentialities, that is, as a spatialized world, by temporally transcending the immediacy of this environment. This implies the opening of a temporal horizon in which the simple flowing passage of change is transformed by defining the present as that in which past tensions or desires have been satisfied or frustrated, and in which there are existing tensions or desires which may be satisfied in the future. This subjective space-time is not to be counterposed to real space-time. It is a real emergent order. The organism in its environment thereby becomes an embodied subject in a world, a world which is constituted in progressively more complex ways as it strives to come to terms with its environment. That is, organisms conceived in terms of theoretical biology based on process metaphysics can be understood to be essentially as they have been described by philosophical biologists.

While being less concerned with the nature of subjectivity than philosophical biologists, ethologists' conceptions of life have generally accorded with their ideas and the process view of the world while allowing for more detailed analyses of the diversity of life-forms. The initial direction of ethology was given to it by Jacob von Uexküll who analysed the constitution by animals of their worlds, focussing on how the perception world and the action world of organisms are related through function circles (for food, for enemies, and so on) to constitute first their surrounding worlds, and then through the coordination and relating of perception and action in different function circles, to inner worlds. By studying the function circles of each organism he revealed the distinctive worlds of different organisms, showing how 'there are as many surrounding worlds as animals.'[66] While few ethologists share von Uexküll's vitalism and anti-evolutionism, the effect of his influence has been that in their study of the nature of action and perception, ethologists have examined and come to understand the vast variety of life-worlds of organisms, and the diverse means by which these are constituted. They have defined awareness and thought in terms of such constitution, and in this way they have revealed the various stages which have led to the complex, social, open textured worlds constituted by humans.[67]

One of the most fruitful theoretical analyses of the stages of development of forms of action and cognition is that of Piaget. Piaget's developmental epistemology of humans was an extension of his original studies in biology, and he returned in later years to apply the ideas of developmental epistemology to the study of the development of cognition in organisms.[68] The basis of his theory has been the conception of the cognitive function as an extension of organic regulations, constituting a differentiated organ which regulates exchanges with the environment. The principle of this organization is the generalization of schema or structures of interpretation and action from one situation to another, assimilating the environment to the organism's schema, and at the same time accommodating these schema to the environment.[69] New developments can also be made through association and integration of such schema, a process which Piaget illustrated:

> ... the edible snail Helix Pomatia L. lays its eggs in the ground a few centimetres below the surface. Not having much intelligence, it is doubtless incapable of foreseeing the advantages of behaving in this way; so we cannot point to any anticipation in what it does. However, (a) it takes shelter from the sun and cold beneath stones, etc.; (b) it is capable of generalizing this

66. Jacob von Uexküll, *Theoretical Biology*, London: Kegan Paul, Trench, Trubner & Co., 1926, p.176.

67. See Konrad Lorenz, *Behind the Mirror: A Search for a Natural History of Human Knowledge*, tr. Ronald Taylor, London: Methuen, 1977; W.H. Thorpe *Animal Nature and Human Nature*, London Methuen, 1974; and Donald R. Griffin, *The Question of Animal Awareness: Evolutionary Continuity of Mental Experience*, N.Y.: The Rockefeller University Press, 1976; *Animal Thinking*, Cambridge Mass.: Harvard University Press, 1984; and *Animal Minds*, Chicago: University of Chicago Press, 1992.

68. Jean Piaget, *Biology and Knowledge: An Essay on the Relations Between Organic Regulations and Cognitive Processes*, tr. Beatrix Walsh, Chicago: University of Chicago Press, 1971.

protection schema in times of intense cold to the point where it will even bury itself in winter; (c) it has a tendency, no doubt hereditary, to hibernation, and shuts itself up in its shell, blocking the entrance with some epiphregmatic secretion (accumulated mucous); (d) moreover, it lays eggs, and one can well imagine that it will never confuse them with any excretion, so that, however rudimentary its perceptions may be (proprioceptive as well as exteroceptive), it takes these eggs into its sphere of conservation as soon as it lays them. Thus the tendency to lay eggs below the ground could be seen as the result of coordination or assimilation of the laying schema into the schema for self-protection or sheltering in the ground.[70]

Piaget's work raises the question of the nature and ontological status of such schema. This is a difficult concept, and like the concept of 'field' in nineteenth century physics, is still in the process of being elaborated. Schema are generally defined as cognitive structures, and thereby as self-regulating systems of transformations which are neither reducible to their constituents, nor characterizable in terms of executive agency controlling constituents. However when defining the ontological status of such schema, there is a clear failure to distinguish between what is potential and what is actual, and then to treat potentialities as actualities. Thus schema are treated as entities which assimilate, or accommodate to, other entities (environmental data). This reification leads to such problems as accounting for how any organism can attain any awareness of what is not assimilated to schema, and thereby how it is possible for schema to develop.

Such problems can be avoided if schema are conceived to be 'structures' as previously defined; that is, as ordered potentialities - the potentialities to order the interaction of the organism with its environment, where such ordering activity involves the capacity to apply to new situations a transformation of the relationships between what is elementized in previous cognitive activity. There are then two actual processes involved in cognition, each to some extent immanent causes of their own activities, one in which cognitive potentialities are realized in particular situations, and another of relatively much longer duration whereby cognitive potentialities are created, maintained, developed and integrated into hierarchies. These must be understood in terms of radically different temporalities. But then what is to be made of the notion that cognitive schemas are generalized, through assimilation and accommodation, from situation to situation? This terminology can be retained so long as the notions of assimilation and accommodation are reinterpreted as the activity of ordering the interaction of the organism with its environment, and the activity of developing the nature of this ordering and the potential for future ordering. This then avoids the dualism on which possible objections to Piaget's position could be based. It now becomes a matter of considering the nature of the different types of ordering involved.

How then is such cognitive development related to physiological development? The central nervous system is an essential means by which multi-celled organisms regulate interaction and exchanges with the environment and is the precondition for the emergence of consciousness, but consciousness is more than the central nervous system. It is the emergent ordering which actually constrains the functioning of the nervous system. One version of this view has been argued for by Roger Sperry:

> ... conscious awareness, in the present view, is interpreted to be a dynamic emergent property of cerebral excitation. As such, conscious experience becomes inseparably tied to the material brain process with all its structural and physiological constraints. At the same time the conscious properties of brain excitation are conceived to be something distinct and special in their own right. They are 'different from and more than' the collected sum of the neurophysico-chemical events out of which they are built... Although the mental properties in brain activity, as here conceived, do not directly intervene in neuronal physiology, they do

69. The most thorough analysis of the concepts used by Piaget remains Hans G. Furth, *Piaget and Knowledge: Theoretical Foundations*, 2nd ed. Chicago: Chicago University Press, 1981.

> supervene. This comes about as a result of higher level cerebral interactions that involve integration between large processes and whole patterns of activity. In the dynamics of these higher level interactions, the more molar conscious properties are seen to supersede the more elemental physio-chemical forces, just as the properties of the molecular supersede nuclear forces in chemical interaction.[71]

However while Sperry tries to represent consciousness as an emergent feature of the functioning of the brain, the position defended here is that consciousness is only intelligible as an emergent feature of the organism with a central nervous system in interaction with its environment, involving a complexity of temporalities. While only offering some brief comments on the emergence of the mind, C.H. Waddington offered a more satisfactory theoretical starting point when he suggested that:

> ...if you think of the brain as a system of sets of circuits through which currents are passing, this concept involves both the past and the future, since the loops can control the incoming signals which go into the brain and thus influence the effect they will have on future actions. We therefore seem to have, even in the simplest act of perception, both the past and the future incorporated into an active participation with the outside world.

Whether or not he was influenced by Waddington, this very basically is the position developed in depth with great subtlety by Gerald Edelman.[72]

Once awareness and purposeful behaviour have been reintroduced as an intelligible emergent feature of the world, another dimension can be added to evolution. This is the dimension recognized by Baldwin and Lloyd Morgan at the turn of the century and rediscovered more recently by Hardy and Waddington. These biologists recognized that in evolution form follows function, and function is established through the initiative of the organism. Hardy illustrated this with the example of blue-tits which learnt to open the tops of milk bottles with their beaks, a skill which spread rapidly throughout Europe. Hardy pointed out that if the bottles were to be provided with successively thicker tops, those tits with more effective beaks for opening the bottles would be more likely to survive. In this way there would be an evolution within the tit population towards specialized tin opening beaks. Hardy argued on this basis that it is not random mutation and selective pressure which are the main causative factors in evolution, but:

> ... the restless, exploring and perceiving animal that discovers new ways of living, new sources of food, just as the tits have discovered the value of the milk bottles... It is adaptations which are due to the animal's behaviour, to its restless exploration of its surroundings, to its initiative, that distinguishes the main diverging lines of evolution; it is these dynamic qualities which led to the different roles of life that open up to a newly emerging group of animals in that phase of their expansion technically known as adaptive radiation - giving the lines of runners, climbers, burrowers, swimmers, and conquerors of the air.[73]

However organisms do not struggle for survival in isolation, but as members of communities, as members of species and as members of ecosystems. Each of these has irreducible dynamics which must be taken into account by evolutionary theory. In relation to communities it is necessary to consider the forms of communication and cooperation which have developed. In relation to

---

70. Piaget, *Biology and Knowledge*, p.240.
71. R.W. Sperry 'A Modified Concept of Consciousness' in *Psychological Review*, Vol.76, 1969, p.533f.
72. Edelman's arguments are summarized in *Bright Air, Brilliant Fire: On the Matter of the Mind*, London: Allen Lane, 1992. His most important ideas are more fully developed in *The Remembered Present*, N.Y.: Basic Books, 1989.
73. A.C. Hardy, *The Living Stream*, London, 1965, p.192f.

species it is necessary to consider the various forms of reproduction which have emerged to produce phenotypes able to survive within various environments or to transform these environments to facilitate their survival. And then it is necessary to consider the various forms of dependence and interdependence within ecosystems and between ecosystems which generate the conditions which enable individual organisms, communities and species to survive. Each of these dynamics is irreducible to any other, yet each is the conditional cause of the others. It is the dynamics of eco-systems which have been studied in greatest detail.

## *Ecology*

Ecological theory is in a fairly chaotic state.[74] It consists of 'several disparate bodies of theoretical ecology stemming from roots in pre-ecology and early formal ecology which are not well integrated with each other.'[75] Nevertheless, there has been a tendency for ecological theory to oscillate between a holistic organicism and an individualistic reductionism. In the first decades of the century ecology was predominantly organismic, with interactions among members of an association of organisms being compared by C.C. Adams in 1913 to 'relations existing between the different cells, organs or activities of a single individual.' He went on: 'The physiological needs and states of an association have as real existence in individual animals as similar needs in the cell or cells which compose the animal body.'[76] Such views were further developed by Clements and his followers under the influence of the biogeography of Humboldt and Grisebach and Herbert Spencer's scheme of evolution through successive stages of differentiation and integration. After the Second World War this organismic approach was transmogrified into systems ecology and supplemented with notions taken from thermodynamics and information theory. However the use of organic analogies in ecology had been attacked in the 1930's by Gleason, who summed up his position in 1975: 'Far from being an organism, an association is merely the fortuitous juxtaposition of plants. What plants? Those that can live together under the physical environment and under their interlocking spheres of influence and which are already located within migrating distance.'[77] While such arguments were ignored at the time, after the Second World War Gleason's approach, elaborated by population biologists, came to predominate. This triumph has been represented by Simberloff as the triumph of materialism and probabilism over essentialist idealism.[78]

However both these branches of ecology have been brilliantly attacked by Richard Levins and Richard Lewontin in 'Dialectics and Reductionism in Ecology.'[79] Here they argued for a position which 'views the whole as a contingent structure in reciprocal interaction with its own parts and with the greater whole of which it is a part. Whole and part do not completely determine each

74. For the history of ecology see Donald Worster *Nature's Economy: A History of Ecological Ideas* [1977], Cambridge: Cambridge University Press, 1985; Sharon E. Kingsland, *Modelling Nature: Episodes in the History of Population Ecology*, Chicago: University of Chicago Press, 1985; or Robert P., McIntosh, *The Background of Ecology: Concept and Thoery*, Cambridge: C.U.P. 1985. For an excellent short history see Robert P. McIntosh, 'The Background and Some Current Problems of Theoretical Ecology', *Synthese*, Vol.43, 1980, pp.195-255.

75. MacIntosh, 'The Background and Some Current Problems of Theoretical Ecology', p.212.

76. C.C. Adams, *Guide to the Study of Animal Ecology*, N.Y., Macmillan, 1913; cited by MacIntosh, p.209.

77. H.F.. Gleason, 'Delving into the History of American Ecology', *Bulletin of the Ecological Society of America*, Vol.56, 1975, p.i.

78. D.S. Simberloff, 'A Succession of Paradigms in Ecology: Essentialism to Materialism and Probabilism', *Synthese*, Vol.42, 1980, pp.3-39.

79. This paper originally appeared in *Synthese*, Vol. 43, 1980, in reply to the reductionist attack on organic holism by Simberloff in 'A Succession of Paradigms in Ecology.' It has been reprinted in a revised form in Levins and Lewontin, *The Dialectical Biologist*, Cambridge: Harvard University Press, 1985, pp.132-160. Levins has recently published a book with the Harvard University Press developing further this position and its mathematical formalism.

other.' The ecological community is 'an intermediate entity, the locus of species interactions, between the local species population and the biogeographic region.'[80] Levins and Lewontin assign five general properties to ecological communities. First, the community is a whole in interaction with the lower- and higher-level wholes, while not being completely determined by them. Second, some of the properties at the community level are definable for that level and are interesting objects of study in their own right. Third, the properties of communities and the properties of constituent populations are linked by many-to-one and one-to-many transformations. This means that there are many possible ways in which the integrity of the whole can be maintained, and many ways in which the parts can adapt to the conditions created by the dynamics of the whole. Fourth, law and constraint are interchangeable. While in physics the boundary conditions within which lawful action is manifest are generally ignored as irrelevant, in ecology the boundary conditions are just as much the object of interest as the lawful behaviour. Fifth, species interact, either directly, as in the predator-prey relation, symbiosis or aggression, or indirectly through alteration of the common environment.

Levins and Lewontin were both participants at the Belagio Conference organized by Waddington. While they were inspired primarily by Engels' *Dialectics of Nature* to stress emergence, partial autonomy and interdependence, the way they have developed their ideas leads to a complete accord with the process conception of being. The forms of relationships they have focussed on can be seen as instances of the complex relationships between immanent and conditional causation. Their work both develops and facilitates the clarification of the process conception of the world.

However process philosophy also points towards other lines of research. To begin with, it is a simple matter to extend the analyses of Levins and Lewontin to include the world ecosystem as a whole, the 'biosphere' as first Eduard Suiss, then Vladimir Vernadskii called it, or 'Gaia' as James Lovelock and Lynn Margulis more recently have called it.[81] If there is anything distinctive about the biosphere as an ecosystem, it is the extent to which interaction is based on alteration of the common environment. Suiss, Vernadskii and Lovelock have been pre-eminently concerned with geological, chemical and atmospheric transformations of nature by life processes. The biosphere is taken to include all those geological, atmospheric and biological processes and cycles through which organisms maintain and transform the conditions for life on Earth, and Lovelock has argued that to conceive the biosphere in this way requires that Earth be thought of as a living organism. Beyond this, when the focus is on self-maintaining order, a freer notion of what there is, is possible. It is no longer necessary to think of the object of analysis as a discrete entity consisting of parts. It is possible to acknowledge that organisms are simultaneously or at different times participants in a number of ecological processes. Another dimension to this anti-reductionist position can then be added by considering the different temporalities of the different processes of life and their relationships. Some processes, for instance the development of the composition of the atmosphere with its layer of ozone, or the reproduction of certain species of trees, require very long durations in comparison with other life processes, and this is significant for understanding their relationships of autonomy and interdependence. Recent work has been undertaken along these lines by R.V. O'Neill, D.L. DeAngelis, J.B. Waide and T.F.H. Allen.[82]

---

80. *The Dialectical Biologist*, p.136.

81. The best overview of the Gaia hypothesis is contained in Peter Bunyard and Edward Goldsmith eds, *Gaia: The Thesis, the Mechanisms and the Implications*, Camelford, Cornwall: Wadebridge Ecological Centre, 1988. This contains a history of the concept by Jacques Grinevald, 'Sketch for a History of the Idea of the Biosphere' and a number of contributions by Lovelock and Margulis.

82. R.V. O'Neill, D.L. DeAngelis, J.B. Waide and T.F.H. Allen, *A Hierarchical Concept of Ecosystems*, Princeton: Princeton University Press, 1986. See especially Chapter 5: "Some Elements of Hierarchy Theory" which considers the theoretical issues, and Chapter 7 "Ecosystems as Hierarchies of Species" in which the theory is applied to ecological systems.

Allowing for such non-reductionist dynamics and complex relations in DNA production, in epigenesis, in cognitive development and in ecosystems, a basis is provided for explaining the punctuated equilibrium in species evolution as revealed by palaeontology. In times of stress the organism as a whole can affect its DNA to produce rapid increases in mutation rates. Epigenesis and cognitive development involve dynamics which limit the possibilities of transformation and guarantee that transformations will be in quantal leaps. While mature ecosystems are capable of preventing any new lines of development establishing themselves, they are subject to collapse after which rapid speciation involving quantal leaps can occur. These principles can operate from the smallest ecosystems to the biosphere.

## *The Becoming of Life*

With such emergent dynamics and inter-dependence it is necessary to redefine the very meaning of evolution. Evolution can no longer be conceived to be simply about differences between organisms and the variety of their adaptive characteristics. It is necessary to consider evolution as part of a general theory of life and its distinctive processes, ranging from the biosphere to particular biochemical processes. Along these lines Brian Goodwin has called for and outlined 'a new conceptual scheme from which both evolution and development emerge as essential aspects of biological process.'[83] According to this scheme:

> The actualization of specific morphological and behavioural patterns in organisms by the action of particular genes and environments on the space-time order of the developing organism described by the laws of organization of the living state is the biological process of creation. The exploration of the potential set of forms defined by these laws, by changes in genes and in the environment, is the process of evolution; while the generation of individual entities of specific form from this set is development. A biology based upon a generative paradigm focuses on these processes of biological creation as the central and distinctive features of the living condition, and sees the actual history of organism (their contingent evolution) as intelligible only in relation to the logic of creative process.[84]

It is within the creative process of becoming of the biosphere that organisms have evolved and developed their awareness of the world and themselves. This cannot be conceived simply in terms of individual organisms, but must be seen in terms of life, the complex of dissipative structures emerging from the thermodynamically far from equilibrium situation maintained by the sun, the development of ecosystems sustaining diversities of species within which awareness has emerged and developed, first through species and communities and then through individual members. Humanity has emerged as part of this creative becoming of life.

---

83. Goodwin, 'Changing from an Evolutionary to a Generative Paradigm', p.99.

84. Ibid. p.118. For the further development of this approach, see Goodwin, *How the Leopard Changed its Spots*.

# 15

# HUMANITY AS AN EMERGENT PHENOMENON WITHIN NATURE

Over the last three hundred years humans have come to be conceived of as either mechanical parts of a mechanical nature, or as virtually super-natural beings who live in a world of culture.[1] The conception of humans as mechanisms derives originally from Hobbes and has been developed in political, ethical, psychological, and most importantly, in economic theory. Counterposed to this has been a tradition which in various ways has striven to fill the gap in our understanding of social life between the State and the individual, the neglect of which, it is argued, has rendered political, ethical, psychological and economic thought superficial.[2] It is this 'humanistic' tradition which has stressed culture, human creativity and 'meaning', but then for the most part ignored the natural conditions of life. There are exceptions to this beginning with J.G. Herder who upheld the notion of creative humans as part of nature by proposing an anti-mechanist conception of nature; but such thinkers have not been taken seriously. Now the mechanistic conception of humans is invalidated by the failure of mechanistic materialism in the natural sciences, while a theory of knowledge adequate to the physical sciences together with the process conception of being, provides a foundation for the humanistic tradition. As Ortega y Gasset asserted: 'In order to speak then, of man's being we must first elaborate a non-Eleatic concept of being as others have elaborated a non-Euclidean geometry. The time has come for the seed sown by Heraclitus to bring forth its mighty harvest.'[3] Humans can now be conceived as culturally constituted creative agents *within* nature.

The humanistic tradition can best be understood against the background of the achievements of Hegel in synthesizing all previous social and political theory through developing a coherent foundation for the conception of humans as creative participants in the becoming of the world; and the subsequent disintegration of his system.[4] This provides an historical perspective in which the relationships between different approaches to the study of humanity can be seen as either one-sided developments of Hegelian ideas, or as reactions to Hegel's limitations. My contention is that process philosophy provides an interpretation of these developments, enabling them to be evaluated and re-integrated into a unified research programme, bridging the gap between the natural and the social sciences, the sciences and the humanities, knowledge and evaluation, and the objective realm and the subjective realm. Humans will be able to see themselves as self-creative participants in the becoming of nature and society, and the development of their understanding as the world becoming conscious of itself. As Marx prophesied: 'Natural science will... subsume the science of man just as the science of man will subsume natural science: there will be one science.'[5]

To accord with the Heraclitean conception of being, humanity must be understood as an emergent process or complex of processes within nature, as part of the biosphere, the complex of

1. This dualism even exists in geography where one would expect the opposition to be overcome. See J.J. Johnston, *Geography and Geographers*, 2nd ed., London: Edward Arnold, 1983, esp. p.175.

2. On the origins of this anti-mechanistic tradition, see B.A. Haddock in *An Introduction to Historical Thought*, London: Edward Arnold, 1980.

3. Jose Ortega y Gasset, *History as a System*, tr. Helene Weyl, [1941], N. Y.: Norton, 1962, p.203.

4. A similar argument, but with different emphases, has been made by Rudiger Bubner in 'On Hegel's Significance for the Social Sciences' in *Hegel and the Sciences*, ed. Robert S. Cohen and Marx W. Wartofsky, Dordrecht: Reidel, 1984, pp.143-160.

5. Karl Marx, 'Economic and Philosophical Manuscripts', in *Karl Marx: Early Writings* tr. Rodney Livingstone and Gregor Benton, Harmondsworth: Penguin, 1975, p.355.

dissipative structures which has emerged in the thermodynamically far from equilibrium situation maintained on earth by the sun. Living entities are processes which define their environments as their worlds, worlds in which they are then sensuously engaged - attracted and repulsed by it, taking it in, incorporating it and excreting it, transforming it and being transformed by it. This characterizes both human individuals and human societies. As Richard Adams wrote: 'societies operate as dissipative structures; they are continuities of form that are constituted by the very flow of energy that is expended (i.e. converted) in the process of acting out the behaviours and doing the work (from both human and non-human sources) that is carried out in the context of social relationships.'[6] So, as Serge Moscovici argued:

> Man's single-handed conflict with nature should be seen as a confrontation *within* nature... The notion that nature is inhuman and man unnatural is totally invalid. No part of man is or ever was closer than any other to an ever-changing nature.[7]

However while humanity is a form of life, not all life is humanity. So what is distinctive about humanity? Humans cannot be distinguished from other animals by their using tools or having a culture which develops from generation to generation. Ethologists have shown that many kinds of animals have these characteristics.[8] The evolution of humanity has involved the simultaneous emergence of a complex of interdependent processes and structures.

## *The Hegelian Concept of Humanity*

The importance of Hegel is to have characterized the most distinctive features of this complex. Hegel rejected Kant's notion of the preformed ego, the 'I' represented as a pure unity relating to itself. Instead Hegel portrayed the ego as the result of the development, from immediate sensitivity to self-awareness, then to self-consciousness gained through a reciprocity of perspectives in interpersonal relationships, and finally to universality through participation in ethical and cultural life. He characterized this formative process as part of three interdependent dialectical patterns: symbolic representation which operates through the medium of language; interaction on the basis of reciprocity which operates through moral relations; and the labour process which operates through the tool.[9] It is through participating in these dialectical patterns of culture that human organisms transcend their particularity and unite with the universal to gain an identity. As Hegel put it in *The Phenomenology of Mind*:

---

6. Richard N. Adams, *Paradoxical Harvest*, Cambridge: C.U.P., 1982, p.17. Adams has further developed his ideas on this in *The Eighth Day: Social Evolution as the Self-Organization of Energy*, Austin: University of Texas, 1988.

7. Serge Moscovici, *Society Against Nature: The Emergence of Human Societies* tr. Sacha Rabinovitch, Hassocks: Harvester Press, 1976, Introduction.

8. See for instance Jane van Lawick-Goodall, 'The Chimpanzee' in Vanne Goodall ed., *The Quest for Man*, London: Phaidon, 1975.

9. See G.W. F. Hegel, *System of Ethical Life and First Philosophy of Spirit*, tr. H.S. Harris and T.M. Knox, N.Y.: S.U.N.Y. Press, 1979, pp.205-253. These lectures, given at Jena in 1803/4 and again in 1805/6, have been critically examined by Jürgen Habermas in 'Labour and Interaction: Remarks on Hegel's Jena *Philosophy of Mind*, in *Theory and Practice* [1971] tr. John Viertel, London: Heinemann, 1974, pp.142-169. The threefold division derives ultimately from the Pythagorean division between lovers of wisdom, lovers of honour and lovers of gain. In later works Hegel subordinated this trichotomy to the division between Subjective, Objective and Absolute spirit, but the earlier division was not totally abandoned.

> ... self-consciousness is only something definite, it only has real existence, so far as it alienates itself. By doing so, it puts itself in the position of something universal, and this its universality is its validity, establishes it, and is its actuality.[10]

While most philosophical anthropologists abjure the terms used by Hegel, they have nevertheless acknowledged the validity of the duality within human awareness between the immediacy of engagement in the world and the transcendence of this immediacy.[11] For instance Helmuth Plessner distinguished humans from non-humans by their eccentric positionality. Humans take up a position in the world as do other organisms to become embodied subjects, but as subjects they also take a perspective outside their bodies to experience themselves as physical objects.[12] Along similar lines, but emphasising the social nature of this eccentric perspective, George Herbert Mead argued that human becoming is characterized by a continuous dialectic between the 'I' as creative subject, and the 'me' which derives from appropriation by individuals of the perspective of the 'generalized other' towards themselves.[13]

The possibility of this duality can to some extent be explained in naturalistic terms compatible with the process conception of being through the genetic epistemology of Jean Piaget. Piaget was concerned to explain the development of cognition from early childhood to adulthood, with particular concern to explain the emergence of the capacity to do science. To do this he represented the development of cognition as the adaptation of structures or schema of interpretation and action to assimilate environmental data in order to engage effectively in the world, producing a hierarchy of cognitive structures, with higher levels in this hierarchy operating on the lower levels of cognitive activity. Each structure was represented as a self-regulating system of transformations which compensates for internal and external imbalances and develops beyond itself into more advanced structures. Piaget traced this development through the elaboration of the most basic forms of sensori-motor intelligence tied to the content of specific sensory inputs and motor actions, through pre-operational intelligence in which schema are dissociated from particular content, through concrete operational intelligence in which schema develop to allow for operations independent of environmental interaction, to formal operations in which operations are performed on operations, as occurs in mathematical thought. This whole process was represented as taking place through the continuous development, differentiation and integration of schema which leads from a subjective, unintegrated, body-centred activity to a practical separation of means and ends and the development of a logic of action, to the capacity to retrace a cognitive route (to see that if a liquid is poured from one container to another that it can be poured back again, and that therefore there must be a conservation of liquid), to the capacity to think mathematically. This development was seen to involve a growing decentration from immediate experience, that is, from the experience of immediate engagement in the world, which is itself to some extent transformed by this decentration. The process of this development of cognition was described by Piaget in the terminology of Waddington's theory of epigenesis: '... intellectual growth contains its own rhythm and its "chreods" just as physical growth does.'[14]

A number of criticisms can and have been made of Piaget's ideas (apart from the obvious one forcefully made by Vygotsky that it is asocial, and the one that was made in the previous chapter,

---

10. G.W.F. Hegel *The Phenomenology of Mind*, tr. J.B. Baillie [1931], New York: Harper & Row, 1967, p.514f.

11. For an account of the development of philosophical anthropology see Axel Honneth and Hans Joas, *Social Action and Human Nature*, [1980] tr. Raymond Meyer, Cambridge: Cambridge University Press, 1988.

12. See Helmuth Plessner, *'De Homine Abscondito'*, *Social Research*, 1969, Vol.36, pp.497-509; and *Laughing and Crying*, tr. James Spencer Churchill and Marjorie Grene, Evanston: Northwestern University Press, 1970, Ch.1.

13. George Herbert Mead, *Mind, Self and Society*, ed. Charles W. Morris, Chicago and London: University of Chicago Press, 1934.

14. Jean Piaget, *Biology and Knowledge*, tr. Beatrix Walsh, Chicago: Uni. of Chicago Press, p.21.

that Piaget has hypostatized structures rather than treating them as ordered potentialities).[15] It appears from empirical studies that cognitive development is less 'tidy' than Piaget implies, with different levels of intelligence co-existing at any time. And a central tenet of Piaget's doctrine, that all abstract thinking is built on structures developed through practical engagement in the world, is refuted by examples of people with severe cerebral palsy who have shown themselves capable of a high levels of intellectual achievement. Such examples suggest either that the developments described by Piaget have to a considerable extent been 'genetically assimilated', or that there is more than one way for intelligence to develop. More fundamentally, Piaget has not understood the cognition of children in their own terms but as defective stages on the way to scientific cognition - which itself is understood in an excessively formalistic manner. As a consequence of this, he has focussed on cognitive structures in abstraction from the child's fragmentary, but global experience of the world. This is associated with basic omissions from Piaget's conceptualization of cognition. Without seeing such achievements as articulations of this global experience, Piaget is left with no means of understanding the relationship between each achievement, the relationship between abstract thought and global experiences characteristic of emotion, empathy and imagination, the use of metaphor and metonymy and the construction of narratives. He has ignored forms of intelligence beyond 'formal operations' (exemplified by mathematical thinking), namely 'dialectical' thinking - the capacity to question assumptions, to consider alternative assumptions, to use metaphors to see the world in entirely new ways, to change focus from parts to wholes and from wholes to parts so that they are seen relationally, to produce and understand narratives, and perhaps most fundamentally, to recognize explicitly the global experience of the world which is always assumed implicitly as that which is articulated by all particular determinations, whether concrete or abstract.

These criticisms can be accommodated, and it is possible to reformulate Piaget's ideas so as to avoid the hypostatization of 'structures'. Reformulated, it is the organism in interaction with its environment which develops the capacity to generalize types of ordering activity to new engagements with the environment, while at the same time developing the potential of this activity. In the case of the emergence of new levels, this involves the development of the capacity to order the ordering activity associated with more immediate involvement in the world. This reformulation at the same time has the advantage of emphasising the contingency of the world the organism is attempting to come to terms with and the limitations of all cognitive activity and explicit knowledge in this regard, and allows Piaget's concepts to be refined by taking into account the durational nature of any ordering activity and the different spatio-temporal orders associated with different types of ordering activity.

The durational nature of cognition is particularly important in explaining decentration. The concept of decentration cannot be made sense of if the stream of consciousness is conceived as a linear sequence of events. Consciousness involves a multilinear becoming and requires the ontology of process philosophy to be made intelligible. For instance the sensori-motor differentiation of actions into means and ends implies a durational cognitive activity transcending this differentiation. More significantly, reversibility of operations generates the capacity to constitute the environment as entities which are instances of classes, enabling the world to be constituted as the enduring background of ordered relationships between entities to all particular perceptions and actions. This ability implies the emergence of a new spatio-temporal order, which is then consolidated with the development of formal operations. As Nathaniel Lawrence argued:

> The smooth slippage of closed events in a continuous progression along a time line is not adequate to the facts. Consciousness accumulates large patches of temporality into a variety of 'nows' of many sizes. It synthesizes them in a great many ways ... and thereby generates the raw materials from which many abstract meanings for time can be derived: mathematical,

15. Many of these have been accepted by Piaget himself, who until is death in 1980 was continually revising his ideas. On this, see Hans G. Furth, 'Piaget's New Model,' *Piaget and Knowledge*, 2nd ed., Chicago: University of Chicago Press,

> physical, perceptual etc. In short, the conveyor belt metaphor of temporal sequence does not accommodate to the multiple modes of arrest and synthesis by which consciousness establishes both its open-ended quasi-identity and the continuous summation of the world-in-relation-to-consciousness.[16]

The subject is not an enduring substance, but a process of becoming continually forming itself within the context of nature, culture and social forms, in which every act must be supported by a self-conception which is an enduring expectation which will only be confirmed by what the subject discovers itself to have been standing for.

Cognitive development only takes place in the context of social relations of some sort, yet it is the precondition for the complex and diverse forms of human social relations. It is through these social relations that the individual is constrained to achieve this decentering of consciousness, and it is through them that some degree of reintegration of the individual becomes possible. This brings us to the dialectical patterns of cultural development.

!! With the characteristic penchant for reductionism of the Western intellectual tradition, each of the three dialectical patterns of culture identified by Hegel has been used by different thinkers as the sole basis for explaining the development of society. As Jürgen Habermas pointed out:

> *Ernst Cassirer* takes the dialectic of representation and makes it the guiding principle of a Hegelianized Kant interpretation, which at the same time is the foundation of a philosophy of symbolic forms. *Georg Lukács* interprets the movement of intellectual development from Kant to Hegel along the guide-line presented by the dialectic of labour, which at the same time guarantees the materialistic unity of subject and object in the world-historical formative process of the human species; finally, the neo-Hegelianism of a thinker such as *Theodor Litt* leads to a conception of the stepwise self-development of spirit which follows the pattern of the struggle for recognition.[17]

However such thinkers have succeeded in advancing our understanding of different aspects of these dialectical patterns, and to capture their achievements, these will be redefined in broader terms: as the dialectic of orientation, as the dialectic of recognition and as the dialectic of power.

Before examining these dialectical patterns of culture in detail, a number of points can be made about the unique nature of such processes. These patterns are dialectical because they are based on people as conscious agents creating themselves. As such, they cannot be understood simply in terms of individuals, nor as emergent processes transcending individuals, but must be understood as processes through which individuals emerge to become semi-autonomous participants in the on-going creative becoming of these patterns, which are semi-autonomous from these individuals. Furthermore, individuals are struggling for goals which are neither final ends nor simply potentialities for achieving these, but are simultaneously both ends desired and potentialities for pursuing further ends. Orientation, recognition and power thus have, as Derrida has noted in relation to desire in general, a deferred quality; it is never possible to actually achieve these as final states, as final resting points. The potentialities produced are potentialities both of the dialectical patterns themselves and of individuals participating in them, and the becoming of the patterns and of the individuals who emerge in this becoming is endless. Associated with this, dialectical patterns have no definite boundaries, either temporal or spatial. Although there is considerable spatial differentiation of social activity insulating people from each other, all dialectical activity relates itself, even if only through exclusion, to all potential participants. Finally, dialectical activity

---

1981, Ch.15.

16. Nathaniel Lawrence, 'The Illusion of Monolinear Time' in James M. Edie, Francis H. Parker and Calvin O. Schrag ed.'s, *Patterns of the Life-World*, Evanston: N.W.U.P., 1970, pp.298-314, p.309f.

17. Habermas, 'Labour and Interaction', in *Theory and Practice,* p.157f.

carries with it the possibility of critical reflection and transcendence. To be participating in these dialectical patterns is to be at least provisionally committing oneself to certain evaluative stances within these patterns, and to be at least tacitly aware that such stances are incompatible with other possible stances, and that one's own stance is therefore questionable. So as Hegel saw, the dialectic of representation carries with it the tendency for people to transcend limited, one-sided forms of thinking and replace them with forms of thinking which come nearer to grasping the whole in its complex diversity, the dialectic of recognition tends to reciprocity, carrying with it a tendency to generate social relations which extend recognition and respect to more and more people, and the dialectic of labour tends to generate more effective technologies and organizations.

## *The Dialectic of Orientation*

The most influential anti-mechanist social theory in recent years has been associated with the attempt to explain society in terms of the dialectic of symbolic representation. However this project has fragmented with various facets being examined in isolation, delimited as distinct and self-contained fields of study. Those dominated by the mechanistic world-orientation tend to focus on the power of language to designate things or to represent states of affairs; those inspired by the tradition deriving from Herder and von Humboldt have focussed on the creative expression of the individual subject (or as in the case of Heidegger, of the world); those inspired by the tradition of hermeneutics have focussed on the process of understanding, while the structuralists have focussed on conventional sign systems. Concern with what is expressed, with advances in comprehension and its relation to representation has been for the most part been the preserve of the philosophy of science. It is to reconcile these various approaches that it is suggested that the dialectic of symbolic representation be reconceived as the dialectic of orientation.

So conceived it is possible to see how this is generated and maintained. The decentering of experience at the pre-operational level of intelligence is associated with the emergence of imagination, the capacity to produce and think in signs which facilitate cognitive activity independent of the immediate situation, and associated with this, a growing awareness of others with a different perspective on the world. This leads children to distinguish their own perspectives (and intentions based on these) from those of others and to distinguish all perspectives from the world itself, revealing the questionability of their own viewpoints, and raising questions about the relationship between these to the viewpoints of others. This engenders (or augments) a curiosity, and an impetus, facilitated by the development of the capacity to communicate by signs, for children to express themselves to validate their own perspectives and experiences in the eyes of others and to relate their own perspectives and experiences to those of others.[18] Through participating in language and other sign systems, appreciating the expressions of others and gaining affirmation of their own expressions, the surrounding world comes to be experienced as common world, a reality shared with others about which stories can be told and about the nature of which people can speculate. Children are induced in this way to participate in and to contribute towards defining a social imaginary world transcending their immediate experience in terms of which they can locate themselves and which can serve as a reference point for discourse, or at least for the achievement of a reciprocity of perspectives. Further decentration leads at least some people in some societies to a conscious struggle to explore the limits of their understanding and the means to achieve it, and to strive to articulate a conception of the world valid from the perspective of a 'generalized other', to strive to see the world from a perspective shared not only with those around one, but with all anonymous contemporaries, with all predecessors and successors. It is through the reproduction of this struggle and through the ensuing communicative activity, that a community

---

18. This process has been analysed by Stephen Strasser in *The Idea of a Dialogal Phenomenology*, Pittsburgh: Duquesne University Press, 1969.

beyond the immediate experience of each individual becomes imagined as reality, providing the context of 'subject positions' within which each communicative act can take its place and be made sense of.

Signs are both the condition of and are generated, reproduced and developed in this struggle for orientation. Peirce defined a sign as 'something which stands to somebody for something in some respect or capacity.'[19] However what a sign stands for is never simply given, but is in some sense a construct. According to Whitehead 'the mind is functioning symbolically when some components of its experience elicit consciousness, beliefs, emotions, and usages, respecting other components of experience. The former set of components are the "symbols," and the latter set constitute the "meaning" of the symbols.'[20] A sign can be a thing (structure), event or process encountered in nature or society, it can be a communicative act, or it can be an entity designated or produced by such an act. It stands for, reveals or draws attention to, or creates or elicits beliefs, emotions or evaluations concerning components of the world, emotions, thoughts, beliefs or speculations about the world, or imaginative constructions or fictions. According to Peirce, signs can be classified (not necessarily exclusively) into indexes, icons and symbols.[21] An index refers to that which it denotes through being causally related to it, as smoke is causally related to fire or a footprint to an animal. An icon refers to that which it denotes merely by virtue of its own character which it possesses whether or not the object denoted actually exists. Examples are images, diagrams and metaphors. A symbol is a conventionally defined sign which would lose its character as a sign if there were no interpretant.

The structuralists have focussed their attention on sign systems, the conventional codes which specify the relationships between sets of perceptually distinct phenomena to enable the production and interpretation of communicative acts. One of their most important achievement was to have shown how many other sign systems than language are involved in communication; that all actions and the material products of actions, 'that all the various non-verbal dimensions of culture, such as styles in clothing, village lay-out, architecture, furniture, food, cooking, music, physical gestures, postural attitudes such as buildings, gardens, forms of dress and so on are organized in patterned sets so as to incorporate coded information in a manner analogous to the sounds and words and sentences of a natural language.'[22] Structuralists have also revealed the ordered nature of sign systems, the patterns of oppositions between signs, and the relationship between and role of metaphor and metonymy - or as Lévi-Strauss reformulated these, paradigmatic associations and syntagmatic chains, in communication. But they have tended to reify the order they have found, treating it not as potentialities facilitating communication and action but as something existing in its own right which not only delimits what can be expressed and understood, but which determines what people say and do. As Lacan argued that people do not speak, they are spoken; they do not think, they are thought.[23]

Poststructuralists in the last two decades have attacked this reification of sign systems. However the underlying principles on which the structuralist reification is based had already been effectively criticised in Russia in the 1920's by Bakhtin, Medvedev and Volosinov, without this attack having led to the relativistic bind of post-structuralists such as Derrida. Volosinov opposed Saussure's abstraction from language of a synchronic system of signs, arguing that language is not an inert system of norms to which a speaker must conform to be understood. What is of interest to the speaker is the adaptability of linguistic forms to express new meanings in concrete contexts, while

19. *Philosophical Writings of Peirce*, ed. Justus Buchler, New York: Dover, 1955, p.99.

20. Alfred North Whitehead, *Symbolism: Its Meaning and Effect*, [1927], N.Y.: Fordham University Press, 1985, p.8.

21. One of the problems with the whole field of semiotics is the absence of any consensus on definitions. On this see Umberto Eco, 'Symbol' in *Semiotics and the Philosophy of Language*, Bloomington: Indiana University Press, 1986.

22. Edmund Leach, *Culture and Communication*, Cambridge: Cambridge University Press, 1976, p.10.

23. Jacques Lacan, *Écrits*, London: Tavistock/Routledge, 1980, p.69. For a critique of this, see J.P. Sartre, 'Replies to Structuralism' in *Telos*, Vol.9, 1971, p.113.

understanding the speaker's utterance is not simply the recognition of form but involves understanding, from within particular concrete contexts, of its meaning. As he put it:

> ... the constituent factor for the linguistic form, as for the sign, is not at all its self-identity as signal but its specific variability; and the constituent factor for understanding the linguistic form is not recognition of 'the same thing,' but understanding in the proper sense of the word, i.e., orientation in the particular, given context and in the particular, given situation - orientation in the dynamic process of becoming and not 'orientation' in some inert state.[24]

This criticism can be generalized to all other sign systems and in essence corresponds to Bourdieu's criticism of structuralist anthropology.[25] For Bourdieu, the patterns of oppositions evident in the practices and products of societies are not fixed structures which organize the way people act, but are the outcome of people's creative efforts to act from situation to situation in accordance with their *habitus*, that is, their:

> *dispositions*, structured structures predisposed to function as structuring structures, that is, as principles of the generation and structuring of practices and representations which can be objectively "regulated" and "regular" without in any way being the product of obedience to rules, objectively adapted to their goals without presupposing a conscious aiming at ends or an express mastery of the operations necessary to attain them and, being all this, collectively orchestrated without being the product of the orchestrating action of a conductor.[26]

The patterns of oppositions noted by structuralists are simply by-products of such practical efforts.

Structuralists and poststructuralists alike have given scant attention to what must be recognized as a central aspect of the dialectic of orientation: what is being communicated (etymologically: made common) in communicative activity. John Austin pointed out that only a minor part of communication is stating what is the case. Communication is articulating the world into consciousness to create a meaningful public space, and is in part creative of relations between people, between individuals and society, and between humans and their environment. Among other things, communication involves defining immediate situations, including the relationship between those engaged in communication, defining, questioning and redefining the broader context of such situations, producing or reproducing narratives, speculating, expressing emotions, attitudes and intentions, forming more enduring relationships (as in making a promise, swearing allegiance, or simply becoming friends), and negotiating, arguing a case, or drawing attention to the communicative act itself or to assumptions (such as the meaning of terms or the conventions of narrative construction) which make communication possible.[27] Where communication is concerned with the nature of the world and with revealing its significance, then the speculative attainment, development, affirmation and criticism of shared assumptions, interpretive schemes and ideals and showing when their deployment is appropriate is more fundamental than, and is the condition for, reporting states of affairs. Such interpretative and evaluative schemes range from those associated with body schema and practices, the habitus, to explicitly formulated conceptual frameworks, and from schemes associated with comprehending particular situations to world-orientations and grand narratives.

---

24. V.N. Volosinov, *Marxism and the Philosophy of Language* [1930] tr. Ladislav Matejka and I.R. Titunik, N.Y. and London: Seminar Press, 1973, p.66.

25. See especially Pierre Bourdieu, *Outline of a Theory of Practice,* Cambridge: Cambridge University Press, 1977, p.22ff.

26. Ibid., p.72.

27. Austin suggested that there might be as many as 10,000 different linguistic functions performed by speech acts.

Communication characterizing the nature and significance of the world is usually associated with practical activity and often is only understood practically within particular situations, facilitating the coördination of people's actions and lives and the comprehension of each other's situations, actions and significance. It is primarily through the metaphorical generalization of schemes of interpretation from context to context, from society to nature and back again and their embodiment in the physical world, in social relations and in people's *habitus* that cultures gain their coherence, a coherence which is usually reinforced through a dominant metaphor or thematic motif integrated into a dominant narrative. However there are few cultures which are so primitive or degenerate that at least some of its members do not devote at least some of their time to criticising and trying to overcome the contradictions and limitations of received beliefs, interpretations, forms of thinking and narratives, and to explicit efforts to construct alternatives to define and express themselves and their relationship to the world.[28] It is through such efforts that we have gained a heritage of a diversity of speech genres, worlds of mythology, song, dance, poetry and novels, of sculpture, architecture and other artworks, and critical traditions of history, philosophy, logic, mathematics and science.

JürgenHabermas has argued that there are three (or four) validity claims implicitly raised and reciprocally recognized with every speech-act - in relation to cognition that the propositional content of a speech-act is true, in relation to interaction that the performative component is correct, and in relation to expression that intentions are being expressed sincerely.[29] Occasionally he also includes the claim that what is said is intelligible. These validity claims, Habermas argues, are an inescapable aspect of all communicative acts, although in non-verbal acts they are less well defined. However beyond these validity claims (if these claims are indeed universal) there are more basic claims. To speak is to give expression - though never complete expression - to a tacitly presupposed order of meanings associated with a mode of being in the world and a form of life with standards defining what is the appropriate way for people to live and to act.[30] It is being implicitly claimed that the schemes of interpretation and associated modes of being, forms of life and standards presupposed by and expressed in such speech-acts are appropriate and adequate to the situation, and that they are consistent with other interpretative schemes accepted as valid. While schemes, modes of being and standards can be questioned, they can only be transcended by being replaced. The total abandonment of all standards is inconceivable. Nihilism is itself the product of standards (for instance, of what is to count as an 'objective' attitude to the world).

The relationship between individual efforts at orientation, communicative acts, speech genres, texts, cultural fields, discursive formations, narratives, various types of media. systems of signs and enduring schemes of interpretation, involve multiple spatio-temporalities. Expressive acts (although not necessarily expressions, e.g. writings) are of a short duration by comparison with the evolution of stories, genres, cultural fields, discursive formations, schemes of interpretation and sign systems. By participating in dialogue or in other forms of communication people are constrained by past communications, narratives, speech genres, existing cultural fields and discursive formations and a hierarchy of enduring interpretive schemes and sign systems which make it possible for them to communicate, while participating in the spatio-temporal order of the evolution of each of these.

---

28. On this, see Geoffrey Samuel, *Mind, Body and Culture*, Cambridge, C.U.P., 1990.

29. See Jürgen Habermas, *Communication and the Evolution of Society* [1976] tr. Thomas McCarthy, London: Heinemann, Beacon Press, 1979, p.58.

30. On this, see Charles Taylor, *Human Agency and Language: Philosophical Papers 1*, Cambridge: C.U.P., 1985, 'Language and Human Nature', pp.215-247.

## *The Dialectic of Recognition*

While it has been less common, efforts have also been made to theorize social dynamics solely in terms of the dialectic of recognition.[31] Many of the symbolic interactionists inspired by G.H. Mead saw people's behaviour as being virtually determined by the criteria of acceptability of the reference groups and significant others in terms of which they defined themselves and their significance. Thus the criminologist Edwin H. Sutherland argued that 'A person becomes delinquent because of an excess of definitions favourable to violation of law.'[32]

As with the other dialectical processes, the dialectic of recognition is engendered by the development of the individual within a social context.[33] To begin with, infants do not conceive themselves as separate beings at all. As Heinz Remplein argued of the original condition of children: 'Above all, there is lacking the split between I and you that gives a characteristic tension to the experience of the adults.'[34] As the original fragmentary consciousness of the child's body becomes integrated to form a precise corporeal schema there emerges a global consciousness of the body's position in the world. This self-awareness immediately creates an imbalance in experience which leads to the recognition of others as autonomous beings which enables the child to see an image of itself in the responses of others to it, and to identify with this image. Characteristically, where such pronouns are available, children first refer to themselves by name, then by the pronoun 'me'. The use of the pronoun 'I' or its equivalent is a later stage of development and indicates an individuation (the nature and extent of which varies between cultures) of the experience of becoming consequent to the reflexive constitution and recognition by the child of itself as one embodied consciousness among others.[35]

This individuation is, and usually remains precarious since it is founded on the development of a conceptualized self which derives from and is dependent upon recognition and affirmation by others, but involves the assertion of independence against these others. This generates the original desire to be recognized by others which expresses itself in simultaneous wilfulness and the quest for attention. The conceptualized self and the conception of others develops reciprocally by relativizing particular others in relation to others in general. The 'you' first becomes 'mother', then becomes 'a' mother along with other mothers. The 'you' which was unique in the original dyad becomes 'the' other in reference to 'me'. By a process of successive identifications, the struggle for recognition is then generalized from significant others to reference groups, and with some people, at least in some societies, to a generalized other, the point of view which is defensible in an open court of reason. This struggle for recognition engenders the participation by individuals in the 'moral order', the order of symbols, status relations and moral notions through which people, the roles they play and their actions are recognized as significant and are granted respect or disdained.

---

31. See Tamotsu Shibutani, 'Reference Groups as Perspectives' in Jerome G. Manis and Bernard N. Meltzer ed.'s. 2nd ed., Boston, Allyn and Bacon, 1972, pp.160-170; and Howard S. Becker *The Outsiders*, Glenco, Illinois: The Free Press, 1963 for two of the best symbolic interactionist analyses. These works are developments of the research programme of G.H. Mead as formulated in *Mind, Self and Society*, ed. Charles Morris, Chicago: Chicago University Press, 1934.

32. E.H. Sutherland, *Principles of Criminology*, 5th ed. Chicago: Lipincott, 1955, p.78. See also Daniel Glaser, 'Criminality Theories and Behavioural Images' in Jerome G. Manis and Bernard N. Meltzer, *Symbolic Interaction*, 2nd ed. Boston: Allyn and Bacon, pp.482-497.

33. For one of the best analyses of this see M. Merleau-Ponty's study 'The Child's Relations with Others', in *The Primacy of Perception*, ed. James M. Edie, Evanston: North Western University Press, 1964.

34. Heinz Remplein, *Die Seelische Entwicklung des Menschen Im Kindes - Und Jungendalter*, 14th ed. Munich, 1966, p.184; translated and cited by Strasser, *The Idea of a Dialogal Phenomenology*, p.84.

35. On this see Rom Harré, *Personal Being*, Cambridge Mass.: Harvard University Press, 1984, Ch.4. Harré refers to the individuated, embodied subject as the 'self.'

MacIntyre argued that people can only orient themselves in society through being told stories which enable them to understand and take up their positions within the stories which are being living out. As he put it:

> I can only answer the question "What am I to do?" if I can answer the prior question "Of what story or stories do I find myself a part?" We enter human society, that is, with one or more imputed characters - roles into which we have been drafted - and we have to learn what they are in order to be able to understand how others respond to us and how our responses to them are apt to be construed. It is through hearing stories ... that children learn or mislearn both what a child and what a parent is, what the caste of characters may be in the drama into which they have been born and what the ways of the world are. Deprive children of stories and you leave them unscripted, anxious stutterers in their actions and in their words.[36]

As participants within this moral order and having attained some degree of identity as a consequence, some individuals is some societies are launched on a quest for coherence in their judgements and actions. This involves individuals in representing themselves to each other and to themselves as unfinished autobiographies or narratives. In formulating these autobiographies people define themselves in terms of commitments to a hierarchical order of projects, ranging from short term projects such as fulfilling the expectations of the role or roles they are immediately engaged in, to the projects through which they define the significance of their lives. In this way people's autobiographies are related to the biographies of others and the histories of social formations: families, communities, organizations and cultural, social and political movements, and at least tacitly, to the narratives through which classes, nations and civilizations define themselves and their place in the world.[37] Through the quest for coherence in their own lives people are aroused to search for coherence in the moral order, and beyond this, in the history of their families, communities, classes and nations, and in the history of civilization and humanity itself.

Through this some people to acquire and develop the capacity to question and reformulate this moral order and the narratives of the social orders which represent and legitimate it. While it is possible for a society to be composed of institutions, organizations and groups embodying different and incommensurable ideals and values so that there is no coherent moral order, the tension generated within individuals struggling for coherent identities guarantees that in all but the most oppressive societies there will be at least some impetus towards achieving such cultural coherence. It is to this impetus that in Western societies we owe a heritage of universalist moral notions, a history of ethical thought devoted to refining and justifying these notions (although it is only with modernity that these notions have been abstracted from politics and theology), histories of classes, nations, civilizations and of humanity, a number of competing grand narratives of progress, and an array of institutions and organizations which incorporate such notions at least as ideals.[38]

As with the dialectic of orientation this dialectic of recognition also involves the emergence of a spatio-temporally transcendent order, or complex of orders, created and sustained by the struggle for recognition and respect. Participation in these enables people to transcend their immediate being in the world, enabling them to achieve the reflexivity required to integrate their disparate engagements in the world into the unity of themselves as unfinished stories or biographies, and thereby to become active moral agents. By internalizing the viewpoint of different reference groups, individuals who remain with their own subjective, immediate stream of time consciousness

---

36. Alasdair MacIntyre, *After Virtue*, 2nd ed., (Notre Dame, Notre Dame University Press, 1984), p.216.

37. The most profound work on narratives comes from Paul Ricoeur. On the relationship between narrative and identity see Ricoeur *Oneself as Another*, tr. Kathleen Blamey, Chicago and London: The University of Chicago Press, 1992.

38. Of course this is contested by Nietzsche in *The Genealogy of Morals*. However it is possible to accept the argument of Nietzsche and his followers that *ressentiment* of the weak has been a driving force behind the development of ethics without accepting the argument that this has been the only driving force.

simultaneously incorporate the intersubjective time dimensions of these orders.[39] Their actions and lives are in this way raised from their particularistic immediacy to become part of the spatio-temporally transcendent processes of becoming of different moral orders, and in this way, formulated as narratives, they take on an objective significance transcending the contingent existence of their biological existence. The sense of being someone with a significance which endures through diverse activities in diverse contexts is only attainable at the point of intersection of such multiple spatio-temporal systems of social experience.

## *The Dialectic of Power*

The most enduring form of anti-mechanistic social theory has been based on the elaboration of the dialectic of labour. As Habermas pointed out, Lukács, and following Lukács, many of the Western Marxists influenced by Marx's early works, have seen humanity as creating itself through its humanization of nature. Developing control of nature, developing 'the forces of production', involves not only developing technology, that is, knowledge, skills and apparatus, but also forms of social organization to coordinate people's activities, distribute products, educate people to participate in such organizations, and controlling people to ensure they play a productive role in all this. To capture all these dimensions it is necessary to reformulate the dialectic of labour as the dialectic of power. Under this rubric it is necessary to consider both theories of technology and theories of power. This covers an enormous range of issues and debates, not all of which can be considered here.[40] The central problem in all these is defining what power is.

In terms of process philosophy all processes manifest power, as both a potential and in their activity, in maintaining their existence. It is the capacity to produce, and the production of, additional ordering in the world, and is the very be-ing of any process. The dissipative structures which develop in thermodynamically far from equilibrium situations are particular types of self-ordering activity in which power is the capacity to order and the ordering of the flow-through of usable energy and materials (that is, stable forms of energy). Living beings as complex dissipative structures are unique in that their self-ordering activity involves defining their environments as worlds in relation to themselves and correspondingly, involves the development of awareness, appetites and aversions, and the power to order their engagement with their worlds accordingly. The distinctively human form of power is essentially socio-cultural (presupposing and involving both of the other dialectical processes - without being reducible to them), and it can only be understood in relation to (although it is not entirely reducible to) institutions.

The theory of power which comes closest to acknowledging all this is that offered by Richard Newbold Adams. According to Adams:

> Everything in the environment of man is composed of energy forms and processes and can be measured in terms of the amount of energy that is potentially available for conversion or is being converted. ... In dealing with social power ... we are concerned not so much with the rate of flow or conversion as with *the control that one actor, or party, or operating unit*

39. See Katherine Mary Tillman, 'Temporality and Role-Taking in G.H. Mead' in *Social Research*, Vol.37, 1970, pp.533-546. Mead's analysis of this problem, which is indebted to both Hegel and process philosophy, overcomes many of the problems the French post-structuralists (notably Derrida and Lacan) have grappled with. For an account of problems of the post-structuralists in this regard, see Peter Dews, *The Logics of Disintegration*, London: Verso, 1987, particularly Ch.3.

40. On the philosophy of technology see the anthology *Philosophy and Technology*, ed. Carl Mitcham and Robert Mackey, N.Y.: Free Press, 1983 (which does not include any work of Martin Heidegger or Don Idhe). On the range of theories of power see Stewart R. Clegg in *Frameworks of Power*, London: Sage, 1989.

> *exercises over some set of energy forms or flows*, and, most specifically, over some set of energy forms or flows *that constitute part of the meaningful environment of another actor*.[41]

In later work Adams emphasised that the control that matters most is control over the triggers which begin processes of energy transformation.[42] Control was defined as 'making and carrying out decisions about the exercise of a technology', and technology defined as 'a set of knowledge, skills and materials ... necessary to alter the order (i.e., space-time relations) of some set of energy forms or achieve an energy conversion.'[43]

While having the virtue of identifying the central features of power and what is of central importance in power struggles, this theory takes as unproblematic the existence of forms of energy, and also the operating units as centres of action. Martin Heidegger in his study of technology defined technology as a way of revealing, criticising modern technology for revealing nature, and ultimately people themselves, as nothing but standing reserves, as merely things or forces to be controlled or utilized for controlling something else. Contrasting this with the ancient Greek understanding of technology, he argued that modern technology blinds people to the responsibility of nature for the bringing forth of products.[44] Adams has not entirely freed himself from this perspective despite his conception of humans as themselves part of nature as energetic processes. It is necessary to recognize that nature itself is active, bringing into being the forms and flows of energy, including humans, which together generate the products associated with human agency. But before humans can play a part in this they must be formed through social relations. Only through socialization (itself a transformation of energy) do individuals become effective agents. As Stephen Clegg argued, developing an insight of Foucault:

> ...all forms of agency will be an achievement of control produced by discipline. Consistency, coherence and memory of self as such are not given but learned and accomplished. The agency of a person is no less an achievement of discipline than is that of an organization.[45]

In fact both the ability and the desire to achieve power is engendered by the symbolic constitution of the individual in the context of social institutions. It is within a social context the child develops its own capacity to manipulate the world and to decentre itself from its immediate involvement in the world. In so doing it develops the capacity to recognize the outcome of its actions, to use tools and to create things - while losing the experienced unity with its mother's power. It then becomes aware that its activities, creations and its very being as an entity within the world have a symbolic dimension and are subject to the interpretation and action of others. The child is thereby made aware of its own contingency and the limitations of its power. Others not only threaten the child and limit it physically, particularly its access to what it desires, but have the capacity to reduce it to an instrument of their own projects. However at the same time the child is socialized, trained and educated into an inter-world of shared praxis, of tools and other instruments - knives, hammers, shovels, roads, buildings, weapons, machines, factories, processes of production - together with codes of conduct, social roles, institutions and organizations which constrain and thereby coordinate individual actions and activities, and a shared social imaginary through which all these are defined as such and understood. These enable the child at least to some

---

41. Richard Newbold Adams, *Energy and Structure: A Theory of Social Power*, Austin: University of Texas Press, 1975, p.12.

42. Richard Newbold Adams, *The Eighth Day: Social Evolution as the Self-Organization of Energy*, Austin: University of Texas Press, 1988, p.47ff.

43. Adams, *Energy and Structure*, p.13f.

44. Martin Heidegger, 'The Question Concerning Technology' in *The Question Concerning Technology and Other Essays*, N.Y.: Harper, 1977.

45. Clegg, *Frameworks of Power*, p.188.

degree (as the precondition of its staying alive and later, reproducing) to participate in society's power - its capacity to regulate (and its activity of regulating) in precise ways the transformations of potent energy. The individual in being designated a particular subject gains access to and is able to appropriate the products of this activity - the reordered and accumulated potent energy, particularly food, clothing and shelter, and instruments - associated with these transformations. In this way a possibility is opened up by society, and at the same time an impulsion is generated in the individual, to realize this possibility, of participating in the control not only of the surrounding world for the immediate future, but of the conditions of life. The constant reproduction of the quest by people for such power generates the production, development and transformation of not only means to live and instruments of production, but also technological know-how and organizations which order the interactive processes between people and with nature, thereby maintaining and developing enduring social structures of power.

As in the other dialectical processes, individuals who emerge through their participation in the processes of controlling the world then become active agents in the transformation of structures of power. There is more potential for conflict in the dialectic of power than in other dialectical processes - over who will have access to the means of production, in the relations of power between people, who will do the work and who will get the products and other benefits of organized action, over who will have the opportunity and means to reproduce themselves, over how things will be done, over whose aspirations and goals will be taken most into account in decisions, over which roles people will occupy and especially who will occupy the main positions of power in organizations, over what issues will be raised and considered when decisions are made, over how people will be organized and which power structures will prevail, over what channels of communication will be created, who will control access to these channels, who will define the agenda and who will be granted the means to develop ideas and be granted the authority to define reality. These conflicts spill over into and profoundly affect the other dialectical processes.

The forms of power achievable by individuals or organizations are also radically different. There is a vast difference between being able to use tools or other instruments, having skills in interpersonal relationships, being able to influence the actions of others, having privileges and access to products of consumption, having symbolic power - being able to command respect for one's views, and having political power - being able to participate in the decision-making of the community, of organizations or of the State. Then there are complex power relationships and struggles between different organizations and between individuals and organizations, from primary groups to nation States to transnational companies. These can be very complex. Within individual organizations, even those committed to well-defined goals, there are invariably sub-groupings to some extent in conflict with each other, and within any society there are vast numbers of organizations with varying degrees of stability and permanence, often with overlapping memberships, struggling to maintain themselves and to define and realize their goals.

Through the diversity of power struggles there is at least some impetus towards a general augmentation of the power by people over the conditions of their existence, or at least some aspect of the conditions of their existence. Since to be engaged in such a dialectic is to be committed to achieving power, the forms of power which augment everyone's power will generally meet with less resistance than other forms, and those organizations which develop their power will tend to prevail over those which do not. The dialectic of power is then, like the dialectics of orientation and of recognition, a social phenomenon which must be understood in terms of a relationship between individuals and instruments, organizations and complexes of organizations, which durationally transcend to various degrees the particular activities and lives of individuals, being both the condition and the result of their struggles.

## *The Inter-Relationships Between Dialectical Processes*

While each of these dialectical processes has its roots in the diremption within social relations brought about by the growing decentering of experience, and all dialectical processes are involved in each and every action, expression and creation of each and every person, each dialectic has its own unique dynamics irreducible to the dynamics of the others. Each of these can be seen as a conditional cause of the others, thereby making possible a multiplicity of complex dynamic relationships. This provides a research programme of tracing these interdependencies and their developments, and much of Hegel's *Phenomenology of Spirit* can be understood as undertaking this programme. For instance Hegel's most famous analysis: that of the relationship between Master and Slave in Ancient societies, begins as an account of a struggle for recognition.[46] The Master subdues the Slave, forcing him to recognize and subordinate himself to the will of the Master. In this way the Master should be successful in his struggle for recognition. However in reducing the Slave to a thing and treating him as an instrument, the recognition obtained is deprived of any significance. The Slave on the other hand can see in the Master something to aspire to. But beyond this, the Slave in constant fear of death is shaken from concern with his particular existence to take the point of view of the universal, and at the same time in being forced to work for the Master gains mastery over nature and impresses himself upon it. By creating a standing reflection of himself as a universal being, the Slave becomes such a being and gains self-substantiation in a way which is denied to the Master, whose relationship to nature is mediated by the Slave. Through such analyses, which were augmented by the division between Subjective, Objective and Absolute Spirit, Hegel tried to interpret history as the progressive actualization of the World Spirit in which it struggles, through a series of forms of Objective Spirit, to create the material and social conditions to develop consciousness of itself in art, religion, and finally philosophy - the realm of Absolute Spirit. And in the process of developing this research programme Hegel formulated his ethics and political philosophy.

However the consequence of rejecting Hegel's general Neoplatonic scheme has been that very few thinkers have tried to consider all dialectical processes simultaneously. Most of those examining the relationships between dialectical processes have been concerned with explaining one scheme, and occasionally two, in terms of another. The most thoroughly analysed relationship has been that between the struggle for power and symbolic activity, a relationship examined by first by the Marxists, and then by a diversity of schools in a diversity of countries. For instance the Hegelian Marxist, Lucien Goldmann, examined literature against the background of the dialectic of labour and the class conflicts associated with this.[47] Berger and Luckmann developed a phenomenological approach to analyse the struggle for power as primarily a struggle to define reality in general and situations in particular.[48] Marshall Sahlins examined the dialectical relationship between power relations and symbolic action, drawing on the work of the structuralists in an effort to transcend the tendency towards a reductionism to practical interests by Marxists, ecological anthropologists and Berger and Luckmann, while Abner Cohen examined the same relationship to transcend the reductionism of the structuralists.[49] Foucault's examination of the relationship between power and knowledge is a further example of the study of this relationship. Much of the work of the symbolic interactionists is an attempt to explain conceptions of the world

46. Hegel, *The Phenomenology of Mind*, p.234ff. This is developed further in *Philosophy of Mind*, §433-§435.

47. See Goldmann, *The Hidden God* for a study of the work of Pascal and Racine as expressions of the world-view of members of the *Noblesse de Robe* during the early stages of capitalism.

48. Peter L. Berger and Thomas Luckmann, *The Social Construction of Reality*, [1966] Penguin, Harmondsworth, 1972.

49. Marshall Sahlins, *Culture and Practical Reason*, Chicago: University of Chicago Press, 1976; Abner Cohen, *Two Dimensional Man*, Berkeley and Los Angeles: University of California Press, 1976.

in terms of the struggle for recognition,[50] anthropologists have analysed the struggle for power as a means to gaining honour and Rom Harré has argued that in advanced capitalist societies, the struggle for respect is irreducible to practical concerns.[51] Pierre Bourdieu's research program is essentially a reductionist study of cultural activity in terms of the dialectic of power mediated through the dialectic of recognition, but he then provides a place for the dialectic of orientation as an emergent through his concept of the cultural field. It is rare for social theorists involved in such studies to acknowledge that the different dialectical processes have their own autonomy, and very rarely do they recognize more than two semi-autonomous dialectical patterns.

However it is not the limitations of social theory following the breakdown of Hegel's system which are most important, but the achievements which in one way or another transcend Hegel. Two traditions which originated in Hegel have gone beyond his achievements. The first is the Marxist tradition and the second the existentialist. In advancing beyond Hegel, these traditions have to a considerable extent contracted their field of comprehension and lost some coherence in doing so.

## *Marxism and Emergent Social Dynamics*

As was pointed out in Chapter 9, Marx was not an entirely consistent thinker. He mediated between different traditions and never managed to formulate his ideas in terms of a coherent conception of being. In considering Marx as an advance on Hegel it is those aspects of Marx consistent with a process view of the world which I am concerned to defend. But to defend these aspects of Marx's thought it is necessary to unravel the incompatible strands in his work and to show which are the most significant ideas. To begin with, Marx belonged to the Young Hegelian movement which reformulated Hegel's system to unleash its critical potential. The Neoplatonism of this early phase was partially transcended by adopting the economic reductionism of the Scottish school of historians, and then both these positions were transcended in Marx's most important achievement, his analysis of capitalism. But what was involved in this study of capitalism? What is always taken to be Marx's central thesis is that in some sense or other the economy is basic to understanding society. But corresponding in part to the confusion of ontologies underlying Marx's work, he formulated this thesis in three distinct ways, two of which are blatantly incompatible.

The first way in which the economy is held to be basic is in the sense that the labour process is 'the necessary condition for effecting exchange of matter between man and Nature; it is the everlasting Nature-imposed condition of human existence, and therefore is independent of every social phase of that existence, or rather, is common to every such phase.'[52] As such the productive process is the metabolism of society, and as with the study of organisms, everything else must be understood in relation to it. The second sense in which the economy is basic is clearly distinguished from the first and pertains fully only to capitalism. As Marx wrote:

> It is not the unity of living and active humanity with the natural, inorganic conditions of their metabolic exchange with nature, and hence their appropriation of nature, which requires explanation or is the result of historic process, but rather the separation between these inorganic conditions of human existence and this active existence, a separation which is completely posited only in the relation of wage labour and capital.[53]

---

50. See papers by Blumer and Shibutani in the anthology edited by Arnold M. Rose, *Human Behaviour and Social Processes*, Boston: Houghton Mifflin, 1962.

51. Rom Harré, *Social Being*, Oxford: Blackwell, 1979.

52. Karl Marx *Capital* Vol.1, tr. Samuel Moore and Edward Aveling [1887], Moscow: Progress Publishers, 1974, p.179.

53. Karl Marx *Grundrisse*, tr. Martin Nicolaus, Harmondsworth: Penguin, 1973, p.489.

This is the sense in which the market is seen to have developed as an emergent process to transform people, reproducing the relations of production which reduces people to labour power to be bought and sold as a commodity, and then to have developed according to its own immanent dynamics to dominate the whole of society, and ultimately, of the world. The third sense is a form of technological determinism. As Marx argued in *The Poverty of Philosophy*: 'In acquiring new productive forces men change their mode of production; and in changing their mode of production, in changing the way of earning their living, they change all their social relations. The hand-mill gives you society with the feudal lord; the steam-mill, society with the industrial capitalist.'[54]

This third sense of the primacy of the economy, the technological determinist sense, the sense which is generally taken as synonymous with Marxism, provides a general scheme of history which is designed to explain both developments in different types of society and the movement from one type of society to another.[55] It is the standard reductionism deriving from the Scottish historians and is ultimately rooted in mechanistic materialism. It presupposes that the egoistic conception of humans deriving from Hobbes is valid for all societies. But if all history could be explained so simply, then there could be no emergent dynamics. There would be no object, no autonomous dynamics of capitalism to be explained by Marx in his major work: *Capital*. Furthermore this reductionism is inconsistent with the conception of humans as creative social beings which underlies Marx's critique of capitalism and which is required to justify any optimism about the future. Thus the third sense in which the economy is held to be primary is incompatible with the second sense, which is the central theme of Marx's work, and therefore must be rejected by anyone who accepts Marx's central arguments, quite apart from all the empirical evidence against it. This leaves the first and the second sense to be considered, each of which fully accords with the process view of humanity as an emergent process within nature and of society itself as consisting of emergent processes.

The first sense in which the economy is held to be primary does not contradict Hegel's philosophy. Marx's position in this regard can be seen as a development within the framework of the Hegelian system which underplays the dialectic of recognition and the dialectic of representation in favour of the dialectic of labour (and is associated with efforts to explain the dialectic of representation reductionistically in terms of the dialectic of labour). It is the second sense of the primacy of the economic in which Marx transcends the framework of Hegel's analysis because it implies that capitalism is developing according to laws transcending the dialectical rationality of human becoming. These laws describe the tendencies of a process which emerges from and then constrains the dialectical processes. Although, as was pointed out in Chapter 9, there is a dialectic of economic categories presupposed in the development of capitalism which is associated with the development of contradictions in the economic system, the laws of capitalist development are more like the laws of the physical world than dialectical patterns of becoming. The dynamics of the economy confront people as a second nature, and the tendencies described by these laws could just as well lead to the destruction of humanity as to the realization of humanity's highest potentialities. It was merely a contingent fact that the tendencies in capitalism at the time in which Marx was writing were producing the conditions which could have facilitated the creation of a new form of society in which people's alienation from their creative powers and from society could have been overcome. The fact of the emergence of a process within and transcending the dialectical patterns of becoming of society makes it impossible to accept the teleological view of history of Hegel since, as was suggested in a previous chapter, emergence implies a genuinely open future - since what emerges is not completely determined by the conditions of its emergence.

But if there is one emergent process within culture, there is no reason why there cannot be others. This is one of Georg Simmel's central insights:

---

54. Karl Marx, *The Poverty of Philosophy*, [1947] Moscow: Progress Publishers, 1973, p.95.

55. For a relatively rigorous defence of this vulgar Marxism see G.A. Cohen, *Karl Marx's Theory of History - A Defence*, Oxford: Clarendon, 1978.

> Whenever life progresses beyond the animal level of culture, an internal contradiction appears... We speak of culture whenever life produces certain forms in which it expresses and realizes itself... But although these forms arise out of the life process, because of their unique constellation they do not share the restless rhythm of life, its ascent and descent, its constant renewal, its incessant divisions and reunifications... They acquire fixed identities, a logic and lawfulness of their own; this new rigidity inevitably places them at a distance from the spiritual dynamic which created them and which makes them independent... This characteristic of cultural processes was first noted in economic change.[56]

Simmel's research programme involved identifying and analysing the nature, generation and reproduction of these forms. William Mc'Neill's analysis of the emergence of 'microparasitism' and 'macroparasitism', Lewis Mumford's analysis of the emergence and dynamics of cities, Robert Michels' analysis of the iron law of oligarchy in political parties, Michel Foucault's identification of emergent discursive formations: the asylum, the clinic, the prison and so on, Bourdieu's analysis of the dynamics of cultural fields, the work of various Marxists who have identified and revealed emergent tendencies in both late capitalist and non-capitalist socio-economic formations, the work of Wallerstein and his colleagues in describing the concentration of economic and political power and the differentiation of the world-system of capitalism into cores, semiperipheries and peripheries, and the accounts of Flannery, Rapaport and Bunker of the tendency of dominant social systems to 'hypercoherence', to increase control, to use up more and more available energy, until a stage is reached where they have so much power that they can survive while contributing little or nothing to the systems on which they are dependent - until they destroy these systems, the conditions of their own existence,[57] can all be interpreted as studies of emergent social forms or processes in accordance with this research programme. And by so interpreting these analyses and their theoretical objects it becomes possible to overcome difficulties within these analyses and to show their relevance to each other. For instance it is possible to account for the identity of discursive formations over time - something which was a major problem for Foucault, to represent the differentiation of the world-system as only a tendency of one process among others - thereby allowing for the vast variety of responses to the expansion of capitalism by different regions, and to allow for greater complexity in the economy itself than Marx or all but a few of his followers have considered - allowing for the partial autonomy of and interaction between local, national and international economies, for the emergence of new semi-autonomous forms of State and non-State institutions associated with modern capitalism, and so on. All these emergent processes can then be evaluated according to their effects on other processes.

To explain such emergent processes it is necessary to refer back to the three dialectical processes as conditional causes. Ultimately it is because the world exists in a state of far from thermodynamic equilibrium, and because this has given rise to a world ecosystem which maintains the conditions for human life that complex social structures have been able to form, and all emergent social processes are dissipative structures reproducing themselves by maintaining, and being able to maintain, a flow-through of useful energy and materials. However it is because such emergent processes provide and reproduce the conditions for at least a large number of people to orient themselves, to gain a sense of their own significance and to gain some control over their lives that people accept and conform to the constraints imposed by these emergent processes. Such conformity can be reinforced by the differentiation of people's situations within these processes.

56. Georg Simmel, 'The Conflict in Modern Culture', in Donald N. Levine ed. *George Simmel on Individual and Social Forms*, Chicago: University of Chicago Press, 1971, pp.375-393, p.375f.

57. Kent V. Flannery, 'The Cultural Evolution of Civilizations', *Annual Review of Ecology and Systematics*, Vol.3, 1972, pp.399-426 based on the work of Rapaport. This notion has been further developed by Stephen Bunker in *Underdeveloping the Amazon* and used to characterize the core zones of Western capitalism. It could be equally applied to the Soviet bureaucracy.

For instance in capitalism, the bourgeoisie are provided with the best means to orient themselves, to gain respect and to control the conditions of their existence, but must maximize the profits of their enterprises to avoid declining into the proletariat, who in turn must work hard to avoid ending up in the reserve army of unemployed. Where some people refuse to conform, there are always others lower down striving to move up in society willing to conform to and defend the system in their place. But emergent processes are not entirely explicable in terms of their material and environmental causes. They must be to some extent recognized as immanent causes irreducible to the conditions of their emergence, and they must to some extent be explained in their own terms, as Marx attempted to explain capitalism as an emergent, self-reproducing ensemble of social relations based on the universalization of the commodity form to produce and reproduce capital and wage-labour.

The conception of society in which a number of semi-autonomous processes are recognized leads to the problem of understanding the relationship between these diverse processes, which in turn has led to the study of different spatialities and temporalities associated with these processes.[58] Most notable in this regard have been the historians of the Annales school who have emphasised the distinction between, as Braudel described it, 'the conspicuous history which holds our attention by its continued and dramatic changes - and that other, submerged history, almost silent and always discrete, virtually unsuspected either by its observers or its participants, which is little touched by the obstinate erosion of time.'[59]

Althusser in his effort to transcend the limitations of Hegelian Marxism also acknowledged these different temporalities in history, writing: 'As a first approximation, we can argue from the specific structure of the Marxist whole that it is no longer possible to think the process of the development of the different levels of the whole *in the same historical time*... On the contrary, we have to assign to each level a *peculiar time*, relatively autonomous and hence relatively independent, even in its dependence, of the "times" of the other levels.'[60] But in his proposed scheme for examining society Althusser simply accepted the traditional scheme of orthodox Marxism, which may have been valid when applied to nineteenth century capitalism, as timelessly valid for all forms of society. He went on to argue that: 'we can and must say: for each mode of production there is a peculiar time and history, punctuated in specific way by the development of the productive forces; the relations of production have their peculiar time and history punctuated in a specific way; the political superstructure has its own history...; philosophy has its own time and history...; scientific formations have their own time and history, etc...'[61] But it makes virtually no sense to distinguish between mode of production, relations of production and political superstructure in, for instance, the Kabyle studied by Bourdieu, and it is highly suspect in late twentieth century capitalist societies where political organization and relations of production are so intimately involved in much of the advanced forms of production.[62] And the distinction between philosophy and science is of recent origin. There was nothing like it in the seventeenth century.

When fully developed, the possibility of emergence of processes within social dynamics must lead to a rejection of such preconceptions about the differentiations within society. The specific nature of such differentiations themselves have histories which must be examined in each society, showing the relationships between each semi-autonomous process at different levels, from small groups to the dynamics of civilizations. As Braudel wrote: 'History accepts and discovers

---

58. The importance of spatio-temporal order has long been recognized in geography and architecture. Recently Anthony Giddens has developed ideas on this subject. See Anthony Giddens, *The Constitution of Society*, Cambridge: Polity Press, 1984, Ch.3.

59. F. Braudel, *The Mediterranean and the Mediterranean World in the Age of Philip II*, (2 volumes) Vol. 1. tr. S. Reynolds, London: Collins, 1972, p.16.

60. Louis Althusser and Etienne Balibar, *Reading Capital* [1968] tr. Ben Brewster, 2nd ed., London: N.L.B., 1977, p.99f.

61. Ibid. p.99f.

62. As Boris Frankel has argued in *Beyond the State?* London: Macmillan, 1983, Ch.'s 1,2 & 3.

multidimensional explanations, reaching as it were, vertically from one temporal plane to another. And on every plane there are also horizontal relations and connections.'[63] And elaborating on this elsewhere: 'Some structures, because of their long life, become stable elements for an infinite number of generations: They get in the way of history, hinder its flow, and in hindering it shape it. Others wear themselves out more quickly. But all of them provide both support and hindrance. As hindrances they stand as limits ("envelopes," in the mathematical sense) beyond which man and his experiences cannot go.'[64]

Elsewhere he recognized a multiplicity of spatial orders inter-related with such temporal orders.[65] Thus a society must be understood more as an ecosystem of processes (and the structures maintained by them) with analogous relations to those revealed in ecology by ecologists such as Levins and Lewontin. Such processes incorporate ways of conceiving the world in terms of which people define themselves and act purposefully, frequently develop according to dynamics which transcend and constrain the dialectical processes, and at the same time are processes within nature and must be understood in relation to geographical and ecological conditions of humanity. These processes are often in conflict with each other, and such conflict can eventually lead to the destruction of one process by another which is dependent upon it for its very existence. The concepts of conditional and immanent causation provide a means to understand and clarify such a multiplicity of relationships of partial dependence and autonomy, and often partial conflict, between the different human processes and between these and other natural processes; and also what a spatio-temporal order is (an order of potentialities for coordinated interaction such that this facilitates and is constrained to maintain these potentialities), how different processes generate different spatio-temporal orders, and the significance of this for understanding the inter-relationships between processes.

## *Existentialism and the Individual*

The development of the notion of emergent processes presents the problem of what is the relationship between these emergent processes and the underlying dialectical struggles of and between people. The fact of emergence, by undermining the notion of history as a teleological unfolding of an inner essence, whether of the World Spirit or of humanity, suggests a different conception of the place of the individual in the world than that implied by Hegelian or Hegelian Marxist thought. Individuals can no longer be reduced to vehicles of this unfolding moved by the 'cunning of reason.' Neither can they be reduced to cyphers of social structures constructed by a process of 'interpellation' as Althusser and his followers (including Foucault in this regard) have represented them. Individuals can be seen as emergent processes from, and within, nature, culture and society, and as participants in the process of becoming of the world. Marx's critique of Hegel is intimately related to the existentialist critique, and each can be seen to be compatible with the other when interpreted from the perspective of process philosophy.

The existentialist critique of Hegel began with Schelling's later philosophy and was further articulated by Kierkegaard, who had attended Schelling's lectures (along with Engels, Burckhardt and Bakunin, among others) in 1841. Kierkegaard was troubled by how in Hegel's system 'the existing subjectivity tends more and more to evaporate.'[66] Consequently he focussed on the

63. Braudel, *The Mediterranean and the Mediterranean World in the Age of Philip II*, p.16.

64. Fernand Braudel, 'History and the Social Sciences' in *On History*, tr. Sarah Matthews, Chicago: Chicago University Press, 1980, pp.25-54, p.31.

65. Fernand Braudel, *Civilization and Capitalism 15th-18th Century*, (3 volumes) Vol. III, *The Perspective of the World*, tr. Siân Reynolds, London: Fontana, 1985, I, 'Divisions of Space and Time in Europe', pp.21-88.

66. *Kierkegaard's Concluding Unscientific Postscript*, tr. David F. Swenson and Walter Lowrie, Princeton: Princeton University Press, 1968, p.112.

individual as a contingent subject perpetually becoming, with all the uncertainty and anxiety entailed by this. Rejecting Hegel's faith that the finitude of existence could be transcended by taking the perspective of the Absolute, that philosophy could escape 'from the weary strife of passions that agitate the surface of society into the calm region of contemplation...'[67] Kierkegaard argued:

> The principle that the existing subjective thinker is constantly occupied in striving, does not mean that he has, in the finite sense, a goal toward which he strives, and that he would be finished when he had reached this goal. No, he strives infinitely, is constantly in the process of becoming.[68]

Similar sentiments were expressed by Nietzsche.

While Kierkegaard was responding to Hegel, his abstraction of the individual subject from the world reflects the underlying dominance of the mechanistic world-view.[69] In terms of process philosophy, the individual as a process of becoming is intelligible as an emergent process within the world. Through participation by the sensitive organism in the dialectical processes of becoming and the various semi-autonomous processes of society, the organism is individuated as a subject, and this individuation consists in the emergence of the capacity, inherent in the nature of the different dialectical processes, to reflect on the conditions of its existence, to take responsibility for its conception of the world, to choose which others to regard as significant, and to strive to live life accordingly, modifying or transforming relationships of power in the process. That is, the individual has the capacity (cultivated in some societies, suppressed in others) to develop a mind.[70] The mind is not a substance. To make up one's mind is to interpret one's situation, to commit oneself to projects accordingly, and to have gained sufficient self-mastery to persist against obstacles in the effort to realise these projects. To have a mind of one's own is to have developed one's understanding, to have established one's convictions about the nature of the world and oneself and what projects are worth striving to realize, and to be able to formulate effective projects of action in accordance with these convictions and to be prepared to act accordingly. 'Mind' so conceived, is in accordance with common usage and the ontology of process philosophy, a structure, the potential to order activity in a way which cannot be entirely understood from the physical, biological, cultural and social conditions of one's existence since it involves new constraints on activity not present in these conditions. Freedom as the potential for self-determination is a function both of the development of mind and the nature of the individual's situation, and there is no guarantee that it will be achieved. Children are born in chains, and the challenge of life is liberation; but this liberation is always socially, biologically and physically situated.

The nature of human consciousness has been examined most systematically by the phenomenologists and those influenced by them, including the poststructuralists and 'hermeneuticists' (although their work builds upon the work earlier philosophers such as Dilthey, Bergson and William James). The founder of the phenomenological movement, Edmund Husserl, was concerned to transcend both naturalistic reductionism and relativism by developing philosophy into a rigourous science. This was to be devoted to obtaining apodictic knowledge by applying a presuppositionless method to examine and describe lived experience. This science was to be more

---

67. G.W.F. Hegel, *The Philosophy of History*, tr. J. Sibree [1899] New York: Dover, 1956, p.457.

68. *Kierkegaard's Concluding Unscientific Postscript*, p.84.

69. This is argued by Karl Löwith in 'Man Between Infinities', in *Measure: A Critical Journal*, Vol.1, 1950, pp.297-310. See also Hans Jonas, 'Gnosticism, Existentialism, and Nihilism', *The Phenomenon of Life*, Chicago: Chicago University Press, 1966.

70. One of the most disorienting aspects of the modern Western world is the abandonment of a culture which promoted the development of mind and its replacement by a culture which severely suppresses its development.

fundamental than the natural sciences, and to reveal the natural sciences as just one creation of the Spirit among others. In this project, Husserl failed. But in doing so he transcended mechanistic categories and developed a set of concepts based on a view of human consciousness as intentional (as always consciousness of something) and as temporal, as a process of becoming inseparable from its world which is constituted by it and which is transformed as part of its own development. These concepts, and the research program they engendered, enabled the existentialists to examine in a systematic way themes which had only been touched on fragmentarily by Kierkegaard and Nietzsche.

In developing this research programme, phenomenologists have investigated the lived experience of being in the life-world - that is, the *umwelt* (the surrounding world), the *mitwelt* (the world shared with other people), and the *eigenwelt* (the 'self-world'), and in terms of these concepts, have examined what it means to be embodied, the spatiality and temporality of being-in-the-world, what is involved in being with, confronting and forming relations with other people, the experience of meaning in the world and of the associated claims of the world upon one, the nature of acting and being engaged in action, both as an individual and with other people, and the nature of emotions, imagination and self-deception.[71] They have analysed the structures of the socially created, 'inter-world', the world of physical constructions - buildings, roads, furniture, instruments, works of art, and so on, of the meanings sedimented in these creations, of the typified expectations and responses of people, of designated roles and statuses, of institutions, rules, regulations and laws, that is, the world within which people are habitually engaged; and they have examined the complex spatio-temporal organization of this life-world and its impact on individuals. These analyses have facilitated the study of the contradictions in the social world and the experience of alienation, the study of social commitments, joint praxis, the formation of groups and revolutionary movements and the crystallization of institutions.[72] Ideas developed in such investigations have been further elaborated in psychology, psychiatry and the social sciences.[73] Such work has produced a notion of humanity as essentially creative, characterized not so much by the ability to produce a culture but by the ability to transcend old cultural forms. As Merleau-Ponty argued, 'What defines man is not the capacity to create a second nature - economic, social or cultural - beyond biological nature; it is rather the capacity of going beyond created structures in order to create others.'[74]

A deeper understanding of this creativity has been achieved by Paul Ricoeur through his work on narrative. For Ricoeur, narrative is the fundamental structure of the experience of time; its ultimate referent is lived time. There are three dimensions, or forms of mimesis, in narrative. Firstly life itself is an inchoate narrative. It is for this reason that we have a pre-understanding of what human action is, of its semantics, its symbolism, its temporality. The second aspect involves the representation of action according to specific rules of emplotment, that is, the making of a structure to interpret and organize this pre-understanding. Through the activity of emplotment a quasi-world of action and characters is generated. Innovations are made by inventing plots by

---

71. Maurice Merleau-Ponty's *Phenomenology of Perception*, [1946] tr. Colin Smith, London: Routledge & Kegan Paul, 1962 is one of the most profound of these works, but there are many others, including those of Sartre, Ingarden, Schutz, Strasser, Ricoeur, Natanson, Laing and Esterson.

72. See Jean Paul Sartre, *Critique of Dialectical Reason*, [1960] tr. Alan Sheridan-Smith, London: New Left Books, 1976. For a lucid account of this see R.D. Laing and D.G. Cooper, *Reason and Violence: A Decade of Sartre's Philosophy*, London: Tavistock Publications, 2nd ed. 1971.

73. See for example Rollo May et. al. ed.'s, *Existence: A New Dimension in Psychiatry and Psychology*, N.Y., Basic Books, 1958 for a representative selection of existential phenomenological psychiatry. For a review of such psychiatry see Herbert Spiegelberg, *Phenomenology in Psychology and Psychiatry*, Evanston: N.W. Uni. Press, 1972. For the whole range of the human sciences see Maurice Natanson ed. *Phenomenology and the Social Sciences* 2 volumes, Evanston: Northwestern University Press, 1973.

74. Maurice Merleau-Ponty, *The Structure of Behaviour* [1942] tr. Alden Fisher, Boston: Beacon Press, 1967, p.175.

means of which 'goals, causes, and chance are brought together within the temporal unity of whole and complete action.'[75] A complete action can consist of a number of other actions, and it can be the action of an individual - from some particular achievement to having lived a whole life, or of a group, such as winning a war, founding a nation or establishing or destroying a civilization. The third aspect is the reception and actualization of that structure. People are confronted with and drawn into the quasi-world, distancing them from their own life-worlds, revealing and challenging their taken for granted horizons of expectations. They are provided with room to manoeuvre, to think about the way they live and to appropriate the new structure to organize their own actions and lives.

Process philosophy (which through the indirect and direct influence of Bergson was one of the most important starting points for the development of phenomenology, and also the hermeneutics of Ricoeur) provides a naturalistic and physicalist justification for, and interpretation of, the concepts developed by the existential phenomenologists and hermeneuticists, and reunites these ideas with the natural sciences and the human sciences to conceive humans as conscious participants in the process of becoming of nature and society, simultaneously obviating the problems in both Anglo-American and French philosophy of mind.[76] Accordingly process philosophy justifies in a naturalistic way the existential philosophy expounded by Merleau-Ponty:

> As its name suggests, existential philosophy consists of taking as one's theme not only knowledge or consciousness understood as an activity which autonomously posits immanent and transparent objects but also existence, i.e., an activity given to itself in a natural and historical situation and as incapable of abstracting itself from that situation as it is of reducing itself to it. Knowledge finds itself put back into the totality of human praxis, as it were, given ballast by it. The 'subject' is no longer just the epistemological subject but is the human subject who, by means of a continual dialectic, thinks in terms of his situation, forms his categories in contact with his experience, and modifies this situation and this experience by the meaning he discovers in them. In particular this subject is no longer alone, is no longer consciousness in general or pure being for itself. He is in the midst of other consciousnesses which likewise have a situation; he is for others, and because of this he undergoes an objectivation and becomes generic subject... Man no longer appears as a product of his environment or an absolute legislator but emerges as a product-producer, the locus where necessity can turn into concrete liberty.[77]

However while recognizing both that humans are part of nature and that they have very distinct qualities which make them significant beings for the world as a whole, humans cannot be represented as the end product of evolution. As was pointed out in the previous chapter, the concept of evolution is itself problematic, and cannot be conceived of as a process of development to higher and higher levels. Evolution involves the development of ecosystems, ranging in size from those associated with microscopic environments to the world as a whole, consisting of from a few to vast diversities of species, many of which play essential roles in maintaining these ecosystems. Such developments frequently lead to dead ends, catastrophes and reversals in the

75. Paul Ricoeur, *Time and Narrative*, (3 volumes), Chicago: Uni. of Chicago Press, Vol.I, ix.

76. While Anglo-American philosophy of mind generally takes mechanical nature as the point of departure and then struggles to find a place for the conscious subject, French philosophy of mind has traditionally taken the unitary ego as the point of departure and then criticised this, trying to find a place for desire, the unconscious, and occasionally nature. For an approach which does formulate the philosophy of mind in terms of a theory of being which allows for real emergence, which takes the individual as emerging from both nature and social relations, and which critically examines both French and Anglo-American philosophies, see Rom Harré, *Personal Being*, Oxford: Blackwell, 1983.

77. Maurice Merleau-Ponty, 'Marxism and Philosophy' in John O'Neill ed. *Phenomenology, Language and Sociology: Selected Essays of Maurice Merleau-Ponty*, London: Heinemann, 1974, pp.174-185, p.182f.

fortunes of different life forms. The average life span of each species in this process is about three million years, and there is no reason why humanity should not be eliminated in due course. Many species become extinct because they destroy the environmental conditions of their existence. The only unique characteristic of humanity is that people are capable of understanding and changing the processes through which they are destroying the conditions of their existence. Unlike other species, the extinction of humans will be their own responsibility.

indwelling

# 16

# ETHICS, POLITICAL PHILOSOPHY AND THE SOCIAL SCIENCES

In the early chapters of this work it was shown how it is impossible to even think clearly about environmental problems from within the framework of concepts prevailing in Western societies. It was shown that Marxism does provide a framework for analysing the cause of environmental destruction (despite many of Marx's own views), that Marxists are correct to identify the immanent dynamics of world capitalism as the immediate source of most of the world's present environmental problems; but in practice, the failures of orthodox Marxists have revealed the extent to which Marx failed to fully transcend the forms of thinking of Western civilization, and of capitalist society in particular. Neoplatonic and mechanistic themes within Marxism have negated much of its liberating potential. Something more is required. Process philosophy provides such a new starting point - for understanding the world, for judging the significance of life, for deciding how to live and how to act, for evaluating and creating institutions and for working out political goals and strategies.

According to this philosophy, human subjects are socio-cultural beings, part of and within the world, some of the beings through which the world has attained and is attaining consciousness of itself. The goal of enquiry is understanding, an 'indwelling' in the world such that the world becomes intelligible. The importance of abstract forms of thinking, the development of which has been a major achievement of the culture of Western civilization, is recognized; but the nihilistic effects of ignoring the level of abstraction involved and taking abstractions for reality, the 'fallacy of misplaced concreteness', are avoided by reconceiving what it involves. Rather than being seen as a transcendence of the changing sensible world to arrive at knowledge of what is eternal - whether of forms, of the laws of nature, or of facts and logical relations, abstraction can be seen as part of the process of creating the means for deeper understanding of the world. There is no reason why understanding so conceived should not lead to an appreciation of the world's significance, and to an appreciation of the relative significance of its different constituents.[1] And where the primary focus is on the becoming of all that is, it is impossible to understand beings without appreciating their intrinsic value. From the 'universe of death', as Coleridge described the world of mechanistic science, a science based on process philosophy will lead closer to the way the world was experienced by Wordsworth when he wrote:

> ... all
> That I beheld respired with inward meaning.

The framework of mechanistic concepts has not only been effective as a means to understand the world. The metaphor of mechanism has also provided the ideal for people to conform to, despite the dogma of the disjunction between questions of how the world is and how the world ought to be. In arguing for a process view of the world based on an auditory analogy, it is not only being argued that this is the best metaphor to make the world intelligible; it is being argued that this should serve as the root metaphor of the ideals for people, society and humanity to strive for. Music should replace the machine as the dominant thematic motif of civilization. However with this new metaphor there is no longer any reason for regarding its dual role as the basis for interpretation and the basis for evaluation as inconsistent. Advancing understanding is itself

1. On this, see Michael Polanyi and Harry Prosch, *Meaning*, Chicago: University of Chicago Press, 1975.

participation in the creative becoming of the world, while the way the world is understood orients people for action in relation to this becoming. This process involves the development of concepts which then mediate people's interactions with each other and with the rest of nature, and is thereby a major aspect of their self-creation. Human societies are seen as processes of becoming within nature, and individuals are seen as becoming autonomous selves through their participation in the cultural dynamics of their societies. Individuals emerge as capable of critically reflecting on and thereby developing their cultural heritage and of acting according to their subsequent convictions; and like melodies in a symphony, the contribution people make to society, to humanity and to nature remains a part of these even after they have ceased to exist as active individuals. With each thought and action people are creating themselves, their community and the world; and the lives they lead are an indelible contribution to the becoming of the world.

The version of process philosophy proposed here is not being presented as the eternal truth, but as the means for the fullest comprehension of the world of the present age, of its achievements, problems and limitations and of the possibilities open to it. It is presented as itself historically situated, as a contribution to an on-going dialogue, providing a provisional orientation to the world which must continually be tested, both as the basis for extending our understanding of the world and as the basis for action, and which at least in its present form will itself be transcended in the future. The basic scheme of a philosophy of process has been outlined in previous chapters, and it has been shown how it is required to overcome the fragmented nature of modern science and how it provides the basis for a new conception of life and humanity. In the final two chapters this scheme will be articulated, showing how it can provide the foundations for a new ethics, political philosophy and science of humanity, an orientation for living, for social, political and economic action, for a world-wide environmental movement, and ultimately, for a new, post-European, post-nihilistic world civilization. It provides a basis for articulating the aspirations of people able to contribute to the achievement of this new world order, affirming the most important ideals of Western societies and of the tradition of Marxism: the heroic moralism and the unfettered search for truth of the West and the quest for a just social order within which people will be able to reappropriate their creative powers, the basic ideal of Marxism, while at the same time undercutting the opposing tendencies of both, the tendencies toward domination, purely instrumentalist thinking and nihilism which have been generated by the pursuit of these ideals. By facilitating this, process philosophy provides a starting point for confronting environmental problems. However this will require the transformation of these ideals, the way culture is divided into its different realms of discourse, and the meaning of many of the most significant terms in common discourse.

Some idea of what a world founded on process philosophy would be like can be gained from an existing society in which people already conceive the world as a process of becoming, the Fipa of Tanzania.

## *The Fipa of Tanzania*

The Fipa worked out in practice the implications of a process conception of being for life and have embodied this as a habitus. Although they are a relatively small society, they provide an image of a real alternative to the prevailing forms of human life. Their achievements are sufficient to reveal the potential for humanity if such a world-orientation were to be adopted.

Based on an underlying metaphor of the struggle to control a python, the Fipa see the universe as a multidimensional structure bound together through the common theme of a unitary process of the inner darkness of the non-intellectual self and the outer darkness of wild nature being changed by being brought into relation with Fipa humanity, which is itself changed in this never ending process. In this, the development of the individual and human society are seen as interdependent aspects of a single process, central to which is the development of understanding through communication. As the ethnographer of the Fipa, Roy Willis, wrote: 'In speech the self emerges as originator and constructor - of meaning. Which is to say that in the process of verbal communication the human individual achieves self-definition. In the act of giving which is the speech-performance, the giver also receives - of himself.'[2] Through this speech there is a continuous expansion of common understanding which unites humanity. As Willis observed:

> The Fipa intuition of the world and human nature as essentially process... has the consequence that the intellectual picture of the universe is always provisional... Instead of the maintenance and extension of social distinctions and cognitive categories, we find Fipa constantly seeking to subsume existing discriminations and categories within more inclusive and fundamental concepts. The constant expansion of intellectual apprehension into the opaque areas without human society and within the human individual tends to unify the individual and collective experience and transcend differentiating characteristics of human beings and external nature.[3]

To maximise the potential for this communication, the Fipa have organized their villages in concentrated, but formally unstructured settlements which increases physical proximity between people.

However it is not only through speech that the individual achieves self-definition. It is also achieved through the work by which nature is continually in the process of being domesticated. The inspiration to work is neither simply self-interest nor moral obligation, since Fipa see themselves as participants in a community of reciprocal interests. This view of things has produced a strong work ethic, but it is very different from the work ethic developed in Western Europe. Willis contrasted the two:

> Calvinism partakes of the dualism inherent in Western culture in opposing its ultimate value, the spiritual salvation of the individual, to the individual's social action in the world, which is seen as a means to this ultimate end. This dualism, is, however, transformed by historical development into its opposite, in which a dominant rational materialism encroaches into a diminishing area of human 'spiritual autonomy'. In contrast the monistic Fipa world view sees the development of the individual and human society as interdependent aspects of a single life process; there is thus no possibility of a structural transformation of the Fipa world view towards a domination of human beings by reified abstractions, such as Western man has notoriously suffered. Instead we see, in the nineteenth century apogee of Fipa culture, peace and industry in association not with a grim-faced Puritanism but with a vivacious and sociable populace.[4]

While this work ethic involves a striving for control over life, the aim is not to subjugate the world. The idea of reducing nature to a mechanical order is totally alien to the Fipa. The process of domestication of the world is an unending one. As Willis pointed out: 'The python image represents an immortal antagonist without and within; it also appears as a giver and creator of life... meaning emerges endlessly from the process of interaction between the known and the unknown, intellect and force, familiar and strange.'[5] The Fipa have never believed they could transcend this becoming. As one of their sayings points out, while you are making your clothes, the clothes you are wearing are wearing out.

The Fipa notion of human community as in the process of becoming has led to a refusal to make blanket judgements about foreign ethnic groups or to judge individuals by the external marks of

---

2. Roy Willis, *Man and Beast*, Frogmore, Paladin, 1975, p.89.

3. Ibid. p.123.

4. Ibid. p.127.

5. Ibid. p.124.

ethnic identity, and by the recognition of strangers as potential contributors to the on-going dialogue by which community is formed. On the other hand when they have been attacked, the Fipa have responded courageously. In the nineteenth century they were sandwiched between two expansionist African imperialisms, Bemba and Nyamwezi. They willingly made the sacrifices necessary for self-defence, but did not develop a chauvinistic hostility to their opponents. They combined 'the maintenance of territorial security through a strong military force with a consistently non-aggressive foreign policy.'[6]

The explanation for the extraordinary qualities of the Fipa lies in their way of conceiving themselves. Willis considered the possibility that Fipa society could be explained in terms of environmental conditions, but pointed out that the Nyamwezi who live in a similar physical environment and have a similar millet based economic system have a different social organization. He concluded: 'Our analysis leads us to suppose that these facts reflect basic values projected by the structure of Fipa cosmology, rather than any innate ethical superiority in Fipa humanity.'[7]

## *From Instrumental to Creative Rationality*

To begin the reorientation involved in conceiving humanity as a creative participant in the becoming of the world it is necessary to reconceive the nature of human action. The concepts in terms of which people have come to define themselves are such as to make it difficult to conceive of effective action which is not based on domination of nature and of people. Thus Habermas argued in opposition to Marcuse's proposal for a non-oppressive science and technology: 'The idea of a New Science will not stand up to logical scrutiny any more than that of a New Technology, if indeed science is to retain the meaning of modern science inherently oriented to possible technical control. For this function, as for scientific-technical progress in general, there is no more "humane" substitute.'[8] Process philosophy has provided the basis for a new science. It will now be shown how it can provide the basis for a new conception of action and technology.

People comprehend the world and define their situations by means of concepts. Most of these concepts are simultaneously evaluative and descriptive. 'Yellow' is unusual in being merely descriptive, while 'good' is very unusual in being purely evaluative. Concepts, such as 'chair', 'table', and 'boat', evaluate as they describe.[9] To refer to something as a chair, for instance, is to designate it as something good to sit on. There are also evaluative concepts which define people, their relationships and their actions or activities, and the basic structure of the ethical process through which people accord and are accorded recognition, are respected or disdained, is an order of such concepts. For instance the concept of 'ship's captain' is not only linked to other concepts (such as 'ship', 'shipping company', 'crew', 'cargo', 'passengers'), facilitating the achievement of a common orientation, the coordination of action and the creation and sustaining of organizations and institutions by defining ends to be achieved and revealing how to achieve these, but also implies status and what actions are appropriate for justifying this status. If someone is a ship's captain, he ought to maintain order on ship, ensure the ship's safety, and so on, even under adverse circumstances. There are also concepts, including the concept of 'concept', which enable individuals to reflect on and evaluate the adequacy of the concepts and conceptually constituted social processes within which they are participating. The concept of 'good' is the most important of these.

---

6. Ibid. p.127.
7. Ibid. p.127.
8. Jürgen Habermas, *Toward a Rational Society*, tr. Jeremy J. Shapiro, London: Heinemann, 1971, p.88. This conviction underlies all Habermas's work.
9. This point has been well made by Julius Kovesi in *Moral Notions*, London: Routledge & Kegan Paul, 1967.

Such concepts are underpinned, though not necessarily in an entirely coherent way, by more basic concepts and by the general world-orientation dominating society. In capitalism it is through money that people, roles and actions are designated as significant. The role of captain is important because ships make a profit, which means that it is worth exchanging money for labour-power which can function in the role of captain. The status of money in society in turn is sustained by the conception of humans as egoistic individuals who only enter into association with others because it is in their selfish interests to do so, and by the notion of economic progress as improved efficiency engendered by the struggle between egoists mediated by a monetary economy. 'Economic progress' is then sustained in a broader context in which it is seen as part of 'evolutionary progress'.

Practical reason is essentially bound up with such concepts, and always involves simultaneous participation in each of the dialectical processes of culture. People act by defining themselves within situations or negotiating such definitions in terms of the concepts available to them and then responding to the experienced claims made by these situations upon them by formulating projects which they then strive to realize. This generally involves acting in accordance with the implications of these concepts, becoming through their actions and achievements what they have defined themselves as being. For instance ships' captains are expected to put the safety of their passengers before themselves. For a captain to define such a situation involves experiencing this claim upon him calling for the appropriate action. The captain who subordinates his concern for his own safety to that of his passengers in a situation of great danger thereby becomes a 'real' captain. To fail to so act would be to become a coward, and thereby a 'poor excuse' for a captain. Simultaneously, people are defining themselves through narratives: as unfinished autobiographies formulated in terms of such conceptually defined roles and evaluations, relating themselves, their histories and their ambitions and projects to the unfinished biographies of others and to the histories and goals of social formations - from families to civilizations, which are also constituted by such concepts and narratives.[10]

Such autobiographical, biographical and historical self-definition is generally defined in relation to some general ideal of good order in the world. The ideal in most business enterprises in Western countries is to control everything, to make everything, both nature and people, serve as predictable instruments for achieving extrinsically defined ends.[11] This ideal is an expression of the metaphor of a machine. In all machines the whole is explained by the motion of the parts, while at the same time parts and their movements are evaluated according to their degree of subordinated to the ends to be achieved by the machine. The actions of a ship's captain should be directed towards the subordination of both himself, the crew and the ship to the goal of transporting cargo or passengers, moving them from one location to another. It is by virtue of the efficiency achieved by such subordination that economic enterprises are seen to maximize profits and so survive, grow and to contribute to economic progress - essentially the total instrumentalization of the world for the maximum production of commodities.

This is not to say that people function as cyphers of their cultures, acting out the logical implications of concepts embodied in institutions and society. Fulfilling the expectations made upon them requires effort, and people succeed in mobilizing themselves to different degrees. As Aristotle argued, the degree of success is largely a function of upbringing, of how habits have been inculcated in people. People embody ways of conceiving the world and orientations for action as a habitus. They must then struggle to maintain the integrity of this habitus in an active world shared with others, and there is an inevitable creativity involved in the application of concepts to new

10. David Carr has analysed the close relationship between the structure of human action organized by projects and the structure of narratives in *Time, Narrative, and History*, Bloomington: Indiana University Press, 1986.

11. On how this ideal is put into practice in ITT under the management of Harold S. Geneen see Richard Tanner Pascale and Anthony G. Athos, *The Art of Japanese Management*, [1981] Harmondsworth: Penguin, 1986, Ch.3. This is contrasted unfavourably with Japanese management practices as exemplified by Matsushita (based on a world-orientation closer to process philosophy).

situations, in the negotiation of shared definitions, and in the way individuals relate to the organizations of which they are part and to their own actions. As social beings choice is almost unavoidable because people are active in different roles which make competing claims upon them (for instance, in being both a ship's captain and a father), there is almost always a dissonance between conceptualizations pertaining to spatio-temporally broader contexts and those associated with more immediate situations which must be reconciled (such as between factors pertaining to honour and those pertaining to physical well-being), words can be understood differently by different 'reference groups', there are always rival ways of conceptualizing the world and rival definitions of each situation and of each organization and institution, and there are always contradictions in the culture with which individuals must come to terms (for instance between the ideal of getting rich and the ideal of upholding the standards of one's profession). Also, it is to some extent open to individuals to decide which others and whose definitions of reality and of themselves they will take seriously, and which of their actions to identify with - whether to regard particular actions as fully expressing what they are, as means to be able to do what they most identify with, or as merely play-acting. Finally, concepts define reality in opposition to other possibilities, and in doing so reveal these possibilities, thereby freeing individuals to reject all claims made upon them by situations as conventionally defined simply to express their autonomy.

Beyond this, concepts are never entirely adequate to grasp or define the complexity and emergent novelty of the experienced world. It is possible that all proposed definitions of reality are radically defective, and people may experience all sorts of meanings and engender effects, either within themselves, in others or in the world, which are unanticipated and incomprehensible in terms of the concepts by which they have defined the world and themselves. In response to such situations people are able to critically reflect on received ways of understanding the world and to redefine old or develop new concepts, and by defining social relations in terms of these concepts, to bring into life new social forms. This is what occurred both when law and when money were first instituted. By instituting 'law' in the early Middle Ages as a signification having a common meaning, it became possible to reformulate social relations, to see in social conditions the need for legal codification and alteration, and then to institute a manifold of reorganizations, redeterminations and reformations of already present social significations in society.[12] The same sort of process occurred with the introduction of money - and we are still wintering the extension of the commodity form associated with this institution. Such reconceptualizations are not confined to social relations. When nature came to be defined as an economic resource, a whole new set of relations between humanity and the world was brought into being.

In earlier chapters it was shown how environmental problems within Western civilization have revealed the radically defective nature of the concepts institutionalized or 'incorporated' within it. Having established an alternative metaphysical basis for understanding the world, and thereby having provided an alternative thematic motif to unify culture, these defective concepts can be replaced by alternatives which explicitly acknowledge the creativity involved in human becoming and the becoming of the rest of nature. When the world is conceived of in terms of an auditory analogy as a durational process of becoming, the end can no longer be thought of as what comes at the end of history. The good to be aimed at by individuals and society must pertain to the whole duration of becoming, whether this be of an individual's life, of a society, of humanity, or of nature.[13] If the notion of progress is maintained, then this must be understood in relation to the improvement of the spatial and temporal whole, just as each instrument and each note or melody in a symphony must be evaluated by a composer in terms of both its intrinsic quality and whether it contributes to the whole piece of music. This is inimical to the reduction of any part or any stage in

12. The creative emergence involved in this has been stressed by Cornelius Castoriadis in *The Imaginary Institution of Society*, tr. Kathleen McLaughlin, Cambridge: Polity Press, 1987.

13. This extends Aristotle's argument in the *Nicomachean Ethics* that 'Happiness ... requires ... a complete lifetime.' 1100a4-5. Also 1098a18.

this extensive durational becoming to a mere means to an end to come later. The nihilism which, as Nietzsche noted, is the eventual outcome of such an instrumentalization of the present, of defining the significance of life in terms of a purpose to be realized in the future - which is forever put off and which eventually fades into nothingness, is thereby avoided.[14] This change in thinking must be articulated into everyday life, into interpersonal relations, and into productive activity. It is no longer acceptable to think of action in terms of a sharp division between means and ends; defining situations and acting on the basis of such definitions, but must be seen as self-creation, a contribution to the world along with the end products of such activity.[15] And the end products of activities themselves must not be taken as what is valuable in action, but, as Marx argued in the *Grundrisse*, as new potentialities, the significance of which are only realized in later activity by being consumed, used or appreciated.

This is not to say that all activity is on one plane of becoming. Some activities participate only in short durational processes, while other activities also participate in long durational processes of greater significance to the becoming of the world. But no plane of becoming can be reduced to nothing but an instrument of another (for instance biological becoming to cultural becoming) without corrupting it.

In this scheme of things the instrumentalist notion of rationality must be rejected and replaced with a 'creative rationality'. If the world is a process of becoming consisting of a multiplicity of inter-dependent, semi-autonomous sub-processes of becoming, treating it as a collection of predictable objects to be used efficiently is to fail to acknowledge the reality of creative becoming and of the processes which maintain the ordered potentialities which people identify as objects. It involves a failure to see that one's projects, or one's society's projects, are at the same time part of the becoming, or at least affect the conditions for becoming, of other processes with some autonomy of their own, and that one, or one's society, can be a constituent of these processes. By contrast creative rationality involves recognizing that in one's thoughts and actions one is creating oneself as a participant in the becoming of a world consisting of self-creating processes with various degrees of autonomy, stability and dependence. In defining the world in terms of concepts one has consciously committed oneself to, one is forming a relationship and thereby contributing to the world's becoming. To conceive the world as a mere instrument is in fact to create a relationship between oneself and the rest of the world which debases it to a mere instrument; a debasement which is likely to have unforeseen and unfortunate consequences. Practical rationality must be understood in relation to such defining, as establishing a 'ratio' between each situation defined and the rest of the world, between the concepts in terms of which the world is defined and rival concepts, and between the different projects revealed as possible through defining situations in terms of these concepts. Being rational is deliberately defining the world and the potentialities and significance of the co-becoming participants associated with one's own self-creation in terms of the most discursively defensible concepts presently available, and acting accordingly, thereby 'realizing' these concepts in one's action and life. This requires a recognition of the continuously creative nature of this becoming in which one is participating, and of the possibility and the likelihood of emergent novelty. So rather than treating actions, objects and events as simply means for attaining distant ends, actions must always be seen as changing the conditions for the becoming of processes in the future, opening up and closing off different potentialities of one's own and of other processes.

Once rationality is understood as creative rather than as instrumental, the idea of power and control can be redefined, and it can be seen to make sense to say that total control is not a desired end. Gaining total control over the world would mean destroying its autonomy and creativity. For instance it is imaginable (although highly unlikely) that the self-regulating and creative dynamics of

14. See Friedrich Nietzsche, *The Will to Power*, §666.

15. Cornelius Castoriadis has argued for such a change in the concept of technique in *Crossroads in the Labyrinth*, [1978] tr. Kate Soper and Martin H. Ryle, Brighton: Harvester, 1984, pp.229-259.

the world ecosystem by which its stability is maintained could be replaced by artificial mechanisms - and in fact it was seen in a previous chapter that some Soviet thinkers called for such control. This would mean that the continued survival of humanity and other life forms in the world would be dependent on the continual monitoring and manipulation by humans of the conditions required for this survival. On a smaller scale this is in fact the situation which has been produced with the development of forms of agriculture which are dependent on farmers to continually control levels of water and fertilizer and to administer pesticides. This is the enslavement of people to their control mechanisms rather than an augmentation of their power. It is better to live in a world which is not under such instrumental control, which has dynamics of its own to maintain the conditions favourable to human life.[16] The control to be aimed at by creative rationality then should not be seen as the reduction of the world to a mechanical order to serve human purposes, but as the creation of the structures which will facilitate the shaping by people of their lives. To have power is to have the means to develop ones understanding of the world and oneself, and to be situated within structures through which this understanding can be spontaneously and creatively expressed.

It is in terms of these new notions of action, of rationality, of progress and of power or control that ethics, political philosophy and the struggle for the liberation of life must be reformulated.

## *A New Ethics*

In the epilogue to *The Phenomenon of Life*, Hans Jonas argued that:

> Ontology as the ground of ethics was the original tenet of philosophy. Their divorce, which is the divorce of the "objective" and "subjective" realms, is the modern destiny. Their reunion can be effected, if at all, only from the "objective" end, that is to say, through a revision of the idea of nature. And it is becoming rather than abiding nature which would hold out any such promise.[17]

In this work such a reunion has been attempted by defending and elaborating a process view of the world, of life and humanity. The implications of this reunion for ethics can now be spelt out.

In formulating ethics in terms of process philosophy, the very nature of ethics must be reconceived. Within the framework of mechanistic materialism the individual consciousness is seen as an inexplicable intrusion into a meaningless world of moving matter. Almost all ethical thought since the seventeenth century has been coloured by this way of viewing things. Consequently ethics has come to be conceived in terms of an opposition between morality, justified by the reason or feelings of individual subjects, and self-interest understood as the natural tendency of a self-reproducing mechanism to reduce everything to instruments for its survival and for the satisfaction of its appetites. For the most part, this has led to the separation of ethics from other realms of discourse and to an almost exclusive concern with the rightness or wrongness of particular actions or kinds of action. With the conception of humans as creating themselves through appropriating and developing their cultural heritage it should be clear just how pathological is a society which assumes that people are moved by appetites and aversions and which takes concern for others as problematic. A process view of the world situates people within a meaningful world. It does not separate ethical discourse from other realms of discourse, and its primary focus is on the life of people as a whole. The fundamental ethical questions become: What is a good life? What sort of contribution is it best to make to the unfinished becoming of society, humanity and the world?

---

16. This point has been well argued by Stephen R.L. Clark in 'Gaia and the Forms of Life' in Robert Elliot and Arran Gare, *Environmental Philosophy*, St. Lucia: University of Queensland Press, 1983, pp.182-200.

17. Hans Jonas, *The Phenomenon of Life: Towards a Philosophical Biology*, [1966] Chicago: University of Chicago Press, p.283.

What sort of being is it most worthwhile for individuals to strive to become? Hence, ethical action cannot be treated as something separate from economic or political action. Furthermore it is not sufficient to provide merely abstract determinations of what is the good life. People are always already participating in an institutionalized moral order which defines the significance of their actions and lives, and it is necessary that this be taken as a starting point. Ethical theory must concern itself with the way people and actions are accorded recognition and respect or disdain within society, with how structures of recognition are maintained and how they can be changed. Ethics immediately raises the political question: Does the existing social order, including the structures of recognition, facilitate the attainment of the highest forms of life? Ethical philosophy cannot be detached from political philosophy, or from economic, social or political science.

If it is possible to give a simple answer to the question What is a good life? it would be 'a fulfilling (or fulfilled) life'. But what is a fulfilling life? No one could possibly think of his or her life as fulfilling unless it had some meaning.[18] As Nietzsche succinctly put it: 'If we have the why of life, we can put up with almost any how. Men do not strive for happiness; only Englishmen do that.'[19] Process philosophy allows that the world and people's lives as part of this world can have meaning. Through their participation in the dialectics of orientation, recognition and power, through their struggle to understand the world and their place within it, to achieve relationships of mutual recognition, and to gain control over their destinies and to live according to their convictions, people are becoming part of a temporal order transcending their organic existence, thereby raising the immediacy of their situations to a different plane of becoming to achieve identities as significant human beings within the world. Such a conception of humans implies an abandonment of the opposition between self-interest and social responsibility. The self only emerges through relations to others, and these social conditions are logically prior to self-interest. Self-formation and commitment to others are indissociable. As Rabbi Hillel put it:

> If I am not for myself, who will be for me?
> If I am for myself only, what am I?
> If not now, when?

Where the becoming of humanity is understood in terms of a creative rationality, the project of finding algorithms for deciding correct courses of action must be abandoned. What is required is a return to the ethics of virtues, as called for by Alasdair MacIntyre,[20] with the main task being the development of a framework of concepts, defining what is virtuous and vicious, by which people can orient themselves in their self-creation. Such a framework can be developed by taking existing concepts and redefining them to accord with the process view of the world. As defining the quality of actions or of life these are not to be conceived of as imperatives in the sense of constraints on self-interest, but modes of being or becoming required to live a good life. At the same time these should not be seen only in relation to individuals, but should be seen as candidates for defining a new moral order. Three concepts in particular can be redefined and developed for this purpose: justice, duty and integrity.

## *Justice and Injustice*

Justice can be defined as the appropriate recognition and acknowledgement, in action, thought and feeling, of the nature and thereby the meaning and significance of all beings. This is a

---

18. As Viktor Frankl has argued in *Man's Search for Meaning*, N.Y.: Pocket Books, 1984.

19. Friedrich Nietzsche, *Twilight of the Idols*, [1889] tr. R.J. Hollingdale, Harmondsworth: Penguin, 1968, 'Maxims and Arrows', §12.

20. Alasdair MacIntyre, *After Virtue*, 2nd ed. Notre Dame: University of Notre Dame Press, 1984.

development of the ethical philosophy of William Wollaston as formulated in *The Religion of Nature*. Wollaston argued: 'That whoever acts as if things were so, or not so, doth by his acts declare, that they are so, or not so; as plainly as he could by words, and with more reality.' and that: 'No act (whether word or deed) of any being, to whom moral good or evil are imputable, that interferes with any true proposition, or denies any thing to be as it is, can be right.'[21] For instance to punish a person who is innocent is, by that action, to imply that the person is guilty. This contradicts the true state of affairs and is therefore wrong. Injustice, as a failure to acknowledge the nature and significance of beings affected by one's actions, always involves such falsehoods. Similarly, to take the property of another without reason is by that action to define the other's property as one's own, denying the true state of affairs.

However justice should not be thought to pertain only to action. It should extend to what is thought and what is felt. Injustices are committed merely by failing to recognize the true nature of beings, quite independent of any action towards them - which is why clearing the name of a dead person can be a legitimate struggle for justice. Drawing out the implications of this, justice requires of people that they critically examine their conceptions of the world, particularly those conceptions which are institutionalized, to ensure that they do justice to everything. Then it is necessary to have the appropriate emotional responses to be just. To take pleasure in the undeserved failure of another, or to resent their deserved success, is also unjust. And considered as a virtue, being just requires the capacity to work out compromises between opposing claims of justice, and to give equal consideration to and to keep everything involved in situations in proportion. As the ancient Greeks recognized, proportion or balance (*sophrosyne*) is of paramount importance for justice. Without such proportion, the quest for justice can easily turn into oppression. Yet it is impossible to provide purely formal criteria for achieving such balance.

This notion of justice captures the essence of rival theories of justice without being reducible to them. It acknowledges Plato's view of justice as each thing keeping to its appropriate place since this must follow from actions based on the appropriate recognition of all beings. It encompasses Aristotle's definition of justice as that which preserves and promotes the well-being of the social and political community,[22] and it accords with Thomas Aquinas' definition of justice as 'a habit whereby a man renders to each one his due by a constant and perpetual will.'[23] Rights claims associated with contracts, explicitly formulated or implied, can be acknowledged as part of justice as defined above, as can non-contractually based legal rights; but these cannot be the whole of it. If contracts are made, these must be recognized by relevant actors, but claims for justice can still be made upon people without contracts having been made, while legal rights other than contracts must be embedded in some notion of justice transcending the notion of rights to have any moral force. Kant's criterion for defining justice, i.e., that: 'Every action is just that in itself or in its maxim is such that the freedom of the will of each can coexist together with the freedom of everyone in accordance with a universal law'[24] can also be acknowledged to have some validity. However it provides only the negative conditions for freedom. Such a formalistic criterion based on the acceptance of a total separation between knowledge about the nature of the world and practical reason, cannot capture every aspect of justice.

Justice so conceived goes beyond these doctrines, requiring of people sensitivity, consideration, imagination and compassion to understand the situations and perspectives of other beings - whether human or non-human, and breadth of understanding to appreciate the past causes and present dynamics responsible for existing conditions and to appreciate all the effects of actions. It also

---

21. William Wollaston, 'The Religion of Nature Delineated', Section I, 1025-6 in L.A. Selby-Bigge ed. *British Moralists*, Oxford, Clarendon Press, 1897, p.362. For an account of the misrepresentations of Wollaston see Joel Feinberg, 'Wollaston and his Critics', *Journal of the History of Ideas*, Vol.XXXVIII, No.2, 1977, pp.345-352.

22. Aristotle, *Nicomachean Ethics*, Bk V, 1129b18-19.

23. Thomas Aquinas, *The Summa Theologica*, II-II, Question 58.

24. Immanuel Kant, *The Metaphysical Elements of Justice*, [1797] tr. John Ladd, Indianapolis: Bobbs Merrill, 1965, p.35.

requires insight to avoid the distorting effects of self-interest, jealousy, resentment, malice, envy, arrogance and laziness, of projecting onto others the dissociated elements of oneself or one's group, of transferring onto others one's past forms of personal or social relationships, of using unjust acts of others to legitimate one's own injustices, of defining others to effect rigid boundaries in group experience, and so on.[25] And it requires judgement to balance different claims to justice, taking into account different social pressures to distort judgements.

It is the notion of justice which Simone Weil upheld when she pointed out the radical difference between calls for justice and assertions of rights.[26] The connotations of claims to rights reveals the meaning context within which the modern concept of rights was developed, a society of egoistic individuals in commercial relationships.[27] To call for justice for oneself, on the other hand, is to request that what one is, what one's situation in the world is, what are one's needs, what one has suffered, what efforts one has made and what are one's potentialities, particularly one's potentialities to be hurt on the one hand, and on the other to contribute to 'the common good of one's communities',[28] be understood in all their uniqueness, appreciated, and taken fully into account. Similarly when calling for justice for other people, for one's community or for other life forms. And while demands for rights are assertions of the primacy of the individual over the community, calls for justice affirm the reality of community, including the community of members of ecosystems. As John Finnis has pointed out: 'the objective of justice is ... the common good, the flourishing of all members of the community'.[29] To ignore a claim to a right is an offence against the individual only; to ignore a claim to justice is an offence against the entire community.

A number of points have been raised against Wollaston's views, and to defend the notion of justice presented it is necessary that these be examined. Joel Feinberg argued that Wollaston has provided no basis for distinguishing the significance of the falsehoods implied by different actions.[30] For instance no distinction is drawn between treating a person as a post and treating a post as a person. While the latter might appear inappropriate, it would not appear to be morally wrong except where such treatment resulted in failure to act appropriately elsewhere. For this reason it is necessary to have an underlying epistemology and ontology which allows for distinctions of significance to be made between kinds of beings. This was provided by the philosophies of Plato and Aristotle, but not by empiricism and mechanistic materialism. However process philosophy also provides a basis for such judgements. Where knowledge implies understanding, that is, 'indwelling' in the world and in the specific entities understood, and these entities are understood as processes of becoming, it is impossible to understand the world without appreciating its intrinsic significance and of each entity within it, of the differences between non-life and life, between plant life and animal life, and between animal life and human life; and thereby the difference between action on a post and action on a person. This does not mean that the conception of entities (humans, organisms or ecosystems) following from a process world-orientation enables people to deduce in any but a general way how people should act towards them. Rather, as explanations within the sciences are not determined by metaphysical assumptions but are

25. Projection and transference are recognized by psycho-analysts as playing a major role in psychodynamics. However they are just as pervasive in the cultural dynamics of groups, as for instance when powerful, aggressive nations see quite powerless nations as threatening their security. The phenomenon and consequences of maintaining strong experiential boundaries have been described by Mary Douglas in *Purity and Danger: An analysis of the Concepts of Pollution and Taboo*, London: Routledge & Kegan and Paul, 1966.

26. See p.46 of this work.

27. The evolution of the Roman and medieval notion of *'jus'* to the notion of 'right' developed by Hobbes, from 'the fair' to 'a liberty to do something', is described by John Finnis in *Natural Law and Natural Rights*, Oxford: Clarendon, 1984, p.206ff.

28. Ibid. p.164.

29. Ibid. p.174.

30. Joel Feinberg, 'Wollaston and his Critics', *Journal of the History of Ideas*, Vol.XXXVIII, No.2, 1977, pp.345-352.

only acceptable if they are intelligible in terms of generally defensible metaphysical schemes, so the more particular concepts by which people define their situations and orient themselves for action are only just if they accord with the basic nature of entities as comprehended in terms of the most defensible metaphysical scheme. Thus, practices or actions which conceive people as mere objects to be manipulated or as nothing but labour power to be bought and sold, deny people their essential humanity. From the perspective of process philosophy, they do not do justice to their potentialities and are therefore unjust.

An older argument against Wollaston comes from Hume who argued that:

> ...there is an evident reasoning in a circle. A person who takes possession of *another's* goods and uses them as his *own* in a manner declares them to be his own, and this falshood [sic] is the source of the immorality of injustice. But is property, or right, or obligation intelligible without an antecedent morality?[31]

But assuming a pre-existing morality is not a problem in itself. All human activity and ethical theorising originates from within a cultural tradition containing a moral order. This was only seen as a problem as such by Hume because of his basic commitment to a view of humans according to which the existence of such an order is unintelligible. However Hume has pointed to a real difficulty with Wollaston's approach: that he has provided no basis for critically evaluating the received moral order or for resolving conflicts between opposing ways of conceiving things. But again, by defending a dialectical theory of knowledge in opposition to both logical empiricism and relativism, construing the goal of enquiry as understanding, and providing a theory of being which allows that beings in the world have different significance and that humans have potentialities worth realizing, such a basis is provided. If there is any dispute over evaluative ethical concepts, the dialectical approach implies that it is enough to settle arguments that reasons can be provided to convince people to choose between accepting or rejecting their validity, or that one definition or application is superior to another, while the process view of the world provides a framework and ultimate reference point for such arguments. Ultimately, dispute resolution requires the construction of a narrative from the perspective of one ethical position which reveals both the achievements and failings of rival ethical positions.

On such a basis it is also possible to criticise the institutions of society for being unjust. While forms of life which ascribe property rights in such a way that nature is reduced to a mere instrument and people are defined in terms of their ownership of property can be regarded as just if the world is nothing but a Darwinian struggle for survival, these must be condemned as unjust if the process view of evolution is successfully defended. All life forms must then be ascribed intrinsic significance with a dynamics of their own which should be respected, and people treated as creative processes of becoming with the potential to form communities based on mutual recognition of each other's significance. Similarly, if socio-biologists are right then it is proper to maintain gender relations which deny 'femininity' and thereby respect to women who strive to develop their full potential to participate in economic, political and cultural life, but totally unjust if the process view of humans is correct. Such institutional criticism is central to Marx's *Capital* where the categories defining right economic behaviour and constituting the forms of life in capitalist society were implicitly, but nevertheless savagely criticised on the basis that they define humans, who Marx conceived to be creative social beings, as nothing but labour power to be bought and sold as a commodity. Marx's analysis provides a model for further critiques, particularly of the assumption by economic institutions of economic categories which do not do justice to nature, to those excluded from the economic system and to future generations. A socio-economic formation in which nature and people are defined by institutions as nothing but resources to be used efficiently is essentially unjust.

---

31. David Hume, *A Treatise of Human Nature*, London: Dent, 1962, Bk III, Pt. 1, Sec. 1, Vol.2, p.171n.

## *Duty and Corruption*

This brings us to the notions of duty and corruption. Most of the more important actions within societies are undertaken by people acting in the context of and as representatives of institutions or organizations. These always embody ways of defining the world, ideals and goals to be striven for; and institutional roles are defined in relation to these. The most important ethical concepts in relation to institutional or organizational behaviour are those of duty and corruption. In accordance with process philosophy the notions of duty and dutiful can be redefined to imply a less moralistic and more activist stance than is usual. 'Duty' has unfortunate connotations of being an obligation which must over-ride self-interest. To avoid this, duty can be redefined as the behaviour required to become a 'real' member of one's profession (for example, putting one's passengers before oneself to become a real ship's captain), with what is required extended to taking responsibility for the institutions within which one is participating. Rather than 'dutiful' simply defining individuals as those who fulfil, or at least strive to fulfil, the expectations of their roles, it can be redefined to require that they also understand or strive to understand and evaluate the significance of their roles within institutions. 'Corruption' can be defined as the failure of people to do their duty. Action as a participant in an institution or organization is corrupt not only when just role expectations are not conformed to, but also when these role expectations and the goals and ideals of the institution or organization have not been questioned by individuals. Action deliberately not conforming to role expectations and institutional or organizational ideals and goals which are seen as unjust is not corruption but subversion. The ideals and goals of institutions and organizations are always open to revision, and there should be a constant struggle between different people to define or redefine their ideals and goals. Subversion can at the same time be a constructive reformulation of institutions or organizations around different ways of conceiving the world and around different ideals and goals.

## *Integrity and its lack*

Finally we come to the concept of integrity. Integrity means wholeness. It is the measure of the coherence or 'narrative unity' one's life gains through striving as far as one's abilities will allow to be just to the world and to oneself in action, thought and feelings, and through one's commitment to justice however adverse the circumstances.

Justice, duty and integrity are closely related concepts. The notion of justice pertains not only to one's relationships others, but also to oneself. To do oneself justice one must accord in thought and practice appropriate recognition of what one is, of one's human nature, of one's needs, of one's appetites and aversions and of one's unique abilities. This requires the appropriation and participation in the development of one's cultural heritage, the fullest possible development of one's understanding and awareness of the world, of oneself and of one's particular situation, and action on this basis to make the fullest contribution to the becoming of the world. Succeeding in this, creating in oneself in the duration of one's lifetime a process of objective significance, is achieving integrity.

Since a life of integrity is a genuine form of emergence within the world, it is not easy to convey an understanding of what it is to live such a life. Perhaps one of the best efforts in this direction was made by Erik Erikson who wrote of the person with integrity:

> Although aware of the relativity of all the various life styles which have given meaning to human striving, the possessor of integrity is ready to defend the dignity of his own style against all physical and economic threats. For he knows that an individual life is the accidental

coincidence of but one life cycle with but one segment of history; and that for him all human integrity stands or falls with the one style of integrity of which he partakes.[32]

However this needs to be complemented by an account of what it means to lack integrity. This has been superbly characterized by Miroslav Holub in his poem *Polonius*:[33]

Behind every arras
he does his duty
unswervingly.
Walls are his ears,
keyholes his eyes.

He slinks up the stairs,
oozes from the ceiling,
floats through the door
ready to give evidence,
prove what is proven.
stab with a needle
or pin on an order.

His poems always rhyme,
his brush is dipped in honey,
his music flutes
from marzipan and cane.

You buy him
by weight, boneless,
a pound of wax flesh,
a pound of mousy philosophy,
a pound of jellied
flunkey.

And when he's sold out
and the left-overs wrapped
in a tasselled obituary,
a paranoid funeral notice,

and when the spore-creating mould
of memory
covers him over,
when he falls
arse-first to the stars,

the whole continent will be lighter,
earth's axis straighten up
and in night's thunderous arena
a bird will chirp in gratitude.

32. Erik H. Erikson, *Childhood and Society*, [1965] Frogmore: Triad/Paladin, 1977, p.241f.

33. From *Miroslav Holub: Selected Poems*: tr. Ian Milner and George Theiner, Harmondsworth: Penguin, p.73f.

Unlike the notion of self-actualization, integrity cannot be construed to justify treating the rest of the world as a means to one's own development. In this regard the process view of integrity is entirely in accordance with the ideas of Viktor Frankl who argued:

> By declaring that man is a responsible creature and must actualize the potential meaning of his life, I wish to stress that the true meaning of life is to be found in the world rather than with man or his own psyche, as though it were a closed system. By the same token, the real aim of human existence cannot be found in what is called self-actualization. Human existence is essentially self-transcendence rather than self-actualization.[34]

While integrity involves developing one's potentialities, this must be in response to the claims of the world upon one, as a significant contribution to the becoming of a world which must be understood, both in practice and on reflection, to have a significance beyond one's own life. The aim in life should be to find a goal worthy of one's abilities.

The quest for integrity is always undertaken in a world of institutions with pre-defined roles, ideals and goals, and in such institutional contexts, integrity and duty are indissociable. Institutionalized roles embody ideals, and some minimal integrity is required to live up to these ideals in the face of outside pressures or in the face of problematic situations. However such embodied ideals may be indefensible, and questioning these ideals and living one's life according to one's judgements, struggling against the pressures of established definitions and enduring the ensuing retribution, social invalidation and hardship, requires considerable courage, effort and fortitude. Acting and living with integrity requires a struggle for self-mastery, strength of character and the cultivation of that strength. It requires the development of the ability to measure oneself not against those around one but against the 'generalized other', perhaps totally unembodied in the present, at least among one's acquaintances, and then to live according to this measure despite the opinions of those around one.

The impulse to achieve integrity can be identified with conscience. The etymological meaning of conscience is 'with knowledge' or 'with deliberation' and implies the claim of the world revealed by understanding and deliberation. In relation to the 'court of conscience' of the casuists in the Middle Ages it was associated with the effort to direct action in accordance with the fullest possible knowledge. With the Reformation, conscience was internalized as a part of the heroic moralism of Western culture. This conscience reached its highest development in the ethical thought of Rousseau and Kant. But associated with the advance of nihilism and the decline of this moralism, conscience has been redefined as the subjective experience of constraint produced by the accidents of one's upbringing. The notion of conscience, like that of integrity, seldom enters into the discourse of moral philosophers. With process philosophy both the rational, emotional and the social dimensions to conscience are restored. It can be understood as the impulse to live in accordance with justice and to thereby attain and maintain one's integrity. It is the impulse to become human.

The quality of integrity is then a function of the extent of the context people take into consideration in defining themselves and choosing how to live. As Voznesensky wrote in *Antiworlds*:[35]

> In finding their truths, lives vary in daring:
> Worms come through holes and bold men on parabolas.

---

34. Viktor E. Frankl, *Man's Search for Meaning*, [1963] tr. Ilse Lasch, N.Y.: Pocket Books, 1984, p.175.

35. 'Parabolic Ballad', tr. W.H. Auden, from *Antiworlds: Poems by Andrei Voznesensky*, ed. Patricia Blake & Max Hayward, London: O.U.P., 1967, p.49.

People who define their lives only in relation to their place of work, a local group or community and who strive for integrity within this context without any concern for the relationship of this community to the rest of the world can achieve only a very limited integrity. The highest degree of integrity requires a struggle to consider what contribution one's life is making not only to one's immediate community, but also to one's society, to humanity, to life itself and the whole of nature, understood not only in terms of one's contemporaries, but also in terms of the entire history and the entire past and future of the world, and then to live in the light of this understanding. Striving for greater integrity will inevitably make life far more difficult; it will bring one into conflict with those around one, it will involve more failures and detours, and in terms of the prevailing criteria, one's life will appear far less successful than it would otherwise be; but then one's life will take on a greater meaning in relation to the broader, longer durational and more significant processes within which one will be authentically participating.

Achieving integrity requires all that justice requires - consideration, compassion, sensitivity, imagination and perspective, and almost always - courage. It is by recognizing that one's integrity is one's authentic contribution to the becoming of the world, and seeing one's present life and actions from the perspective of the end of one's life, and one's whole life from the perspective of the totality of the world's becoming, that such courage can be gained. Integrity therefore requires above all else the development of one's understanding of the world and of oneself.

## *Ethics and the Environment*

The concepts of justice, duty, integrity and their opposites finally provide a language for bringing questions about our relations to other life forms, ecosystems and future generations, the relationship between the wealthy and the poor of the world, the nature of built-up environments, and so on, into the realm of rational ethical discourse. It has been argued that underlying the environmental crisis is the domination of Western society by a mechanistic world-orientation, that mechanistic materialism is invalid and that the world can best be understood as a process of creative becoming within which we are semi-autonomous participants. Underlying the environmental crisis is the basic injustice of falsely assuming in the way society is organised, in its major institutions and in people's most important activities, that the world is a mechanical order of things. It is this which Peter Singer was reacting against when he protested against the treating of animals 'like machines that convert low-priced fodder into high-priced flesh...'[36] But this injustice is also evident in treating life forms (individuals, communities, species, ecosystems) as though they existed in isolation without intrinsic significance, rather than as intrinsically valuable participant processes in inter-dependent, self-stabilizing communities and ecosystems. Further injustices are perpetrated by regarding people of other nations as nothing but competitors in a struggle for survival and the poor of the Third World world as merely the losers in this struggle, in denying the significance of different cultural traditions throughout the world and seeing them as merely obstacles to progress, in acting as though future generations were merely the collection of people who might exist in the future, and in creating forms of life which define people as egoists whose ultimate end is nothing more than satisfying their appetites, social climbing and being entertained.

As institutional actors, those who have the courage to re-evaluate the state of the world must confront the corruption of the dominant institutions of society, and then must strive to reorganize them - particularly those associated with the economy. In terms of mechanistic materialism the economy is the circulation of money through which goods and services are exchanged for the factors of production, and progress is anything which increases the number of goods and services involved in this exchange, while in terms of process philosophy the economy of society is its 'household management', the organization of the metabolism of society, especially its interaction

36. Peter Singer, *Animal Liberation*, N.Y.: Avon Books, 1975, p.94.

with its environment, and progress is improving the conditions for civilization, for the highest forms of relationships between people and for the life of culture, while at the same time preserving and contributing to 'the integrity, stability, and beauty of the biotic community'.[37] It is necessary to evaluate the functioning of the economy according to whether it is based on a just conception of all elements of and in the environment, of the participants in the production process and their relationships, of people of different locations, nations and regions, and of future generations. The economic system of capitalism is based on unjust conceptions of all of these and is having disastrous effects as a consequence.

However the concepts proposed here are not only means to enable people to define what is right and wrong, or even to evaluate institutions. They are proposed as the basis for an alternative moral order and as the foundation for an alternative social order. Part of the function of such a moral order is to enable individuals to define the significance of others and to work out who to align themselves with and who to oppose. But at least as important, especially in the face of a society hostile to one's ideals, an alternative moral order provides one with the means to define the significance of one's own life and actions independently of the opinions of those surrounding one. Environmentalists in the modern world are now in a somewhat similar situation to Hamlet - aware that something is radically wrong but confronted by a general consensus that everything is in order. Herbert Marcuse wrote of the modern condition:

> A comfortable, smooth, reasonable, democratic unfreedom prevails in advanced industrial civilization, a token of technological progress. Indeed, what could be more rational than the suppression of individuality in the mechanization of socially necessary but painful performances; the concentration of individual enterprises in more effective, more productive corporations; the regulation of free competition among unequally equipped economic subjects; the curtailment of prerogatives and national sovereignties which impede the international organization of resources.[38]

The reasonability of this is vouchsafed by the dominant world-orientation, grounded in the mainstream of science, embodied by individuals as a habitus and in the major institutions of modern society, and providing the concepts which mediate people's relationships and in terms of which they define their goals. Consequently all that appears to be important is the comfort, wealth and entertainment provided by technical progress. But a vast range of apparent problems suggest something is rotten in the state of the world. Examination of each of these problems reveals them to be interconnected, and deeply connected to the mechanistic world-orientation which denies their significance. The situation confronting the affluent is whether to drift through life along the easiest path, or whether to look behind particular problems to their deeper causes and to critically examine the beliefs and attitudes which have come to be taken as self-evidently valid. Ultimately the question is whether they will remain cyphers for prevailing social forces, or whether they will live their lives with integrity. Confronted with this choice, those who have faced up to environmental problems might well sympathise with Hamlet's lament:

> The time is out of joint; O cursed spite,
> That ever I was born to set it right![39]

An alternative moral order based on concepts such as justice, duty and integrity is required to give people the strength to attempt this task and begin the struggle to create a new social order.

---

37. Aldo Leopold, *A Sand County Almanac*, Oxford: Oxford University Press, p.224.
38. Herbert Marcuse, *One Dimensional Man*, Boston: Beacon Press, 1966, p.1.
39. Shakespeare, *Hamlet*, I.v.188.

## *Political Philosophy*

In the present age the liberal political philosophies on which Western political institutions were originally based have lost their relevance. The development of the world economy with its transnational corporations transcending the control of national governments together with the complexity of and inter-relationships between communities, economic organizations, the consciousness industry and military, legal, penal, educational, welfare and political institutions have left the concepts of liberal democratic thought - 'public realm versus private realm', 'freedom', 'democracy', 'liberty' etc. - virtually without content,[40] while the States of most countries are no longer capable of dealing with the social, economic and environmental problems confronting them. This has been recognized by Marxist theorists of the State, but such theorists have simply analysed these problems as 'the crisis of the State'. They have not proposed any solutions. This reflects one of the great defects of Marxism - its absence of a political philosophy.[41] But the conscious regulation of material production according to a settled plan called for by Marx and his followers can only mean that economics should be subordinated to politics. It is the failure by Marxists (apart from Habermas and Bobbio) to realize this and to think through its implications which more than anything else is responsible for the Marxist tendency towards authoritarianism.[42] The proposed solution by some anarchists and environmentalists to the failures of both liberal-democratic and Marxist practices - that the State be abolished and society broken up into small, independent, self-subsistent communities is totally unrealistic in the light of present problems, the present population of the world and the power structures already in existence.[43] As Boris Frankel has cogently argued, what is necessary to confront current problems is not the contraction of States, but their expansion - albeit in a quite different form than at present.[44] The question which must be faced is how to organize political, social and economic institutions and processes so as to decentralize power and avoid the tendencies of organizations to become self-serving at the expense of the people they were designed to serve - while still dealing with issues transcending local concerns. The philosophical problem is to reformulate or create new political concepts to enable people to think about the political problems facing the world.

There have been five great political philosophers in European history: Plato, Aristotle, Thomas Aquinas, Hobbes and Hegel. Hobbes was the political philosopher who provided a new starting point to replace the synthesis of Platonic, Aristotelian, Stoic and Judaic thought - of which Aquinas had been the foremost exponent - and provided the starting point for modern social contract theories of rights, utilitarianism, economic theory and Social Darwinism. Hegel is the philosopher who, by incorporating ideas from Montesquieu, Herder, Rousseau and Kant (the four next most significant political philosophers in European history), produced an historicist reformulation of Platonism (incorporating some elements of Aristotle's philosophy) to meet the challenge of

---

40. This is not to say that these concepts do not have rhetorical force; they underlie and are used with considerable effect by the New Right - but not to gain anything which could be called freedom, democracy or liberty, but to effect the subordination of everyone to an international economy dominated by giant transnational financial, manufacturing, and agribusiness organizations.

41. Marx argued that: 'The life-process of society, which is based on the process of material production, does not strip off its mystical veil until it is treated as production by freely associated men, and is consciously regulated by them in accordance with a settled plan.' (*Capital*, Vol.1, Moscow, Progress Publishers, 1974, p.84.) How do these radical individualists settle on a plan?

42. This tendency was anticipated by Michael Bakunin in *God and the State* published in 1882. Norberto Bobbio is one of the few Marxists to address the issue. See *The Future of Democracy*, tr. Roger Griffin, ed. Richard Bellamy, Cambridge: Polity Press, 1987.

43. See Rudolf Bahro, 'Dare to Form Communes', *Building the Green Movement*, London: Heretic Books, 1986, pp.86-91.

44. Boris Frankel, *Beyond the State?: Dominant Theories and Socialist Strategies*, London: Macmillan, 1983.

Hobbesian philosophy. This he defended firstly through his metaphysics, and then through a narrative of world-history formulated from the perspective of this metaphysics in terms of which the achievements and limitations of all past political thought and political forms were evaluated. Rejecting the atomic individualism of social contract theorists and utilitarians, Hegel argued that humans are essentially socio-politico-cultural beings, that societies embody a rationality and that individuals only become fully human, only become rational, free individuals and recognize themselves as such, through participating in the ethical life of society. In a modern society such freedom is gained through the family in which the ethical spirit has its immediate substantial existence in its natural universality, then in civil society, the realm of formal universality in which people produce and exchange goods to satisfy their needs. Hegel argued that while this is an order of interdependence in which the self-interested pursuit of each contributes to the welfare of all, the free market would tend to concentrate wealth and pauperize large sections of the population if left to itself. It must be constrained by corporations organized on the basis of each trade to give isolated and competing producers the chance of a communal life and a recognized trade. However corporations themselves are not enough, and civil society, along with the family, has to be ordered into a larger, more cohesive unity: that of the State, the self-conscious ethical substance in which the family principle and civil society are unified and particular self-consciousnesses are raised to consciousness of their universality. In terms of the concepts elaborated above, the State is the ordering of society by which justice is made to prevail over particular interests, the ordering which embodying a commitment to justice, recognizes the significance of each individual, takes the perspective of the whole society, and then strives to realize the common good.

Given that the uncontrolled operation of markets will lead to the destruction of the world ecosystem and that efforts to replace markets by planning have failed, it is this Hegelian model of an economy subordinated to justice which must be aimed at. While Hegel's basic Neoplatonic framework and some details of his political philosophy are open to question, the great achievement of Hegel was to have shown the way to reconciling Herder's notion of life as social self-expression with Kant's notion of the autonomous rational will, while still granting a place to the functioning of the market.[45] But from Karl Marx to Karl Popper, Hegel's political philosophy has been attacked for its theoretical assumptions and has been identified with oppressive developments in politics. This has led to a failure to appreciate Hegel's achievements, and it is this more than anything else which has contributed to the triumph of Hobbesian thought. What is proposed here is that the theoretical attacks on Hegel can be obviated and those aspects of his thought which might give sustenance to oppressive political tendencies avoided - while at the same the problems and complexities of the modern world can be confronted and his ideas extended to deal with the environment, by reformulating his political philosophy through process philosophy.

One of the main problems in Hegel's political philosophy is that it provides no way to evaluate the forms of thinking embodied in existing political institutions. In this regard Hegel left people in the lurch, claiming that the philosopher is only able to reveal the rationality of history after the dust had settled. Hegel's followers who did grant a place to reason in guiding reformist or revolutionary action failed to provide an attractive vision of the future. Either they confined reason to a purely critical role, or less commonly, represented this end as static and formal. The effect of their ideas was to lead to all the past and the present being viewed as mere instruments for the realization of an ideal. By contrast Aristotle's approach to political philosophy provides a way of evaluating the institutions and organizations of and forms of thinking embodied in societies, and thereby for developing programmes of political reform. For Aristotle, ethics and politics are indissociable. His *Nicomachean Ethics* was devoted to working out what is the highest good for humans, the ultimate end which is desired for its own sake and for which all other ends are means, while his *Politics* was devoted to working out how societies should be organized to enable people to realize the highest

45. For a defence of Hegel on these grounds, showing his relevance for the present, see Charles Taylor, *Hegel and Modern Social Theory*, Cambridge: C.U.P., 1979.

good. While one might disagree with Aristotle's conclusions as to what the highest good for humans is, it is difficult to conceive of a better formulation of the relation between ethics and politics, and how to conceive the fundamental problem of political philosophy.

The answer given to the first and most fundamental question: What is the ultimate end of life? will depend on what conception of humans and their place in the world is argued for. To begin with, Aristotle argued that the ultimate end of life is spiritual well-being (*eudaimonia*) which is achieved by the 'activity of the soul in conformity with excellence or virtue, and if there are several virtues, in conformity with the best and most complete.'[46] On the basis of his metaphysics and corresponding conception of the nature of humans, he argued that the highest virtue is the activity concerned with theoretical knowledge or contemplation. In relation to politics he then argued that the ideal polis is one 'which has virtue sufficiently supported by material resources to facilitate participation in the actions which virtue calls for.'[47] In terms of the metaphysics and corresponding conception of humans defended here, people are striving to orient themselves, to live and act in a way which deserves and receives recognition and respect from people who are themselves worthy of respect, and to gain sufficient control over the conditions of their existence to shape their lives according to their understanding and convictions. If the process view of the world is valid, societies should be organized to facilitate the achievement of these ends.

Societies should therefore be organized to promote cultural life, justice and liberty. Cultural life should be understood as the communicative activity in which, through dialogue, literature, art, drama, architecture and other forms of communication, ways of understanding, experiencing, modes of being in the world and forms of life are revealed and appreciated, tried out and questioned, further developed or replaced, and problems of localities, organizations, nations, humanity and life are defined and projects of action are publicly formulated, elaborated and evaluated. It is through such cultural life that people orient themselves. The most important measure of success in this is the degree to which people are able to construct coherent and convincing grand narratives which relate all particular orientations and projects, and to define their own lives in relation to these. Concomitantly, achieving justice should be understood as each individual, whether human or non-human, being given appropriate recognition in thought and action, in social practices and institutions. Cultural life is a condition for achieving this, but it also requires empowerment of people, the security to pursue justice and the means to gain redress against injustices. Liberty should be understood as the condition in which people can live with integrity. This requires not only freedom from constraints, but also the means for people to form relationships and make commitments to others, to reproduce themselves and to bring up their children, to develop their abilities and their understanding of the world, and to participate as fully as their abilities will allow and in accordance with their convictions, in the economic, political and cultural life of society. For there to be liberty, societies must provide their members with economic security, with the education necessary for them to be able to participate in the cultural life of society, with media to communicate their ideas, with occupations in which they can realize their highest potentialities to contribute to society and the world, and with the means to participate in defining and redefining the goals and values of the social formations within which they are participating.

This notion of liberty is opposed to the doctrine of negative liberty formulated in terms of mechanistic materialism by Hobbes who argued that: 'Liberty, or Freedome, signifieth (properly) the absence of Opposition'[48]. It accords with the notion of positive liberty proposed by Montesquieu (and then taken up and developed by Rousseau, Kant, Hegel and Marx) who argued

46. Aristotle, *Nichomachean Ethics*, I, vii, 1098a16.

47. Aristotle, *Politics*, VII i 1323b38.

48. Thomas Hobbes, *Leviathan*, Harmondsworth: Penguin, 1968, Ch.XXI, p.261. For an examination of the relationship between Hobbes and his predecessors in this regard see Quentin Skinner, 'The Idea of Negative Liberty', in *Philosophy in History*, ed. Richard Rorty, J.B. Schneewind and Quentin Skinner, Cambridge: C.U.P., 1984, pp.193-221.

that political liberty: 'does not consist in an unrestrained freedom. In governments ... liberty can consist only in the power of doing what we ought to will'.[49] Negative liberty is important not in itself but as a condition for achieving positive liberty. Cultural life, justice and liberty must be seen as mutually dependent, though irreducible to each other. Existing institutions should be evaluated and preserved, transformed or abolished according to whether and how much they facilitate cultural life, justice and liberty.

With this conception of politics, the environment must be given central place: as the condition for the continued maintenance and reproduction of society and for the realization of humanity's highest ends, and as consisting of non-human life forms with a significance in their own right which, if justice is to be done, must be appropriately recognized. The most important form of justice in terms of which any society and every institution in society must be evaluated is in its relation to its environment

## *Generativity and Decadence*

However there is an important insight embodied in Plato's philosophy which was to some extent lost sight of by Aristotle. This is that people are moved to action by having an image of how they should be, and some notion of what it means to fall away from this ideal. For Plato the good polis or society, that is, the form in which all societies are striving to participate, is the just society, one in which those dominated by their intellect rule over those dominated by their spirit, who in turn rule over those dominated by their appetites. While few are attracted to the static ideal portrayed by Plato, his description in Book VIII of *The Republic* of what is involved in falling away from justice, in the advance of decadence, has been one of the most powerful images affecting European political life (with later Rome generally being taken as the model of decadence).[50] Plato's account of the difficulty experienced by those oriented towards achieving higher ends when confronted with the low cunning of the street-wise, his account of the development of militarism as those dominated by intellect are displaced by those dominated by spirit, of their replacement in turn by those questing for wealth and the corrosive effect this has on people's attitudes to life, of the rejection of all constraints when those who are dominated by their appetites reject all discipline paving the way for the triumph of tyranny, should not be taken as a description of reality (as Aristotle took it to be) but as a powerful analysis of a very real tendency.

Ideals of how societies should be have almost always been represented as static. But all static societies are repulsive, and the greatest oppression in the world has resulted from the tendency to see the present as a mere instrument for some future state. Hegel attempted to solve this problem by historicising Plato. Following Herder, he represented people as having their national genius manifest in their religion, their polity, their ethics, their legislation, and their science, art and mechanical skills. People are inspired to bring to fruition the potentiality of their nation, to realize freedom by recognizing, believing in and willing what is common to the whole (in effect, participating in and living according to Rousseau's General Will). This freedom is objectified in the State which unifies and directs the nation. But people lose their dedication to the State as its contradictions and irrationalities are revealed. By the time the ideals underlying the State have been brought to full consciousness by philosophers they are no longer able to inspire people. The society becomes decadent and a new nation invigorated by a new, as yet inarticulate vision comes to dominate the stage.

Hegel rejected the idea that philosophy could play any part in this process, and without a new vision for society being provided by philosophers, economists and Social Darwinists have been able to foist on people their vision of the ideal society as a perfect machine. Decadence and social

49. Baron de Montesquieu, *The Spirit of Laws* tr. Thomas Nugent, New York: Hafner, 1966, XI, 3, p.150.

50. Contributors to the notion of decadence include Vico, Montesquieu, Gibbon, Hegel, Spengler and Toynbee.

vigour have come to be understood simply in terms of the opposition between self-indulgence on the one hand and militarism and machine-like efficiency on the other. The only mobilization of people's potentialities conceivable has come to be the mobilization to conquer and dominate other people and the mobilization of people for industry. Vigour has come to be identified with the growth of Gross National Product and the rise of economic power to dominate other nations, and at least in Anglophone nations, decadence is equated with failure to reduce everything to instruments for economic development. Plato has been well and truly stood on his head. The ideal has been equated with what for Plato was the triumph of the most base, and people have been blinded to the possibility of anything beyond this. And these values are driving humanity inexorably towards the complete destruction of the environment.

The process view of the world provides the possibility of constructing an alternative ideal to judge societies by and for accounting for tendencies to decadence while avoiding the Platonist tendency to represent the ideal as a static form to be realized in the future. To begin with it is necessary to acknowledge that the world is a process of creative becoming without any definite end, and then to reformulate political ideals on this assumption. I propose that the 'generative' society by taken as the ultimate ideal. The generative society is not a static end state, but the structure which cultivates and provides the conditions for the fullest development of the potentialities of its members to participate in the creative becoming of society, of culture, of humanity and of nature. A generative society is a society in which people have liberty, which has an active cultural life as people struggle to orient themselves, and in which people are successfully struggling to make justice prevail, a society in which people have the conditions for and are struggling to deepen their understanding, heighten their awareness and extend their consciousness of the world, to confront society's and the world's problems and to express this in their work and lives. This end is not a future state, but the quality of the whole, unfinished duration of society's and the world's becoming. The present as part of this whole duration cannot be reduced to a mere means for realizing this end.

The progress of decadence can be described in opposition to this as beginning with the decay of dialogue and the disintegration of narratives, particularly broader narratives defining the history and goals of humanity, of civilization or of the nation as people contract their horizons, both spatially and temporally, cease to strive for an orientation to life beyond their immediate situations, and cease trying to understand and justify what they are doing - becoming hostile to any fundamental questioning of their lives, goals or ways of thinking. In intellectual life, metaphysics is replaced by sophistry, scholasticism, or analytic philosophy, the quest for understanding is replaced by the meaningless accumulation of facts and the quest for technical control, and the struggle to organize experience into coherent narratives is abandoned. Following this, the actions through which people strive to attain a sense of their significance cease to be defined from the perspective of the 'generalized other', or in relation to a grand narrative and come to be measured in terms of their impression on others. People strive for status rather than to live worthwhile lives. As people lose their sense of justice and injustice, what is and is not corrupt, political decisions come to be based on compromises between people with power - and the powerless are forgotten and trodden under foot. Institutions and organizations cease to be questioned and evaluated for their contributions to life in general and become progressively more self-serving. Since status without any general perspective to justify it can only be defined in opposition to those who are deprived of it, people's struggles for recognition take the form of dividing people into winners and losers. This generates increasingly complex interpersonal, social and political games which are usually unproductive and frequently destructive. As social relations become increasingly disaffirming and frustrating, games are oriented towards achieving power over others, and as a consequence, power, wealth and income are concentrated. Under these circumstances in which people are no longer able to fulfil themselves, they become more aggressive and violent, particularly towards those designated as 'pollution' by exclusive groups. Deviousness, 'rat cunning' and moral cowardice become habitual. As people lose sight of even the most pressing problems of their society, social crises proliferate. In this final state of decadence, people's creative potentialities cease to be cultivated, and no other potentialities are acknowledged than the most basic capacities to consume, to serve as instruments and to win out in power struggles, either civil, economic or military. In such circumstances cynicism appears clever, and idealism as a sign of feeble-mindedness. Those who do manage to rise above the prevailing condition, who do strive to orient themselves to a broader perspective and who struggle to meet the challenges confronting their societies, are isolated. Demagoguery, scheming and brute force become the order of the day.

With the conception of humans that has been defended the tendency noted by Hegel for major societies to embody an ideal which people strive to realize, and for successive social orders to embody forms of thinking which are more rationally coherent and which acknowledge a progressively greater proportion of the population as free, can be explained. People do require ideals to orient themselves, and are inspired by ideals which provide an orientation for action which enables them to achieve a sense of their significance. The revelation of contradictions is disorienting, preventing people attaining the sense of the unambiguous significance of their lives for which they are striving. Under these circumstance fewer people will be inspired to serve the institutions representing such ideals. And a society sinking into decadence will be less able to survive challenges to its power. On the other hand new groups of people struggling for power will usually only be successful against established power groups when they are able to formulate their struggles in terms of more coherent visions of the world which acknowledge the significance of more people than the world-orientation of their opponents.

But there is more to it than this. Ideas only begin to become important forces when disparate groups are struggling to overcome the conflicts which divide them in order to challenge the power of others. This is what was shown to have been the case in early medieval Europe and in early modern Europe, and at various times in the history of Russia. Furthermore, while the incoherencies of ideals can count in part for the decay of societies, there are also tendencies within all human organizations towards corruption and decadence quite apart from the inadequacies of the ideals which they incorporate. In other words, while ideals must be recognized as important, the tendencies towards generativity and decadence in societies are more complex than Hegelians have allowed. While from the perspective of process philosophy it is possible to explain the tendency towards greater rationality and freedom in society, there is no justification for believing in the necessity of such advances, nor for the belief in a final end state for which all previous history is only the means. It is likely that there will be periods of chaos and violence between generative eras, and there is no guarantee that on the collapse of one generative era a new generative era will emerge from such chaos. The identification of the sequence of social formations with the march of divinity, or humanity, towards its final self-actualization, must be rejected.

## *World Politics and the Problem of Representation*

One of the central problems of political philosophy, particularly in the present, is what is to be taken as its object of analysis. Both Plato and Aristotle took the polis as their object for political philosophy, Aquinas took the whole of Christendom and the relationship of this to kingdoms, while Hobbes and Hegel took the autonomous nation-State. This reflects the context within which these political philosophers were developing their ideas. Focussing on the complexity of modern institutions and the way people are controlled by them Foucault argued for a rejection of the traditional notion of sovereignty and called for political activity to be directed to local sites rather than to control of the State,[51] while Marxists and environmentalists have revealed political problems which transcend all national boundaries, implying that only by addressing the global situation can

51. Michel Foucault, *Power/Knowledge*, ed. Colin Gordon, Brighton: Harvester Press, 1980, esp. 'Truth and Power', pp.109-133.

political action be of any significance. Such a global outlook was originally taken by Kant who, in his *Perpetual Peace*, called for a 'league of peace', and this call for an internationalist orientation was revived by one of Hegel's students, Friedrich Carové. Carové argued that the ultimate realization of rationality in history was not the nation-State of Prussia (or USA) but an international State, and in particular an international legal system in which every individual in the world is recognized as a free agent. Beyond this he argued that the ultimate actualization of the ideal of an ethical community in which the free self-conscious Spirit would feel entirely at home demands the absorption of the political State into the association of humanity in a divine, fraternal community, involving the creation of a unified, world-wide public consciousness which would allow each and every individual to comprehend the whole variety of human expression as revelations of people's own divine faculties, capacities and powers.[52] It seems that Foucault and Kant, Carové and Marxist world-systems theorists are all correct in identifying political problems at different levels than the nation-State, although not in their assumption that the domains which they have identified could be the sole locus of political and cultural action. By formulating political philosophy in terms of process philosophy a basis is provided for dealing with politics (which can then be conceived as the process of defining and redefining the goals, ideals and values to be realized by and within any social formation) at a multiplicity of levels without assuming that any one level is pre-eminent. However to relate each level to each other and all to the world community as a whole it is necessary to work out how to represent people at different levels of organization.

One of the most important problems arising from this is to work out what is representation and how effective representation can be achieved in a world of enormous complexity. In *The New Science of Politics*, Eric Voeglin defined a representative as 'a person who has the power to act for a society by virtue of his position in the structure of the community, without specific instructions for a specified business, and whose acts will not be effectively repudiated by the members of the society.'[53] This definition leaves it open how such representation is possible and what it means to be properly represented. To comprehend this, representation must be seen as simultaneously involving each of the dialectics of culture: that of orientation, of recognition and of control. To begin with, representation is an essential part of the struggle for orientation and for recognition, and must be evaluated in terms of its success in this regard. Representation is part of the process by which groups of people are defined and define themselves as being communities with specific problems, aspirations and significance. It is the condition for a potential community to become a reality. However to succeed in becoming a reality, the community must be represented as part of the whole of reality, as part of a general order of things so that they can identify their own lives within, and orient themselves to, the world at large. And as Kenneth Boulding (following Fred L. Polak) has argued: 'there is a great deal of historical evidence to suggest that a society which loses its identity with posterity and which loses its positive image of the future loses also its capacity to deal with present problems, and soon falls apart.'[54] Future and past generations must be represented to properly represent people in the present. Similarly people require an identity with the environment, particularly in the immediate vicinity, and should be represented as part of local, regional and global ecosystems.

Having representatives will not in itself guarantee proper representation. For this to be achieved, representatives must have a perspective on the world (which must include accurate knowledge and continued access to appropriate information - but which is not reducible to these) which defines the past and articulates the concerns and aspirations of the individuals or groups represented and which can be expressed and integrated into whatever decision-making or actions

52. For a study of Carové see John Edward Toews, *Hegelianism*, Cambridge: Cambridge University Press, 1980, pp.134-140.

53. Eric Voeglin, *The New Science of Politics*, Chicago: The University of Chicago Press, 1952, p.37.

54. Kenneth E. Boulding in Herman E. Daly (ed.) *Toward a Steady-State Economy*, San Francisco: Freeman, 1973, p.129.

the representative is involved in. This requires of this particular perspective that it be able to be related to the broader perspectives on which political decisions are made. Relating perspectives to decision-making, and relating perspectives to each other is achieved by constructing narratives - histories defining the achievements and failures of past projects, defining the problems of the present, and projecting a future to be realized. The development of perspectives and their integration into narratives is then the most important condition for achieving real representation. The condition for the development of such perspectives is not just free speech, but unbiased media able to support and communicate the development and criticism of perspectives, educational institutions which take as their prime goal not vocational training but the development of understanding of the world - of people's ability to define themselves by appropriating and participating in the development of their cultural heritage, and an active cultural life in which the general public is engaged in defining itself historically, questioning, developing and replacing prevailing perspectives and the projects based upon them.

Representatives must also be effective, they must have the power to ensure that all they represent is taken into consideration, that their perspectives are incorporated into political decisions. The problem is to ensure that representatives have some redress when they perceive their representees to be unjustly done by, while limiting their power to impose unjust decisions on others. That is, the structure of the community must be such that all can be effectively represented. So long as one group, for instance large, transnational corporations, are able to hold societies to ransom, then large numbers of people in society will be inadequately represented.

To achieve representation in a complex world requires the encoding of perspectives in impersonal laws - as Rousseau, Kant, Hegel and Carové among others have argued. Such laws represent the people whose interests are taken into account by them, and who can then make claims which will be backed up by the community. But quite apart from these laws providing means for those with political power to oppress people, legal systems tend to become self-serving, and to subordinate people to their own ends. The only hope of checking such corruption is by keeping alive the idea of justice as something to which all government and legal processes must always be subordinated, and maintaining a critical process of review supported by an active and critical cultural life to expose when and where government and legal processes are unjust; and only those purported laws which are just should be accepted as laws.[55] This requires the subordination of the abstractions of law to the narratives by which people define themselves and their struggle for justice.

However even if all these conditions were met, there would still be no guarantee that people would be justly represented. So long as there are representatives there will always be a tendency towards corruption. To begin with, representatives have a tendency to usurp symbolic power from the groups they purport to represent. Such purported representatives may then not only fail to express the concerns of the group from which their symbolic power derives, but may contribute to decisions and processes directed against the concerns of this group. Such usurpation can be institutionalized and then disguised, producing a form of fetishism.[56] For instance priests usually do not define themselves as representing the people who believe in their religion, but as representing God, although their status as priests would amount to nothing if nobody but they themselves believed in the doctrines of their religion. Such fetishised usurpation is present where-ever social processes are reified, where civil servants present themselves as representatives of the State, where the ruling class present themselves as representatives of the Nation, or where revolutionaries present themselves as representing History or the Proletariat. Such usurpation is usually associated

55. Aquinas wrote: 'As Augustine says: "that which is not just seems to be no law at all", wherefore the force of law depends on the extent of its justice.' *Summa Theologica*, I-II, Qu.95, Second Article.

56. The nature of this has been brilliantly analysed by Pierre Bourdieu, 'Delegation and Political Fetishism' in *Language and Symbolic Power*, tr. Gino Raymond and Matthew Adamson, ed. John B. Thompson, Cambridge: Polity Press, 1991, pp.203-219.

with an impersonalization of the symbolic role. Thus the Pope presents himself not as exercising his power but as the medium through which God expresses Himself, the bureaucrat presents him or herself as a mere instrument of the State, and the revolutionary as an instrument of history or of the proletariat. In a society where such fetishised representation is widespread, there is a downgrading of individuals, and with this, of the life of dialogue essential to achieving and maintaining genuine representation. If individuals purport to represent only themselves, they are seen as representing no-one, and the significance of inquiry and dialogue through which people are struggling to develop their understanding of the world, the essential condition for the development of adequate perspectives and for the critical review of institutions, is denied proper recognition - or worse, if this involves questioning of those whose symbolic power is fetishised, as anathema.

The structure of representation which is most likely to be successful and to avoid such corruption is one which decentralizes power so that laws are enacted and decisions made at the most spatially and/or functionally proximate centre of decision-making at which all those most affected by decisions can be represented, yet which can at the same time effectively represent particular concerns at 'higher' levels of organization when necessary (where 'higher' simply designates the broader scope to be considered). And it is necessary to acknowledge once and for all the correctness of Montesquieu's view on the need for a division of powers, to have a plurality of structures which can act as checks on each other to counter the tendency for the most ruthless to take control of organizations and for organizations to become self-serving and oppressive. To succeed in this, it is necessary to go far beyond the division between the executive, the legislature and the judiciary and to develop federated systems of government, different forms of economic enterprise (as market socialists such as Alec Nove have proposed), politically oriented trade and professional unions, media free from control of governments, press barons and advertisers, open civil services where civil servants are free to publicly criticise their superiors and where policy proposals are published - as in the Swedish civil service, education systems with autonomy from government and economic pressures - preferably ones in which a diversity of institutions compete with each other as in nineteenth century Germany, an independent legal system recognizing the subordinate status of laws to justice and which is really accessible to all, and so on. These should be organized so that it is possible and in the interests of members of each institution to expose the corruption of members in both their own and in other institutions.[57] To facilitate decentralization it is necessary to have procedures for putting forward ideas and problems for consideration and for challenging decisions and censuring corrupt representatives at each level of organization both from higher levels (which represent broader interests), and lower levels (which represent more particular concerns). Also, to avoid the tendency for energy rich regions to dominate energy poor regions (as has occurred in Brazil, for instance), it is necessary that the personnel and funding for organizations in any region come directly from that region, and that these organizations have the power to stop economic enterprises based elsewhere operating in their region. The challenge is then how to design organizations which decentralize power and provide for initiative and review from different levels, while retaining the capacity to coordinate lower levels or related organizations to respond to more universal and longer term problems. Again, the most important means for achieving this challenge is an active cultural life in which shared perspectives are developed and maintained, perspectives formulated as narratives which show how the present has developed from the past and which articulate particular concerns and relate these to broader concerns of society and humanity so that people, at each level of an organization and in related organizations, can understand each others' points of view, ambitions and projects and balance claims to justice.

With this notion of representation it is possible to further elaborate on the nature and conditions for generative and decadent societies. A generative society can be seen as one which justly and effectively represents through the narratives being lived out its individual members, groups of

57. On the conditions which prevent or facilitate corruption, see Stephen Bunker, *Underdeveloping the Amazon*, Ch.7, 'Collaboration, Competition, and Corruption in Two Colonization Projects', pp.190-197.

people sharing significantly similar situations, the society as a whole, future generations, humanity and the biotic community in the general functioning of organizations and in political decision-making processes of the society, so facilitating the fullest development of the potential of the society, its individual members and the rest of the biotic community. Decadence corresponds to a failure of representation, which manifests itself when people in society, particularly those committed to justice and to living with integrity, are not represented or can no longer identify with those who purport to represent them and can no longer get their particular concerns taken into consideration and catered for, and when future generations, the rest of humanity and the environmental conditions for life are not taken into consideration. Genuine political struggles can be seen as essentially struggles for representation (rather than merely the struggle for power within the existing order), and the rise and decline of societies is the consequence of both the success or failure of different people in these struggles and of the structures of representation which they create. The struggle against decadence requires the formulation of perspectives on the world and the integration of these into narratives in which people can see themselves represented, and leaders who, embodying such perspectives, can effectively articulate the interests and aspirations of people and inspire them to struggle to realize the goals projected by these narratives in practice, to crystallize these narratives in institutions and thereby to transform society. At present, the problem is that, through a process of political integration and exclusion in the core zones of the international capitalist system, less and less people are being effectively represented by the narratives of progress and the institutions which dominate the world, while future generations and the environment are scarcely represented at all. What is now required is a world-wide struggle to represent the entire population of the world, together with all future generations as a community within the biotic community of which humans are part, and through the construction of a new grand narrative, the articulation of this into a sufficient number of levels and divisions to effectively represent each individual, each local community, each group, nation and region in the context of this global community, both human and non-human.

## *The Human Sciences*

In working out how to act and how to live, in challenging and attempting to alter the existing orientational, ethical and power structures of society, and in formulating political goals and planning political action, it is necessary to consider what is possible. To reveal the possibilities of making justice prevail, of achieving proper and effective representation, it is necessary to understand what the present state of affairs is, how existing orientational, ethical and power structures are maintained and reproduced by the complex of social practices, institutions and economic, social and political processes already involved in the dynamics of the social world, ranging from the local to the international level, and the relationship between each of these and the present state and dynamics of the rest of nature. It is necessary to understand these dynamics to reveal when, where and by whom action to improve the world can be effective. Being effective should not be understood simply in terms of gaining power, but in terms of what relationships between people could be made to prevail. It is necessary to consider not only what oppressive forms of relationships could be overcome, but also what structures of orientation, recognition and power can be created and maintained and which of these would be most likely to ensure that inquiry and communication would be cultivated, justice achieved, effective representation gained, worthwhile ends realized and tendencies towards corruption minimized. Success will require the creation of an image of the future together with the specific goals which must be attained to realize it, based on an understanding and critical evaluation of existing processes and structures making up society. This critical understanding of the world should enable individuals and groups to define their problems and aspirations, to consider each structure of communication, recognition and power and each emergent social process and complex of processes, in terms of whether and how they facilitate or prevent the achievement of their own particular goals, the goals of their community and the goals of humanity.

To this end, ethics and political philosophy should be integrally related to efforts to understand these complex relations, to orienting people for action and for life, to providing the means by which individuals could be understood and could understand themselves in relationship to the complex order of society. There are two rival ways in which people attempt to deepen their understanding of society, the tradition of historical and fictional narrative construction which proceeds by attempting to construct coherent narratives about agents, both individual and collective, and the social sciences which attempt to predict and explain social phenomena. What is required is a transcendence of this opposition, and this is made possible by process philosophy.

As shown in the previous chapter, process philosophy provides a conceptual framework for overcoming the divisions between the humanities and the sciences, between the social sciences and the natural sciences, and between theory and practice. By conceiving the goal of science as the development of understanding, and by providing a unified conceptual framework for analysing the relationships between the dynamics of the physical world, the biological world and the complex of processes which make up the social world, and which can then situate conceptually and analyse the structures of the life-worlds of people, process philosophy provides a basis for a thermodynamically and ecologically based socio-political economics which could relate all world problems into an integrated historical narrative, while situating people as individual agents within this narrative. Through such a narrative, people, conceived of as situated, but partly self-creative processes of becoming within the becoming of the world, as participating in this becoming with each thought and action, could be provided with the means to extend and deepen their understanding of themselves as potential agents of this becoming.

This historical social science would transcend the disciplinary boundaries which at present dominate the study of humanity. While it is necessary to abstract out individual social processes to understand their particular dynamics, process philosophy requires that such abstraction always be acknowledged as such, and that abstract models never be identified with social reality. The disciplinary boundaries formed by such abstractions are so at odds with the complex interdependencies within society that the general population are being blinded by prevailing social science rather than informed by it. As James O'Connor argued: 'as social theory becomes more specialized, the economy, society, and polity become more unified... Hence, never before has it become so essential to invent, however crudely and provisionally, a method which combines historical interpretation, ideology critique, political economy, economic sociology, and political sociology.'[58] At most, three broad disciplinary boundaries might be regarded as acceptable within the human sciences: the study of culture, a broadly conceived human ecology, and psychology, although even these would be related through philosophical anthropology and be in constant interaction; and all studies of humanity would be historical, while all history would be theoretically informed. Since the defining feature of humans, being the condition of both the existence of complex institutions and their individualism, the study of culture or cultures must be regarded as the pre-eminent human science, 'human ecology' dealing with the structures or institutions and emergent processes associated with people's transformations of their physical, biological, socio-cultural environments, encompassing geography, political economy, sociology, politics and law,[59] would assume a conception of humans and then continually revise this in the light of advances in the study of culture and psychology, while psychology would conceive its object of study, the individual subject, as being essentially biological, cultural and social as well as personal, and therefore incapable of being totally abstracted from the study of culture and the dynamics of

58. James O'Connor, *Accumulation Crisis*, Blackwell: Oxford University Press, 1984, p.vii.

59. The study of money-making and spending - chrematistics - must be placed in a broader context and evaluated in terms of this broader context. For some advances in this direction see André Gorz, *Critique of Economic Reason*, tr. Gillian Handyside & Chris Turner, London: Verso, 1989.

societies.[60] These sciences would conceive people, from individuals to humanity as a whole, in the broader perspective of the world ecosystem as a complex of dissipative structures ultimately maintained by the condition of far from thermodynamic equilibrium produced by the sun, in which all power is ultimately control over the transformations of usable energy, and historically as a complex of narratives, institutions, traditions and emergent social processes whereby humans have been formed and have transformed themselves and their environments to create the present world-order.[61]

Respecting such interdependencies would not involve reducing the complexity of social reality to manifestations of one holistic process. Social reality cannot be reduced to a single plane of becoming. As Foucault argued (reflecting the influence of Braudel, and ultimately, of Bergson):

> It's not a matter of locating everything on one level, that of the event, but of realising that there are actually a whole order of levels of different types of events differing in amplitude, chronological breadth, and capacity to produce effects. The problem is at once to distinguish among events, to differentiate the networks and levels to which they belong, and to reconstitute the lines along which they are connected and engender one another.[62]

It is consequently impossible for social science to give a transparent representation of social reality as a totally predictable order. And it cannot presuppose a privileged perspective within this process of becoming of humanity. In place of prevailing economics which tacitly presupposes the perspective of governments and businesses in the economic centres of the world, the new historical political economy should enable people to define and orient themselves to the world from their own particular situations, whether they be businessmen, workers, peasants or unemployed, males or females, representatives of governments in core, semi-peripheral or peripheral regions of the world, representatives of international or local organizations, or whatever.

Presupposing that humans are cultural beings and that social science itself is cultural activity, this social science would be explicitly evaluative, with evaluation being grounded in the conception of humans and their place in nature assumed as the hard core of a research program. But developed according to a dialectical theory of knowledge, this conception of humanity would be seen to be open to question, to revision or to replacement, rather than, as with the assumptions about humanity of prevailing social science, being presupposed. This would reincorporate questions of evaluation into the realm of rational discourse. With the process conception of humanity as the reference point for evaluation, in place of 'economic man' and the Social Darwinian notion of progress through the survival of the fittest, social science would firstly evaluate social formations in terms of their contribution to the stability and resilience of the world's ecosystems, their sustainability, and then in terms of the quality of the life-worlds generated by them. Social relations, institutions and emergent social processes would be judged in terms of the justice of the conceptualizations of the world embodied and reproduced by them, in terms of how they facilitated

---

60. This is the form of psychology argued for by Rom Harré in *Social Being*, Oxford: Blackwell, 1979 and *Personal Being*, Oxford: Blackwell, 1983 - anticipated to some extent by Vygotsky.

61. Such an ecological economics has been in the making for over a century without gaining institutional recognition. See Juan Martinez-Alier, *Ecological Economics*, Oxford: Blackwell, 1987. A milestone in its development was Nicholas Georgescu-Roegen's work, *The Entropy Law and the Economic Process*, Cambridge: Harvard University Press, 1971. See also Herman E. Daly, *Steady-State Economics*, San Francisco: Freeman, 1977, and Narindar Singh (*Economics and the Crisis of Ecology*, 2nd ed. Delhi: Oxford University Press, 1976), and Herman E. Daly and John B. Cobb, Jr. *For the Common Good: Redirecting the Economy Toward Community, the Environment, and a Sustainable Future,* Boston: Beacon Press, 1989. I am at present attempting to reformulate and synthesize ideas from institutional economics, political economics, global economics and bio-economics through process philosophy.

62. Michel Foucault, *Power/Knowledge*, tr. Colin Gordon et. al. Brighton: Harverster Press, 1980, p.114.

or failed to facilitate the attainment by people of recognition and respect, and in terms of the conditions being provided to people to participate in the shaping of their destinies.

Such a social science would also orient people for action. Georg Lukács argued:

> As long as man concentrates his interest contemplatively upon the past or the future, both ossify into an alien existence. And between the subject and the object lies the unbridgeable 'pernicious chasm' of the present. Man must be able to comprehend the present as a becoming. He can do this by seeing in it the tendencies out of whose dialectical opposition he can make the future. Only when he does this will the present be a process of becoming, that belongs to him.[63]

According to the process view of the world the complete separation of theory from practice is impossible. Social science should facilitate deeper 'indwelling' within the world so that its significance and the significance of different possible projects can be judged. Beginning with the assumption that science is part of the orientational structure, the on-going dialogue through which people are developing their understanding of themselves and their place in the world, process philosophy supports Jürgen Habermas's contention that it is impossible to comprehend the social world without evaluating the validity claims being made by social actors.[64] A social science and history based on process philosophy would be concerned to reveal unjust forms of thinking and the forces engendering and reproducing them to liberate people from ideological mystification. Its aim would be to provide people with better means to understand themselves and their motives, to reveal what ends are worth striving for, and to provide people with the means to articulate their aspirations. It would aim to enable people to better comprehend the different tendencies within the world, the extent to which their own ends are being frustrated or facilitated by these, and what part they could play in furthering or inhibiting these tendencies. Trying to illuminate the present in the light of the past, it would aim to contribute to the construction of the future. The development of such a social science would be part of the self-formation of humanity.

This would require more than just exposing the failings of the existing order. One only refutes what one replaces. And as Rom Harré pointed out: 'people create themselves and their patterns of interaction by virtue of the psychological and social theories to which they subscribe.'[65] This new social science would be providing people with new ways to conceive themselves, their society and the world to replace those being revealed as defective. It would replace the categories of existing economic theory, the 'forms of existence' of capitalist societies by concepts consistent with a process view of the world.[66] 'Labour-power' would be replaced as the dominant concept defining work relationships by concepts which acknowledge the full needs and potentialities of people as creative, social agents, and the dynamics and intrinsic value of other forms of life. Podolinski's energy theory of value and the associated reformulation of the concept of 'surplus value' would replace the neo-classical concept of exchange value; and Daly and Cobb's concept of 'Index of Sustainable Economic Welfare'[67] and Oldak's concept of 'gross social wealth', or some equivalent, would replace the notion of 'gross national product' as the ultimate reference points for evaluating economic performance, thus situating the monetary system within the environment and bringing into focus the real contribution of economic activity to the conditions of life - both human and non-human. With such concepts, mining, cutting down trees and the destruction of agricultural land

63. Georg Lukács, *History and Class Consciousness*, tr. Rodney Livingstone, London: Merlin Press, 1971, p.204.

64. Jürgen Habermas, *The Theory of Communicative Action*, [1981], tr. Thomas McCarthy, (2 vols), Vol.1, Boston: Beacon Press, 1984, pp.102-141. This is also argued by Charles Taylor, 'Understanding and Ethnocentricity', *Philosophy and the Human Sciences: Philosophical Papers II*, Cambridge: C.U.P., 1985, Ch.4.

65. Rom Harré, *Personal Being*, Oxford: Blackwell, 1983, p.24.

66. For some developments in this direction see Daly and Cobb, *For the Common Good*.

67. Ibid. 'Appendix'.

would be recorded as costs and loss of wealth, while activities which are at present excluded from national accounts would be accorded value. A sharp distinction would be drawn between regenerating sources of usable order - such as sunlight, climatic systems, rivers, species, ecosystems, people etc. which alone should be designated as resources (from the Latin *resurger* - to rise again), and usable order which has been saved up - such as concentrations of minerals, oil etc., are rightly designated 'reserves' (from the Latin *reservar* - to save up), and currency given to the concept of ecocide - the destruction of resources and dissipation of reserves so defined. Other concepts would then be reformulated to accord with this new way of thinking.

However it is not only particular concepts which would be transformed, but along with these, the image of society. There will always be an image of society dominating any community, and this will always function to some extent as an ideal. Process philosophy would replace the analogy of the machine which underlies prevailing economic thought with an auditory analogy to enable society to be understood as a creative process of becoming within nature. At the same time it would promote some variant of Wallerstein's notion of world-system, since quite apart from its role in revealing the causes and extent of economic exploitation, political oppression and environmental destruction, such an image of the entire world is required to construct a world community.

Through analysis of the tendencies within the existing societies, such a social science could mobilize people to replace prevailing concepts and images by revealing the commonality of interests between those who are oppressed by the present system, by presenting an image of the future worth striving for, and by giving some idea of the paths which people, individually and collectively, could take to help realize this future. And as Marx argued, the validity of social theories can only be judged by whether people take them up and define the world accordingly, and then by whether the promise of these theories, the potentialities they purport to reveal, are realized in action: 'Man must prove the truth, i.e., the reality and power, the this worldliness of his thinking in practice.'[68]

## *Policy and Strategy Formation*

One of the most important requirements for representing people and for transforming society is systematically formulating and evaluating political and economic policies and programmes. Formulation of policies and programmes is usually based on mechanistic assumptions in terms of neo-classical economic theory in which the economy is treated as a closed system, driven by greed, tending towards equilibrium, and in which nature is treated as nothing but a passive resource. Evaluation is generally based on some version of cost-benefit analysis. Cost-benefit analyses were criticised in Chapter 2 for their assumption of a mechanistic conception of the world, and deriving from this, their assumption that the world can be understood as the sum of all its states of affairs and events. Self-organization processes with their immanent dynamics are ignored. Such analyses cannot take into account the complex interdependencies of reality and replaces democratic procedures by a managerial approach to decision-making in which decisions are taken on the basis of pseudo-scientific quantification procedures. But an alternative strategy and policy-making procedure has been developed which accords with the conception of people orienting themselves primarily through narratives, which assumes a dynamic, active world, and which tends to democratize decision-making rather than concentrating it in the hands of 'experts'. This is 'retrospective path analysis' developed by Cliff Hooker.

Retrospective path analysis consists in firstly the selection of macro-economic goals by considering a variety of end-points forty to fifty years in the future, and then secondly examining various paths to the desired future state. However there is no reason why this cannot be extended to considering goals for the whole of civilization several centuries into the future, and considering a

---

68. Karl Marx, 'Theses on Feuerbach', 2.

variety of sub-goals for achieving this. This procedure departs from the normal approach in calculating a course of action retrospectively from some future date, specifying 'those key transitions in social structure and functioning generally which, taken in proper sequence, will lead from the present to the desired future social condition.'[69] Such an approach focuses attention on the conditions necessary for achieving the desired future states, on the tendencies inimical to the realization of such ends, and on the crucial societal decisions at the branchpoints of different possible paths of development.

Retrospective path analysis accords with the way people generally formulate and commit themselves to projects. Projects formulated and acted on in this way have essentially the same structure as narratives and allow for a complex structure of sub-projects as sub-narratives. Formulating such projects would provide people with unfinished stories or complexes of stories to situate themselves within as creative agents. Decision-making would require recognizing the limits of knowledge, taking into consideration how much room for manoeuvre is given to different actors during the process of reaching desired ends. Decisions would be constantly open to re-examination and reformulation. Furthermore, since the way people think can and should be included as one of the ends to be aimed at, people's way of thinking and relating to each other and to the world could be incorporated into the path analysis. This means that retrospective path analysis would avoid the tendency to reduce other people who are to be involved in striving to realize ends into instruments. It would open up for democratic discussion the question of what sort of future we want, and open to question what sort of people we wish to become and what sort of relationships between people we should be developing. This would require a fundamental questioning of what kind of beings we are, what is our place in the cosmos and what are our potentialities. Such a decision procedure would contribute to transforming people's attitudes from a mechanistic world-orientation to a process world-orientation, from seeing themselves as beings standing outside the world trying to control it to experiencing themselves as processes of becoming actively participating as cultural beings in the becoming of the world. People would become responsible agents creating themselves through forming and reforming the narratives defining themselves and their place in the world.

Retrospective path analysis accords with and would reinforce the need for a new science of humanity based on process philosophy. While cost-benefit analyses implicitly assume an instrumentalist form of rationality and a crude positivistic theory of science in which knowledge amounts to the ability to predict the probabilities of the occurrence of different future states and events, retrospective path analysis is consistent with the notion of creative rationality and the ethical notions associated with it, and requires the development of the form social science being proposed - one which facilitates analyses of the diversity of and complex inter-relationships between processes, and which subordinates such analyses to historical narrative. The full development of such a social science would provide the means for situating policy analyses within the broader socio-cultural dynamics of particular societies, of civilizations and of humanity as a whole over different durations, and take into account, consider and balance the different claims to justice of acting upon such policies. It is such a form of policy formulation and the human science which is required to confront the present environmental and cultural crises in all their complexity.

---

69. Clifford A. Hooker, 'Scientific Neutrality versus Normative Learning: The Theoretician's and Politician's Dilemma', in David Oldroyd ed., *Science and Ethics*, Kensington: New South Wales University Press, 1982, pp.8-33, p.17.

# 17

# TOWARDS AN ECOLOGICALLY SUSTAINABLE CIVILIZATION

Given the dynamics of the existing economic and political organization of the world, putting a stop to environmental destruction and maintaining a sustainable world-economy can only be achieved by creating a new social, political and economic world order. In the light of the analyses offered in this work it should now be a straightforward matter to describe in broad outline what kind of new order will be required: a drastic reduction of social inequality throughout the world and in each country, the decentralization of political power, and a radical revaluation of nature and community. The world-system of regional exploitation needs to be destroyed, and international relations rebuilt on the basis of justice in the relationships between people and between humanity and nature. Population growth needs to be checked by eliminating the poverty, insecurity and ignorance which generates it. Sustainable life-styles and forms of agriculture should be preserved or developed to replace forms which degrade the environment. For those in the economic core zones where people have achieved the material conditions for a decent life, lifestyles which use up the minimum amount of reserves and which preserve resources, which slow down the dissipation of entropy rather than hasten it, need to be promoted. This will require the transformation of the moral structures of societies so that people are accorded recognition when they contribute towards such changes and participate in such lifestyles, and despised otherwise.

The biggest problem in achieving this will be to overcome the autonomous dynamics of international capitalism, to liberate the Third World from its subjugation and exploitation and to develop new politico-economic structures throughout the world which redistribute power. Markets need to be insulated from each other in order to prevent regional exploitation and to undermine the dynamics of international capitalism, and it will be necessary to put an end to or prevent the formation of markets altogether in those areas of the world only capable of supporting in a sustainable way subsistence modes of production. Breaking the domination by the economic core zones, transnational finance, agribusiness and global manufacturing organizations and the comprador classes who serve as agents for them, while at the same time representing those interests and concerns which transcend national boundaries, will require the unification of major regions of the world.[1] Within these regions each nation needs to centralize power to control the market and to deal with those issues affecting the nation, the broader region and humanity, while decentralizing power to ensure against the blindness of bureaucracies and the tendency for metropolises to exploit peripheral regions.

While the unique histories and qualities of each locality, country and broader region need to be taken into consideration, the kind of economic system most likely to enable people in the industrialized West to control their destinies in accordance with the long term interests of humanity and nature, is some variation of the market socialism argued for by Alec Nove - with economies consisting of centralised state corporations (which should include all military equipment manufacture), socialised enterprises (state or socially owned with full autonomy and with management responsible to the work-force), co-operative enterprises (of which Mondragon is an exemplary example),[2] small-scale private enterprises subject to clearly defined limits (which would

---

1. The need for such regional struggles for independence as a means of overcoming international exploitation has been argued for by Dudley Seers in *The Political Economy of Nationalism*, Oxford: O.U.P., 1983.

2. For an excellent study of Mondragon, see Henk Thomas and Chris Logan, *Mondragon: An Economic Analysis*, London: George Allen & Unwin, 1982.

include the family farm), and individuals (e.g. artists, freelance journalists and plumbers).[3] However there should be far more regional containment of markets, particularly for capital, than advocated by Nove, and markets should never be more, or be conceived as more, than instruments to decentralize power and decision-making and to provide enough competition and freedom for individual enterprise to stimulate creative effort and guard against corruption and 'bureaucratization' of organizations.

It should be recognized that the market mechanism can only function properly for some goods and services, and that the market by itself is not an efficient or proper means for allocating returns to factors of production. The market mechanism is blind to the medium and long-term future, and blind to the intrinsic value of nature and people. An unhindered pricing mechanism will not lead to the best use of reserves and resources because people in the future, let alone plants, animals and ecosystems, cannot bid on the market. If it is allowed to operate unhindered for labour it debases people, reducing them and their creative activity to commodities, it creates insecurity and it leads to the concentration of income and wealth, all of which corrupt the ethical and political life of society. And if the pricing mechanism is allowed to operate unhindered for capital it is temporally unstable, tending to concentrate income, wealth and power, producing cycles of booms and depressions, and spatially unstable, tending to concentrate the means of production in small regions which is disastrous for the people outside these regions and catastrophic for the world's environment dominated by these regions. The operations of the market will not support efforts to address long-term problems which will only benefit future generations - whether these be efforts to conserve reserves and preserve resources, to reafforest land, to reduce pollution, or advance our understanding of the world. The common good, the distribution of and rewards to factors of production and meeting long term problems, need to be recognized as ethico/political problems. The market should always be subordinated to ethically based political institutions with the power and the will to take longer-term perspectives and to ensure that justice prevails in people's relationships to each other and to nature.

Developing such a new ethical, political and economic order will involve a long and complex struggle. To achieve the necessary changes, a fundamental, world-wide cultural transformation will be required. The former premier of the Soviet Union, Mikhail Gorbachev, arguing that the development of armaments is no longer the means to security, pointed out: 'This is a totally new situation which signifies a break with the traditions, the way of thinking and the patterns of behaviour, which have developed over centuries, and even over millenia.'[4] These same traditions, ways of thinking and patterns of behaviour also have to be changed to overcome the environmental crisis. Changing the way people understand themselves and incorporating a new way of thinking into society, as both the condition for addressing the major problems of the age and the condition for changing the social and political order of the world, is the most difficult task of all. However while it is almost unimaginably difficult and will take perhaps centuries to achieve, it is a task which should now be regarded as absolutely essential if humanity, and most other life-forms on earth, are to survive.

To begin the struggle for such a massive transformation it will be necessary to work towards the establishment of what the Italian Marxist Antonio Gramsci called an alternative hegemonic culture to oppose the hegemony of the increasingly nihilistic culture of international capitalism. It will be necessary to make the immediate economic and political crises afflicting capitalism a central issue in the struggle to establish and develop this alternative hegemony, since the effects of such crises are so all pervasive that no social movement which fails to confront them can be taken seriously. However these crises should be shown to be related to the broader context of the environmental crisis, and the alternative hegemonic culture, to be effective, needs to be based on a new world-orientation articulated into a new grand narrative transcending Eurocentricism and

---

3. Alec Nove, *The Economics of Feasible Socialism*, London: George Allen & Unwin, 1983.

4. Michail Gorbachev, *Selected Speeches and Articles*, 2nd ed., Moscow: Progress Publishers, 1987, pp.213-224, p.219.

anthropocentricism, a grand narrative which redefines the past and projects a new future, and relates all individuals, all organizations, all communities and all societies to the struggle to realize this future. Furthermore, a movement devoted to reforming the world needs to gain more than the mere allegiance of people willing to fight for new institutions and new power relations. Through this struggle it is necessary for people to change the way they experience the world, the way they understand themselves and their place in the world, how they relate to each other, the way they live and the way they organize themselves. The new way of conceiving things, the new world orientation and new grand narrative need to be incorporated into the autobiographies and broader narratives by which individuals and communities define themselves. Ultimately they need to be incorporated into their mode of being in the world as a habitus which can challenge the prevailing habitus with its mechanistic world-orientation. It is necessary to begin the process of embodying a new world-orientation into social relations, organizations, institutions, the built-up environment, and language itself.

## *Hegemony and Alternative Hegemony*

The concept of hegemony is one of the most fruitful and influential concepts developed within Marxism.[5] It was originally used by Plekhanov and other Russian Marxists in the 1880's in their call to the working class to lead an alliance with the peasantry to overthrow Tsarism. This involved transcending limited economic concerns and developing a national approach to fight for the liberation of all oppressed nationalities, classes and groups. The strategy was taken up and developed by Lenin in opposition to the passive 'economism' and 'class reductionism' of the Mensheviks, and the success of the Bolsheviks under the leadership of Lenin was based on this strategy. However it was Gramsci who in the 1920's transferred a term which had only been used in formulating strategy into a concept of analysis, and developed the notion of cultural hegemony.

In developing this concept it is likely, although difficult to prove, that Gramsci was influenced by the ideas of Bogdanov.[6] Bogdanov had set up workers' academies in Italy between 1909 and 1911, following which, Tasca, Gramsci's early mentor in the Socialist Party, advocated a program of education and cultural development for the working class. In 1919, paralleling the *Proletkul't* movement in the Soviet Union, Tasca, Gramsci and Togliatti founded a journal, a weekly review of socialist culture. To highlight the importance of culture, Gramsci extended the term 'hegemony' to include all the practices of the capitalist class in attaining and maintaining State power. He argued that in class rule, force is only the last resort, that a class can only gain and retain power by leading ideologically and politically. Hegemony is then a relation not of domination by means of force, but of consent by means of political and ideological leadership. It is the organization of consent.

Developing the concept of class hegemony enabled Gramsci to reveal how entrenched the organization of consent can be. Hegemony is not achieved through a few intellectual ideas, but is integrated into people's lives through civil society. As one interpreter summed up Gramsci's notion of hegemony:

---

5. For the concept of hegemony see Antonio Gramsci, *Selections from the Prison Notebooks*, ed. and tr. Quentin Hoare and Geoffrey Nowell Smith, New York, International Publishers, 1971, passim. See also Christine Buci-Glucksmann, *Gramsci and the State*, tr. David Fernbach, London: Lawrence and Wishart, 1980; Roger Simon, *Gramsci's Political Thought*, London: Lawrence and Wishart, 1982; Carl Boggs, *The Two Revolutions*, Boston: South End Press, 1984; Ernesto Laclau & Chantal Mouffe, *Hegemony & Socialist Strategy: Towards a Radical Democratic Politics*, tr. Winston Moore and Paul Cammack, London: Verso, 1985; and Joseph V. Fermia, *Gramsci's Political Thought*, Oxford: Clarendon Press, 1987.

6. On this, see Zenovia A. Sochor, 'Was Bogdanov Russia's Answer to Gramsci?', *Studies in Soviet Thought*, Vol.22, 1981, pp.59-81, esp. p.60.

> It is a whole body of practices and expectations, over the whole of living: our senses and assignments of energy, our shaping perceptions of ourselves and our world. It is a lived system of meanings and values - constitutive and constituting - which as they are experienced as practices appear as reciprocally confirming. It thus constitutes a sense of reality for most people in society, a sense of absolute because experienced reality beyond which it is very difficult for most members of the society to move, in most areas of their lives. It is, that is to say, in the strongest sense a 'culture', but a culture which has also to be seen as the lived dominance and subordination of particular classes.[7]

The social relations of civil society, embodied in the great variety of organisations making up civil society, are at the same time relations of power just as much as, though in a different way than, the coercive relations of the State. The State is then redefined as civil society plus political society; in other words, hegemony protected by the armour of coercion. On the basis of this analysis, Gramsci argued that the tasks ahead of Marxists in Western Europe were considerably more difficult than those which had faced the Bolsheviks in Russia. Western Marxists have to overcome not only the coercive State, but civil society through which capitalists organize consent and disperse their power. Thus, in comparing Tsarist Russia and the West, Gramsci wrote: 'In Russia the State was everything, civil society was primordial and gelatinous; in the West, there was a proper relation between State and civil society, and when the State trembled a sturdy structure of civil society was at once revealed.'[8] What is required in the West is not a war of manoeuvre, but a war of position, demanding enormous sacrifices by infinite masses of people. This involves developing an alternative culture to the hegemonic culture of the ruling class.

In developing this point, Gramsci took Lenin's rejection of economism and class reductionism further, along lines already chartered by Bogdanov, arguing that in its struggle for hegemony the proletariat must undergo moral and intellectual reform and develop an ideology to bind together diverse social elements. It should combine the interests of other classes, groups and movements with its own interests so as to create a national-popular collective will. To do this it needs to overcome all the narrow, corporate prejudices of a fundamental class and make all necessary compromises in political and economic programmes in order to build up and sustain a bloc of social forces with a common world-view. There must be 'a cultural-social unity through which a multiplicity of dispersed wills, with heterogeneous aims, are welded together with a single aim, as the basis of an equal and common conception of the world.'[9]

Such a unity could not be attained by adopting Marxism in a pure form. It was seen to be necessary to formulate a more complex synthesis of class objectives with popular-democratic themes that have arisen out of the unique and original history of each country. To achieve this it is necessary to engage in critical reflection on the existing ideological complex. Such reflection should not be left to groups of intellectuals. It is something that everyone should be involved in. As Gramsci argued:

> It is essential to destroy the widespread prejudice that philosophy is a strange and difficult thing just because it is the specific intellectual activity of a particular category of specialists or of professional and systematic philosophers. It must first be shown that all men are 'philosophers', by defining the limits and characteristics of the 'spontaneous philosophy' which is proper to everybody.[10]

---

7. Raymond Williams, *Marxism and Literature*, Oxford: Oxford University Press, 1977, p.110.

8. Gramsci, *Selections from the Prison Notebooks*, p.238.

9. Ibid. p.349.

10. Ibid. p.323.

The question is whether individuals are to allow their philosophy to be imposed on them, or whether they are to consciously and critically work out their own conception of the world and 'take an active part in the creation of the history of the world...'[11]

Such philosophizing should not be seen as a contemplative activity, 'but also and above all as a cultural battle to transform the popular "mentality" and to diffuse the philosophical innovations which will demonstrate themselves to be "historically true" to the extent that they become concretely - i.e., historically and socially - universal.'[12] Gramsci believed that it is in situations of crisis and engagement that people are most able to overcome their intellectual passivity and to work out their own conception of the world. Correspondingly, philosophizing should not aim to make a fresh start, but should begin by differentiating and changing the relative weight of the elements of the old ideology, while reorganising the new ideological system around a different central unifying principle to form a coherent, critical conception of the world. If the old ideology was genuinely popular, then it is necessary to preserve at least some of its elements in the new system, even if slightly altered in the process. Only in this way is it possible for the ideas and aims of a revolutionary class to become deeply rooted among the people. Political action can only be successful by drawing on the cultural heritage of the nation. However, unlike Bogdanov, Gramsci did not offer an alternative cosmology which could achieve this, and simply took for granted Marx's grand narrative of proletarian liberation.

## *Gramsci Today*

Gramsci's main work was written in prison, and his writings were reflections on the failure of Marxists and on the success of the fascists at a crucial conjuncture of history. They were meant to provide guidance for the future. We are now in a similar, though more significant conjuncture to that of the 1920's and early 30's. To begin with, the world is facing an economic crisis. Unemployment has already risen dramatically over the last two decades, although this has been disguised by its irregular growth, with big increases occurring at approximately eight year intervals. Unemployment in the O.E.C.D. countries rose in the recession of 1967 to 5 million, in the recession of 1973-75 to 15 million, and in the recession of 1982 to 32 million.[13] We can expect unemployment to go far higher, as it is already in the Third World. Morocco, for instance, already has an unemployment rate among able-bodied men between the ages of 15 and 64 of over 60%.[14]

There are a number of causes of this state of affairs. To begin with, there has been a revolution in technology which has generated in advanced capitalist nations both a big increase in productivity and increased unemployment. The founder of the science of cybernetics, Norbert Weiner, anticipated that the development of information technology would cause a depression more severe than that of the 1930's. As he argued:

> Let us remember that the automatic machine ... is the precise economic equivalent of slave labour. Any labour which competes with slave labour must accept the conditions of slave labour. It is perfectly clear that this will produce an unemployment situation, in comparison with which the present recession and even the depression of the thirties will seem like a pleasant joke.[15]

---

11. Loc.cit.
12. Ibid. p.348.
13. Andre Gunder Frank, *Critique and Anti-Critique*, N.Y.: Praeger, 1984, p.231.
14. See Susan George, *A Fate Worse Than Debt*, Harmondsworth: Penguin, 1988 for a description of the economies of the Third World nations.
15. Norbert Weiner, *The Human Use of Human Beings: Cybernetics and Society*, Boston: Anchor Press, 1954, p.220.

So far only a small proportion of the potential for saving labour through computer-chip technology has been exploited. It has been calculated by a computer scientist at Carnegie-Mellon University that by the year 2010 the number of people employed in manufacturing in the United States will drop from 26 million to 3 million.[16]

However the development of technology is only part of the problem. John Kenneth Galbraith concluded his book *The Great Crash of 1929* published in 1954 by listing the five weaknesses of the US economy in the 1920's which had an especially significant bearing on the ensuing depression. These were the growing inequality of income distribution, the bad corporate structure (due to the growth of holding companies and investment trusts), the bad banking structure, the dubious state of the foreign balance, and the poor state of economic intelligence.[17] All these weaknesses, which were patched up during the 1940's, 50's and 60's, are emerging again. There has been a massive redistribution of income and wealth to the wealthy, with the richest 1% of the US population increasing their share of national wealth from 20.8% in 1949 to 34.3% in 1983, compared with 36.3% in 1929.[18] Through a spate of takeover activity on a colossal scale, corporate structures are weak and there has been a decline in productivity.[19] International finance has undermined almost all the controls on banking which were put in place after the Great Depression. In his book *The Financial Revolution*, Adrian Hamilton described how 'larger and larger institutional savings are chasing fewer and fewer investment outlets. The major manufacturing industries are contracting. The Third World has been shut off from new funds. The funds within the system are moving in faster circles, chasing the marginal profit that they can glean from their own movement.'[20] John Maynard Keynes, pondering on the causes of the Great Depression, had noted that 'Speculators may do no harm as bubbles on a steady stream of enterprise. But the position is serious when enterprise becomes a bubble in a whirlpool of speculation. When the capital development of a country becomes a by-product of the activities of a casino, the job is likely to be ill-done.'[21] Hamilton has shown that the whole world economy has become one great casino for the super-rich. One outcome of this has been the corruption and collapse of Thrifts in USA which it is estimated will cost the public from $US500 billion to $US1.4 trillion over the next 40 years.[22] There are also massive trade imbalances between nations exacerbated by the absence of stable exchange rates and by the debt crisis in the Third World (with Third World debt in 1988 standing at $US1.2 trillion). This debt, which is forcing countries to compete with each other to increase exports, thus forcing down prices, is having much the same effect as Germany's reparation payments in the 1920's which undermined not only Germany's economy, but the economies of the victors whose industries were undermined by cheap imports from Germany. And to top all this off, the discipline of economics has come to be dominated by the same sort of pre-Keynesian neo-classical ideas fetishizing the market which prevailed in the 1920's. As Lester Thurow concluded

---

16. David Dickson, *The New Science of Politics*, 2nd ed., Chicago: The University of Chicago Press, 1984, p.51.

17. John Kenneth Galbraith, *The Great Crash 1929*, [1954], Harmondsworth: Penguin, 1984, p.194ff.

18. See Ravi Batra, *The Great Depression of 1990*, Schwartz and Wilkinson, 1987, p.118, and Lester C. Thurow, *The Zero-Sum Solution*, Harmondsworth: Penguin, 1987, p.122ff.

19. The deleterious effects of corporate takeovers have been examined in Alan Auerbach ed., *Corporate Takeovers: Causes and Consequences*, Chicago: University of Chicago Press, 1988; and David Ravenscroft and F.M. Scherer, *Mergers, Sell-Offs and Economic Efficiency*, N.Y.: Brookings Institute, 1988. See also 'Corporate Finance: Do Mergers Work?', *The Economist*, 17-23 December, 1988, pp.68-70. On the general effects of this on the U.S. economy see S. Melman, *Profits Without Production*, N.Y.: Knopf, 1983.

20. Adrian Hamilton in *The Financial Revolution*, Harmondsworth, Penguin, 1986 p.235. See also 'A Survey of America's Capital Markets' in *The Economist*, 11-17th June, 1988, after p.56.

21. John Maynard Keynes, *The General Theory of Employment, Interest and Money*, [1936], London: Macmillan, 1964, p.159.

22. For a review of studies of this, see Michael M. Thomas, 'The Great American Shambles', *The New York Review of Books*, Jan. 31st, 1991, Vol.XXXVIII, No.3, pp.30-33.

his study of the discipline: 'Economics is in a state of turmoil. The economics of the textbooks and of the graduate schools not only still teach price-auction model but is moving towards narrower and narrower interpretations. The mathematical sophistication intensifies as an understanding of the real world diminishes.'[23]

These problems are reinforced by fundamental transformations in the international economic order.[24] Transnational business organization have grown to such an extent that they are not merely uncontrollable by national governments, but through their control of media and investment are able to dominate governments. One consequence of this is the breakdown of the international regulation of trade which has left each country competing with every other country to increase its exports over imports, and forced nations to compete with each other to deflate their economies to reduce imports. The success of Japan, West Germany and Taiwan in this struggle relative to the United States, Britain and more significantly, almost all the Third World has led to enormous instability and will prevent a repetition of the Keynesian strategy of President Reagan where massive expenditure on armaments lifted the world out of the recession of 1982. Secondly, growth of transnational corporations has forced nations and workers throughout the world to compete with each other to reduce taxes and wages to retain investment, while the growth of international finance has virtually destroyed the power of governments to regulate their economies.[25] Many States have now lost or abandoned sovereignty over their national economies, and the welfare organizations built up after the Second World War are being dismantled.[26] Thirdly, while there was much poverty in the 1920's and 30's one could still believe that this could eventually be overcome through continued economic growth. The environmental crisis has undermined this assumption. As Dudley Seers pointed out:

> We are entering a period in which resource limits can no longer be ignored, nor can the interests of different sections of the world be assumed compatible: to solve one country's problems may well be to aggravate those of another... The [economic] crisis is not just a cyclical downturn nor even ... the slack phase in a hypothetical Kondratieff cycle... [It is] the culmination of a period of increasing strains on the world's productive structures, natural resources, and political systems. Thus a swift rise in world output would soon reveal shortages in oil, various minerals, and food, and increased international tension...[27]

This prediction is clearly borne out by boom in commodity prices from 1987 to mid-1988, a period in which metal prices more than doubled.

Finally, as in the 1920's there is a cultural crisis (but without generating the corresponding intellectual and artistic creativity). The nihilism of Western culture is increasingly manifesting

23. Lester C. Thurow, *Dangerous Currents: The State of Economics*, Oxford University Press, 1983, p.236. See also Paul Ormerod, *The Death of Economics*, London: Faber and Faber, 1994.

24. Charles P. Kindelberger in his book *The World in Depression, 1929-1939*, Harmondsworth: Penguin, 1986 focussed on the instability of the international economic order rather than on the conditions within countries. However the list of causes of the depression he gave in the concluding chapter also characterizes the present situation. A good overview of the state and trends of the world-economy can be gained from R.J. Johnston and P.J. Taylor eds, *A World in Crisis*, Oxford: Blackwell, 1986 and Susan Strange ed., *Paths to International Political Economy*, London: George Allen & Unwin, 1984.

25. In effect the predictions of Stephen Hymer made in the early 1970's have proved correct and the consequences have been even worse than Hymer predicted. See Stephen Hymer, 'Internationalization of capital and international politics: a radical approach', in Edward J. Nell ed., *Growth, Profits & Property: Essays in the Revival of Political Economy*, Cambridge: Cambridge University Press, 1980, pp.189-203.

26. See Scott Lash & John Urry, *The End of Organized Capitalism*, Cambridge: Polity Press, 1987, Ch.7 onwards for a detailed study of this.

27. Seers, *The Political Economy of Nationalism*, p.1. Seers is not alone in holding this view. See papers in Daniel Yergin and Martin Hillenbrand eds, *Global Insecurity*, Penguin: Harmondsworth, 1983.

itself, expressing itself in decadence, the growth of organized and unorganized crime, an inability to face up to and mobilize against the problems of society, exhaustion and fragmentation of intellectual life, and the embracing of simplistic, irrationalist ideologies. The stresses of this cultural crisis have in turn undermined the ability of individuals to cope with life. As James O'Connor argued:

> Capital's passion, money in search of more of itself, is unregulated by public conscience, institutionalized morality, or the state. The individual is bereft of a trustworthy social superego; neither capital nor the state can administer the passions and conscience. The individual is thus isolated, not merely materially and socially, but emotionally, a 'stranger in the crowd.'... In this cauldron of uncertainty and insecurity, a world where most people are encouraged to aspire to the banal, the routine, the scheduled, personality crisis erupts.[28]

The booming drug culture is a manifestation of this crisis.

## *Revamping Gramsci: The Environmental Crisis and Process Philosophy*

In the face of this crisis, Gramsci's project of developing an alternative hegemonic culture to unify opposition to the existing order should be recognized as more important than ever. However, in the light of what has been argued in this work, it is necessary to re-evaluate the whole focus of those struggling against the oppression of the existing socio-economic order. It is necessary to go beyond Gramsci in what is to be made the central unifying principle around which the culture of the alternative hegemony is to be organized. What is required is a return to the project of Bogdanov and his supporters. If the problems of and oppression within the world are to be effectively confronted, then the environmental crisis should be the focus of a world-wide alternative hegemonic civilization, and my contention is that the unifying principle of this alternative culture must be a new metaphysics and cosmology, that of process philosophy. And for this to be articulated into guidance for action, it is necessary to elaborate in terms of it a new grand narrative projecting a new future.

The environmental crisis has destroyed the central tenet of those apologists for the existing order, that present suffering is necessary for economic progress which will eventually make everyone better off. There is no reason at all to believe that the present era of economic crises will usher in a new era of prosperity. The pressure on individuals and societies to increase production to overcome unemployment and international debt are not only oppressive, they are driving humanity to the destruction of the conditions of its continued existence. Environmental degradation is implicated in all oppression in the world and vice versa; changing our relationship to the environment to overcome the environmental crisis will only be possible by overcoming all major forms of economic, social, political and cultural oppression. It is no longer the expropriation of surplus value from workers which is the most oppressive aspect of capitalism, but its monopolization of control over the world's reserves and resources, its wasting and destruction of these combined with the exclusion of more and more people both from access to them and from participation in economic life.

It is now essential that present economic policies extolling the unleashing of market forces be abandoned and that economies of all nations be brought under democratic control. Nations peripheral to the world economy in particular need to liberate themselves from the economic core regions in order not only to overcome the oppression of their people, but also to conserve and preserve the world's environment, while people in all nations need to struggle against the fetishism

28. James O'Connor, *The Meaning of Crisis*, Oxford: Blackwell, 1987, p.177.

of commodities to create environmentally sustainable forms of life. Stephen Bunker has made this point well:

> Dominant classes depend on their societies' total environment; in this sense they depend on the organization of other classes' adaptation to the environment. The clearest lesson of class relations in the Amazon is that dominant groups which impoverish the rest of society ultimately impoverish themselves. Only when human communities with balanced exchange relations exist is it possible for social organization to adapt to its total environment in ways which sustain both human community and the ecosystem itself.[29]

Environmentalism as the struggle against ecocide thus can unify all struggles against oppression. It is simultaneously a symbol for the untenability of the existing economic and political organization of the world, a symbol against oppression throughout the world, both within and between nations, a symbol for the inter-relatedness and interdependence of the human community and of other life forms, and a symbol affirming all life, providing the foundation for a new vision of the future. It is this which the West German Greens recognized, and which made their achievement important for the rest of the world. As Werner Hülsberg wrote in the conclusion to his study *The German Greens*:

> The real contribution of the West German Greens, ... is that they understood and grasped the ecological question not just as another question and not as a political neutral task but rather as the decisive question, the acid test, of *left-wing politics*.... The ecological question has become today a symbol for the general dissatisfaction with a model in which traditional politics is only capable of following the dictates of economic interests and in which science has become a whore on sale to the highest bidder. The ecological question presents us with the need for a new emancipatory model of eco-socialism.[30]

## *Nationalism versus Globalism*

In a world in which even most national governments in the economic centres have failed to effectively confront their own environmental problems it is hardly likely that actions taken by international organizations will have much success unless backed by local organizations. Correspondingly, while it is necessary for people to act locally, purely local action ignoring the broader context affecting local issues is unlikely to do more than slow down the rate of environmental destruction. Those struggling against global environmental problems can only succeed by developing strong organizations committed to the conservation and preservation of the environment which can effectively represent local environmental concerns within broader national, regional and international forums. Environmentalists will have to struggle for representation of the environment in local, national and regional politics, and then use the environment as a focus to mobilize people to liberate themselves from the destructive imperatives of the world economy. This will require the fostering of an environmentalist nationalism.

The promotion of such nationalism is also required to recreate the sense of community and personal identity required for effective action, to overcome the rootlessness of people which is depriving them of the will to struggle for anything. In the modern world radical political movements, including Marxist movements, have only ever been successful where they have been more or less explicitly fused with local cultural traditions as nationalist movements, and

---

29. Stephen Bunker, *Underdeveloping the Amazon*, Urbana and Chicago: Illinois University Press, 1985, p.252.

30. Werner Hülsberg, *The German Greens*, London: Verso, 1988, p.219f.

nationalism has been central to the struggles of every other country which has been at all successful in overcoming its subjugation by the world-market. Only by cultivating nationalist sentiments will it be possible to mobilize people to bear the costs of the struggle for regional control over economic life, to generate concern by people for justice for their compatriots, to inspire them to develop more austere forms of life which conserve reserves and preserve resources, and to develop institutions powerful enough to tackle broader environmental problems. Only through nationalistic struggles will the hold of the consumer oriented culture of the economic cores be broken, forms of relationships between people transcending commodity fetishism be developed, the organizational basis and the cultural conditions for confronting the long-term problems associated with the environment be created, and the possibility of transcending capitalism altogether be revealed. The point is that except for intellectuals, pure cosmopolitanism is too rarefied an orientation in the struggle for justice. Most people need to feel that they will be recognized and taken into account at a more immediate level before they will define their own lives in terms of this struggle.

To begin with, nationalism should continue to be fostered in the Third World. Third World people need to struggle through local, national, and also regional organizations to preserve their own environments from exploitation by the economic core regions. What is particularly required in the unification of regions, such as Latin America and Africa, to oppose domination by North America and Western Europe. Given the location of most environmental destruction in the Third World, this struggle for liberation should be recognized as the most important struggle of all.

What role, then, should environmentalists prescribe for the affluent nations of the world in the world-economy? Should their main concern be to provide for the impoverished of the Third World by striving to make the world into one vast Welfare State as Willy Brandt has proposed?[31] Or should their main concern be to orient themselves towards preserving their environments by reducing economic output and reducing their consumption? In the tradition of most humanitarian thinkers Ervin Laszlo made the point that: 'The World Bank calculated that hard-core world poverty could be erased by an investment of one dollar per barrel of oil used between now and the end of the century. Some one billion people would be lifted from abject and inhuman conditions to a life worthy of human beings. Such funds are comparatively modest and they could easily be raised. They equal the yearly incremental expenditures of the world's privileged classes on tobacco, alcohol, and cosmetics.'[32] But thinking in such terms offers further legitimation for the drive for continued economic growth in the affluent nations, since this can then be represented as the means for overcoming poverty in the Third World. And apart from the unlikelihood of further economic growth leading to any greater generosity towards the Third World from the affluent countries, there is no reason to believe that government aid to Third World countries will solve the problem of poverty.

In most cases, foreign aid from the governments of the core zones to the Third World has further entrenched existing oppressive power structures, while solutions foisted on Third World governments by benevolent international agencies have been singularly unsuccessful. When affluent people try to extend their own organizations and policies into areas which are socially simple, energy poor and devoid of organizations and institutions which can match the organizational strength of such agencies, these agencies have facilitated their own and the peripheral societies' permeability to and exploitability by nationally and internationally dominant classes. Those people who have overcome their poverty have been those who have relied on their own efforts, and people outside these regions could not have directed their struggles. As Denis Goulet argued on the basis of a study of strategies for development in Guinea-Bissau: 'Paradoxically, the lesson of greatest

31. See Willy Brandt et. al. *North-South: A Programme for Survival*, London: Pan Books, 1980. Also Willy Brandt, *World Armament & World Hunger*, London: Gollancz, 1986.

32. Ervin Laszlo, *The Inner Limits of Mankind*, Oxford: Pergamon Press, 1978, p.46.

importance is that the best model of development is the one that any society forges for itself on the anvil of its own specific conditions.'[33] So, as James Lamb argued:

> Development should be a *struggle* to create criteria, goals, and means for self-liberation from misery, inequity, and dependency in all forms. Crucially, it should be the process a people choose, which heals them from historical trauma, and enables them to achieve a newness on their own terms.[34]

This has been the secret of the success of Kerala in India, Algeria, Eritrea and Zimbabwe.[35] For such reasons Dudley Seers who spent much of his life as an adviser to Third World governments, opposed foreign aid to developing countries.[36] The only aid likely to be effective is aid put at the disposal of the poor; that is aid 'disposed of locally, by the poor countries' poor themselves, on their own terms and in support of local work, education and the meeting of basic needs, thus benefiting development from below.'[37] Only such aid should be supported.

The primary goal of people in the affluent core economies should be the termination or radical reduction of the economic links between their countries and Third World economies, and an end to the exploitation of Third World reserves and resources, to the importation of agricultural products and timber. As Stephen Bunker argued: 'Ultimately the need is to slow the flow of energy to the world centre.'[38] This struggle should not be thought of in purely altruistic terms. It should be seen as part of and linked to the struggle to stop the bankrupting of farmers, the deindustrialization of regions, unemployment and the impoverishment of people in these core zones. To avoid exploitation, to escape the vicissitudes and pressures of the international capitalist system and to gain democratic control over their economies, to turn the advantages of improved technology to bettering the conditions of life, regions, whether local, national or broader geographical areas, should as far as is possible aim at economic self-sufficiency.[39] As John Maynard Keynes argued in 1933: 'Ideas, knowledge, science, hospitality, travel - these are the things which should of their nature be international. But let goods be homespun whenever it is reasonably and conveniently possible, and above all, let finance be primarily national.'[40] With recent developments in technology, such localization of production for most goods is now more possible than ever. The task ahead of people in the economic core zones and semi-peripheries is to create steady-state economies for their own benefit, for the benefit of people in the Third World, for the benefit of all future generations and for the benefit of all other life-forms.

---

33. Denis Goulet, 'Looking at Guinea-Bissau: A New Nation's Development Strategy', Occasional Paper no.9: Overseas Development Council, March 1978, p.52, cited by Charles K. Wilber and Kenneth P. Jameson in 'Paradigms of Economic Development and Beyond' in Charles K. Wilber ed. *The Political Economy of Development and Underdevelopment* 3rd ed., N.Y.: Random House, 1984, pp.4-25, p.22.

34. James J. Lamb, 'The Third World and the Development Debate' *IDOC - North America*, January-February 1973, p.20, cited ibid. p.23.

35. On the success of Zimbabwe see Paul Harrison, *The Greening of Africa*, London: Paladin, 1987, pp.87-97.

36. Seers, *The Political Economy of Nationalism,* p.156. Some detailed support for Seers' views is provided by Marcus Linear, *Zapping the Third World: The Disaster of Development Aid*, London: Pluto Press, 1985 and Jon Bennett in *The Hunger Machine*, Cambridge: Polity Press, 1987, Ch.3, 'Aid: the Poisoned Gift?'. As Bennett argued, 'Aid as a political weapon is not always covert: indeed, it could be argued that all *official* aid is simply an extension of foreign policy and a means of entrenching economic dependency.' (p.84.) and 'Aid is, in fact, a transfer of money from Northern tax-payers to Northern private companies...' (p.104f.) Bennett's criticisms are not directed against private aid organizations.

37. Erik Dammann, *Revolution in the Affluent Society*, tr. Louis Mackay, London: Heretic Books, 1984, p.44.

38. Bunker, *Underdevelopment in the Amazon*, p.253.

39. This has been forcefully argued by Herman R. Daly and John E. Cobb, Jr. in *For the Common Good*, Boston: Beacon Press, 1989.

40. John Maynard Keynes, 'National Self-Sufficiency' in *The Yale Review*, Vol.22, 1933, p.758.

Nationalism is the only ideological weapon with the potential to combat the forces of international capitalism to achieve such control.[41] Justice will be achieved through the development of a reformulated nationalism, or it will not be achieved at all and victory will pass to those demagogues of the extreme right willing to incite exclusive groups to violent struggle for what reserves and resource are left.

## *What is Nationalism?*

Nationalism has a bad name among radicals. In the past it has been linked with imperialism, wars of aggression and with the persecution of minorities, and at present it is associated with the ethnic violence in the Balkans and Eastern Europe. So just what is nationalism? And how can it be utilized by environmentalists?

Nationalism is essentially 'a territorial ideology',[42] while modern States are territorial political institutions. Nations are both cultural and political phenomena. As Benedict Anderson has argued, the nation is 'an imagined political community' which is created through being imagined.[43] There are three component parts to the doctrine of nationalism: that the people of a region should be self-determining, that they have a unique national character which should be fully expressed, and that each nation contributes its special genius to the common fund of humanity. Historically, the 'nation', or rather 'community of nations' has succeeded ethnic groups, and then world religions, as the focus of social integration beyond the biological family, and nations are now the major actors in grand narratives of humanity's progress. Nationalism, affirming the community of people in a region, relates their traditions and their future, provides people with an identity and forges a common destiny for its members. Nationalism, as a narrative or unfinished story of the people of a region, serves to coordinate people's actions and lives, to mobilize them for action, and to legitimate the institutions of the State. It serves the State by strengthening the institutional relationships between the State and civil society, by furthering the internal unification of culturally and economically diverse regions into a more homogeneous State territory, and it divides one political community from another, in many instances determining the geographical boundaries of the State. Conversely, by affirming the existence of a community, nationalism legitimates claims by people for just representation by the State, that the State will itself be just and that it will put to rights injustices perpetrated against its members. As John Breuilly argued on the basis of his exhaustive examination of the history of nationalism: 'an effective nationalism develops where it makes political sense for an opposition to the government to claim to represent the nation against the present state.'[44]

Nationalism has taken a variety of forms.[45] Nationalism emerged in Spain, England and France as the merchant classes of these societies struggled for political representation in the new absolutist States which had emerged from the late feudal kingdoms. While this nationalism developed through the cultural unity engendered by the development of new print-languages, culture was not an issue in its development, and the nations involved were assumed to correspond to the territorial boundaries of the State. Such nationalism was identified with the democratization of government. The growth in power of England and France was a stimulus for the development of two other forms of European nationalism - the separationist nationalism of Ireland, Belgium and Norway, and the

---

41. As Dudley Seers has argued in *The Political Economy of Nationalism*.

42. This definition is developed by James Anderson, 'Nationalism and Geography' in *The Rise of the Modern State*, ed. James Anderson, Brighton: Wheatsheaf Books, 1986, pp.115-142, p.116.

43. Benedict Anderson, *Imagined Communities: Reflections on the Origin and Spread of Nationalism*, London: Verso, 1983, p.14.

44. John Breuilly, *Nationalism and the State*, Manchester: Manchester University Press, 1985, p.382.

45. See ibid.

unificationist nationalism of Italy, Germany and Poland. In each of these cases the promotion of national cultures was central and preceded the emergence of national States. Outside Europe, nationalism first developed among European colonies in the struggles for independence, then a 'reform' nationalism developed in countries threatened by European imperialism, notably in Japan, Turkey and China. Towards the end of the nineteenth century and culminating in the Second World War, European nationalism was extended to include the working class in the community of the nation, but at the same time, at least among the major powers, it became more authoritarian and expansionist. It came to be associated with a rejection of the right of every nation to political self-determination and independence and the assertion of the privileged position of one's own nation - the 'chosen nation'.[46] However there were other European States in which a new reform nationalism developed to mobilize people against the vicissitudes of the world capitalist economy without this chauvinist quality. Such reform nationalism, associated with the development of the social or liberal corporate States, became increasingly influential among small European nations after the Second World War.[47] There also emerged at this time a post-colonialist nationalism in the Third World, in some cases serving to unify inherited political divisions, in others to separate regions from inherited State structures or to combine regions across State boundaries. A new set of nationalist movements have emerged within Europe striving for independence from old States, for instance Scotland, Flanders. Croatia and Estonia are successful examples of this. And finally there has developed in the Third World a new form of nationalism aimed at uniting broader regions such as South America, Africa and the Islamic countries into a united struggle against domination by the First World.

While the fostering of nationalism has led to greater justice and vast improvements in the quality of life of those who are united by it, nationalism is also associated with tendencies to deny justice to racial minorities and outsiders. It is these destructive tendencies of nationalism which have been used to justify the claim of cosmopolitan intellectuals that nationalism is essentially pathological. But are double standards and their consequences inevitable? Benedict Anderson has argued that they are not. He reminds us that far from being a concomitant of nationalism, racism is a throwback to notions of class: 'The fact of the matter is that nationalism thinks in terms of historical destinies, while racism dreams of eternal contaminations, transmitted from the origins of time through an endless sequence of loathsome copulations: outside history. ... The dreams of racism actually have their origin in ideologies of class, rather than in those of nation: above all in claims to divinity among rulers and to "blue" or "while" blood and "breeding" among aristocracies.'[48] The ideology of nationalism is more consistent with quest for universal justice. One case which illustrates this relationship is the nationalism promoted in the 1930's in Sweden.

Up until the 1930's, Sweden was dominated by the export oriented sector of the bourgeoisie. This group was defeated during the Great Depression when the workers in alliance with farmers' interest groups and the home market fraction of the bourgeoisie gained power. In 1932, the Social Democratic Party gained control of parliament, and until recently retained this almost continuously. But even more importantly, the Swedish Confederation of Trade Unions first united the working class, then transcended its narrow concerns with working class incomes to represent the interests of all those oppressed by economic developments. It was the prototypical case of the success of a Gramsci type alternative hegemony, based on a fusion of socialist and traditional ideas, succeeding in becoming the dominant hegemony. This success was achieved by forging of a left-wing form of nationalism.

Winton Higgins described the response of the Swedish labour movement to the Great Depression: 'Alone among Western labour movements, the Swedish movement took the Depression

---

46. This development was recognized by Rodolf Hilferding. See *Finance Capital*, [1910] tr. Morris Watnick and Sam Gordon, London: Routledge & Kegan Paul, 1981, esp. p.335.

47. On this see Peter Katzenstein, *Small States in World Markets*, Ithaca: Cornell University Press, 1985.

48. Benedict Anderson, *Imagined Communities*, p.136.

as the cue to mount an all-out assault on the organising principle of bourgeois politics, the theory and practice of economic liberalism.'[49] It then began developing general policies for the whole nation based on maintaining full employment, equalizing wages, and controlling levels of investment: 'It has developed a practice of national policy formation and implementation outside the framework of the state, and thus is also partly extra-parliamentary party and partly alternative state apparatus.'[50] Transcending working class particularism, the labour movement recast Swedish political culture through the concept of 'peoples' home' (*folkshemspolitik*). Gören Therborn described the role of this concept:

> The Peoples' Home had an implicit connotation of 'family' - rather than 'house' - of family community and equality with 'no favourites or stepchildren'. It connoted common concern and caring for each other and had its focus on society rather than on the state and particular institutions. It is noteworthy and testifies to the tactical skill and success of the SAP, that the notion turned out quite compatible with a reaffirmation of classical working class demands in the fields of social policy.[51]

This universalism transcended the notion of individuals' social rights, and replaced it with a *Weltanschauung* of national solidarity. In accordance with this, Alva Myrdal presented the case for social security not as a question of social insurance, but as a question 'of social policy, as a productive social policy - as common investment by the nation in its future welfare - with its accentuation of family policy and of preventative measures.'[52]

Associated with this internal policy, the Swedes under the Social Democratic Party aligned themselves with the oppressed of the world, and more consistently and successfully addressed environmental problems than virtually any other nation (with the possible exception of Denmark). Despite their cold climate, they now use only one half as much energy per head of population as people in USA.[53]

## *Environmental Nationalism and Process Philosophy*

As we have seen, the development of nationalism first requires a struggle for cultural independence and a sense of cultural identity in the regions in which people live. For such reasons Dudley Seers argued for the development of national cultures. He called for education in the traditional arts subjects, arguing that 'the centre-piece of education is history, the history of the nation *in relation to its continent and the world*, ranging right up to the present.' And he argued for 'making people familiar with their nations cultural heritage - myths, fables, songs, dances, carvings and sculpture, buildings, etc. - which expresses national experience and can help inhibit the growth of cultural dependence.'[54] But while these are important, such an education by itself is too passive. There is no reason to suppose that the culture of any region in its existing form will be

---

49. Winton Higgins, 'Unemployment and the labour movement's breakthrough in Sweden' in Jill Roe ed. *Unemployment*, Sydney: Hale & Iremonger, 1985, pp.105-124, p.105.

50. Winton Higgins, 'Working Class Mobilization and Socialism in Sweden', Paul Boreham and Geoff Dow eds, *Work and Inequality*, Melbourne: Macmillan, 1980, p.156.

51. G. Therborn, 'The working class and the welfare state', 5th *Nordic Congress of Research in the History of the Labour Movement*, Murikka: Finland, August, 1983, p.20.

52. Cited ibid. p.23.

53. On the Swedish effort to address environmental problems see Peter Bunyard, 'Sweden: Choosing the Right Energy Path?' *The Ecologist*, Vol. 16, No.1, 1986, pp.24-28. In 1988 laws were passed to phase out battery hens, and to improve the living conditions of cattle and pigs.

54. Seers, *The Political Economy of Nationalism*, p.105.

adequate for what is required of it. Education should induce people into an on-going dialogue so that they can become critical participants in the development of their cultural heritage, and define their own lives in terms of this participation.[55] To achieve this it is necessary to see cultural development and nationalism in terms of a theory of culture, a grand narrative, a cosmology, and a general philosophy.

In terms of process philosophy, the culture of a region is part of the process of a people's self-creation, part of their on-going struggle (with varying degrees of success) to orient themselves, practically and theoretically, in relation to nature, to each other, to their society and to outsiders; to recognize and appreciate nature's, their own and each others' significance and potentialities, and to realize these potentialities. The primary means by which people do this is through the construction of narratives. Cultural diversity is required to appreciate unique situations, to explore diverse possibilities and reveal the limitations of different modes of existence. Each culture is a contribution to life and to the culture of humanity, as part of the world's and humanity's self-creation, not entirely determined by past and present conditions, yet dependent on environmental and material conditions, including the cultures of others in the past and in the present. The unique significance of each local culture and all its subcultures (and the people embodying and developing this culture), can be fully appreciated, but still criticised from this perspective, allowing individuals to assimilate aspects of other cultures to their own. The study of the local environment is part of the development of culture, and by fostering a recognition of the relationship of society to its environment in the past and present, nationalism can be fused with the commitment to conserving and preserving the integrity of this environment. By seeing cultural development in terms process philosophy, the struggle for national independence can be seen as a struggle within nature and for nature, as part of the world's struggling to become conscious of itself in all its diversity to reveal and realize its potentialities.

Process philosophy also provides a means to integrate cultures. It provides a framework of concepts which can facilitate far more efficiently than prevailing concepts an understanding by individuals of their place within the world. It enables individuals to easily comprehend the major advances in the natural sciences, allowing them to understand their place in the natural world, enables them to grasp the complexities of societies and the international socio-economic order, and legitimates the central place of narratives in orienting people. In this way process philosophy should enable people to see through the illusions purveyed by the priests of the hegemonic culture, the 'scientific experts', whether these be orthodox natural scientists, economists or experts in cost-benefit analysis. Most importantly, it should enable the members of a region to see their common interests, the relationship between these and the future of their environments, and between their own environments and the world ecosystem.

Nationalism can then be redefined as the commitment by a regional community to justice within the region, to preserving and developing its potentialities. Above all, as the ultimate condition of all potentialities, nationalism should involve a commitment to preserving and conserving the local environment. For these commitments to mean anything they must be incorporated into the narrative defining the nation - the unfinished story which provides the ultimate reference point for all its communities, organizations and institutions to define and legitimate themselves and their projects.

By providing a way of thinking about one's place in the world which neither atomizes the world nor dissolves each part into the totality, process philosophy makes it possible to formulate a multi-levelled nationalism, to acknowledge the significance and partial autonomy of one's local community while seeing this as participating in a national community which itself has a partial autonomy, which is in turn participating in a broader regional community (for example, Western Europe, the Islamic world, Africa or Latin America) with some partial autonomy, which again is

55. For an illuminating study of education and its problems and what is required to overcome them and to induct students in this way see Basil Bernstein, 'On the classification and framing of educational knowledge' in *Class, Codes and Control*, St Albans: Paladin, 1973, pp.227-256.

participating in a world community which is more than the sum of all the particular communities which compose it. Individuals can then be simultaneously nationalistic in relation to their local region, to their country, and to a major region of the world, while at the same time being committed to international justice and to the subordination of national interests to the interests of humanity and to the health of the bio-sphere. The question then is one of justice, of appropriately acknowledging the uniqueness and significance of each level of the community. Environmental nationalism should then be seen as the struggle to maintain, to transform or to create power structures, from the local to the international level, which appropriately recognize the people of regions, their future generations, their non-human forms of life and the general environment, and to effectively articulate the needs, concerns, potentialities and aspirations of all these life forms.

## *Cultural Inertia and Creative Rationality*

The transformation of society from one socio-economic formation to another on an international scale will be a long drawn-out process in which opposing social forms will co-exist, and in which there will be failures, retreats and regressions as well as successful advances by those struggling for a more just order. As Marx argued in *Capital*: 'epochs in the history of society are no more separated from each other by hard and fast lines of demarcation than are geological epochs.'[56] A world-wide social and cultural transformation is something which will have to be struggled for over centuries, and in which even successful struggles in any individual's lifetime can only be regarded as contributions to this struggle. Furthermore, it is only through people recognizing this, and recognizing that life in the present cannot be reduced to a mere means for the realization of a new world-order that this struggle is likely to succeed.

One of the major concerns of this work has been to reveal the nature and dynamics of cultures, and in so doing, to reveal their inertia and what is involved in major cultural transformations. It has been shown how the Christianity which developed in the early Middle Ages was built on previous modes of being in the world, and the intellectual revolution associated with the development of mechanistic materialism was already foreshadowed by, and was actually an articulation and an expression of, previous developments in social practices. In Russia where more radical changes were made over a relatively short period, these required a tremendous effort. Through the whole of the nineteenth century the élite of the intelligentsia struggled to develop the world-orientation and associated mode of being in the world required to overthrow Tsarist rule, and much of the turmoil following the Revolution was produced by the struggle to change the habitus of the rest of the population. And to a considerable extent this transformation, which was not entirely successful, was only possible because of the resonance between the Orthodox Christianity of Russia and the Neoplatonic aspects of Marxism, the pre-existing model of Western European dynamism continually brought home by the threat of domination by Western Europe, and the propensity of Russian culture to invert itself. The main reason for the difficulty in effecting cultural change is the way particular modes of conceiving the world are embodied in practices and institutions, with all practices in societies resonating with and thereby supporting each other, requiring of individuals that they conceive themselves and their relationships in a certain way to get on in the world. Ways of conceiving the world are embodied not only by individuals and their social relations and practices, but also by modes of production, institutions, organizations, and by the transformations of the physical world.

However once the instrumentalist form of thinking deriving from the mechanical world-orientation is abandoned there appears to be grounds for hope that radical cultural changes can be effected, at least in the long-run. Instead of focussing solely on gaining positions of political power and achieving specific, pre-defined political goals, an approach is revealed in which a broader front

---

56. Karl Marx, *Capital*, Vol.1, Moscow: Progress Publishers, 1974, p.351.

is available for action. By changing focus in this way, it becomes clear that far more people can and must play a part and be involved in a wide variety of tasks to effect the requisite changes in society.

The essence of the conception of power which emerges from process philosophy is that it is both the potentiality for and the actuality of self-creation as co-becoming with a multiplicity of other inter-dependent, semi-autonomous processes in the process of becoming of the world. To be in control of the world is not to reduce everything to instruments. It is to be able to live and act rationally, where rationality is understood as creative rationality, the ordering principle of the self-formation of people. Creative rationality involves striving to think and act justly, recognizing in thought and action the nature, dynamics and significance of all processes related to one's life. The ends of actions should not be defined in abstraction from these other processes and should always take into consideration the conditions being created or destroyed for other actions and for other processes. In effect, one (or one's group, organization, nation etc.) should see oneself as a participant within an ecosystem, a system of 'homes' of all individuated processes of becoming which make up the process of becoming of the world, and all one's actions in terms of what difference they make to this system.

This means that activity directed to changing the world should not be conceived as an engineering task to erect a preconceived model of how things ought to be; it is activity aimed at establishing and increasing the power and influence of practices, social relations, institutions and the products of activities embodying one form of thinking over those which embody another. Firstly it is necessary to challenge and replace the dominant stories defining individuals, communities, organizations, nations and civilization. It is necessary for individuals to change their habitus in the same way as the Russian intelligentsia forged a new habitus in the nineteenth and early twentieth centuries. Beyond this, it will be necessary to look for, or create niches within which the theoretical ideas, interpersonal relationships, practices, ways of living, relationships to the physical environment, and organizations embodying the process conception of the world can be established and made to flourish, and which in so doing can provide further niches for other process oriented research, relationships, practices, lives and organizations to establish themselves. The aim should be to develop these in such a way that they eventually undermine and displace the practices, relationships, ways of living, institutions and organizations embodying the mechanistic world-orientation.

This will be a multi-dimensional struggle. All action is simultaneously a participation in a multiplicity of processes - natural, biological, social and cultural - each of which has a different temporal rhythm. Each action and each life has significance far beyond what is generally recognized by one dimensional instrumentalist rationality. A life, or even an individual action, can at the same time be personal event, a political event, and a creative participation in the long durational dynamics of a culture through its symbolic significance, particularly if by such a life or such an action the way reality is normally defined is challenged. Through their actions and by the lives they lead, people are defining and redefining the meaning of history. This is what Merleau-Ponty was trying to convey when he argued:

> History is the judge - not History as the Power of a moment or of a century - but history as the space of inscription and accumulation beyond the limits of countries and epochs of what we have said and done that is most true and valuable, taking into account the circumstances in which we had to speak.... What [people] expect of the artist or politician is that he draw them towards values in which they will only later recognize their own values. The painter or politician shapes others more than he follows them. The *public* at whom he aims is not given; it is a public to be elicited in his work. The others of whom he thinks are not empirical 'others', nor even *humanity* conceived as a species; it is others once they have become such that he can live with them. The history in which the artist participates ... is not a power before which he must genuflect. It is the perpetual conversation woven together by all speech, all

> valid works and actions, each according to its place and circumstance, contesting and confirming the other, each one recreating all the others.[57]

To free themselves from the prevailing social perspectives and to reveal to themselves and to others the possibility of reconceiving the world, it is important that people actively participate in social struggles for a better world. The very fact of being part of a political struggle makes possible changes in perspectives and attitudes, especially if the struggle is well chosen and well organized. This seems to have been one of the main reasons for the success of the German greens. People should not be disheartened by the limited chances of achieving any particular goal. Actions, or even lives, which at the time appeared to have failed totally, as examples have entered the transcendent temporal order of the symbolic realm, acquired a symbolic significance which has influenced people for thousands of years, and changed the course of history.

## *Alternative Policies: Towards a New Grand Narrative*

While it is necessary to think of the struggle for a new order as a long term endeavour, to become a political force in the present it will be necessary to articulate the problems and aspirations of people in the short term with those of the intermediate and long term, the problems and projects of local areas with broader regional and world problems and projects, the immediate problems and goals of individuals and groups with national problems and goals and with the problems and goals of humanity. What is required is the construction of a new grand narrative, a new story of humanity's transformations in which people can identify themselves in history relation to the rest of the world, including the environment, and take up a position in the struggle to realize short, intermediate and long term goals of their communities, organizations, their nation and humanity.[58] The grand narrative should evaluate people of the past in terms of the nature of their relationship to their environment and project a future of an environmentally sustainable civilization, while showing how problems confronting people in the present are related to environmental degradation, and how overcoming this relates to their own aspirations. It should be elaborated so that all actors, whether individuals, organizations or nations, can identify and situate themselves, and evaluate all other actors.

To develop this it is necessary to use retrospective path analysis and the concepts of the ethics, political philosophy and science of society based on a process view of the world. In terms of these it is necessary to work out what form of world-society and civilization could provide the best conditions for people to live in while preserving the environment. This can be posited as something to be aimed at several hundred years in the future, and the paths which need to be taken from our present situation to realize that goal can then be examined. In the light of this general project, it should be possible to consider each region in the world and for each nation to consider in more detail what paths they need to take if humanity is to achieve a sustainable world-order. In this way it should be possible to formulate national and broader regional goals for a hundred years or so into the future, and then to consider how different locations, institutions and organizations can be developed to realize these goals. These goals should be developed as part of nationalistic endeavours to provide the conditions within each country for people to contribute as much as their abilities will allow to their nation, to humanity and to life itself.

It is then necessary to examine the immediate problems of societies and of individuals from within this general framework to relate these to the broader problems of civilization and the world

---

57. Maurice Merleau-Ponty, *Humanism and Terror: An Essay on the Communist Problem*, tr. John O'Neill, Boston: Beacon Press, 1969, p.188f.

58. On what is required to construct a new grand narrative see Arran Gare, *Postmodernism and the Environmental Crisis*, London: Routledge, 1995, Ch.5.

ecosystem. It is particularly necessary to identify forms of oppression which are preventing the realization of such national goals and the long term goals of humanity so that attempts to overcome these can be integrated with nationalism, with humanism and with the endeavour to preserve the world ecosystem. As each more local and immediate goal and path is worked out, this should then enable broader and longer term goals and their associated paths to be revised. Developing an image of the future in this way should involve a constant shifting of focus between the general and the specific and between the short term and the long term.

However the most important task for the immediate future is to address the existing series of economic crises and their causes, and to formulate solutions to these in accordance with the long run interests of humanity and of life. What is required is nationalist struggles to wrest back control over the economies of regions from the destructive dynamics of international capitalism, to gain democratic control over the financial institutions, transnational manufacturing and agribusiness companies which are at present destabilizing the world-economy, breaking down democratic institutions and creating economic and political pressures which are forcing people to wreck their environments to stay alive. To overcome this crisis in a way which contributes to overcoming the more basic problems of environmental destruction, which contributes to creating a sustainable world-order based on just relations between people and nature, environmentalists should support the erection of economic barriers to break up the world economy and to control the flow of capital, and promote the development of democratic institutions able to plan for the long term future which can subordinate the functioning of the market to long term national and international interests. In this way they can work towards redefining the nature of economics from the promotion of money-making to managing, preserving and developing the national household, with the environment - the foundation of this household - at the centre of concern.

Although what can be achieved in different countries will vary, environmentalists should spearhead the attack on the policies and the institutional changes effected over the last two decades by the champions of unfettered greed. They should identify and strive to unite all those classes and class fractions suffering under the new hegemony of international capitalists. In the immediate future environmentalists should strive for the reduction in power of shareholders in companies relative to stakeholders, which eventually should involve the decentralization and democratization of industry. As technology reduces the requirements for labour, policies should be formulated to ensure that unused labour is employed to improve the environment. To free people from the tyranny of the market, to promote a 'professional' orientation to work (so that people can achieve a sense of identity from their profession rather than from their income, wealth and level of consumption), to put an end to conspicuous consumption and to allow some people to devote their lives to the long-term problems of humanity, general income distribution policies should be formulated with a small range of income distributions, with guaranteed minimum incomes and with maximum incomes - that is, 100% marginal taxation beyond a certain level. The aim should be to reduce the level of material throughputs in the economies of the affluent nations in a way which does not cause hardship to the poor by eliminating wastage, simplifying life, and radically reducing the incomes of the wealthy, particularly of the financial and administrative parasites who now dominate the world.

In this endeavour it is necessary to create and maintain an image of and a sense of belonging to a just community committed to realizing people's potentialities, a community with which people can identify, commit themselves to, and look to for support. In opposition to the ruling hegemonic ideology it is necessary to develop an image of nations as 'people's homes' committed to recognizing the significance of all its members, its future generations, and its environment. Environmentalists should attack the ideological despotism of administrators and pedants and strive to reorganize research and education to counteract the nihilistic decadence which is now undermining civilization. They should launch a sustained attack on all those disciplines where academic power-brokers, politicians and institutions have uncritically assumed the mechanistic world-orientation, which have been indoctrinating students in nihilism and preventing people from reformulating research or teaching in accordance with more defensible metaphysical assumptions. Then in place of the over-specialized, vocationally oriented, soul destroying education of today, education devoted to developing people's understanding and their abilities to participate in the cultural, political and economic life of society, should be promoted. The image of the future articulated in the struggle against the hegemonic culture should be promoted not for a particular class or nation, but as a future for all members of society, for all humanity and for all life.

## *Conclusion*

This work began by showing the extent of the environmental crisis and its roots in the nihilism of European culture. Efforts to confront environmental problems from within the framework of this culture reinforce the forms of thinking responsible for environmental destruction. This culture dominates the minds and lives of people so completely that the view that the world is devoid of significance, that the only end worth striving for is attaining power over the world for the satisfaction of appetites, has come to appear as realism. The Marxism of the Soviet Union was shown to be a response of Eastern society to the more aggressive culture of Western Europe, a response which led it to use Marxism to appropriate the domineering Western orientation to the world. This produced the same problems. It was shown there does not yet exist a fully developed framework of ideas in terms of which environmental problems in all their complex diversity, and the nihilism which underlies the failure to deal with them, can be adequately understood and confronted. To address this, a version of process philosophy has been outlined and it has been argued that if it were fully developed, this could provide such a perspective for individuals, for the environmental movement and for governments, in the West, in the East, and in the peripheries of the world economy. Thereby it could provide the basis for a world-wide cultural revolution beyond European civilization which could serve as the foundation for a new world order.

This does not mean that with the development of this new conception of the world it will be a simple matter to deal with environmental problems. With the entrenchment in society of old conceptions of the world, with the enormity of the problems, the situation could still seem virtually hopeless. The comfort with which the privileged can live if they conform to the system compared to the insecurity of those who take up causes, the high-technology machinery of oppression available to defenders of the status quo, the powerlessness of those who lose their place in the rat race, the likelihood of failure in any particular project, the general discouragement and disdain, or worse, total non-recognition for genuine opposition to the dominant culture, leads easily to the conclusion that it is not worth the effort. But the task remains and will remain, how to transform humanity so that it contributes positively to the life rather than undermining the conditions of its own existence.

Process philosophy reveals the general approach and direction which it is necessary to pursue if there is to be any hope in the long run. And it is the long run which should be considered. If the analysis presented here is right, the world will in the immediate future become increasingly oppressive and violent. This oppression and violence is likely to be with us for a long time. How long will depend upon how long it takes people to change their conceptions of themselves and their place in the world, both in theory and in practice. Efforts in the present, even if they fail in their immediate goal, are contributions towards this cultural revolution.

There is always a tendency to under-estimate the achievements of those working towards a better world. Such people are in opposition to most of the existing power élites, and therefore subject to being defined by the establishment with a vested interest in denying their significance. But also, and perhaps more importantly, creative work is participation in processes of long duration, while destruction tends to be rapid. Consequently what is really creative tends to be invisible against a background of violence, oppression and destruction. Part of the importance of the process view of the world is that it emphasises the durational nature of becoming and thereby reveals more clearly the reality of such long-term creative efforts.

For similar reasons there is also a tendency to overestimate the success of the ruthless, those unhindered by scruples or concern with justice. Machiavelli's case for dismissing justice as the crowning political virtue has come to be accepted as a truly hard-headed view of politics, especially since it has been supported by social Darwinists and vulgar Marxists. This is true not only of the New Right, but also of most of the Left. But the crude Darwinian theory of evolution on which Social Darwinism is based is invalid, and Machiavelli's own life was hardly a great success. The success of most people and societies which have followed Machiavelli's principles have in fact been short-lived. The evolutionary theory deriving from process philosophy implies a different evaluation of ways of living. To begin with, a place is given to choice, and secondly, to emergent levels of ordering beyond their conditions of emergence. In the case of humans, rationality and justice are comprehensible as real features of human becoming. Humans have the choice of living according to justice or living according to egoistic principles in which everything and everyone are reduced to instruments, and they will be selected in the struggle for survival accordingly. If those who choose to live for justice prevail and a world community based on the commitment to justice for all emerges, there is good reason to believe that in the long run humanity will establish a sustainable relationship to its environment. If those who choose to live on Social Darwinian principles prevail, there can be little hope for the long-term future of humanity.

However there are good grounds for believing that those people who do choose to live justly will prevail in the long run and succeed in changing society accordingly. Those who strive for justice are more likely to be able to support each other, while those who struggle for domination will eventually come into conflict and destroy each other. In communities in which justice prevails as a habitus the creative potentialities of its members can more easily be realized and thereby contribute to the society, those with ability are provided with stronger reasons to apply themselves to the benefit of society, and there is far less time wasted in conflict. A community which also accords due recognition to the processes constituting its environment is similarly more likely to endure. It is societies which have been more just which have endured in the past, and where the modern world is concerned, it is not the societies which have extolled the ruthless pursuit of self-interest and which have developed instrumental and mechanistic thinking furthest which have been most successful, even in terms of the criteria of the culture of these societies. Despite their better geographical and strategic positions, since the Second World War the economies of Anglophone nations have been steadily out-performed by those nations founded on more just relations between their members. Sweden had (until economic rationalists gained power) a far healthier economy than USA. So while the future might look bleak in the short term (which of course is no small matter), evolutionary theory based on process philosophy justifies the hope that justice will prevail in the long term.

In conclusion to their work *Order Out Of Chaos*, Ilya Prigogine and Isabelle Stengers remarked: 'It is quite remarkable that we are at the moment both of profound change in the scientific concept of nature and the structure of human society as a result of the demographic explosion. As a result, there is a need for new relations between man and nature and between man and man.'[59] They then went on to point out that the ideas of physical sciences expounded by them, the ideas of instability and fluctuation in a world of processes, were also relevant to the social world. They pointed out that:

> This leads both to hope and a threat: hope, since even small fluctuations may grow and change the overall structure. As a result, individual activity is not doomed to insignificance. On the other hand, this is also a threat, since in our universe the security of stable, permanent rules

59. Ilya Prigogine and Isabelle Stengers, *Order Out of Chaos*, Toronto: Bantam, 1984, p.312.

> seems gone forever. We are living in a dangerous and uncertain world which inspires no blind confidence... [60]

Environmentalism provides the focus for those people rising to the ultimate challenge of the age, and to the greatest challenge in human history. There is reason for hope that the future belongs to these people; but there is always the possibility of total failure, either of a World War which will obliterate all their efforts or the successful entrenchment of a global ruling class committed to augmenting their levels of consumption come what may. In taking up the challenge, in taking the courage to risk their careers, their security, and in some cases, their lives, environmentalists are proving that life is more than satisfying appetites, petty vanities and a grubby struggle for money and status. They are revealing through their own lives the significance of all life. In doing so they are creating a community transcending national boundaries, transcending the confrontation between East and West and between North and South, a community of all those who strive to live for what is highest in life. The gathering strength of this community will, hopefully, transform the world, creating the conditions for all life, human and non-human, to flourish. But even if these people fail, even if the world is reduced by nuclear war to a lifeless desert, their lives will still be an achievement, a great and indelible contribution to the universe.

---

60. Ibid. p.313.

# Index